Smart Innovation, Systems and Technologies

Volume 398

The Smart Innovation, Systems and Technologies book series encompasses the topics of knowledge, intelligence, innovation and sustainability. The aim of the series is to make available a platform for the publication of books on all aspects of single and multi-disciplinary research on these themes in order to make the latest results available in a readily-accessible form. Volumes on interdisciplinary research combining two or more of these areas is particularly sought.

The series covers systems and paradigms that employ knowledge and intelligence in a broad sense. Its scope is systems having embedded knowledge and intelligence, which may be applied to the solution of world problems in industry, the environment and the community. It also focusses on the knowledge-transfer methodologies and innovation strategies employed to make this happen effectively. The combination of intelligent systems tools and a broad range of applications introduces a need for a synergy of disciplines from science, technology, business and the humanities. The series will include conference proceedings, edited collections, monographs, handbooks, reference books, and other relevant types of book in areas of science and technology where smart systems and technologies can offer innovative solutions.

High quality content is an essential feature for all book proposals accepted for the series. It is expected that editors of all accepted volumes will ensure that contributions are subjected to an appropriate level of reviewing process and adhere to KES quality principles.

Indexed by SCOPUS, EI Compendex, INSPEC, WTI Frankfurt eG, zbMATH, Japanese Science and Technology Agency (JST), SCImago, DBLP.

All books published in the series are submitted for consideration in Web of Science.

Priyadarsi Nanda · Sumit Srivastava ·
Vivek Kumar Verma · Pankaj Vyas
Editors

Intelligent System and Data Analysis

SSIC 2023 Doctoral Symposium

Springer

Editors
Priyadarsi Nanda
School of Electrical and Data Engineering
University of Technology Sydney
Ultimo, NSW, Australia

Sumit Srivastava
Department of Information Technology
Manipal University Jaipur
Jaipur, Rajasthan, India

Vivek Kumar Verma
School of Information Technology
Manipal University Jaipur
Jaipur, Rajasthan, India

Pankaj Vyas
Department of Information Technology
Manipal University Jaipur
Jaipur, Rajasthan, India

ISSN 2190-3018 ISSN 2190-3026 (electronic)
Smart Innovation, Systems and Technologies
ISBN 978-981-97-5199-0 ISBN 978-981-97-5200-3 (eBook)
https://doi.org/10.1007/978-981-97-5200-3

This Springer imprint is published by the registered company Springer Nature Singapore Pte Ltd.
The registered company address is: 152 Beach Road, #21-01/04 Gateway East, Singapore 189721, Singapore

Preface

The Doctoral Symposium at the SSIC 2023 conference provided a platform for emerging scholars to present their innovative research and engage with the academic community. This book Proceedings: SSIC 2023 encapsulates the diverse and groundbreaking research presented at the symposium, highlighting the contributions of doctoral candidates to the field of smart systems and computing.

The objective of this book is to showcase the theoretical insights, methodological advancements, and practical applications that emerged from the symposium. By compiling these works, we aim to foster a deeper understanding of the current research landscape and encourage further exploration and collaboration in these areas.

This volume would not have been possible without the unwavering support of many individuals and organizations. We extend our sincere gratitude to the contributors for their outstanding research and commitment. We also thank the reviewers and committee members for their diligent efforts in ensuring the quality of this publication. Special thanks go to our families and colleagues for their constant support and encouragement.

We hope this book serves as a valuable resource for academics, researchers, and students, inspiring new ideas and advancements in smart systems and computing. As you read through these chapters, we encourage you to consider the significant impact of these research contributions on the future of technology and society.

Ultimo, Australia	Priyadarsi Nanda
Jaipur, India	Sumit Srivastava
Jaipur, India	Vivek Kumar Verma
Jaipur, India	Pankaj Vyas

Contents

1 **Employing Deep Learning Techniques and Web Scraping for Enhanced Climate Analysis** 1
Shubham Shetty, Ritik Singh, Gaurav Patil, and Pramod Bide

2 **Federated Learning: Concepts, Application and Future Scope** 15
Pallav Jain and Sanjay Patidar

3 **Systematic Literature Review on Role of AI in IoT-Based Healthcare Solutions** 31
Manisha, Ashish Sharma, and Rekha Jain

4 **A Hybrid Machine Learning Model for Lung Disease Prediction** 39
Poonam Devi and Shrddha Sagar

5 **Analyzing Student Feedback with Natural Language Processing for Enhancing Student Perception** 57
A. Suganya and Ananthi Sheshasaayee

6 **Enabling Technology for Smart and Intelligent Transportation: Analyse the Uses and Impact of Blockchain Technology and Artificial Intelligent on Trucking Industry for Better Results** 75
Mahendra Parihar and Akshaya

7 **Detecting Parkinson Disease Using Various AI Enabled Methods: Recent Advances, Requirements, and Open Challenges** 91
Sourabarna Roy, Tannistha Pal, and Swapan Debbarma

8 **Synergistic Fusion of Steganography and Cryptography in Medical Imaging: A Comparative Analysis to Enhance Image Security** 107
Sharmila Ghosh, Ashim Saha, and Tannistha Pal

9 Detecting Inattentive and Aggressive Driver Behavior Using Deep Learning: Recent Advances, Challenges with Performance Evaluation 125
Tina Debbarma, Tannistha Pal, and Nikhil Debbarma

10 Improved Brain Tumor Classification Accuracy Using CNN Architecture with Efficient Net on Magnetic Resonance Imaging 137
Poonam Jyoti and Makhan Singh

11 Noise Prediction: A Case Study of Gjilan City 149
Buen Bajrami and Kostandina Veljanovska

12 Economic Crimes in Current Era: World and Indian Context 163
Arjun Choudhary, Ashish Tripathi, Meenakshi Punia, Chetanya Kunndra, and Vikas Kumar Kharbas

13 Analyzing the Citation Networks Using Community Detection Approaches: A Review 181
R. Kiruthika and N. Radha

14 A Hybrid Correlation Feature Selection Based Support Vector Machine (CFS-SVM) for Effective Software Defect Prediction 195
D. R. Medhunhashini and K. S. Jeen Marseline

15 Mitigating Risks in Vehicle Fires: An Integrated Approach for Early Detection and Real-Time Alert System 211
Ranjeetsingh Suryawanshi, Harshita Bhagat, Aryan Karande, Parth More, and Vaishali More

16 Multimodal Data Fusion: Combining Big Data and Deep Learning for Enhanced Predictive Models 223
K. R. Sreelakshmi and Sajimon Abraham

17 Analyzing Artificial Intelligence Based Intrusion Detection System in Detecting DDoS Attack 237
Sai Suchith Kurra, Likhitha Rayana, and Mohamed Sirajudeen Yoosuf

18 Preserving Our Heritage: Buildings Deep Learning Solutions for Monitoring Cultural Heritage Structures Using Automated Crack Detection 253
Rajdeep Chatterjee and Biswaroop Nath

19 Insurance_4_YOU: MCDM Equipped User-Friendly Intelligent System for Personalized Life Insurance Recommendations 269
Asha Rani, Kavita Taneja, and Harmunish Taneja

20 Apriori Analysis of Deep Learning Models on a Multi-class Data Set 283
Sai Sathwik Kosuru, Y. C. A. Padmanabha Reddy, Perala Venkata Akanksha, Kotagiri Chaitanya, Snehaja Parsi, and Madhu Babu Chunduri

21 Comparative Analysis of Emission Prediction Using Machine Learning and Deep Learning 297
Nitai Shah, Vimal Kumar Pathak, and Ramanpreet Singh

22 Disease Detection in Beans Leaves Using Transfer Learning 309
Riya Rawal, Anuradha Chug, and Amit P. Singh

23 A Novel Architecture for a Medical Image Recognition System Using Deep Learning-Based Multiple Regression Evaluation 325
Jyoti Parashar, Kamal Upreti, Iram Fatima, Divya Gangwar, Ashok Kumar Saini, and Prashant Vats

24 Mobile Controlled Wheeled Robotic Arm—A Future of Smart Education 343
X. Anitha Mary, Gurram Aakash, Vijila Gnanaraj, Bhupathiraju S. S. M. Karthikeya Varma, and Kocherlakota Snigdha

25 FDA-TO-FCA: Flow Direction Algorithm-Based Task Offloading in Fog-Cloud Architecture 359
Prashant Shukla, Sudhakar Pandey, Shivani Gupta, and Deepika Agrawal

26 Comparison of Segmentation Algorithms for Extraction of Stone from Kidney X-ray Image 373
N. Nisha, Vijaypal Singh Dhaka, and Deepak Sinwar

27 Sentiment Classification in Code-Mixed Indo-Aryan Languages: A Transformer-Based Survey 391
Saikat Roy and Jatinderkumar R. Saini

28 Secure File Sharing System Using Hybrid Authentication and Secured Key Management Algorithm 407
Aryan Dokania, Anshul Bamb, and J. Dheeba

29 IPDD: Intelligent Parkinson Disease Detection System 417
Manish Singhal, Vinita Tiwari, and Ashish Sharma

30 Performance Analysis of the Quantum Schemes in the Resource-Constrained IoT Network 427
Diksha Chawla and Pawan Singh Mehra

31 E-Healthcare Record System Using Blockchain 445
Parthasarathi Pattnayak, Arpeeta Mohanty, Tulip Das, and Sanghamitra Patnaik

32 Navigating the Landscape: A Comprehensive Survey on Healthcare Recommendation Systems 457
Megha Gupta, Shilpa Singhal, and Dr. Abhinav Tomar

33 Multi-class Generative Adversarial Network for Hyperspectral Image Classification for Addressing Class Imbalance Problem 471
Deepak Kumar and Dharmender Kumar

34 A Comparative Review of Resource Optimization and Provision Techniques in Cloud Computing 483
Sanjit Bhagat, Punit Gupta, and Dinesh Kumar Saini

35 The Future of Artificial Intelligence in Fashion: Innovations, Challenges, and Implications 493
Deepak Sharma, J. P. Sampath Kumar, and K. R. Shylaja

36 A Comprehensive Review on Deep Learning-Based Pansharpening Approaches 511
Gurpreet Kaur, Manisha Malhotra, and Dilbag Singh

37 Advancements and Applications of AI and ML Techniques in Breast Cancer Prognosis: A Comprehensive Survey 529
Anurag Jagetiya and Pankaj Dadheech

38 Perspective Analysis Toward Deep Learning Models for Wireless Capsule Endoscopy Images for Gastrointestinal Tract 541
Rakesh Sharma and C. S. Lamba

39 Exploring the Challenges and Software Risks of Agile Methodologies in Software Development 549
Neha Choudhary, Mahaveer Sain, and Sumitra Chopra

40 Guarding Databases: Unraveling SQL Injection Threats with Advanced Detection and Prevention 565
Harshit Soni, Ashish Tripathi, Rohit Kumar Gupta, and Pragati Gupta

Author Index 577

About the Editors

Dr. Priyadarsi Nanda is a Senior Lecturer at the University of Technology Sydney (UTS) with more than 33 years of experience and a strong researcher specialising in research and development of Cybersecurity, IoT security, Internet Traffic Engineering, wireless sensor network security and many more related areas. His most significant work has been in the area of Intrusion detection and prevention systems (IDS/IPS) using image processing techniques, Sybil attack detection in IoT based applications, intelligent firewall design. In Cybersecurity research, he has published over 130 high quality refereed research papers including Transactions in Computers, Transactions in Parallel Processing and Distributed Systems (TPDS), Future Generations of Computer Systems (FGCS) as well as many ERA Tier A/A* conference articles. In 2017, his work in cyber security research has earned him and his team the prestigious Oman research council's national award for best research. Dr. Nanda has successfully supervised 19 HDR at UTS (16 Ph.D. + 3 Masters), and currently, supervising ten Ph.D. students.

Professor Sumit Srivastava has a Ph.D. in the field of Data Mining. He has rich and diversified experience of more than 18 Years in academics, research, and administration as well. He has been associated with Manipal University Jaipur since July 2012. Professor Sumit Srivastava has also been an invited speaker in Algorithms, Machine Learning, Pattern recognition and Information and Knowledge Discovery at various FDP/short-term courses. He has published around 90+ research papers in SCI, Scopus and peer reviewed journals and conferences of international repute. He has guided seven Ph.D. research scholars and 17 M. Tech students for their projects/internships. He is currently the Senior Member of IEEE and professional member of ACM and has been associated with various scientific societies as a member. He is also serving as the reviewer for Springer Journal of Image Processing and Computer Vision, Wiley journal of Computer Vision, Journal of Imaging and Sensing, Journal of Computing and Communication, Expert Systems with Applications- Elsevier and guest editor for International Journal of Artificial Intelligence and Soft Computing (IJAISC).

Dr. Vivek Kumar Verma an esteemed Associate Professor at Manipal University Jaipur, is a prominent figure in the University. He boasts a substantial academic foundation with a Ph.D. in Natural Language Processing and a rich history of scholarly contributions, particularly in areas like Digital Image Processing and Machine Learning. Dr. Verma has published extensively, with his work appearing in 16 Q1 to Q3 journals as per the Scopus index, totaling 61 publications across various levels of academic journals and conferences. He has also been actively involved in editorial duties, having edited and convened numerous proceedings for international conferences, particularly highlighting his leadership in advancing research and academic discourse. His administrative roles at the university underscore as Deputy Director-Research his commitment to fostering an environment conducive to research and innovation.

Dr. Pankaj Vyas is currently the Head of the Department of Information Technology at Manipal University Jaipur (MUJ), a position he has held for the past 6 years. In this role, Dr. Vyas has been instrumental in shaping the department's academic and research initiatives, driving innovation, and fostering a collaborative learning environment. Dr. Vyas completed his B.E. in Computer Science from Amravati University in 1993, followed by an M.E. and Ph.D. in Computer Science from the Birla Institute of Technology and Science (BITS), Pilani, in 2002 and 2015, respectively. With over 30 years of rich academic experience since 1994, he has made significant contributions to the field of computer science education and research. In 2002, Dr. Vyas joined BITS Pilani, where he held various positions within the Department of Computer Science and Information Systems until July 2017. He then joined MUJ as an Associate Professor in the Department of Computer Science and Engineering (CSE), where he continues to inspire students and colleagues alike. Throughout his teaching career, spanning more than three decades, Dr. Vyas has specialized in courses related to programming and software engineering. He takes great pride in teaching 'Object-Oriented Programming' and 'Operating Systems.' Dr. Vyas has guided five undergraduate theses and two higher degree dissertations, reflecting his dedication to mentoring the next generation of computer scientists. His research interests are focused on software quality improvement, safety enhancement of object-oriented systems, software safety analysis techniques, and early software reliability estimation of object-oriented software systems. Dr. Vyas's extensive experience and expertise make him a valuable contributor to the Springer book for SSIC-2023 conference proceedings.

Chapter 1
Employing Deep Learning Techniques and Web Scraping for Enhanced Climate Analysis

Shubham Shetty, Ritik Singh, Gaurav Patil, and Pramod Bide

Abstract Images obtained by the visual sensors of automated vehicles often suffer from diminished quality due to a multitude of atmospheric phenomena such as haze, moisture, rain, and storms. These phenomena usually disrupt the proper functioning of visual sensors of vision-assisted transportation systems or Advanced Driver Assistance System (ADAS) as well as several other surveillance based systems. Removal of weather effects (deweathering) from these images has been crucial and has gotten a lot of attention in order to address these issues. Therefore, in order to accurately trigger the necessary deweathering (removal of haze, moisture, etc.) operation, a solution is needed to automatically ascertain the weather conditions depicted in an input image. This paper presents a novel deep learning-based framework for weather image recognition that takes into account 11 typical weather phenomena, including dew, rainbows, lightning, sandstorms, snow, rain, fog smog, rime, glaze, hail, and frost. For an input image, the proposed solution approach automatically classifies the image into one of the eleven categories (e.g., sunny or others). In the proposed approach, transfer learning on five state-of-the-art models, namely EfficientNetB7, ResNet50, MobileNet, VGG19, and DenseNet201, has been implemented which are pretrained on the ImageNet dataset. Using this approach, an accuracy of up to 92.5% has been obtained.

1.1 Introduction

Weather plays an integral role in the general ecosystem. The weather surrounding us exerts a significant impact on people's decision-making. Improper classification of weather conditions can prove to be hazardous to human life. Foggy and snowy conditions usually result in a lot of traffic in hilly regions. Fatal traffic accidents can

S. Shetty (✉) · R. Singh · G. Patil · P. Bide
Bharatiya Vidya Bhavan's Sardar Patel Institute of Technology (SPIT), Mumbai 400058, India
e-mail: shubham.shetty@spit.ac.in

P. Bide
e-mail: pramod_bide@spit.ac.in

P. Nanda et al. (eds.), *Intelligent System and Data Analysis*, Smart Innovation, Systems and Technologies 398, https://doi.org/10.1007/978-981-97-5200-3_1

be avoided if we correctly identify the weather in a particular region and take the necessary precautions. Apart from traffic accidents, correct climate recognition can also be used in the tourism industry where areas that are experiencing heavy and extreme conditions may be prohibited to visit for some time. There are many other industries that use climate analysis to plan and optimize their workflow or operations. For example, In the agricultural industry, the farmers may require to know the current weather conditions to start or stop their sprinklers. Construction workers need to plan their supplies and gear on a daily basis at the construction site.

At this point in time, there are traditional methods of climate analysis which involve deploying sensors at the site which is mostly accurate but a very costly solution [2]. Other solutions include manually looking out of the window and judging the climate ourselves or counting on forecasts provided on television by the meteorological department.

A computer vision approach can simply help us automate this procedure and tackle the problem of huge costs that sensors present. A variety of approaches that use different computer vision and deep learning architectures have already been proposed to crack this problem, and the aim of the study is to collate, differentiate, and weigh these approaches and use a new approach that uses latest state-of-the-art models to provide a better solution as they have not been attempted yet as a solution to this problem [3]. Many existing approaches use images from vehicle cameras which narrows the scope of the problem and such approaches are limited to specific applications. Hence, there is still a need for a more generic solution that may be used for climate analysis in a wide variety of use cases.

In this study, there is use of EfficientNetB7, ResNet50, MobileNet, VGG19, and DenseNet201 models which are pretrained on the ImageNet dataset and comparison of the results of each in the form of a detailed study.

The rest of the paper has been organized as follows: We briefly cover Related Works in Sect. 1.2. The Proposed Methodology and Experiments are described in Sect. 1.3. Results and Discussion are mentioned in Sect. 1.3.7 and Conclusion in Sect. 1.4.

1.2 Related Works

Over time, numerous image processing methods and architectures have surfaced in the field of image classification. Climate analysis has been a fairly unexplored area in the field of image recognition. Climate analysis has a plethora of applications which haven't been explored in great depth till date.

Ziqi Zhu and colleagues utilize three networks (AlexNet, modified AlexNet, and GoogleNet) to conduct climate analysis. In their investigation, they examine the effectiveness of both fine-tuning and non-fine-tuning strategies. In the fine-tuning step, the models are pretrained on the ILSVRC2012 dataset and then further trained on the actual weather dataset."WeatherDataset" of four classes with 16635 images is constructed and images are collected from the web. The experimental findings demonstrate that the suggested approach is capable of excellent results, with the recognition accuracy rate of 94.5%. The dataset used has only four classes which may be insufficient for practical applications [1].

Wang Zenghui et al. use an approach which is called as SAID of stacked ensemble algorithms. Two sets of classifiers are used. The first set consists of RBF-SVM, Random Forest, and KNN and the second set consists of Random Forest, Native Bayes, and KNN. The dataset used by the authors is of 1125 images constructed by manually labeling images from Google images, flickr, gettyimages, yahoo, etc., as cloudy, sunrise, rainy, and sunshine. The accuracy achieved by the authors is as high as 94.5%. But since the dataset has only four classes, this might turn out to be a little less for practical applications [2].

Yuzhou Shi et al. employ convolutional neural networks in conjunction with the phenomenon of edge deterioration. Mask R-CNN is used to extract the foreground and the foreground edges in the image as these are the regions of interest and superimpose them into the scale matrix which is input to the classification network. The dataset consists of road-traffic images with a total of 276 images divided into four classes, namely sunny, foggy, rainy, and snowy. The accuracy achieved by the proposed methodology is as high as 94.71%. The dataset used by the authors is very small, and it is also imbalanced which makes it difficult to use in real-life scenarios [3].

Li-Wei Kang et al. introduced a weather classification system based on GoogleLeNet, which utilizes four classes: hazy, rainy, snowy, and a category for images that do not belong to any of the three weather conditions. The primary goal of this research is to develop a classifier that can categorize each input image into one of four classes. The output size of GoogLeNet is directly set to 4, and the model was trained with batch stochastic gradient descend to minimize the softmax loss. The dataset used in this paper is called the MWI dataset, a multi-class weather image dataset including 20k images collected from many web albums and films. The methodology proposed by the authors outperforms other models in this paper, and accuracy achieved is 92%. The authors train these models on personal computers and that is the reason why they use a little less complex models [5].

Shweta Mishra et al. have used pretrained models like AlexNet, VGGNET, GoogLeNet, and Residual Networks and have also added softmax layer of (4,1) because the number of classes is 4. The synthetic images were used to train the classifier layers, and for each model, the mean Average Precision (mAP) was determined. The synthetic dataset used by the authors in this paper contained images annotated with four weather classes. The resolution of each image was recorded at 1280×720 and contains over 110,000 images. The mAP for all the architectures varied between 60 and 74%. VGGNEt had the most efficient classification, while ResNet had the lowest accuracy. There is a problem of bias in vision datasets that needs to be tackled [6].

WeatherNet, developed by Mohamed R. Ibrahim et al., consists of four deep CNN models designed to identify various weather conditions such as dawn/dusk, daytime, nighttime, glare, rain, snow, and fog. Models 1 and 3 are trained to output one of three classes, while models 2 and 4 are binary classifiers. For the classification tasks, they used transfer learning via ResNet50, which produced results that were optimized in terms of accuracy and computational speed. The dataset is made up of 23,865 images that were downloaded from the internet, specifically Google Images, which included pictures for every category of weather and lighting. On test datasets, four models had accuracy of 91.6%, 94.8%, 93.2%, and 95.6%, respectively. The models also demonstrated high accuracy, a high $F1$-score, and low false-positive rates of 6 percent or less. Further fine-tuning for the hyperparameters of the convolutional neural networks can be done [7].

Mohammad Farid Naufal et al. utilize convolutional neural networks with transfer learning to classify weather images. They compare Xception, VGG16, MobileNetV2, and DenseNet201. The authors of this paper use five cross-validations and perform the training for 50 epochs for each model. The dataset used in this paper comprises 1816 weather images obtained from multi-class weather dataset which has four classes, namely cloudy, rainy, shine, and sunrise. Xception achieves the highest accuracy rate of 90.2%, followed by DenseNet201 with an accuracy of 89.34%. The dataset used by the authors has very few images, and the dataset is highly unbalanced [9].

Mohamed Elhoseiny et al. have used the concept of deep convolutional neural network inspired by Krizhevsky's CNN. They use a pretrained CNN trained on ImageNet and then fine tune it for weather classification. Hyperparameter tuning is performed by the authors. The authors compare ImageNet CNN with their own weather CNN on the dataset. A weather image dataset is consisting of 10k images which has various classes like sunny, foggy, rainy, etc. The weather CNN gives a normalized classification accuracy of 82.2%. The hyperparameter tuning done by the authors is not that extensive and could be improved for further increase in accuracy [10].

Jingming Xia et al. employ ResNet15, a simplified version of ResNet50, in their study. They add some residual modules as well. The perform weather classification on their own dataset. The authors create their own dataset in this paper called as the WeatherDataset-4 which has 4983 weather images of four classes, namely foggy, rainy, sunny, and snowy. The average accuracy of climate recognition by ResNet15 is 96.03% which is very high. The dataset used by the authors has only four classes which is a little insufficient for usage in practical applications [11].

In their proposal, Ling Tan et al. introduce a convolutional neural network based on ResNet50, utilizing three channels. The Resnet50 branch extracts global weather features from the images. Two other CNN5 branches are created which extract local features. The global and local features are combined through a concatenation function, followed by classification using a softmax classifier. The dataset used by the authors in this paper has 6185 outdoor weather images and is named as WeatherDataset-6. The average recognition accuracy achieved by the authors after using the model on WeatherDataset-6 is 95.81%. The model that the authors have created only work for images taken in daylight and no testing for images taken at night is done [12].

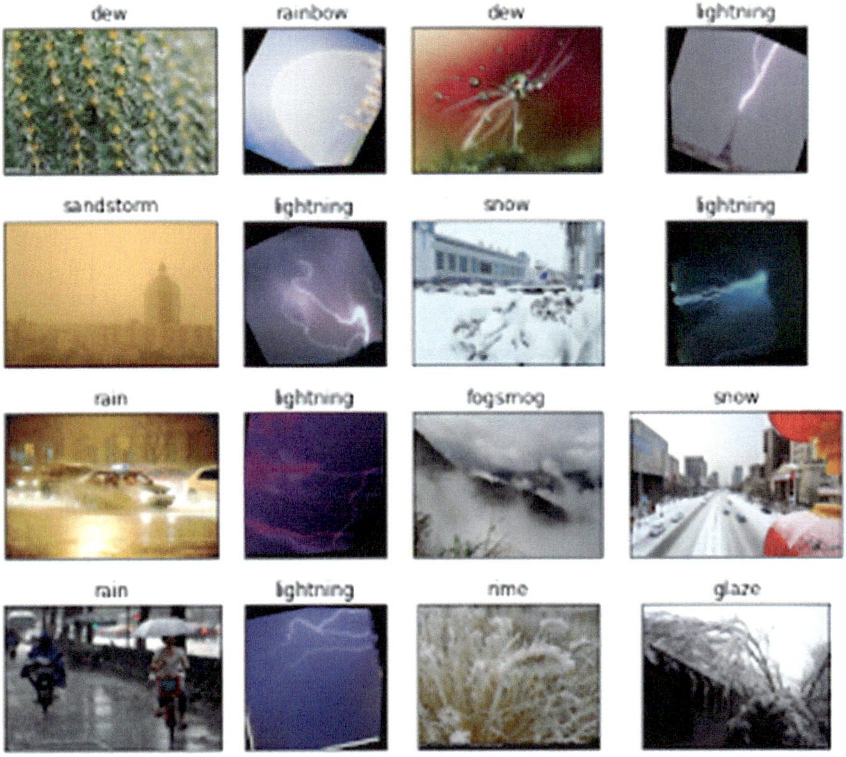

Fig. 1.1 Sample images from the dataset

1.3 Methodology

1.3.1 Dataset Description

For this research, the dataset is created manually by collecting images from the web and manually assigning labels to different classes. A total of 6682 images were collected which span across 11 different classes, namely dew, rainbows, lightning, sandstorms, snow, rain, fog, rime, glaze, hail, and frost. Figure 1.1 depicts a randomized sample of images present in the given dataset. Figure 1.2 represents the distribution of the images in the dataset by class.

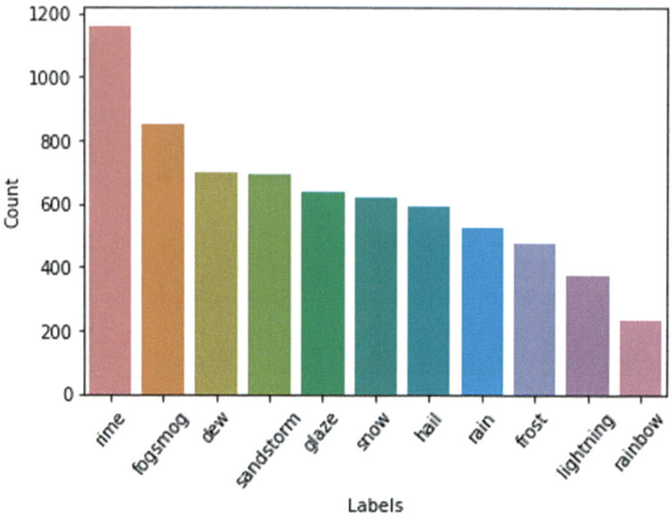

Fig. 1.2 Initial imbalanced dataset

1.3.2 Data Collection Process

The data collection process was a thorough undertaking, involving various searches across websites and online repositories. We prioritized variety and diversity in our image selection, ensuring a comprehensive representation of various weather-related scenarios.

1.3.3 Addressing Data Imbalance

One difficult challenge that surfaced during the dataset creation was the imbalance in the distribution of images across different weather categories. Acquiring an ample number of images for the lightning and rainbow categories, in particular, presented a unique set of challenges. To rectify this discrepancy and uphold the integrity of our dataset, we harnessed the power of data augmentation techniques, strategically targeting these two categories.

1.3.4 Data Augmentation

To rectify the class imbalance, we executed a thoughtfully designed data augmentation strategy. Specifically, the lightning and rainbow categories underwent a series of augmentations, including:

- **Blurriness**: We introduced randomized blurriness to some images, in the view of replicating the natural variability in image quality which is generally encountered in real-life scenarios.
- **Rotation**: Images were subjected to random rotations, spanning up to 15 °C, effectively accounting for variations in perspective.
- **Contrast Adjustment**: Subtle alterations to image contrast were applied to simulate a diverse range of lighting conditions.

1.3.5 Noise Introduction

Considering the origin of our images from the web, a certain degree of inherent noise was inevitable within our dataset. This deliberate retention of noise aimed to authentically reflect real-world conditions. Importantly, the introduced noise was held to a minimum, preserving the genuine essence of the images.

1.3.6 Training

In response to the remarkable success of deep learning in many perceptual tasks, we employed it to do weather image categorization (object identification, speech recognition, and image recognition, among others). One of the most well-known deep learning architectures, convolutional neural networks (CNNs), has shown that it is capable of performing a number of visual tasks with skills comparable to those of humans. But how effectively CNN operates is largely dependent on the volume of training data. A sophisticated deep learning technique called transfer learning enables the reuse of a model for many applications. Given the enormous resources in terms of processing and the time it takes to construct CNNs from scratch, pretrained CNN architectures have become quite popular for computer vision tasks.

This study used five cutting-edge models for transfer learning: EfficientNetB7, ResNet50, MobileNet, VGG19, and DenseNet201. The study makes use of the model weights from training on the ImageNet dataset. According to the requirements of the dataset, a few layers are added to all the models and the top layers of the models are made untrainable. The input layer of our five models was fed-forward with the training and testing images after they were resized to (256,256,3). In addition to the original architecture, there is addition of two dense layers with different neurons but with the same activation function "relu". The output of every dense layer is flattened and fed as input to the next dense layer. The data is categorized using 11 neurons in the last dense layer, which uses the "softmax" activation function. In case of every model, the optimizer used is Adam optimizer with a learning rate of 0.001. Categorical cross-entropy is used as the loss function, and early stopping is enabled with a batch size of 32.

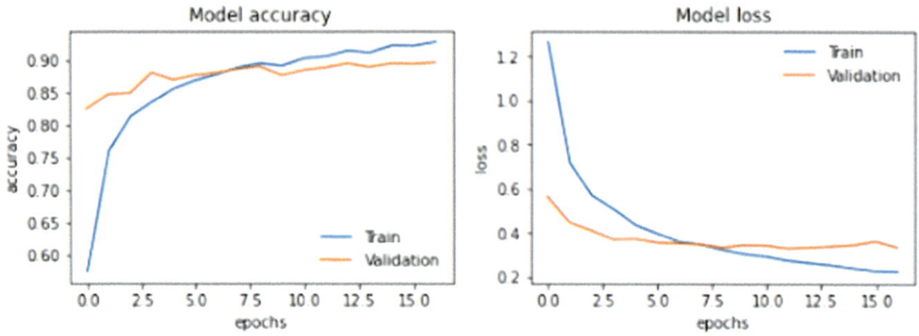

Fig. 1.3 Loss of EfficientDet

The following subsections provide a detailed description of the pretrained weather prediction models:

1. **EfficientDet**: EfficientDet is an object detection model architecture that balances accuracy and computational efficiency. It builds upon the EfficientNet backbone, using compound scaling to optimize depth, width, and resolution. With feature pyramid networks and bi-directional feature networks, EfficientDet captures multi-scale features and efficiently fuses information. It achieves state-of-the-art performance on object detection benchmarks like COCO while being computationally efficient. EfficientDet has gained popularity due to its ability to detect objects across categories effectively. The initial model, known as EfficientNet-B0, serves as the baseline model.
   Building upon this baseline, a process of compound scaling is applied, resulting in the creation of a series of EfficientNet models, ranging from EfficientNetB1 to B7. In this experiment, there is use of EfficientNetB7. After training for 40 epochs, the validation and training losses are decreasing with every epoch and leading to a validation loss of 0.33 and training loss of 0.22 (see Fig. 1.3)
2. **ResNet50**: The winner of the ILSVRC-2015 challenge was Residual Network. It was proposed to add residual blocks in order to solve the issue of a vanishing gradient in ultra-deep CNN. Residual Network (ResNet) was developed in a variety of forms with different layer counts, including 34, 50, 101, 152, and 1202. In this study, the often used ResNet50 was applied. As you can see in the figure, the validation and training losses decreased with every epoch and resulted in validation loss of 0.32 and training loss of 0.29 (see Fig. 1.4).
3. **MobileNet**: MobileNet is a family of lightweight neural network architectures designed for mobile and embedded devices. It reduces computational complexity by using depth-wise separable convolutions, which split the convolution operation into separate layers. This approach decreases parameters and computation, making MobileNet models efficient for resource-constrained devices. MobileNetV2 introduces inverted residuals and linear bottlenecks for further optimization, while MobileNetV3 enhances performance with specialized blocks for specific tasks. MobileNet models are widely used in image classification, object detection, and

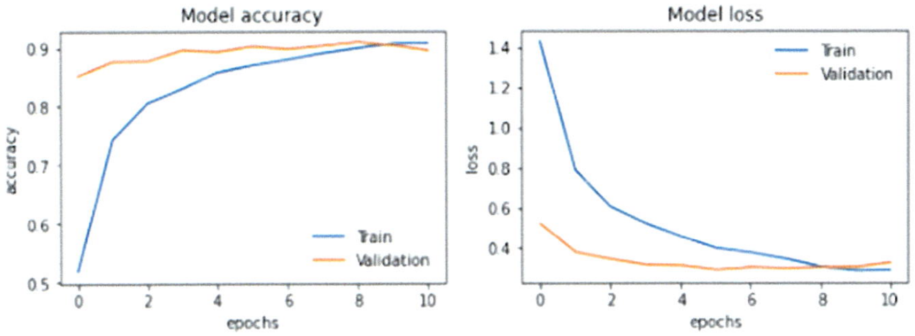

Fig. 1.4 Loss of ResNet50

face recognition, providing a balance between model size, inference speed, and accuracy. After training for few epochs, the loss is decreasing with every epoch and leading to validation loss of 0.32 and training loss of 0.24 (see Fig. 1.5).

4. **VGG19**: VGG19 is a deep convolutional neural network architecture introduced in 2014. With 19 layers, it follows a simple design principle of small receptive fields and fixed-size pooling layers. VGG19 achieved remarkable performance in image classification tasks but is computationally expensive. Nonetheless, it remains a benchmark and reference in deep learning, serving as a teaching tool and inspiring subsequent architectures. After training, the loss decreased with every epoch and finally resulted in validation loss of 0.38 and training loss of 0.30 (see Fig. 1.6).
5. **DenseNet201**: A DenseNet is a special type of convolutional neural network that makes use of dense connections between layers by using Dense Blocks, which link all layers directly with matching feature-map sizes. Each layer receives additional inputs from all earlier layers and transmits its own feature-maps to all later layers in order to maintain the feed-forward character of the system. We have used DenseNet201 version of DenseNet which is 201 layers deep as its name suggested. After training for 40 epochs, the loss is decreasing with every epoch, and there are validation loss of 0.30 and training loss of 0.21 (see Fig. 1.7).

1.3.7 Results and Discussion

The recognition accuracy by each of the models used in this paper varied in the range of 90–93%. Since this study is using transfer learning, each of these models was trained for 50 epochs.

Based on the table, DenseNet201 achieved the highest test accuracy of 92.55%, whereas the lowest test accuracy was that of EfficientNetB7 which was around 87.26%. The other models, namely VGG19, Resnet50, and MobileNet, achieved a test recognition accuracy of 88.25%, 90.49%, and 90.24%, respectively. DenseNet201

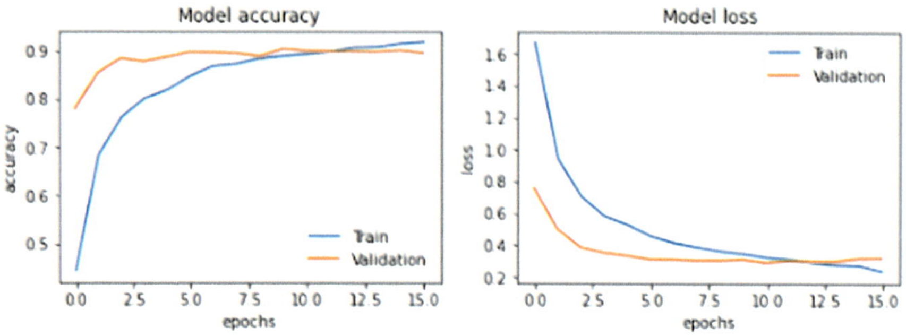

Fig. 1.5 Loss of MobileNet

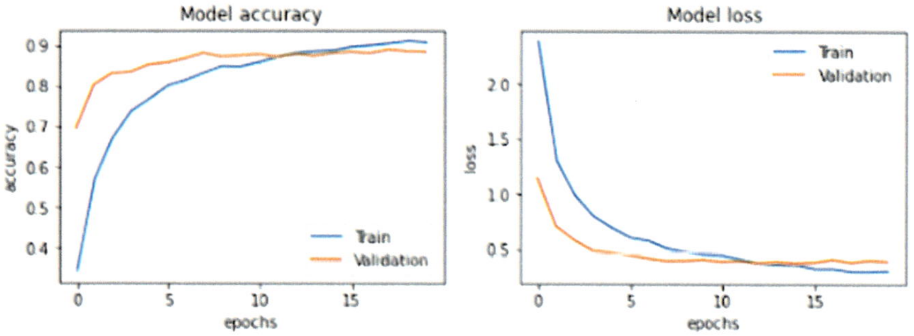

Fig. 1.6 Loss of VGG19

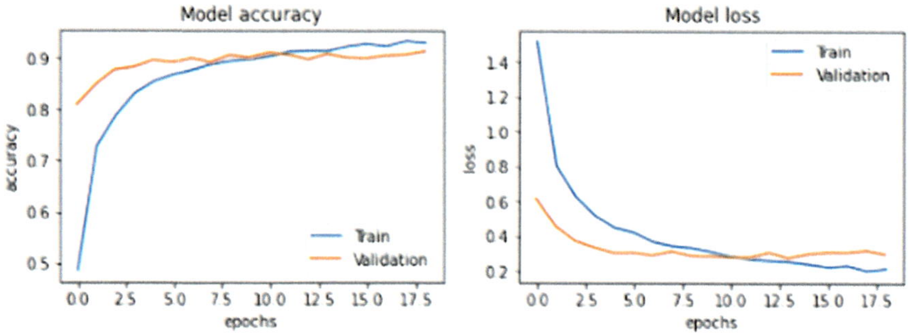

Fig. 1.7 Loss of DenseNet

Fig. 1.8 Output examples of DenseNet201

achieved the best results among all of the tested models. The inference times were also pretty good; hence, DenseNet201 can be used in practical real-life applications. The inferences of DenseNet201 model on the dataset used can be seen in Fig. 1.8. Similarly, the test accuracies achieved by all the state-of-the-art models being compared in this paper are shown in Fig. 1.9.

1.4 Conclusion

Deep learning has evolved tremendously over the past few years as a result of a great amount of research. Multiple new methodologies are being proposed that help in reducing the amount of computation and increasing the speed of giving out results. In this paper, there is an attempt to juxtapose multiple state-of-the-art models mentioned

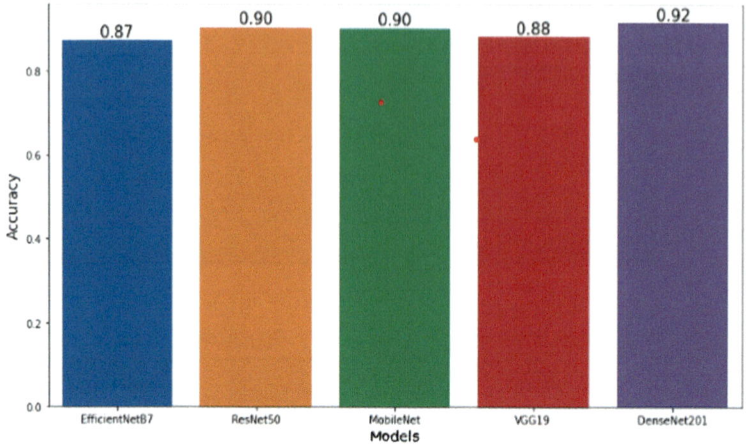

Fig. 1.9 Test accuracy of all models used

above for the task of climate analysis. The findings of this study are encouraging, and there is confidence that they can lay the groundwork for upcoming state-of-the-art models. The proposed method can be applied as a detection algorithm in a variety of industries that use climate analysis.

References

1. Zhu, Z., Zhuo, L., Qu, P., Zhou, K., Zhang, J.: Extreme weather recognition using convolutional neural networks. In: IEEE International Symposium on Multimedia (ISM), 621–625. San Jose, CA, USA (2016). https://doi.org/10.1109/ISM.2016.0133
2. Cao, Y., Yang, H.: Weather prediction using cloud's images. In: 2022 International Conference on Big Data, Information and Computer Network (BDICN), pp. 820–823, Sanya, China (2022). https://doi.org/10.1109/BDICN55575.2022.00157
3. Shi, Y., Li, Y., Liu, J., Liu, X., Murphey, Y.L.: Weather recognition based on edge deterioration and convolutional neural networks. In: 2018 24th International Conference on Pattern Recognition (ICPR), pp. 2438–2443, Beijing, China (2018). https://doi.org/10.1109/ICPR.2018.8546085
4. Mistry, J., Misraa, A.K., Agarwal, M., Vyas, A., Chudasama, V.M., Upla, K.P.: An automatic detection of helmeted and non-helmeted motorcyclist with license plate extraction using convolutional neural network. In: 2017 Seventh International Conference on Image Processing Theory, Tools and Applications (IPTA), pp. 1–6, Montreal, QC, Canada (2017). https://doi.org/10.1109/IPTA.2017.8310092
5. Zhu, Z., Li, J., Zhuo, L., Zhang, J.: Extreme weather recognition using a novel fine-tuning strategy and optimized GoogLeNet. In: 2017 International Conference on Digital Image Computing: Techniques and Applications (DICTA), pp. 1–7, Sydney, NSW, Australia (2017). https://doi.org/10.1109/DICTA.2017.8227431
6. Mishra, S., Kumar, S., Kumar, V.: Multi-view Deep Learning for Weather Recognition (2022). https://doi.org/10.1007/978-981-16-8546-0_7

7. Ibrahim, M., Haworth, J., Cheng, T.: WeatherNet: recognising weather and visual conditions from street-level images using deep residual learning. Int. J. Geo-Inf. **8**, 549 (2019). https://doi.org/10.3390/ijgi8120549
8. Biard, J.C., Kunkel, K.E.: Automated detection of weather fronts using a deep learning neural network. Adv. Stat. Clim. Meteorol. Oceanogr. **5**, 147–160 (2019). https://doi.org/10.5194/ascmo-5-147-2019
9. Naufal, M.F., Kusuma, S.F.: Weather image classification using convolutional neural network with transfer learning. AIP Conf. Proc. **2470**, 050004 (2022). https://doi.org/10.1063/5.0080195
10. Elhoseiny, M., Huang, S., Elgammal, A.: Weather classification with deep convolutional neural networks (2015). https://doi.org/10.1109/ICIP.2015.7351424
11. Xia, J., Xuan, D., Tan, L., Xing, L.: ResNet15: weather recognition on traffic road with deep convolutional neural network. Adv. Meteorol. **2020**, 11 (2020). https://doi.org/10.1155/2020/6972826
12. Tan, L., Xuan, D., Xia, J., Wang, C.: weather recognition based on 3C-CNN. KSII Trans. Internet Inf. Syst. **14**(8), 3567–3582 (2020). https://doi.org/10.3837/tiis.2020.08.024
13. An, N.: Xception network for weather image recognition based on transfer learning. In: 2022 International Conference on Machine Learning and Intelligent Systems Engineering (MLISE), Guangzhou, China, pp. 330–333 (2022). https://doi.org/10.1109/MLISE57402.2022.00072
14. Abdalla, A.M., Ghaith, I.H., Tamimi, A.A.: Deep learning weather forecasting techniques: literature survey. In: 2021 International Conference on Information Technology (ICIT), Amman, Jordan, pp. 622–626 (2021). https://doi.org/10.1109/ICIT52682.2021.9491774
15. Zhao, X., Wu, C.: Weather classification based on convolutional neural networks. In: 2021 International Conference on Wireless Communications and Smart Grid (ICWCSG), Hangzhou, China, pp. 293–296 (2021). https://doi.org/10.1109/ICWCSG53609.2021.00064
16. Chen, Z., Yang, F., Lindner, A., Barrenetxea, G., Vetterli, M.: How is the weather: automatic inference from images. In: 2012 19th IEEE International Conference on Image Processing. IEEE, pp. 1853–1856 (2012)
17. Dannheim, C., Icking, C., Mader, M., Sallis, P.: Weather detection in vehicles by means of camera and LIDAR systems. In: 2014 Sixth International Conference on Computational Intelligence, Communication Systems and Networks. IEEE, pp. 186–191 (2014)
18. Li, Q., Kong, Y., Xia, S.M.: A method of weather recognition Basedon outdoor images. Comput. Vis. Theor. Appl. **2**, 510–516 (2014)
19. Lagorio, A., Grosso, E., Tistarelli, M.: Automatic detection of adverse weather conditions in traffic scenes. In: Advanced Video and Signal Based Surveillance, pp. 273–279 (2008)

Chapter 2
Federated Learning: Concepts, Application and Future Scope

Pallav Jain and Sanjay Patidar

Abstract Artificial is a constantly evolving and progressing technology. In today's society, artificial intelligence models are frequently employed. With today's data and the capabilities of artificial intelligence, we can attain human intellect and, in some situations, considerably more. But also, the rapid increase of data and how it is managed is a cause for concern. Machine learning is an artificial intelligence sub-area that allows a machine to learn from prior data. Nowadays, machine learning is utilized in a broad range of applications. However, it is a tough task to apply machine learning on decentralized data. In the recent years, a new technique known as Federated Learning has garnered considerable attention. It concentrates to train distributed and decentralized data while maintaining its privacy. In our paper we have explained what Federated Learning is, the characteristics of Federated Learning, what are the hardships we face in applying Federated Learning, in which field Federated Learning is currently used and what are the future scope of Federated Learning.

2.1 Introduction

Generalized machine learning algorithms focuses on training data that is centralized. One of Google's most promising cloud infrastructures is a platform for employing machine learning to train protected data. Google claims to have pioneered a novel method for safeguarding smartphones interactions and training associated models via machine learning. This approach is called as Federated Learning [1]. Federated Learning includes blockchain, edge computing and on-device AI. In machine learning, data is gathered from devices such as cell phones, laptops and then it is used in a centralized server. With this data, machine learning systems train itself to predict outcome of new data, and this user data is then collected and sent to servers to train

P. Jain (✉) · S. Patidar
Delhi Technological University, New Delhi, India
e-mail: Pallavkjain@gmail.com

S. Patidar
e-mail: sanjaypatidar@dtu.ac.in

P. Nanda et al. (eds.), *Intelligent System and Data Analysis*, Smart Innovation, Systems and Technologies 398, https://doi.org/10.1007/978-981-97-5200-3_2

models using AI techniques. In Federated Learning, we don't take sensitive user data, instead we train our model locally on devices which are authorized to communicate training datasets to the centralized server. This same procedure occurs on several devices that contain an offline archive of data. The results are then compiled on a centralized server, but this time without any user data. Then we merge the results on a centralized consolidated server.

2.2 Federated Learning (An Overview)

2.2.1 Characteristics of Federated Learning

FL has a lot to do with distributed learning. Distributed computing and storage are two components of a traditional distributed system. The initial model updates of federated learning for Android devices are comparable to distributed computation in many ways. Although Federated Learning placed a strong emphasis on privacy preservation, the recent distributed machine learning research has also focused on privacy-preserving distributed systems. Distributed processing is the use of a communication network to connect several computers in various locations in the control of a centralized server, with every machine doing different pieces of the same operation. As a result, distributed processing is primarily focused on speeding up the processing step, while Federated Learning focuses on developing a collaborative model that does not compromise privacy [2]. Federated Learning has many benefits over classical machine learning techniques in terms of 1. Data security. 2. Data diversity. 3. Real-time continual learning. 4. Hardware efficiency.

2.2.1.1 Data Security

Data security is the process of securing the data from unauthorized access. In classic ML, we share raw data between the systems so that we can train models. This poses a serious security threat as the data shared can be misused. In federated learning, we train our model locally and then the results are shared into consolidated servers so there is a minimum risk of data security violations. The model does not require a data pool because the training dataset is kept on the devices.

2.2.1.2 Data Diversity

It is difficult for companies to club datasets together from such diverse sources including the data security and network unavailability in edge devices challenges. It is easier to access different data using federated learning even though data sources only connect at specific times.

2.2.1.3 Real-Time Continual Learning

In Federated Learning, models are improving continuously via the datasets provided by the client. There is no need to combine data in order to learn continuously.

2.2.1.4 Hardware Efficiency

Federated Learning models do not use single server to train and evaluate data that's why FL models use less hardware than any classic ML models and hence are more hardware efficient (Fig. 2.1).

2.3 Federated Learning (Definition)

Let's say we have X devices, i.e.,

$$(n_1, n_2, n_3, \ldots, n_x) \tag{2.1}$$

and their data is represented by

$$(d_1, d_2, d_3, \ldots, d_x) \tag{2.2}$$

and we want to use their data to construct an ML model. A classic ML algorithm will put all the data together for training the model, i.e.,

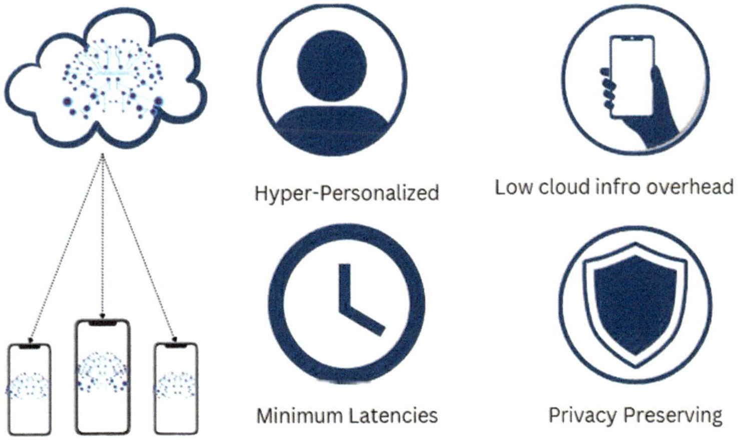

Fig. 2.1 Benefits of Federated Learning

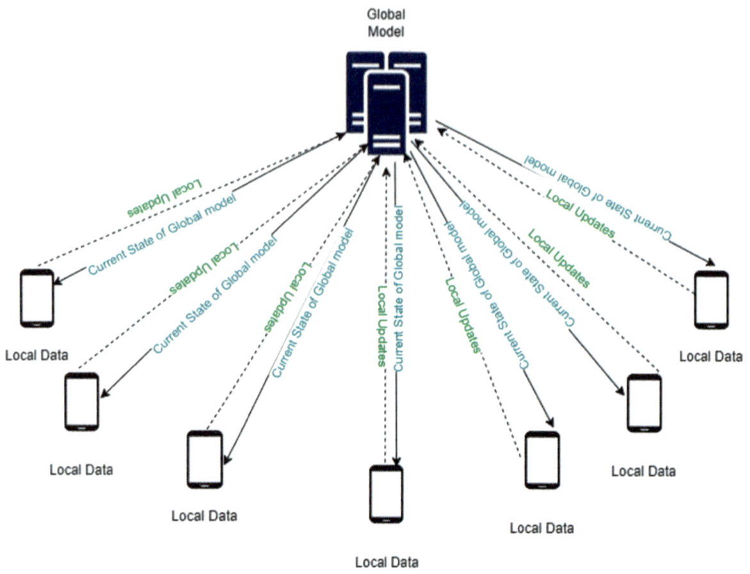

Fig. 2.2 Working of Federated Learning

$$d = d_1\cup, d_2\cup, d_3, \ldots, \cup dx \tag{2.3}$$

Then the data owners will collaborate to train the models in federated learning. Here the owner n_i doesn't disclose their data d_i. Suppose the ML model is M_{sum} and FL model is M_{fed}. Accuracy of M_{sum} is V_{sum} and accuracy of M_{fed} is V_{fed}. The accuracy of M_{fed} must have to be In proximity to M_{sum}, M_{sum}.

V_{fed} Let us consider δ as a non-negative integer, then:

$$|V_{\text{fed}} - V_{\text{sum}}| < \delta \tag{2.4}$$

Here δ is the accuracy loss for Federated Learning algorithm (Fig. 2.2).

2.4 Types of Federated Learning

In a dataset, feature space is just the set of all possible values for a selected collection of characteristics from a piece of data. A sample space is the collection of all possible outcomes and A label is applied by human annotators to a piece of data in machine learning to explain it to the computer and that space is called a label space. Let's take a matrix *Di* which is in '*i*' devices. In the context of datasets, a row typically represents a sample or observation, while a column represents a feature or variable, and the dataset can also have labeled data. Let's consider *X* to represent the feature space, Y to represent the label space and I to represent the sample data ID. The entire training dataset (I, X, Y) is made up of feature X, label Y and sample Ids I.

2.4.1 Horizontal Federated Learning

We share same feature space between datasets in horizontal learning but sample space is different, e.g., there are two banks having different customers, and there are very less number of customers common in both banks, but their business model is similar, i.e., feature spaces of both the banks are same. In [2], the author's proposed a collaboratively deep learning scheme where participants train their model in their local machines and only share subsets of update parameters. Google has also proposed a horizontal Federated Learning solution for Android phone updates [3].

In this, user enhance the model inputs locally and then updates it to a cloud, where the uploaded parameters are used to jointly train the centralized model. To protect the privacy of users, aggregated user updates are added [4]. Homomorphic encryption is added to provide security against any attacks on central server [3].

Hence, horizontal federated learning can be summarized as (Fig. 2.3)

$$X_i = X_j, \quad Y_i = Y_j, \quad I_i = I_j \quad \forall D_i, D_j, i \neq j. \tag{2.5}$$

In a horizontal Federated Learning system, it is typically assumed that the users are trustworthy and the server's security measures are reliable. In a horizontal Federated Learning system, the main concern regarding data participants' data security lies with the server, as it possesses the potential to compromise it. In such works, security evidence has been provided. Another security model that considers bad users was also presented, raising further privacy concerns.

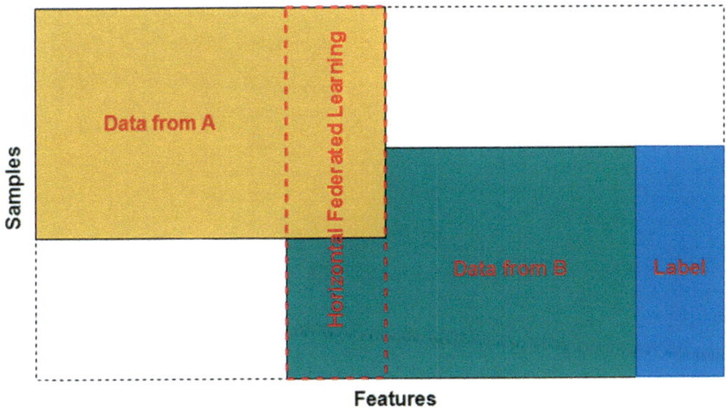

Fig. 2.3 Horizontal Federated Learning [4]

2.4.2 Vertical Federated Learning

It is also called as feature based Federated Learning because the datasets share different feature space and same sample ID [5]. For example, there are two distinct businesses in similar city, an electronic commerce company and a sweet shop, as they are in the same city the customers/users are likely to be same thus there is a large interaction of user space. Suppose the ecommerce website stores the user's expenditure and the sweet shop stores the customer's purchase history, that shows that their feature space is very distinct, suppose both businesses want a prediction model for anything common they can use both dataset having different feature space to train the model.

By merging these many variables and computing the training-loss and gradients in a way that protects privacy, vertically Federated Learning allows for the collaborative construction of models using data from both sides. Under that federal process, the essence and the condition of every indulged parties is the same, and the federal system.

Therefore, in this system we can say (Fig. 2.4)

$$X_i \neq X_j, \quad Y_i \neq Y_j, \quad I_i = I_j \quad \forall D_i, D_j, i \neq j. \tag{2.6}$$

Vertical Federated Learning requires honest participants, so to secure the privacy of the two parties, sometimes they hire a third party called as semi-honest third party (STP). STP doesn't indulge with either party and at the end of training both parties will hold their own data.

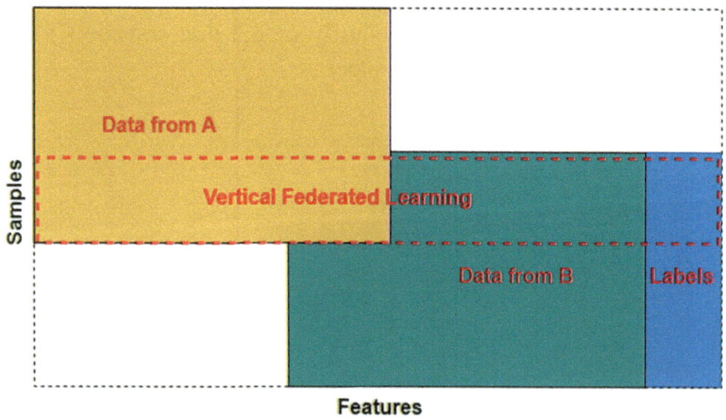

Fig. 2.4 Vertical Federated Learning [4]

2.4.3 *Federated Transfer Learning*

In Federated Transfer Learning, datasets differ in terms of both feature space and samples. Let us take two different entities, a bank and a company, bank is located in Delhi and company is located in Pune, so due to different geographical regions, there is very small intersection of their user group, and also because they have different businesses, only a little part of their feature space intersects, so we can apply transfer learning techniques in such scenarios. Federated Transfer Learning (FTL) is a noteworthy enhancement to existing Federated Learning systems as it addresses challenges that fall beyond the capabilities of current federated learning algorithms.

In FTL, we can say (Fig. 2.5)

$$X_i \neq X_j, \quad Y_i \neq Y_j, \quad I_i \neq I_j \ \forall D_i, D_j, i \neq j. \tag{2.7}$$

2.5 Federated Learning (Objective and Challenges)

Caldas et al. [6] Federated Learning differs significantly from the centralized (cloud-based) machine learning model, posing extra issues in the areas below. The data which is stored on billions of distant devices is used to train the single, worldwide statistical model in classic Federated Learning. In federated learning, typically, the purpose is to reduce the following objective function:

$$\min_w F(w), \quad \text{where } F(w) = \sum_{k=1}^{m} F_k(w) \tag{2.8}$$

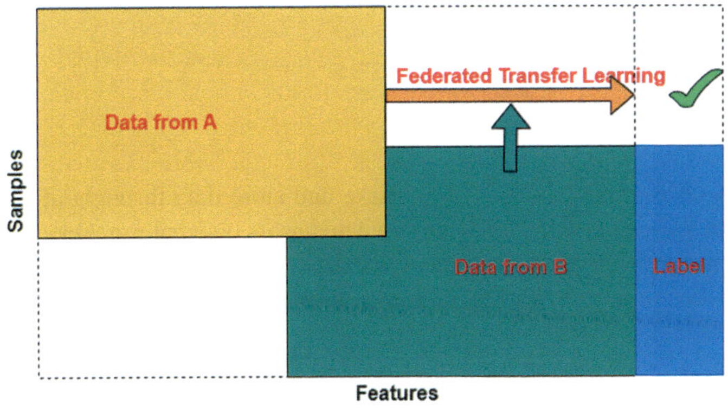

Fig. 2.5 Federated Transfer Learning [4]

m number of devices
F_k objective function for the kth device or empirical risk over local data
p_k relative impact of the kth device.

2.5.1 Challenges Faced by Federated Learning

2.5.1.1 Expensive Communication

A large number of devices connected together make a federated network, and network connectivity between then can be a lot slower than other local processing. In such networks, communication can be much more costly. In order for federated models to perform, it is critical to build efficient communication mechanisms that can transmit model updates or small messages as part of the training process rather than entire datasets over the network [7].

2.5.1.2 Systems Heterogeneity

Due to variations in hardware components such as CPU and RAM, network connectivity options like 3G, 4G, 5G and WI-FI, as well as differences in power supply, federated networks exhibit diverse storage, computing and communication capabilities on each device, including battery levels. Additionally, the number of active devices in a million-device network is typically limited due to constraints related to network bandwidth and system limitations. Only a fraction of the devices, potentially a few hundred, are usually active at any given time. Furthermore, device instability is not uncommon, and it is not unusual for an active device to encounter failures during an iteration. As a result of these system-level characteristics, the occurrence of stragglers (devices with slower processing) and the need for fault tolerance are more prevalent in federated networks compared to traditional data center configurations [6, 7].

2.5.1.3 Statistical Heterogeneity

In a federated network, devices often make and store data throughout the network in a random way. Every device's training data has its own pattern. One client's local dataset is not indicative of total data allocation. Likewise, when customers utilize their amenities or apps to varied measures, their offline datasets tend to fluctuate in size [7].

2.5.1.4 Privacy

In Federated Learning, the data that remains unseen is retained on the edge device since the training process occurs on every individual machine. However, in a collaborative model, as more participants join, the risk of attacks like inference on the training data increases. Privacy-focused Federated Learning techniques, such as safe multiparty computing or differential privacy, often prioritize privacy but may result in reduced model performance or system efficiency, which is a drawback. To create private Federated Learning systems, it is essential to comprehend and strike a balance between these costs, both in theory and through practical experimentation [6].

2.6 Related Work

Federated Learning is an emerging technology which has a lot of potential in different domains. Because of the expanding demand and very good future scope of Federated Learning a lot of surveys, reviews which has been done in the field. In [8], authors have demonstrated the effectiveness of Federated Learning by illustrating an architecture that exemplifies the technology's workings. On-device machine learning, privacy protection technologies, and edge computing all come together in this research to demonstrate Federated Learning. The authors have also simulated Federated Learning task an open-source platform called as TensorFlow federated and simulated learning tasks on MNITST and extended MNIST (E-MNIST) datasets. In [9], the authors presented a unique dynamic fusion-based Federated Learning strategy to increase COVID-19 detection accuracy, communication efficiency, and data privacy. The findings indicated that the suggested method is viable and outperforms the default Federated Learning configuration in terms of model correctness, fault tolerance, resilience and communication efficiency. The authors of [10] undertook a comparative study and experimental analysis of federated deep learning algorithms in the realm of cybersecurity for IoT applications. Industrial Internet of Things, internet of drones, edge computing, Internet of Healthcare Things and Internet of Vehicles are all covered in terms of privacy and security. In addition, an experimental examination of federated deep learning using RNN, CNN and DNN has been conducted. They utilized the MQTTest dataset, the Bot-IoT dataset, and the ToN_IoT dataset.

The authors of [11] have developed a taxonomy of several Federated Learning algorithms aimed at smart IoT applications. They have also highlighted a number of critical flaws in existing Federated Learning techniques, such as robustness, privacy, and excessive communication resource consumption. They concluded that scattered Federated Learning will be an important Federated Learning method for future IoT applications that will involve a large number of devices. In [12], authors have performed a survey, and categorize the Federated Learning approaches in a three-level classification scheme. They also talked about how Federated Learning may be used in networking and communication.

2.7 Current Trends

Due to its status as an emerging technology, Federated Learning offers numerous prospects despite the challenges and limitations it encounters, as mentioned earlier. Recognizing its immense potential, many researchers have delved into related studies and undertaken real-world applications of Federated Learning. In the following section, we will explore the current applications of Federated Learning.

2.7.1 Federated Learning in Mobile Devices

Google proposed the notion of Federated Learning to forecast user input from G'board on Android smartphones [13]. Beside text prediction, emoji prediction is another trending research matter in the branch of Federated Learning. Mobile edge computing (MEC) is a technology introduced in mobile devices in which, rather than incorporating cloud storage and computing into the core network to alleviate network congestion, providers prefer to provide even a service environment close to the customer at the cellular network's edge, but MEC has the drawback of posing a risk of data loss. In [14], author suggested that by combining Federated Learning and 'MEC' we can reduce the risk. Federated Learning and MEC system can be combine with the help of 'In-Edge AI' framework to optimize resource allocation problem. Qian et al. [15] developed a privacy-aware service placement scheme, which gives very good quality service.

2.7.2 Federated Learning in Internet of Things (IoT)

IoT devices use machine learning in a lot of areas such as administration, security, privacy and deployment. In the domain of IoT, there is a lot of very huge training data, which makes machine learning so efficient and effective. However, there are several privacy concerns when employing machine learning in the internet of things. The count of Internet of Things (IoT) devices is experiencing rapid exponential growth and because of that there is an exponential increase in the size of data created due to which it is becoming hard to send to cloud to train the model in a cloud-based environment. Federated Learning provides a simple answer to these problems. End devices (IoT devices) can train models using Federated Learning, so that they can't share their data; instead, they can only communicate model parameters.

2.7.3 *Federated Learning in Cloud Computing*

In classic data analytics, data is collected by smart devices such as IoT devices, smartphones etc. then this data is sent to a data center or server which is cloud-based. In this case, the data undergoes centralized processing. This strategy has two major drawbacks. The first drawback is that owners worry about their data privacy and because of data privacy concern they are hesitant to share their personal information with the cloud. Another disadvantage of using a cloud base system is there is a significant amount if delay in communication because the devices are portable and are widely scattered around the globe, on the other hand cloud servers are located in very isolated places and that's why distance between devices and the servers are very large which causes the delay in communication [4].

Yang et al. [4], the privacy and security issues which are raised by the centralized cloud-based systems can be solved by Federated Learning by allowing on-device joint learning of an ML model without requiring original data for training to be shared with any cloud-based third-party organization. It eliminates the need for original data which is going to an isolated cloud system, instead it require just model parameters to be shared with located close to the parties involved. It is worth noting that communication continues to be a problem in Federated Learning because model upgrades in current deep learning models can be in the gigabyte range, with literally billions of parameters potentially needing to be communicated with the network edge at each iterative process, for a significant number of cycles.

2.7.4 *Federated Learning in Edge Computing*

To overcome the drawbacks of cloud-based systems, edge computing is introduced. In edge computing, edge server's computational and storage capabilities are used in the training of ML models, which are nearer to the origin of data.

Edge computing provides several non-negligible benefits when combined with Federated Learning [16].

2.7.4.1 Security

When opposed to typical cloud-based architecture, edge computing adds some intriguing additional security to the Federated Learning process. In reality, because cloud computing is completely centralized, Federated Learning is particularly susceptible to distributed denial of service (DDoS) attacks and power interruptions On the other side, edge computing distributes data storage over a multitude of devices and data centers hampers attackers' ability to successfully cause any one outage or network shutdown [16].

2.7.4.2 Scalability

Edge intelligence may be leveraged to help organizations scale their businesses. This allows enterprises to deliberately expand their range and abilities of their edge network by placing end devices close to end users thereby avoiding the high expenditures of setting up centralized data centers. Edge intelligence, in conjunction with the related idea of co-location [11], enables businesses to simply and quickly increase the reach and cost-effectiveness of the edge network.

2.7.4.3 Reliability

The possibility of a failure in a network in an outlying data warehouse can be reduced by Edge computing. Edge computing enables the execution of machine learning tasks in proximity to end users, thereby improving the local user experience. Thanks to the mentioned autonomy feature, in the event of a neighboring regional data center failure, the edge devices retain the ability to carry out their operations using their own processing capabilities. The presence of numerous edge data centers and edge computing devices reduces the likelihood of a single failure causing a complete service interruption. As a result, numerous paths are available to redirect the shared model updates to in the event that users' connection to their Federated Learning services is lost.

2.7.4.4 Data Regeneration

If the feature abstraction chores are performed at the edge device level, it becomes possible to recreate an equivalent of the source data at the edge device level without requiring large amount of data from the servers present in cloud. This minimizes the availability consumption and delay of regeneration of data, as well as the overall Federated Learning process [16].

2.7.5 Federated Learning in 5G/6G Services

6G is next in line after 5G. 5G is all about 'connected objects' and 6G will be all about 'connected intelligence'. We can also say that 6G will rely on the foundations of superior AI to create novel and smart services by quickly gathering, communicating and analyzing data at any time and from any location. To deliver on this promise, 6G must be combined with the notion of ubiquitous AI, which will in future replace human centric things rather than machine or application-centric.

The design of classic machine learning is built in a centralized cloud-based system. It does not allow for the realization of the ideal of ubiquitous AI. Due to privacy concerns and a smaller number of communication resources, enabling wireless devices to upload all their data to the cloud for training machine learning models is impractical. Consequently, there is a gradual shift in academic and business efforts toward a collaborative ML framework which allows distant devices to process a collaborative learning model which will not transmit the data. Within this movement, FL remains a central component.

The connection between FL and 5G/6G networks is symbiotic, as Federated Learning offers numerous advantages for the advancement of 5G/6G network adoption, while 5G/6G networks participate in the achievement of Federated Learning. In practice, Federated Learning serves as an effective solution for various challenges within 5G/6G networks.

2.8 Future Scope

With the exponential growth of data comes an increased risk of data theft and privacy breaches. Consequently, Federated Learning holds significant potential for the future. In the following section, we will explore the future prospects and potential applications of Federated Learning.

2.8.1 Self-Driving Vehicles

Self-driving vehicles are the future of automobile industry. As stated in [17], the major problem in self-driving vehicles are the communication overhead and federated learning promises to solve this problem. Authors of [18] has performed an experiment and proved that Federated Learning models are as reliable and even faster than the classic machine learning models in the case of self-driving vehicles.

2.8.2 Wireless Communications

Federated Learning has a privacy-preserving nature, due to which it has a lot of scope in wireless communications. ML have to be introduced in wireless networks for system design, and that's why Federated Learning has a lot of future scope in the field of wireless communication [19].

2.8.3 Blockchain in Edge Computing

Federated learning is mixed with blockchain which is called as FL-blockchain. Blockchain is a digital ledger which works on peer to peer network.

2.9 Conclusion

In today's world, many things are dependent on AI and machine learning, but AI and machine learning has a lot of drawbacks such as data privacy threat, security threat, also it is not feasible to transfer such large volume of data in order to train a model that's why companies are shifting toward different frameworks such as edge computing, distributed machine learning and even though Federated Learning has a lot of shortcomings as discussed in this paper still Federated Learning can be a good option to replace classic machine learning. Researchers are constantly working to minimize the shortcomings of Federated Learning. Federated Learning is now combining with many fields like health care, IoT, networking, communication. There is a lot of future scope for researchers and opportunities in the field of Federated Learning .

References

1. Kairouz, P., McMahan, H.B., Avent, B., et al.: Advances and open problems in federated learning. Foundations and Trends® in Machine (2021)
2. Shokri, R., Shmatikov, V.: Privacy-preserving deep learning. In: Proceedings of the 22nd ACM SIGSAC Conference on Computer and Communications Security (2015). https://doi.org/10.1145/2810103.2813687
3. McMahan, B., Moore, E., Ramage, D., Hampson, S., Arcas, B.A.: Communication-efficient learning of deep networks from decentralized data. Artif Intell Stat 1273–1282. PMLR (2017)
4. Yang, Q., Liu, Y., Chen, T., Tong, Y.: Federated machine learning. ACM Trans. Intell. Syst. Technol. **10**, 1–19 (2019). https://doi.org/10.1145/3298981
5. Nock, R., Hardy, S., Henecka, W., et al.: Entity resolution and federated learning get a federated resolution (2018). arXiv preprint. https://doi.org/10.48550/arXiv.1803.04035
6. Caldas, S., Konecˇny, J., McMahan, H.B., Talwalkar, A.: Expanding the reach of federated learning by reducing client resource requirements (2019). arxiv. https://doi.org/10.48550/arXiv.1812.07210 Focus to learn more
7. Wahab, O.A., Mourad, A., Otrok, H., Taleb, T.: Federated machine learning: survey, multi-level classification, desirable criteria and future directions in communication and networking systems. IEEE Commun. Surv. Tutorials **23**, 1342–1397 (2021). https://doi.org/10.1109/comst.2021.3058573
8. Das, S.K., Bebortta, S.: Heralding the future of federated learning framework: architecture, tools and future directions. In: 2021 11th International Conference on Cloud Computing, Data Science & Engineering (Confluence) (2021). https://doi.org/10.1109/confluence51648.2021.9377066
9. Zhang, W., Zhou, T., Lu, Q., et al.: Dynamic-fusion-based federated learning for covid-19 detection. IEEE Internet Things J. **8**, 15884–15891 (2021). https://doi.org/10.1109/jiot.2021.3056185
10. Ferrag, M.A., Friha, O., Maglaras, L., et al.: Federated deep learning for cyber security in the internet of things: concepts, applications, and experimental analysis. IEEE Access **9**, 138509–138542 (2021). https://doi.org/10.1109/access.2021.3118642
11. Khan, L.U., Saad, W., Han, Z., et al.: Federated learning for internet of things: recent advances, taxonomy, and open challenges. IEEE Commun. Surv. Tutorials **23**, 1759–1799 (2021). https://doi.org/10.1109/comst.2021.3090430

12. Wahab, O.A., Mourad, A., Otrok, H., Taleb, T.: Federated machine learning: survey, multi-level classification, desirable criteria and future directions in communication and networking systems. IEEE Commun. Surv. Tutorials (2021). https://doi.org/10.1109/comst.2021.3058573
13. Yang, T., Andrew, G., Eichner, H., et al.: Applied federated learning: improving google keyboard query suggestions (2018). arxiv. https://doi.org/10.48550/arXiv.1812.02903
14. Wang, X., Han, Y., Wang, C., et al.: In-edge AI: Intelligentizing mobile edge computing, caching and communication by Federated (2019)
15. Qian, Y., Hu, L., Chen, J., et al.: Privacy-aware service placement for mobile edge computing via federated learning. Inf. Sci. **505**, 562–570 (2019). https://doi.org/10.1016/j.ins.2019.07.069
16. Wang, S., Tuor, T., Salonidis, T., et al.: Adaptive federated learning in resource constrained edge computing systems (2019a). arXiv:180405271. https://doi.org/10.48550/arXiv.1804.05271
17. Elbir, A.M., Soner, B., Coleri, S., et al.: Federated learning in vehicular networks (2022). arXiv:200601412. https://doi.org/10.48550/arXiv.2006.01412
18. Zhang, H., Bosch, J., Olsson, H. H.: End-to-end federated learning for autonomous driving vehicles. In: 2021 International Joint Conference on Neural Networks (IJCNN) (2021a). https://doi.org/10.1109/ijcnn52387.2021.9533808
19. Niknam, S., Dhillon, H.S., Reed, J.H.: Federated learning for wireless communications: motivation, opportunities, and challenges. IEEE Commun. Mag. **58**, 46–51 (2020). https://doi.org/10.1109/mcom.001.1900461

Chapter 3
Systematic Literature Review on Role of AI in IoT-Based Healthcare Solutions

Manisha, Ashish Sharma, and Rekha Jain

Abstract There is wide applicability of IoT in many areas of life, one major are of them is Health Care IoT is showing potential application in patient health care. With the rapid advancements in technology and Internet, now conventional techniques of healthcare are vanishing and electronic health care systems are emerging very rapidly. This has increased patient safety, accessibility of health services at remote locations and reduced operational costs. This current study summarizes IoT-based health care solutions with introduction, technology used and application areas. By merging IoT and Artificial Intelligence, systems are becoming more accurate for monitoring and prediction tasks. AI has proven its potential in healthcare area. There is massive data which is continuously coming from various sensors attached to patients' bodies. Machine Learning and Deep Learning models can be used to get insight from this heap of data. Researchers are working on combining IoT and AI to develop more secure, more accurate, more efficient and less costly health care solutions. This paper focuses on AI-ML efforts in this area along with the key findings of various researches. Moreover, various research gaps and challenges has been also discussed to get future directions.

Manisha and Rekha Jain—These authors contributed equally to this work.

Manisha
Department of CSE, IIIT Kota, Kota, India
e-mail: 2021kpad1004@iiitkota.ac.in

A. Sharma (✉)
Department of CSE, Manipal University Jaipur, Jaipur, India
e-mail: ashishsharma.fitt@gmail.com

R. Jain
Department of Computer Application, Manipal University Jaipur, Jaipur, India
e-mail: rekha.jain@jaipur.manipal.edu

P. Nanda et al. (eds.), *Intelligent System and Data Analysis*, Smart Innovation, Systems and Technologies 398, https://doi.org/10.1007/978-981-97-5200-3_3

3.1 Introduction

IoT is having impression in all walks of life, predominantly in the space of health care. The usability of IoT is helping improve quality of life and increasing patient's life expectancy. The exponential rate of innovation is pursing the conventional healthcare techniques to the periphery and is being replaced by cutting edge technology that is heavy on IoT [1]. The result is, there is a paradigm shift in the reach of healthcare services even to remote locations and has ensured higher standards in patient safety. Since such a massive shift has enabled access to services from anywhere, it is also bringing down the cost of service.

There is lot of excitement about. Internet of Things (IoT) as enabler of superior and safer service delivery alternative. The socio-economic benefits can not be ignored, with the emergence of IoT there is a vast pool on employment opportunities that are coming in light [2].

Contrary to the conventional healthcare system that was so dependent of physical examination and long stays at hospitals, that had huge impact on the cost to patients , more so in remote areas. The IoT has presented as a boon by eliminating lot of these needs. Telehealth is the future and with lot of IoT-based devices like wearable tech. Real time monitoring disease diagnosis has become so easy. Home testing kits and quick results are now possible. Health care system has become patient centric from hospital centric. Many clinical tests can be done at home with the help of technology and results can be obtained immediately with precision. These results can be directly sent to hospitals for further analysis from remote areas [3].

IoT: when a network of interconnected electronic devices that have sensors, software and systems are connected over the internet they form their own web called Internet of Things (IoT). These could be simple daily household items to complex industrial utilities. This is a huge ecosystem of interconnected devices that exchange data for a targeted decision making and action. IoT is an advanced system for automation and analytics that is an amalgamation of networking, cloud messaging, artificial intelligence, electronics, etc. The end objective is to deliver expected services with control and transparency. While discussing IoT, there are two important concepts : Communication and Computing.

Communication: Now a days the interconnected web of smart devices can exchange meaningful data for decision making. It has become possible with the help of communication technology. From pace makers to other monitoring devices, are connected through Internet, they can collect and generate meaningful decisions that helps patients. There is a web of actuators, sensors, and devices which are providing foundation to IoT.

Computing: With the onset of IoT, the computing also has taken huge steps forward, computing is no more just real time but it's distributed and driven by an intelligent network of sensor-based devices that are impacting our everyday lives and more so some critical functions like healthcare.

3.2 AI in HIoT

Information Technology is at the epi center of the disruption taking place in expediting health diagnostics and treatment in Healthcare domain. The innovation is focused in offering real time intelligence to enable automation and monitoring of healthcare services. This offers leverage for remote diagnostics and assessment that is impacting significant cost advantage in physician's in person visits and accessibility to advanced services to remote parts.

Not only is the IoT application limited to treatments but with the advancement of wearable tech and sensor based ingestible and embedded technology the patterns of mobility helps build user behaviors [4]. These data points are effectively applied on need based Artificial intelligence driven algorithms to suggest possible health issues and suggests proactive treatments for the future.

Contrary to the traditional cloud based architectures to analyze Big Data that still lacks real time decisions, the IoT applications come to rescue as real time critical decision making is required as we are talking about human life and it's critical body functions. The emergency situations have to be dealt with promptness and swiftly so the cloud architecture that's dependent on bandwidth and lot of external factors is not reliable. Lot of data is generated by wearable, embedded sensors continuously, this data can be collected and effectively processed to identify critical situations with the help of AI and ML-DL algorithms be dealt with promptness and swiftly so the cloud architecture that's dependent on bandwidth and lot of external factors are not reliable. Lot of data is generated by wearable, embedded sensors continuously all the time , this data can be collected and effectively processed to identify critical situations with the help of AI and ML-DL algorithms [5]. These algorithms help in generating intelligent patterns which can helpful in inferencing. Sometimes patients need immediate attention, so these type of algorithm can help in alarming situation to take decisions.

Along with exciting opportunity also comes some major challenges, as the wearable tech and other health-based IoT devices generate huge amount of data "The Big Data" the challenge is to filter through this huge pile of heterogeneous data and build meaning out of it.

There is lot of data redundancy. The mining of such volumes of data becomes foremost important to offer consistent and seamless service. Such endeavors to process and transmit is a very resource heavy task and this needs to be carried out real time [6]. So it is wise to have local processing capabilities and the transmitted data is reduced to bring down transmission costs over the wide networks. Summarized and concise data is more insightful.

The key drivers of big data is the ability to filter meaningful data meant for decision making from the huge set of redundant data. Such methodologies are required to bring efficiency to the ecosystem of IoT as it drastically reduces energy consumption needs needed for sustained long term operations [10]. AI can be advantageous in this situation, ho analyze this heap of data and to generate insightful decisions.

Table 3.1 Comparative study of role of AI in IoT-based health care

S. No.	Author(s), year	Technique	Data	Key findings
1	Kishor and Chakraborty [7]	Support vector machine (SVM), random forest (RF), decision tree, K-nearest neighbor, adaptive boosting	Data collected from hospitals	Proposed a model using machine learning for health care to detect 9 diseases hepatitis, thyroid, heart disease, diabetes, liver disorder, dermatology, breast cancer, surgery, etc. at early RF classifier demonstrated highest accuracy as 97.62%, specificity as 97.81%, sensitivity as 99.67% along with AUC 99.32% of different disease
2	Shu et al. [8]	CNN, LSTM, biomodal deep autoencoder	EEG, ECG, GSR, RSP, EMG, HR signals from sensors	Developed a model for emotion detection based on physiological signals. Complete model for emotion detection, emotion elicitation, feature extraction, classification was developed using input signals from patients to identify depression
3	Aslan [9]	GoogLeNet-based deep learning method	EEG signals, GAMEEMO dataset	A deep learning approach based on GoogLeNet model was proposed to extract the important features of EEG signals and to enhance its performance for the detection of human emotions. Performance of the model was compared with some standard ML classification methods such as KNN, SVN, RF. Accuracy value of this model was found better than existing classification methods. Results of the current work demonstrated that the developed method can be used for automated detection and identification of human emotions in a real-time
4	Yang et al. [5]	GBDT (gradient decision tree)	Metropolitan hospital. Patients used sensors to provide samples	A machine learning model was developed for the early and fast detection of high-risk SARS-CoV-2 by adding demographic features of patients' (age, sex, race), including 27 routine lab tests for detecting SARS-CoV-2 infection status in individual person. This model was developed majorly for front health-line workers
5	Parisi et al. [10]	Lagrangian support vector machine (LSVM) for feature selection and classification	Irvine machine Learning repository of University of California	In case of chronic hepatitis, AI based hybrid model was developed for the prediction of the prognosis of disease in patients. For proposed model area under the receiver operating characteristic (ROC) curve, was 100%

(continued)

Table 3.1 (continued)

S. No.	Author(s), year	Technique	Data	Key findings
6	Janarthanan et al. [11]	Deep learning assisted reconstructed unsupervised coder (UDR-RC)	Wearable sensors	An innovative model was developed using unsupervised deep learning assisted reconstructed coder. This model performs data optimization at the time of preprocessing for on-nodule wearable sensors to find minimum execution time 11.25 ns on test set. It gets better performance for extraction and selection of features in HAT activity in neural networks. The results were promising with accuracy value of 97.5 and 0.52% was the mean squared error
7	Yang et al. [12]	Classification complexity estimating algorithms and principal components analysis	Sensor generated data	Introduced a system for stroke rehabilitation which is IoT-enabled and gets data from an armband which is smart and wearable. This system uses data from armband, 3-D dexterous robot hand and machine learning algorithms. Results for verification were very promising and system identified all gestures with on an average 96.20% accuracy for different cases
8	Malche et al. [13]	ML models	Data of physical activity and voice. Physiological data collected from patients, including ECG, heart rate, body temperature, blood oxygen level respiration rate, blood pressure, and EEG using sensors.	Developed an intelligent remote system for tracking the activity of patient, it can monitor vitals like heart rate, body temperature, pattern of breathing while doing activities and activities like sleeping, exercising, running, and walking remotely using attached sensors. Machine Learning algorithms were used for the detection of different activities and to analyze respiratory health of a patient during these activities. execution of ML algorithms was on local device for the identification of various activities and respiratory health of the patient was analyzed at sensor node using microphone to collect voice signal using ML model
9	Pandey [14]	Logistic regression, SVM	Sensor-based data	Proposed a solution to find whether a person is in stress or not using ML algorithm. Hardware device and software is developed for capturing data and then analyzing it. Finally Logistic regression and SVM were applied for the classification

(continued)

Table 3.1 (continued)

S. No.	Author(s), year	Technique	Data	Key findings
10	Rijwan et al. [15]	CNN	Data from home of patients and hospitals	This work presented effect of ML for healthcare and developed one application called "Medicolite", which comprises different modules to provide convenience for various health related problems. This is smart health care system which detects problems and can prescribe drugs for saving life in case of need. CNN classifier was used for the detection and it gave better performance than SVM and Naïve Bayes. The complete system is based on ML from appointments to diagnosis and prescribing drugs

Now AI is readily available to support healthcare personnel for helping in variety of services from administrative workflow to documentation of clinical processes, patient monitoring remotely and to provide a support in image analysis and medical device automation. Table 3.1 represents the comprehensive review of literature.

3.3 Research Gaps

Health care is fastest growing industry. In healthcare IoT provides many great benefits, and when AI is merged with this, it increases accuracy and precision in results, but there are some challenges also present when developing these systems which needs to be considered. Internet of Things-based health care solutions cannot be considered for designing without considering following challenges:

- Massive data: Data produced by sensors, for single health care application multiple devices are needed. All these devices generated massive data in real time, so organization and extraction of relevant data is a challenging task [16].
- Scalability: Solutions so far generated works good in case of small number of patients, so smart solutions are needed to handle large amount of data [17]. Cloud and AI driven solutions can be helpful in this.
- Availability of datasets for testing and verification of solutions is limited. Patients' data is normally kept confidential, so standard data sets are needed for the better design.
- Security : Patients get lots of benefits from IoT based healthcare, but it also create threats for patients. Medical devices can be an easy target for hackers because these devices are connected to the Internet and they can steal or modify patients' data, which can be dangerous or life threatening for the patients.

- Deep learning and machine learning show a black-box structure, since it is very difficult to understand and interpret the relationships between the inputs and the outputs and explainability is of great importance for making trust between users and models to facilitate the model adoption. So application of XAI concept is needed with the developed models.
- Generalization of the solutions : Every patient is different and shows different symptoms in different situations, so generated solutions are very specific. Hence there is a need of development of generalized solutions, which can cater large number of patients.

3.4 Conclusion

Research and technology innovation has seen leaps of growth over the past decade. When it comes to connecting health professionals, medical devices and sensors, Internet of things (IoT) has shown promising results and taken the landscape by storm. From pacemakers to cochlear implant, IoT has demonstrated what the future holds for human civilization. Distance is just a matter of time, as everything is managed and controlled from far away. Remote monitoring of patients is possible now. This has a significant impact on fatality rate and superior healthcare services.

This review paper starts with the basic introduction of IoT, Health Care, Communication, and Computation. After that merging of IoT and healthcare has been discussed. There are majorly three technologies used in IoT-based health care system, namely Communication Technology, Location Technology, and Identification Technology. Along with this, here in this paper information is explored for the currently available healthcare services which are based on IoT technologies. By incorporating these concepts, this technology has made monitoring, diagnosis and measuring vital information of patients, possible at remote locations. Healthcare services are now patient centric than hospital centric.

AI is boon and it is becoming essential part of human's life. Machine learning is used to create which can learn from labeled or unlabeled data, as well as can learn from their environment. Machine learning already has been shown its potential in various areas and health care is one of them. In this industry, it provides significant advantages by developing decision and prediction methods. Further inclusion of AI in IoT-based healthcare system has been discussed. Table 3.1 shows various AL-ML research efforts which have been developed. Majorly AI-ML is used in patient monitoring, disease identification and prediction of critical situations. At the end research gaps are discussed, which gives some idea about challenges and future directions for further research. Overall this paper comprised introduction of IoT-based health care solutions and various AI-ML efforts continuing in this area.

References

1. Ahmadi, H., Arji,G., Shahmoradi, L., Safdari, R., Nilashi, M., Alizadeh, M.: The application of internet of things in healthcare: a systematic literature review and classification. Universal Access Inf. Soc. **18**(4), 837–869 (2019)
2. Pradhan, B., Bhattacharyya, S., Pal, K.: IoT-based applications in healthcare devices. J. Healthcare Eng. (2021)
3. Islam, M.M., Rahaman, A., Islam, M.R.: Development of smart healthcare monitoring system in IoT environment. Sn Comput. Sci. **1** (2020)
4. Alshamrani, M.: IoT and artificial intelligence implementations for remote healthcare monitoring systems: a survey. J. King Saud Univ.-Comput. Inf. Sci. (2021)
5. Yang, H.S., Hou, Y., Vasovic, L.V., Steel, P.A., Chadburn, A., Racine-Brzostek, S.E., Velu, P., Cushing, M.M., Loda, M., Kaushal, R.: Routine laboratory blood tests predict SARS-COV-2 infection using machine learning. Clin. Chem. **66**(11), 1396–1404 (2020)
6. Saheb, T., Izadi, L.: Paradigm of IoT big data analytics in the healthcare industry: a review of scientific literature and mapping of research trends. Telematics Inform. **41**, 70–85 (2019)
7. Kishor, A., Chakraborty, C.: Artificial intelligence and internet of things based healthcare 4.0 monitoring system. Wirel. Personal Commun., 1–17 (2021)
8. Shu, L., Xie, J., Yang, M., Li, Z., Li, Z., Liao, D., Xu, X., Yang, X.: A review of emotion recognition using physiological signals. Sensors **18**(7), 2074 (2018)
9. Aslan, M.: CNN based efficient approach for emotion recognition. J. King Saud Univ.-Comput. Inf. Sci. (2021)
10. Parisi, L., RaviChandran, N., Manaog, M.L.: A novel hybrid algorithm for aiding prediction of prognosis in patients with hepatitis. Neural Comput. Appl. **32**(8), 3839–3852 (2020)
11. Janarthanan, R., Doss, S., Baskar, S.: Optimized unsupervised deep learning assisted reconstructed coder in the on-nodule wearable sensor for human activity recognition. Measurement **164**, 108050 (2020)
12. Yang, G., Deng, J., Pang, G., Zhang, H., Li, J., Deng, B., Pang, Z., Xu, J., Jiang, M., Liljeberg, P.: An IoT-enabled stroke rehabilitation system based on smart wearable armband and machine learning. IEEE J. Transl. Eng. Health Med. **6**, 1–10 (2018)
13. Malche, T., Tharewal, S., Tiwari, P.K., Jabarulla, M.Y., Alnuaim, A.A., Hatamleh, W.A., Ullah, M.A.: Artificial intelligence of things-(AIoT-) based patient activity tracking system for remote patient monitoring. J. Healthcare Eng. (2022)
14. Pandey, P.S.: Machine learning and IoT for prediction and detection of stress. In: 2017 17th International Conference on Computational Science and Its Applications (ICCSA), pp. 1–5. IEEE (2017)
15. Khan, R., Srivastava, A.K., Gupta, M., Kumari, P., Kumar, S.: Medicolite-machine learning-based patient care model. Comput. Intell. Neurosci. (2022)
16. Selvaraj, S., Sundaravaradhan, S.: Challenges and opportunities in IoT healthcare systems: a systematic review. SN Appl. Sci. **2**(1), 1–8 (2020)
17. Ahad, A., Tahir, M., Yau, K.-L.A.: 5G-based smart healthcare network: architecture, taxonomy, challenges and future research directions. IEEE Access **7** (2019)

Chapter 4
A Hybrid Machine Learning Model for Lung Disease Prediction

Poonam Devi and Shrddha Sagar

Abstract Respiratory disease is common throughout the world, and even during the Covid-19 pandemic, lung ailments were at their peak worldwide. A lot of research has been done and is currently in progress in this field. Some researchers use only "machine learning (ML) or deep learning (DL) techniques", while others use ML classification techniques with feature selection. This paper focuses on developing a hyper-parameterized hybrid model based on the best machine learning classification techniques for lung disease prediction. The big data framework of healthcare and accurate prediction of respiratory disease will help medical practitioners make decisions and cure diseases. In this article, we propose a new framework, a "hybrid machine learning model based on feature selection, feature optimization and classification techniques" named hybrid machine learning framework (HMLF), which will improve upon existing models in lung disease prediction and overcome the problems of underfitting and overfitting.

4.1 Introduction

The type of diseases and their effects on health are increasing rapidly due to climate change, global warming and lifestyle changes. "According to the World Health Organization's (WHO) 2019 report, Chronic Obstructive Pulmonary Disease (COPD) is the third leading cause of death worldwide", and the Lung Cancer Research Foundation's 2022 report indicates that approximately 236,740 people in the USA will be diagnosed with lung cancer. Air pollution and harmful gases are the primary causes of respiratory diseases such as COPD and lung cancer. In India, which is a developing country, the death rate is three times higher than that of developed countries like the

P. Devi (✉) · S. Sagar
Galgotias University, Noida, Uttar Pradesh, India
e-mail: Poonam.20scse3020009@galgotiasuniversity.edu.in

S. Sagar
e-mail: shrddha.sagar@galgotiasuniversity.edu.in

P. Nanda et al. (eds.), *Intelligent System and Data Analysis*, Smart Innovation, Systems and Technologies 398, https://doi.org/10.1007/978-981-97-5200-3_4

USA, according to the WHO 2019 report [1]. In India, where 64.61% of the population is rural, health services in those areas are not up to par, and people living in rural areas have to travel to cities for health check-ups or any health services. Therefore, it is essential to provide remarkable health services and tools that clinical staff can use to perform routine check-ups in rural areas. With this help, they can assist patients if there is a possibility of a critical condition in the near future, allowing patients to receive timely treatment and save lives. Machine learning tools can make all of this possible.

A lot of research has been done on lung disease prediction using machine learning techniques for chest X-ray image analysis processes [2]. However, in India, where more than 60% of the population lives in villages, it is not possible to have "X-ray machines, magnetic resonance imaging (MRI), or computed tomography (CT) scan" machines everywhere. Therefore, it is best to diagnose early stages of respiratory problems by predicting patient symptoms using advanced machine learning tools. This paper will focus on developing a hyper-parameterized prediction model based on the best machine learning classification techniques with feature selection and optimization methods. This will aid in the timely identification of lungs' illness. The purpose of the present research is not to replace advanced tools such as X-ray machines, MRI, CT scans, or specialist physicians but to provide doctors with an alternative option that assists them in all clinics [2, 3].

4.1.1 Feature Selection

The datasets available on the Internet are not processed or organized as per machine learning models. Therefore, to choose the best features is an indispensable data pre-processing activity in which we select only valuable attributes from our dataset that provide a meaningful contribution to our machine learning model [4, 5]. The "filter technique, wrapper method and embedding method" are the three ways for attribute selection [6].

4.1.1.1 Filter Method

The filter method as shown in Fig. 4.1 is a fast and inexpensive data pre-processing technique in which we use a machine learning algorithm to remove redundant and duplicate features from the dataset [7].

Information Gain Method: In the information gain method, we choose the attribute that has the highest information gain value as the splitting attribute.

$$\text{infoGain}(a_i, n) = \sum_{i=1}^{m} \frac{a_i}{n} \log_2\left(\frac{a_i}{n}\right) \tag{4.1}$$

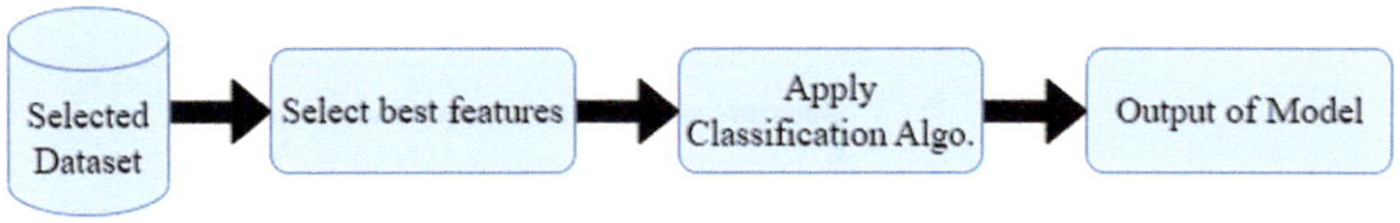

Fig. 4.1 Feature selection process by filter method

where 'a_i' is the number of values in any particular class, 'm' is the total types of class in a particular attribute, and 'n' is the total number of tuples in that particular dataset [8].

Chi-square test: This is a description of a nonparametric test that doesn't require any exact information about the population and is a distribution-free testing method. The method is used to describe the magnitude of the discrepancy between theory and observation [9].

$$\chi^2 = \frac{\sum(O - E)^2}{E} \tag{4.2}$$

Pearson coefficient correlation: The relationship between two attributes is called correlation, and it is measured by the correlation coefficient 'r', which ranges from -1 to 1 or $-1 < = r < = 1$. If both attribute increase or decrease at the same time, the relation is called a positive correlation. If the value of one attribute is increasing while the value of another correlated attribute is decreasing or vice versa, it is called a negative correlation [10].

4.1.1.2 Wrapper Method

The wrapper method as shown in Fig. 4.2 is a greedy method of feature selection that uses both forward selection and backward elimination methods to select the best features from a dataset. It provides a more optimal subset of attributes compared to the filter method, but it is more expensive [4, 11].

Forward selection: In forward selection method, we keep adding the feature which best improve our model and continue this process till an addition of new variable does not improve the performance of the model [4, 12].

Backward elimination: In backward elimination method, we follow an iterative approach where initially we start the process with all attributes, and after each

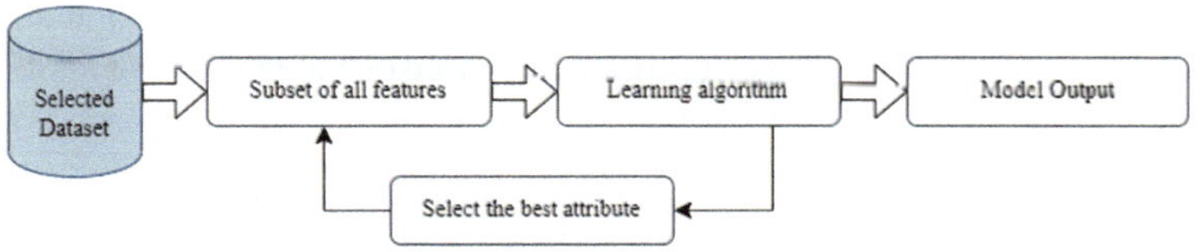

Fig. 4.2 Feature selection by wrapper method

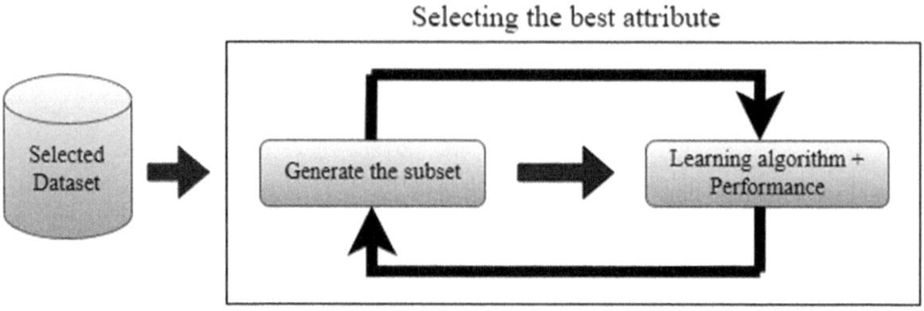

Fig. 4.3 Feature selection by embedded method

iteration, we remove the least significant attribute. We stop the process when no improvement in the model is observed after the attribute is removed [13].

Recursive elimination: This algorithm works on a greedy optimization technique, it recursively fetches out the best features and eliminates the least important attributes from the dataset. To eliminate the least important attributes, it uses an estimator function, and to select the best attributes, it uses an important attribute function [12].

4.1.1.3 Embedded Method

This method as shown in Fig. 4.3 uses its own built-in algorithm to select the best attributes. To create this built-in method, it chooses a subset of attributes and applies all permutations and combinations on them. The best attributes with the highest accuracy are selected [14].

4.1.2 Feature Optimization Techniques

Optimization is a process where we find the optimal values of attributes based on a cost minimization problem, and the selected attributes will fulfil all model requirements [15, 16]. Some selected techniques or algorithms are given below:

4.1.2.1 PSO ("Particle Swarm Optimization")

"Particle swarm optimization is a nature-inspired optimization algorithm proposed by Kennedy and Eberhart in 1995", which is particularly used to resolve computationally difficult optimisation issues. This algorithm is founded on movement and intelligence of swarms, where particles try to reach their best possible solution in each iteration. In each iteration, particles update their position according to their previous position and the relative position of nearby particles. Therefore, this algorithm works based on local and global values. To implement PSO algorithm, we need to use three vectors: "*x*-vector to record the current position of the particle, *p*-vector to record the best solution found by the particle and v-vector to determine the direction of particles for which they travel". All of these vectors update themselves in each iteration [15, 17].

4.1.2.2 GA (Genetic Algorithm)

The GA algorithm is based on Darwin's theory of biological evolution, which is based on the fitness function value. This means that genes whose fitness function value is higher will survive, and those with lower values will be lost. Therefore, it is a biologically inspired evolutionary algorithm that works based on a heuristic approach. This algorithm is used to optimize any problem and also for searching. In this algorithm, we randomly select the best solution out of a set of solutions using an iterative method, and in each iteration, we get a more optimal solution than the previous one [18].

To get an optimal solution, GA uses stages: "initial population, fitness function, selection, crossover and mutation". Let us initially have a population space of 'n' possible solutions, out of which we randomly choose a set of chromosomes that consist of binary values '0' or '1'. The selected chromosomes have values that entirely cover the given search space. These selected chromosomes are then evaluated with the help of a fitness function and rearranged according to their fitness function values. Then, these rearranged chromosomes are passed through genetic operations like crossover and mutation to get modified offspring. We repeat this iteration and modify chromosome values until we get an improved fitness function value [19].

4.1.2.3 Cuckoo Algorithm

It's a metaheuristic optimization-based algorithm given by Xin-She Yang and Suash Deb in 2009. If there are 'n' host nests/solutions in the initial population set, then it chooses a random 'ith' solution based on the performance of Levy flights, hence also called Levy flights cuckoo search algorithm.

Next, it calculates the fitness function '$f(x_i)$' value and again chooses a random nest 'j' among 'n'. If '$f_j > f_i$', then it replaces 'i' by the new solution, otherwise abandons the worst solutions/nests. It keeps the best solution and passes the current best solution to the next generation. The efficiency of this algorithm is higher even though it has fewer control variables [20, 21].

4.1.3 *Hybrid Technique*

Hybrid means a combination of two or more techniques. In hybrid inheritance, we combine two or more inheritances, and in the hybrid technique, we can combine two or more ML techniques to get an optimal model. As we know, sometimes a single technique may not provide a better solution, and at that time, we may use a

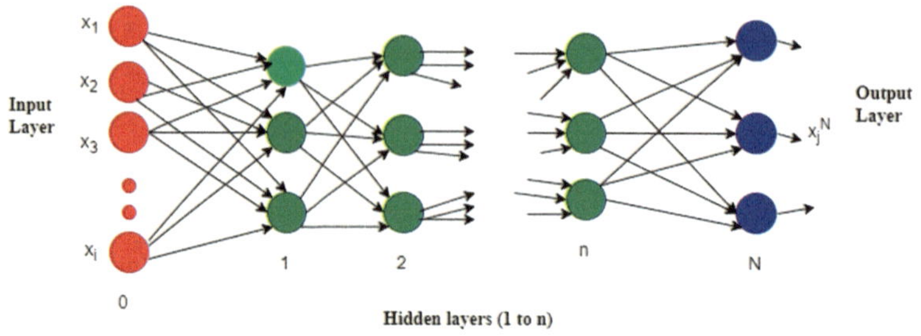

Fig. 4.4 Multilayer perceptron

hybrid approach to get a better result. Even though ensemble techniques are also a combination of several models like bagging, boosting and random forest, the classification techniques used in this model are fixed. However, in the hybrid model, we can combine techniques of our choice [22], such as multilayer perceptron, random forest, bagging and boosting.

4.1.3.1 Multilayer Perceptron

It is a deep learning technique which is also named as feed forward neurological system. It is made up of three layers: "the first is the input layer, the second is the hidden layer (an intermediate layer) and the third is the output layer".

Multilayer perceptron as shown in Fig. 4.4 is a deep learning technique, also known as a feed forward neural system. It comprises of three layers: "the input layer, the hidden layer and the output layer".

In the multilayer perceptron, the input layer (x) is located at the 0th position and its task is to produce output. Layers from 1 to $n-1$ participate in nonlinear mapping of the feature vector x to a new feature x_j, and the final layer, N, is called the productivity layer. In this model, all intermediate layer nodes are interconnected to each other in a way that the production of the first layer is connected to the contribution of the next layer, where each edge has a weight called ($W_{ij}(n + 1)$) [23].

4.1.3.2 Random Forest

A technique where we grow a decision tree from multiple bootstrapped samples by choosing features at random. In random forest, we split the dataset into multiple training sets and make a decision tree of each training dataset, same as the bagging algorithm. But here in the random forest algorithm, we not only bootstrap the training dataset but we also choose a subset of features available to us at every node in our tree. So it will reduce the correlation between the trees. Hence, a random forest technique has low bias and low variance [24].

Working of random forest algorithm as given:

i. If there are N rows and M features in any dataset, split the dataset into T_1, T_2, ...T_n bootstrapped samples or trees.
ii. Choose 'n' and 'm' best features rows at random for each tree where $n < N$ and $m < M$.
iii. Train the decision tree T_1 using selected dataset.
iv. Do row sampling with replacement and train all the trees (T_2, ..., T_n) by following steps ii and iii.
v. At last aggregate all the bootstrapped samples on the bases of majority vote of output.

4.1.3.3 Bagging

Bagging is also known as "bootstrap aggregation". It is an ensemble-based approach comprised of a group of trainees, and these learners make independent errors. In the bagging algorithm, suppose we have a training dataset and 'K' learners; then, the training dataset is also split into K different subsets. The subsets may be small, so there will be high variance, but when we combine all these K learners by bootstrap, variance gets reduced. In the bagging algorithm, we generate multiple models using the same dataset and by applying the same algorithm. However, in the end, we combine all these small models to produce a single one. Bagging is a simple and easy technique that provides better results than the decision tree [25, 26].

4.1.3.4 Boosting

Boosting is an ensemble-based algorithm, also called an iterative-based algorithm, which starts with a uniform probability distribution on the given training dataset. Initially, all training instances have equal weight, but after each iteration, the weight changes. In the 1st iteration, all the 'n' instances have equal probability or weight, i.e. $1/n$. Learner L_1 will apply probability to all these instances, where the learner will label some of the instances correct and some wrong. Now, we will increase the probability of wrong instances and decrease the probability of correct instances. The above process will repeat K times. If we consider each iteration, the error rate is high, so variance will also be high. But at the end, it combines all K learners, which reduces the variance [25, 26].

4.1.3.5 Ensemble Voting (EV)

It is a classification technique in which we can ensemble more than one different classifier to get model output. The model predicts the output based on the highest probability of majority class. It merely accumulates the outcomes of every single classifier which was given to "Voting Classifier" and calculates its final class according to the largest plurality of votes. Instead of developing different specialised framework and evaluating its correctness, we develop just one framework which learns on these algorithms and estimates results primarily on their aggregate plurality of votes for every result class. Voting is of two type

"Hard Voting": It is the most basic kind of voting by majority. We estimate the class label $\hat{O}$ here using the "majority (plurality) voting of each classifier S_i":

$$\hat{O} = \text{mode}\{S_1(x), S_2(x), ..., S_n(x)\} \tag{4.3}$$

"Soft Voting": By "soft voting", we estimate the class labels primarily on the "classifier's projected probabilities P"; however, this strategy is only suggested if the classifiers are fine-tuned.

$$\hat{O} = \text{avg}\max_{j} \sum_{i=1}^{n} w_i Pji \tag{4.4}$$

Here w_i is the weight alloted to the classifier.

4.2 Related Study

Feature selection and optimization have been extensively used in various classification models in order to select the best attributes and achieve more accurate results. This method can use in various models such as ailment analysis, junk mail detection and fake news detection.

For early detection of lung disease, Alshmrani et al. [27] proposed a deep learning model using a combination of VGG19 + CNN techniques. In his paper, the author used multiple class chest X-ray datasets which consisted of lung cancer, pneumonia, lung opacity and tuberculosis (TB). The study evaluated an accuracy of 93.75%, precision of 97.56%, recall of 93.48%, F_1-score of 95.62% and area under curve (AUC) of 99.82% to help physicians diagnose lung disease efficiently and quickly [27].

To evaluate serum for lung disease classification, McDowell et al. [28] collected a multiclass dataset which consisted of asthma patients, COPD patients, lung cancer patients and healthy individuals. The collected samples of concerned patients were sent to a microbiology lab to extract extracellular vesicles (EV) and DNA and perform data coding of samples. The author applied the machine learning methods "generalized linear model (GLM) with and without feature selection, gradient boost machine (GBM), artificial neural network (ANN) and ensemble approach of GBM + ANN" to predict lung disease. The cross-validation method was applied to overcome the problem of overfitting by using applied datasets and training-test sets in a ratio of 70%:30% [28].

Chander Mani Sharma [29] proposed a hybrid model using transfer learning, ResNet-v2 and inception learning. The author used chest X-rays to predict lung disease, for which he used open-source data available on the Internet. The author applied data pre-processing and data augmentation to resize, rescale, crop and scale X-ray images and create a new dataset so that it could fit in a machine learning model. The author also applied "Gaussian blurring and bilateral filtering" image enhancement techniques to make the X-ray picture smoother and eliminate a small amount of noise. The CNN hyper-parameter tuning technique was used to identify the best range of features. The accuracy of the proposed model was 98.66%, and the model was deployed on Amazon IoT core edge environment. The author also conducted a comparative study of his model with other models named "VGG19, MobileNetV2, SqueezeNet and ResNet50" [29].

Sabeena [20] proposed a model named OBEFSs (optimized-based ensemble feature selections) based on a "feature selection, optimization and classification framework for identifying Parkinson's disease by extracting voice". The suggested framework uses an ensemble of feature selection approaches in which the author integrates "FMBOAs, LFCSAs (Levy Flight Cuckoo Search Algorithms) and AFAs (Adaptive Firefly Algorithms) techniques to fetch out essential features from the dataset. The framework uses KPCA (Kernel-based Principal Component Analyses) for dimension reduction and SVM, CNN, FCLSTM-CNN and FCBi-LSTM (Fuzzy Convolution Bidirectional Long Short-Term Memories) for classification to calculate accuracy, F-measure and error rate". In their study, the author found that the FCBi-LSTM classification has the highest accuracy of 98.77% [20].

Almazloum et al. [30] proposed a neural network-based model using multilayer perception for diagnosing pulmonary disease. Here, the authors used 201 records that were split into 103 normal and 98 abnormal cases. The respiratory disease was classified into three classes named obstructive, restrictive and mixed. The proposed model produced accuracy of 92–99% on training data and 87–92% on test data. This would help in clinical decision-making and also help to make pulmonary function test diagnostic software. The author classified 23 selected parameters into three groups based on their characteristics [30].

Harshita [3] proposed a machine learning model for "polycystic ovary syndrome" (which is a hormonal disorder in women which leads to many chronic disease in females) disease prediction using the machine learning algorithms like random forest, SVM and logistic regression. The author used the dataset available on Kaggle. The accuracy of the random forest method is 92.024%, SVM 70.55% and logistic regression 90.18% [3].

Using healthy, pneumonia and tuberculosis patient's dataset, Zak and Krzyżak [31] implemented a deep learning neural network model for detecting lungs' illness using chest X-ray pictures. In which the author used VGG16, ResNet-50 and Inception V3 techniques to classify that model. The author calculated accuracy, precision, F_1-score and sensitivity of segmented and non-segmented X-ray images and compared the execution of the implemented model with the already designed framework and also compared the segmented and non-segmented framework, where he found that the performance of Inception V3 is better than the other one [31].

The work in Allam and Nandhini [32] focuses on feature selection and feature optimization methods using the wrapper method and binary teaching learning method to get better performance of the classification model for breast disease detection. In which the researcher found that the Naïve Bayes classifier provided 98.2% accuracy, SVM classifier provided 98.43% accuracy, and k-NN gave an accuracy of 94.91% [32].

4.3 Proposed Architecture

This paper proposes a new hybrid machine learning framework presented by Fig. 4.5, for predicting lung disease. The proposed model will assist medical practitioners and laboratory assistants in feasible and non-feasible situations for early prediction of lung disease. The proposed model is hybrid in nature which includes data pre-processing, dimension reduction, feature selection, feature optimization and classification techniques, so it may give more feasible results as compared to existing simple models. The nomenclature and figure of the framework are given below:

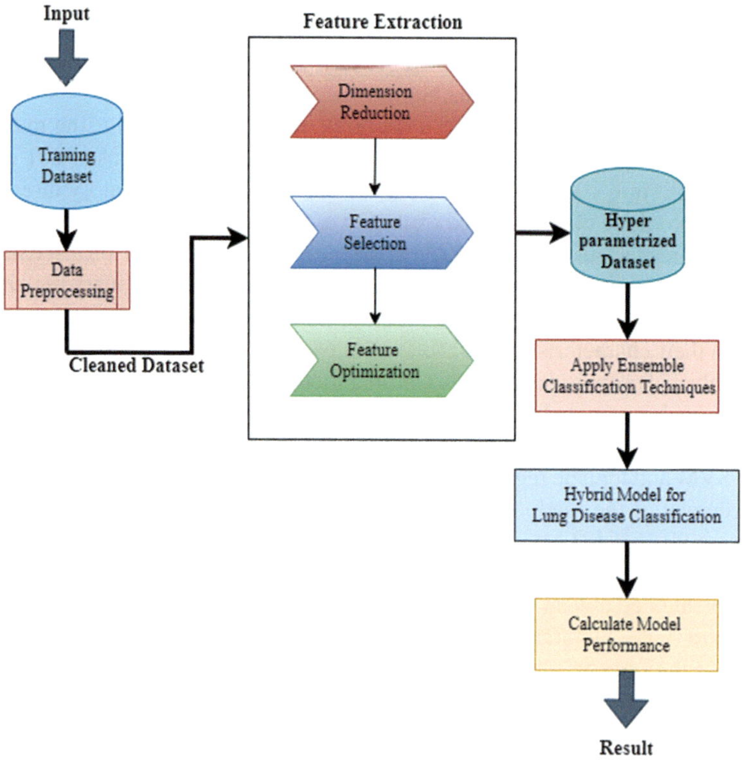

Fig. 4.5 Framework of proposed machine learning model

4.3.1 Lung Disease Training Dataset

It is the dataset that we give to the machine to train it. In simple words, it is the data that is given to the machine as an input. We need to pre-process it to make it understandable by the machine, so human involvement is required here. This data is mostly collected from online open sources or from nearby laboratories in hospitals.

4.3.2 Data Pre-processing

The dataset available on online open sources may have missing values or be in non-machine-readable forms which need to be cleaned and converted into machine-readable formats.

4.3.3 Dimension Reduction

To obtain an optimal dataset, dimension reduction is very important so that we can get optimal results. In large datasets, there are multiple attributes, and if we pass that data to train the machine, it can lead to overfitting problems. To obtain a dataset with optimal features, in the present research, the Principal Component Analysis (PCA) dimension reduction method will be used. Usually, there is high correlation between attributes in any dataset, and the PCA algorithm is used to remove or minimize this correlation.

4.3.4 Feature Selection

In our normalized dataset, there may be lots of features, out of which some features may be redundant, which reduce the performance of the model. So it is good to eliminate those redundant features to get better results. The proposed research will perform with the ensemble of feature selection (EFS) technique to select features for the model.

4.3.5 Feature Optimization

Feature optimization works with hyper-parameter tuning, which minimizes the cost function value and reduces the error rate of any machine learning or deep learning model. This is very significant for a model. The proposed model will use the ensemble of feature optimization (EFO) technique for hyper-parameter tuning of the dataset.

4.3.6 Lung Test Dataset

The test dataset is not the same as the training dataset, but we can say that its schema is the same as the teaching dataset. Using the teaching dataset, a machine learning model is prepared, and using the test dataset, it is tested to determine how accurate or reliable the proposed model is, which depends on its output accuracy. In short, we can say that the training dataset acts as an input, and the testing dataset acts as an output for a model.

4.3.7 Ensemble Method

After data pre-processing, dimension reduction, feature selection and optimization of the hyper-parameterized dataset, it is tested using classification techniques to produce the final result. The classification technique may be one or more than one, but in this proposed model, we will use more than one classification technique, which is why it's called an ensemble method.

4.3.8 Result

This is the last step of the model in which we evaluate the performance of the planned model, which depends on the accuracy of the model. Here we analyse the best output of the proposed hybrid model which may help medical assistants for early and accurate prediction of lung disease.

4.4 Feature Extraction Mechanism

The open-source dataset available online for lung disease prediction is too large and has many attributes, not all of which are useful for detecting lung disease. Using such a large dataset can result in incorrect output, high error rate, longer processing time and higher costs. Therefore, it is advisable to use dimension reduction, feature selection and feature optimization techniques after data pre-processing to obtain a hyper-parameterized dataset that consists of only the most important features.

Figure 4.6 represents the process of "dimension reduction, feature selection and feature optimization". In dimension reduction, the PCA method will be used to reduce

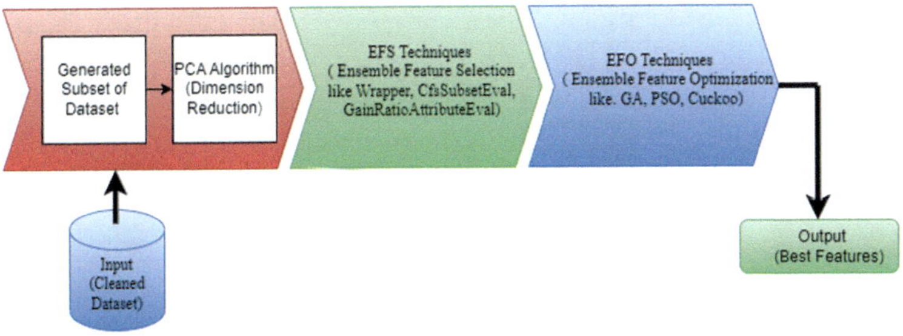

Fig. 4.6 Procedure of feature extraction techniques

irrelevant features from the input dataset. For feature selection, the EFS method will be used, which will eliminate duplicate attributes from the dataset and select the best features. In the feature optimization step, the EFO method will be used, which includes GA, PSO and Cuckoo. After data pre-processing, this is the key process of the model, where a hyper-parameterized dataset is inputted into the classification techniques to obtain the best results.

4.5 Execution Process

Figure 4.7 represents the execution method of the proposed model. In the beginning, we load the lung disease dataset as an input and pre-process it to make it machine-readable. Next, we apply feature reduction, feature selection and feature optimization techniques to get a hyper-parameterized dataset. We then apply classification techniques to assess the performance of the anticipated model. If the results are not good, we apply feature extraction techniques to get the best features and reapply classification techniques to evaluate the result. Finally, after getting the best features, the data is passed through an experimental process to evaluate its performance and calculate the accuracy of the proposed model.

(1) Begin.
(2) Load the lung disease dataset.
(3) Clean the dataset remove its impurities.
(4) Apply feature extraction techniques
    (i) Apply dimension reduction.
    (ii) Apply feature selection.
    (iii) Apply feature optimization.
(5) Remove inappropriate features and keep suitable ones.
(6) Apply ensemble classification techniques on selected dataset.

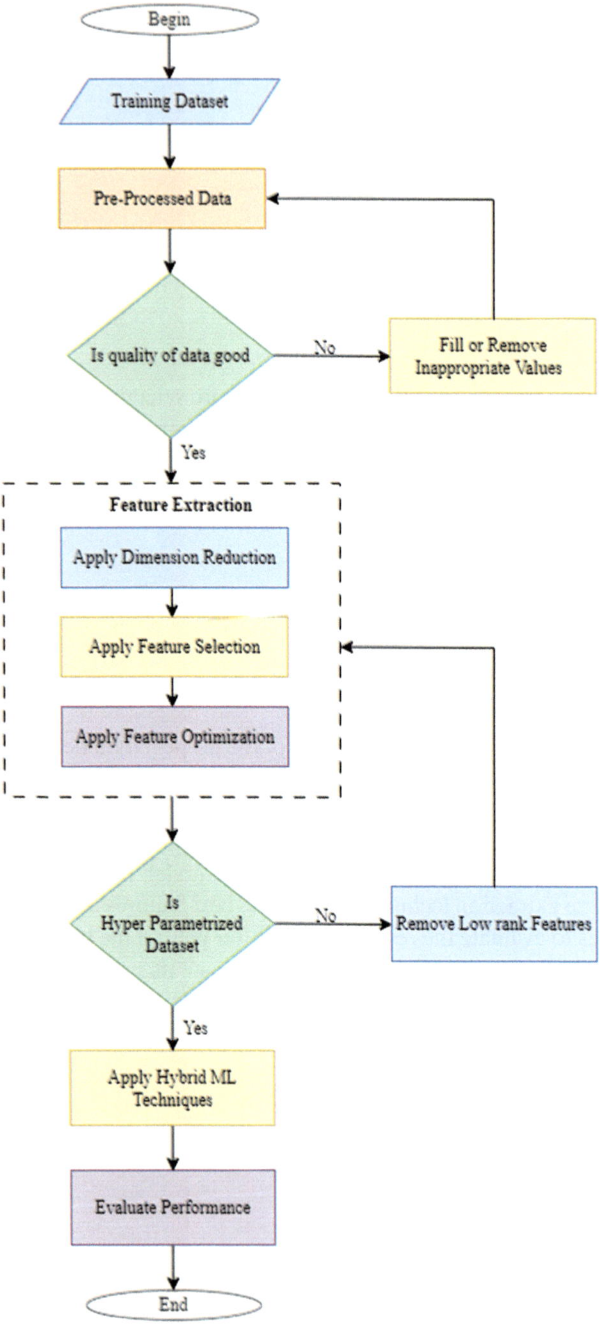

Fig. 4.7 Execution process of proposed model

(7) Evaluate the performance of proposed model.
(8) End.

The textbox above represents the functionality of the proposed model. First, we load the dataset. In the next step, we clean it. After that, we apply feature extraction techniques in three different ways to extract prominent features from the dataset. Finally, we apply ensemble classification techniques and evaluate the performance of the model.

4.6 Conclusion

During the Covid pandemic, lung disease has emerged as a prominent problem in patients, and it is now the third foremost cause of fatality at global level. Early diagnosis is crucial to saving human lives. In this paper, a hybrid machine learning approach is proposed for the early and accurate prognosis of lung disease. The hybrid approach utilizes feature extraction techniques, including PCA for dimension reduction, Wrapper, Pearson coefficient correlation, and Gain Ratio AttributeEval for feature selection, and GA and PSO for feature optimization. The hyper-parameterized dataset is accessed using feature extraction. A hybrid machine learning model, consisting of an ensemble of classification techniques, is used to produce results. The amalgamated approach of the hybrid model surpasses existing models and produces better results, which can help medical practitioners in the early prediction of lung disease.

References

1. https://data.worldbank.org/indicator/SP.DYN.CDRT.IN?locations=IN
2. Goyal, S., Singh, R.: Detection and classification of lung diseases for pneumonia and Covid-19 using machine and deep learning techniques. J. Ambient Intell. Humanized Comput. 3239–3259. https://doi.org/10.1007/s12652-021-03464-7
3. Harshita, B., Leema, N.: DCADS: data-driven computer aided diagnostic system using machine learning techniques for polycystic ovary syndrome. Int. J. Perform. Eng. **19**(3), 193–202 (2023)
4. Devi, P., Sagar, S., Rohil, H.: Prediction of lung disease using machine and deep learning techniques: a review. Int. J. Health Sci. **6**(S2), 7583–7601 (2022). https://doi.org/10.53730/ijhs.v6nS2.6833
5. https://www.simplilearn.com/tutorials/machine-learning-tutorial/feature-selection-in-machine-learning
6. https://www.analyticsvidhya.com/blog/2020/10/feature-selection-techniques-in-machine-learning/
7. Duch, W., Guyon, I., Nikravesh, M., Gunn, S., Zadeh, L.A. (eds.): Feature extraction. Studies in Fuzziness and Soft Computing, vol. 207, pp. 89–117. Springer, Berlin (2006). https://doi.org/10.1007/978-3-540-35488-8_4

8. Shaltout, N.A., El-Hefnawi, M., Rafea, A., Moustafa, A.: Information gain as a feature selection method for the efficient classification of influenza based on viral hosts. Lect. Notes Eng. Comput. Sci. **1**: 625–631
9. Rosidin, S., Muljono, M., Shidik, G.F., Fanani, A.Z., Al Zami, F., Purwanto, P.: Improvement with chi square selection feature using supervised machine learning approach on Covid-19 data. In: 2021 International Seminar on Application for Technology of Information and Communication (iSemantic), pp. 32–36. Semarangin, Indonesia (2021). https://doi.org/10.1109/iSemantic52711.2021.9573196
10. Kumar, S., Chong, I.: Correlation analysis to identify the effective data in machine learning: prediction of depressive disorder and emotion states. Int. J. Environ. Res. Public Health. **15**(12) (2018). https://doi.org/10.3390/ijerph15122907
11. Engin, M., Aras, S.: Detection of lung disorders using embedded and wrapper feature selection methods. Kahramanmaraş Sütçü İmam Üniversitesi Mü hendislik Bilimler iDergisi. **25**(3), 452–460 (2022). https://doi.org/10.17780/ksujes.113837
12. Kancherla, K., Mukkamala, S.: Feature selection for lung cancer detection using SVM based recursive feature elimination method. In: European Conference on Evolutionary Computation, Machine Learning and Data Mining in Bioinformatics, vol. 7246, pp. 168–176 (2012). https://doi.org/10.1007/978-3-642-29066-4_15
13. Balaji, N., Jammu, N., Bhaskara, R.: Automatic diagnosis of pneumonia using backward elimination method based SVM and its hardware implementation. Int. J. Imaging Syst. Technol. **32**(4) (2022). https://doi.org/10.1002/ima.22694
14. Hassan Naqvi, S.Z., Choudhry, M.A.: Embedded system design for classification of COPD and pneumonia patients by lung sound analysis. Biomed Tech (Berl). **67**(3), 201–218 (2022). https://doi.org/10.1515/bmt-2022-0011. PMID: 35405045
15. Kennedy, J., Eberhart, R.: Particle swarm optimization. In: Proceedings of ICNN'95—International Conference on Neural Networks, vol. 4, pp. 1942–1948. Perth, WA, Australia (1995). https://doi.org/10.1109/ICNN.1995.488968
16. https://www.sciencedirect.com/science/article/pii/S2666827022000421
17. Gad, A.G.: Particle swarm optimization algorithm and its applications: a systematic review. Arch. Comput Methods Eng. **29**, 2531–2561 (2022). https://doi.org/10.1007/s11831-021-09694-4
18. McCall, J.: Genetic algorithms for modelling and optimisation. J. Comput. Appl. Math. **184**(1), 205–222 (2005). https://doi.org/10.1016/j.cam.2004.07.034
19. Oh, S., Yoon, J., Choi, Y., Jung, Y.-A., Kim, J.: Genetic algorithm for the optimization of a building power consumption prediction model. Electronics **11** (2022). https://doi.org/10.3390/electronics11213591
20. Sabeena, B., Sivakumari, S., Teressa, D.M.: Optimization-based ensemble feature selection algorithm and deep learning classifier for Parkinson's disease. J. Healthc. Eng. 2040–2295 (2022). https://doi.org/10.1155/2022/1487212, https://doi.org/10.1155/2022/1487212
21. Yang, X.-S., Cuevas, E., Reyna-Orta, A.: A cuckoo search algorithm for multimodal optimization. Sci. World J. 2356–6140 (2014). https://doi.org/10.1155/2014/497514, https://doi.org/10.1155/2014/497514
22. https://www.news-medical.net/life-sciences/Hybrid-Analytical-Techniques-GC-MS-LC-MS-GC-IR-LC-NMR.aspx
23. Almazloum, A.A., Al-Hinnawi, A.-R., De Fazio, R., Visconti, P.: Assessment of multi-layer perceptron neural network for pulmonary function test's diagnosis using ATS and ERS respiratory standard parameters. Computers **11**(9) (2022). https://doi.org/10.3390/computers11090130
24. Jayaraj, D., Sathiamoorthy, S.: Random forest based classification model for lung cancer prediction on computer tomography images. In: 2019 International Conference on Smart Systems and Inventive Technology (ICSSIT), pp. 100–104. Tirunelveli, India (2019). https://doi.org/10.1109/ICSSIT46314.2019.8987772
25. Mamun, M., Farjana, A., Al Mamun, M., Md Salim, A.: Lung cancer prediction model using ensemble learning techniques and a systematic review analysis. In: IEEE Conference World AI IoT Congress, June 2022. (2022). https://doi.org/10.1109/AIIoT54504.2022.9817326

26. Suresh Kumar, M., Perumal, V., Yuvaraj, G., Rajasekar, S.J.S.: Detection of pneumonia from chest X-Ray images using machine learning. Concurrent Eng. **30**(4), 325–334 (2022). https://doi.org/10.1177/1063293X221106501
27. Alshmrani, G.M.M., Ni, Q., Jiang, R., Pervaiz, H., Elshennawy, N.M.: A deep learning architecture for multi-class lung diseases classification using chest X-ray (CXR) images. Alexandria Eng. J. **64**, 923–935 (2023). https://doi.org/10.1016/j.aej.2022.10.053
28. McDowell, A., Kang, J., Yang, J., et al.: Machine-learning algorithms for asthma, COPD, and lung cancer risk assessment using circulating microbial extracellular vesicle data and their application to assess dietary effects. Nat. J. Exp. Mol. Med. **54**: 1586–1595 (2022). https://doi.org/10.1038/s12276-022-00846-5
29. Sharma, C.M., Goyal, L., Chariar, V.M., Sharma, N.: Lung disease classification in CXR images using hybrid inception-ResNet-v2 model and edge computing. J. Healthc. Eng. 2040–2295 (2022). https://doi.org/10.1155/2022/9036457
30. Almazloum, A.A., Al-Hinnawi, A.-R., De Fazio, R., Visconti, P.: Assessment of multi-layer perceptron neural network for pulmonary function test's diagnosis using ATS and ERS respiratory standard parameters. MDPI (Computers) **11**(9) (2022). https://doi.org/10.3390/computers11090130
31. Zak, M., Krzyżak, A.: Classification of lung diseases using deep learning models. In: International Computational Science—ICCS 2020. Lecture Notes in Computer Science, vol. 12139. Springer, Heidelberg (2020). https://doi.org/10.1007/978-3-030-50420-5_47
32. Allam, M., Nandhini, M.: Optimal feature selection using binary teaching learning based optimization algorithm. J. King Saud Univ. Comput. Inf. Sci. **34**(2), 329–341 (2022)

Chapter 5
Analyzing Student Feedback with Natural Language Processing for Enhancing Student Perception

A. Suganya and Ananthi Sheshasaayee

Abstract The growth of technology and digital tools in the educational process results in improved student access, engagement, and outcomes. Aftermath COVID-19, the majority of educational institutions began to use online teaching mechanisms in addition to traditional teacher-centered teaching processes. Thus, the abrupt shift from fully online to a combination of online and offline-based educational instruction has produced a schism in student motivation and understanding of the course. Thus, improving students' overall experience is the primary goal of educational institutions and contemporary research today. The objective of this paper is to examine the factors that most influence students' overall experience toward this shift of online and offline-based education and enhance their retention, engagement, understanding, and motivation toward the current educational experience. A structured questionnaire was sent to higher education students, and 702 valid responses were received. The frequency analysis approach is utilized in this study for extracting the terms with the highest frequency. It will aid in the proposal of a solution to the responses through the use of natural language processing (NLP). The response was validated by NLP tokenization and the findings show that several factors influence students' overall experience and motivation toward the current educational forum, such as teaching quality, classroom environment, curriculum relevance, extracurricular activities, overall interaction, and so on. It is also intended to offer additional insight into the areas that require more emphasis and attention while designing and teaching.

5.1 Introduction

The use of technology in education has transformed traditional teaching and learning techniques in recent years, giving up novel possibilities and chances for both students and instructors. The education system has an extensive repertoire of traditional

A. Suganya · A. Sheshasaayee (✉)
PG and Research Department of Computer Science, Quaid-E-Millath Government College for Women, Chennai, India
e-mail: ananthi.research@gmail.com

P. Nanda et al. (eds.), *Intelligent System and Data Analysis*, Smart Innovation, Systems and Technologies 398, https://doi.org/10.1007/978-981-97-5200-3_5

methods of teaching and learning. Traditional education relates to the traditional form of instruction in which students attend classrooms in person and learn from instructors. It requires face-to-face interaction between instructors and learners [1]. Students learn in an organized atmosphere with a predetermined curriculum [2]. According to research, conventional education can be just as beneficial as online education [3]. Traditional teaching may be boosted with an assortment of techniques, which involves giving students with immediate and helpful feedback [1]. Incorporating innovative methods of teaching and learning, such as blended learning, which integrates traditional and online learning [2]. Critical thinking, creativity, communication, and teamwork are all highlighted as significant factors that can help students acquire the abilities necessary for personal development [4]. After so many decades, the COVID-19 pandemic has produced enormous upheavals in education, necessitating systemic reforms. Both educators and pupils must adapt to an alternate reality as a result of the pandemic, which has culminated in a trend toward online learning and hybrid course delivery [5]. Online learning (OL) is defined as learning in a synchronous or asynchronous setting using various tools, such as mobile devices, PCs, and so on, with an Internet connection. Using these online tools, students may study while interacting with instructors and other classmates from anywhere [6]. In many nations, the education system has been totally transformed into an online manner of teaching and grading.

Online learning delivers an environment for learning and skill enhancement that is accessible anywhere, at any time. It provides learning that is technologically enabled through the utilization of a feasible framework and diverse resources such as text, visual, and audio information [7]. Online educational platforms have grown in prominence in recent years, particularly because of the advent of remote learning as a result of the COVID-19 epidemic. Moodle, Google Classroom, Blackboard, Coursera, edX, Udemy, SkillShare, and Future Learn are other well-known platforms [8, 9]. Student perception is an important factor in determining the standard of instruction delivered by these institutions. It reflects students' complete experiences, opinions, and fulfillment of their objectives, influencing academic achievement, motivation, and overall exposure to the educational environment. Student feedback is an important tool for educational institutions to use in evaluating and improving student perception and overall educational experience [10]. Thus, this research paper proposes an approach that utilizes natural language processing (NLP) techniques, for the effective evaluation of student feedback data. The frequency analysis is used, for identifying reoccurring concepts phrases, or particular areas of concern highlighted by students. This data gives a more in-depth knowledge of the elements that impact the experience with learning, motivation, engagement, retention, and understanding. This automated method allows educators and administrators to swiftly process vast volumes of feedback data. The results of the aforementioned approach show the factors that most influence student perception, such as teaching quality, classroom environment, curriculum relevance, extracurricular activities, overall interaction, and so on, as well as the limitations found in the current system, such as less outdated content, limited interactivity, overwhelming workload, and so on.

5.1.1 Research Objectives

The key objective of the present study is to propose an approach for evaluating student feedback for identifying the variables that influence student experience in terms of curriculum models, as well as discovering shortcomings in present curriculum design and instruction. The precise goals are as follows:

- To gather and analyze student feedback information from a variety of sources, such as questionnaires, websites, or institutional archives.
- To examine the use of NLP tokenization approaches to evaluate feedback from learners.
- To do a frequency analysis in order to uncover reoccurring themes, buzzwords, or topics of concern voiced by students.
- Based on the retrieved data, assess the student's experience and requirements.
- Provide educators with useful insights for improving teaching tactics and student engagement.

5.1.2 Research Questions

- Could natural language processing tokenization, and frequency analysis be used to analyze student comments?
- What exactly are the reoccurring concerns and constraints revealed by the frequency evaluation of student feedback?
- Which of the knowledge gathered from analyzing student comments using frequency analysis be used to improve student motivation, understanding, and overall education reliability?

5.2 Literature Review

The goal of research on the student experience in the current educational system is to enhance the quality of educational forums.

There were several research done on the evaluation of student feedback in which a researcher did a literature analysis on the quality of post-secondary online course evaluation student comments. The authors intended to present a thorough evaluation of studies on the subject. Because of their simplicity and productivity, they discovered that online course assessments are growing increasingly prominent. The authors investigated research that compared the accuracy and dependability of online course assessments to paper-based evaluations. They also looked at research that looked at the elements that impact the quality of feedback collected through online assessments. The authors found that while online course evaluations can give significant input to teachers and administrators, they are accurate and trustworthy [11].

Another researcher recommended using topic modeling, especially the LDA approach, to comprehend the key themes that arise from learners' written responses in course evaluations. When the authors compared their findings to the old approach of manual coding, they discovered that topic modeling was more efficient and successful in identifying themes in the comments [12].

5.2.1 Natural Language Processing in Education

Natural language processing (NLP) includes the application of algorithms and software to comprehend voice and text in order to enhance, scale, and widen the reach of education in society. NLP has been utilized in education to enhance learning environments. Several researchers have used NLP to process educational information and discovered important improvements that may be implemented in educational institutions.

A study of student learning outcomes was addressed by the researcher where the study deals with the creation of a system that analyzes feedback from learners and provides teachers with an understanding of student impressions of their courses using NLP. The report summarizes the findings of pilot research designed to assess the system's effectiveness in enhancing educational results. The study discovered that using NLP-based feedback analysis to improve the learning results of students and provide teachers with substantial insights into their opinions of their courses can be a successful strategy. Overall, the research offers useful insights into the use of NLP to facilitate student feedback analysis and emphasizes the possibilities of this technique to enhance student learning outcomes [13].

One of the researchers compiles data from 107 participants' unrestricted responses obtained via an online survey at the start and conclusion of the sustainability class. Text mining via natural language processing (NLP), principal component analysis (PCA), and overlap analysis of networks was utilized to investigate the changes in learners' perspectives on growth that are sustainable. The study also implies that machine learning and NLP may be used to analyze massive volumes of data from student comments and offer teachers and administrators more accurate and valuable input [14].

Another author gathered 50 responses from students and analyzed them using the frequency of negative feedback sentences. They then utilized support vector machine (SVM) and random forest (RF) algorithms to predict student responses according to criteria such as attendance, involvement, and test results [15].

The researchers created an algorithm that analyzes and provides adaptive feedback based on the data provided by the pupils. The methodology will be evaluated on a large-scale online learning platform, and the results will be compared with student survey-based engagement assessments from the literature. Overall, the work "Adaptive Natural-Language Targeting for Student Feedback" adds to the body of knowledge about NLP strategies for delivering adaptive feedback in online learning environments. It showcases the potential of NLP in improving tutoring software and emphasizes the significance of taking student input into account when designing and improving educational technologies [16].

This study takes data from selected research and codes it using factors such as author, publication year, nation, learning phase, major, asynchronous or synchronous online education, urgent remote learning, sample size, and contentment rate. The research articles included are statistically analyzed to produce effect sizes. The research delves into the utilization of NLP techniques and statistical approaches for analyzing online education feedback, emphasizing the significance of knowing learner attitudes toward online education [17].

5.3 Research Design

To perform the study, a database of student feedback is gathered. Surveys and online platforms feedback gathering techniques in educational institutions were utilized for collecting feedback. A systematic data gathering and processing technique is required when applying frequency analysis to learner feedback. A substantial number of replies should be incorporated into the dataset to provide a precise illustration of student points of view.

5.3.1 Data Collection

The data for the study was gathered from higher education students using a Google Form. Around 748 responses shown in Fig. 5.1 were received based on the questionnaire as shown in Figs. 5.2 and 5.3 and were answered among which 702 genuine replies were obtained from students, and the data was a compilation of bachelor's degrees and advanced degree courses. The responses are compiled in population as per groups in Table 5.1 by demography.

5.3.2 Data Cleaning

The data collected must be cleaned and preprocessed to eliminate noise, unneeded details, and formatting issues which are explained in the results in Figs. 5.5, 5.6, and 5.7. This might entail deleting stop words, grammar, and special characters, as well as standardizing the text format [18].

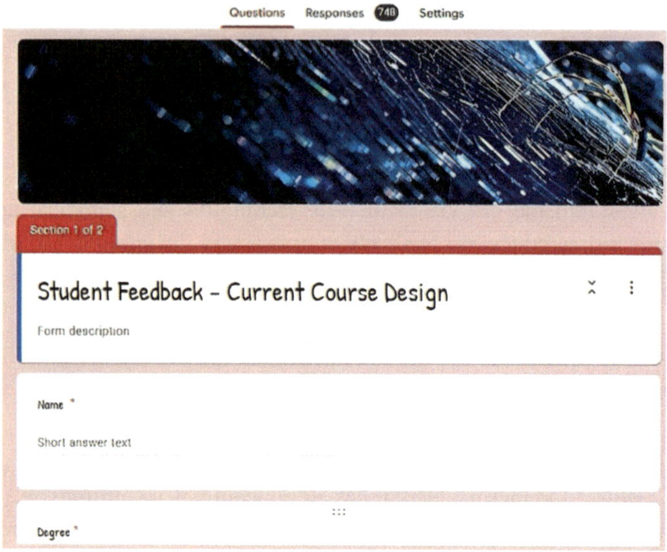

Fig. 5.1 Sentence 748 responses on current course design

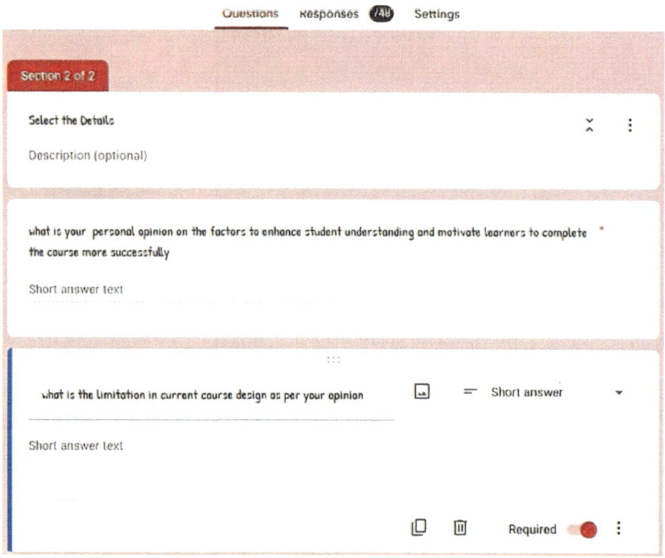

Fig. 5.2 Questionnaire on current course design

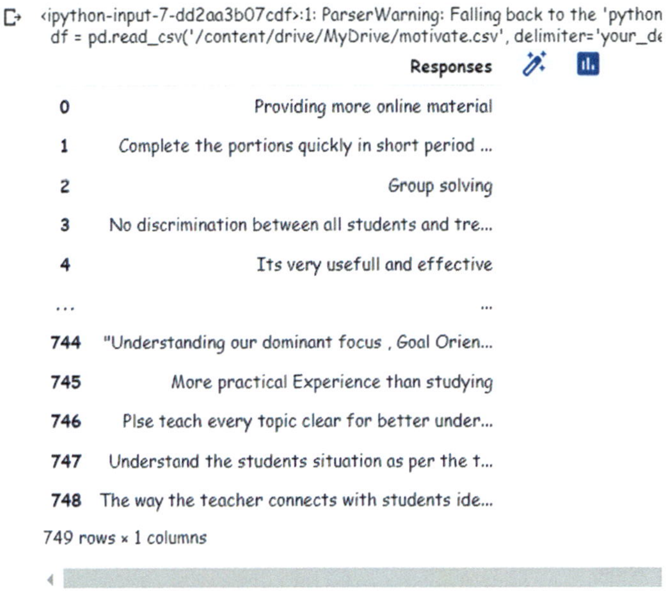

<ipython-input-7-dd2aa3b07cdf>:1: ParserWarning: Falling back to the 'python
df = pd.read_csv('/content/drive/MyDrive/motivate.csv', delimiter='your_de

	Responses
0	Providing more online material
1	Complete the portions quickly in short period ...
2	Group solving
3	No discrimination between all students and tre...
4	Its very usefull and effective
...	...
744	"Understanding our dominant focus , Goal Orien...
745	More practical Experience than studying
746	Plse teach every topic clear for better under...
747	Understand the students situation as per the t...
748	The way the teacher connects with students ide...

749 rows × 1 columns

Fig. 5.3 Sentence into tokens using NLP tokenization

Table 5.1 Demographic distribution of samples

S. No.	Demographic	Population (Total-748)
1	Gender (Male)	403
2	Gender (Female)	345
3	Urban	382
4	Rural	366
5	Bachelor degree	354
6	Advanced degree	292
7	Other degree	102

5.3.3 NLP Tokenization

Tokenization is a method of dividing text into smaller sections known as tokens, which are the fundamental building blocks of NLP [19]. The second stage is obtaining valuable characteristics from student feedback. This is accomplished by dividing a sequence of letters into words as shown in Fig. 5.4 which are the outputs of the code executed in Python, or more specifically, tokens. This is a prerequisite for future investigation. It is impossible to retrieve higher-level data from a document without first recognizing the tokens. For this objective, several tokenization libraries or tools, such as Natural Language Toolkit (NLTK) or SpaCy, were used [20]. In the study, we have used NLP tokenization to find the frequency of continuous words that occur often in the student feedback.

The working of natural language processing in the analysis of the frequency of words is

1. For each row, join all the fields into a single string using a space separator and tokenize the string using a regular expression to extract individual word tokens. Add these tokens to the words list.
2. Loop through the list of words and count their combinations of length n. For each combination, update its count in the word_combinations default dict.
3. Print the repeated combinations of words along with their counts from the word _combinations default dict.

```
['current',
 'system',
 'is',
 'good',
 ',',
 'use',
 'solving',
 'problems',
 'iam',
 'ok',
 'with',
 'the',
 'system',
 'i',
 'want',
 'more',
 'communication',
 'with',
 'teachers',
 'online',
 'learning',
 'is',
 'usefull',
 'need',
 'more',
 'materials',
 'to',
 'learn',
 'online',
 'learning',
 'with',
 'group',
 'solving',
 'was',
 'usefull',
 'need',
 'more',
 'interaction',
 'in',
 'classes',
 'both',
```

Fig. 5.4 Sentence into tokens using NLP tokenization

```
[nltk_data] Downloading package stopwords to /root/nltk_data...
[nltk_data]   Package stopwords is already up-to-date!
providing
online
material
complete
portions
quickly
short
period
give
test
help
understand
concept
clearly
```

Fig. 5.5 Processed words after removal of stop words

```
Stemmed Words:
provid
onlin
materi
complet
portion
quickli
short
period
give
test
help
understand
concept
clearli
```

Fig. 5.6 Stemmed words from the processed words

5.4 Result and Discussion

The proposed methodology was implemented by evaluating the term frequency in the students' feedback for the factors that influence student motivation, retention, engagement, and understanding of the educational forum, as well as the limitations students have regarding the current educational design.

Fig. 5.7 Lemmatization words from the processed words

5.4.1 *Text Preprocessing*

Text preprocessing is an important stage in natural language processing (NLP). Text preprocessing is the process of purging and preparing text data in natural language processing. NLTK and re are prominent Python libraries for addressing a variety of text preparation tasks. Tokenization, stemming, lemmatization, stop word removal, and other techniques are included [21]. Text processing is a process of cleaning the data using these techniques, and the data is received after it is processed as shown in Fig. 5.5, a total of 2715 processed words were received which few words are shown in Fig. 5.5.

After the removal of stop words, stemming and lemmatization are utilized to reduce words to their base or root form. Stemming and lemmatization are two processes that are commonly used to enhance language models by speeding up the procedure for searching [22]. These strategies aid in standardizing word variants and bringing them to a uniform representation, which is especially beneficial for natural language processing (NLP) applications involving keyword analysis and comparison.

Stemmings: The technique of eliminating suffixes from words to acquire the original or base form, defined as the stem, is known as stemming. The outcomes of stems are not inevitably acceptable words, but they reflect the original word's basic sense the processed data are analyzed and the stem words are obtained in which few top words are shown in as shown in Fig. 5.6, the total obtained words are more than 1800.

Lemmatization: The technique that reduces words to their basic or lexicon form, known as the lemma, is known as lemmatization. Lemmatization, contra stemming, assures that the produced lemma is a genuine term with adequate significance in language. Lemmatization-generated words are not like words stemmed it has meaning, if the words are found meaning less without a suffix, then the suffix is not removed as shown in Fig. 5.7.

5.4.2 Tokenization and Frequency Analysis Results

The NLP tokenization technique breaks down the student feedback into individual tokens, such as words and phrases. The tokenization technique made it easier to analyze the frequency of occurrence of certain phrases or concepts in the feedback collection.

5.4.3 Frequency Analysis

Following the tokenization of the feedback data, frequency analysis in natural language processing is performed to assess the prevalence and spatial distribution of tokens among the dataset. Measuring the frequency of each token and constructing a frequency distribution are both part of this process. Frequency analysis can reveal the most frequently used phrases and words by students as shown in Fig. 5.8, showing repeating themes or issues connected to comprehension and motivation [23]. The study uses the term frequency in analyzing the occurrence of repeated words. The term frequency $tf(t, d)$ is the ratio of the frequency of term t inside the document d as shown in Fig. 5.9.

To acquire additional information about the tokenized student feedback data, several data analysis approaches may be used along with frequency analysis.

After analyzing the frequency of each term, we need useful data that indicates the aspects that can improve student motivation, understanding, and overall experience thus NLP. Tokenization is utilized to obtain a combination of two or three continuous words, allowing us to determine what students' requirements. Thus, the data is analyzed and the commonly repeated words are listed in Figs. 5.10 and 5.11.

The frequency analysis indicated the phrases or concepts that were stated the most frequently in the student responses. The study revealed that phrases like "more test," "teacher interaction," "doubt clarification," "classroom interaction," "online resources," and "teacher feedback" were often used in the comments along with various other factors as shown in Fig. 5.12; those consistent indications might give insights into ideas or regions that students are in need of and finds motivating Furthermore, phrases linked with limitations in the current course, such as "adopt technology," "group learning," "teacher interaction," or " teacher feedback" were recognized as common themes along with some more added limitations as shown in

```
Counter({'Responses': 1,
         'Providing': 1,
         'more': 28,
         'online': 20,
         'material': 16,
         'Complete': 1,
         'the': 60,
         'portions': 1,
         'quickly': 1,
         'in': 38,
         'short': 2,
         'period': 2,
         'and': 64,
         'give': 11,
         'some': 10,
         'test': 48,
         'that': 19,
         'will': 12,
         'help': 12,
         'to': 63,
         'understand': 12,
         'concept': 6,
         'clearly': 4,
         'Group': 1,
         'solving': 1,
         'No': 2,
         'discrimination': 1,
         'between': 5,
         'all': 7,
         'students': 44,
         'treàt': 1,
         'a': 20,
         'student': 21,
         'equally': 1,
         'Its': 2,
         'very': 3,
         'usefull': 1,
         'effective': 5,
         'Conducting': 1,
```

Fig. 5.8 Frequency analysis of each token

$$\mathrm{tf}(t,d) = \frac{\text{number of times } t \text{ appears in } d}{\text{total number of terms in } d}$$

Fig. 5.9 Term frequency calculation

```
The repeated combinations of words are:
student all : 3 times
video related : 8 times
will help  : 5 times
study need : 7 times
doubt time : 10 times
concept learn : 8 times
make easy : 10 times
encourage us  : 12 times
better course : 13 times
teaching ways : 13 times
limited notes  : 9 times
test need  : 13 times
discussion done : 14 times
give material  : 5 times
every topic : 12 times
teacher talking : 15 times
our doubt : 15 times
simple way : 8 times
use method : 10 times
offline classes  : 12 times
teacher student : 13 times
my personal : 5 times
opinion things  : 3 times
test improve  : 12 times
group working : 14 times
daily material  : 15 times
easy learning : 12 times
teacher class : 15 times
```

Fig. 5.10 Frequency analysis of student perception toward enhancing education forum

Fig. 5.13 revealing possible areas of improvement. Since there were a lot of repeated words as shown in Figs. 5.10 and 5.11, the top common repeated words are shown in Figs. 5.12 and 5.13.

5.5 Implications and Recommendations

5.5.1 Implications for Education

The outcomes of the NLP tokenization and frequency analysis of student comments have significant educational implications for instructors. The findings can be used to inform instructional design, teaching tactics, and the creation of curriculum. Identifying the sources of student difficulty or confusion can assist instructors in more effectively addressing these gaps. Educators can improve student motivation and learning results by finding the subjects or concepts that students find interesting.

```
The repeated combinations of words are:
less time : 13 times
bore class :9 times
interaction learn  : 15 times
my class : 7 times
learning time : 12 times
new material : 10 times
learn more : 12 times
motivate us  : 10 times
intrest concept : 12 times
teach way : 11 times
more material  : 13 times
test more  : 11 times
new method : 12 times
low learn  : 11 times
slow learn : 13 times
concept difficult : 15 times
doubt time : 10 times
simple technique : 12 times
teach method : 11 times
new concept  : 4 times
teacher time : 10 times
low understand : 15 times
syllabus cover  : 13 times
quiz test  : 10 times
group working : 14 times
online data  : 10 times
more note : 10 times
```

Fig. 5.11 Frequency analysis on limitation in the current education

5.5.2 *Recommendations of NLP in Education*

Several practical recommendations for educators and educational institutions can be offered based on the study's findings: To handle and analyze student feedback data effectively, educators might use NLP tokenization and frequency analysis. NLP libraries and tools can help to automate the process while also providing useful information for student comprehension and motivation. Learner feedback analysis might disclose unique student preferences, abilities, and difficulties. Instructors may utilize this data to personalize lessons, give focused assistance, and alter teaching tactics to meet the requirements of a varied demographic of pupils. NLP can also be utilized in improving the quality of instruction, automated grading, personalized learning, and enhanced student engagement [24, 25]. The implementation of robots in NLP can be used for language learning, student assessment, and teacher training [26].

Further to effectively enhance online and offline learning operations, assisting students in returning to traditional offline learning, utilizing blended learning, reviewing their capacity to work remotely, employing Universal Design for Learning (UDL) ideals, and providing teachers training can be done. Although the transition

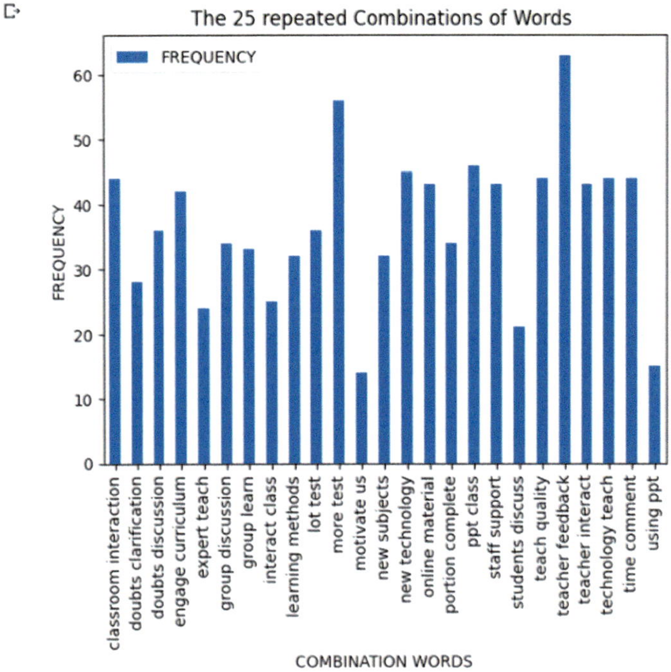

Fig. 5.12 Frequency analysis on student overall perception

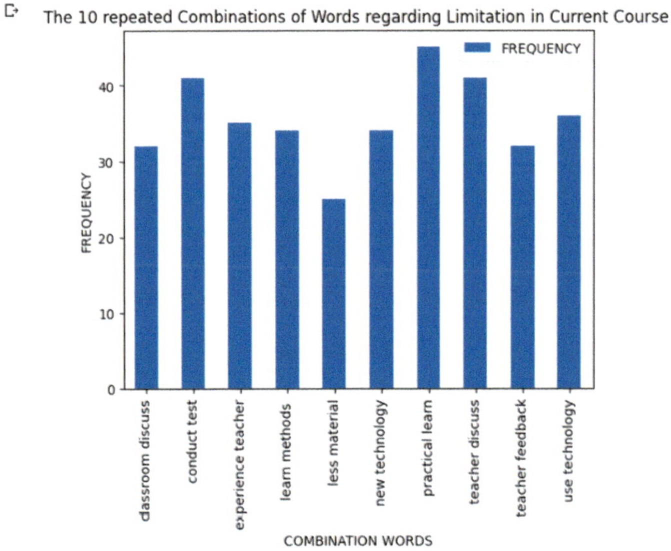

Fig. 5.13 Frequency of limitation in current course

between online and offline education has posed obstacles, it has also provided an opportunity for students to discover novel areas of education and for educators to develop novel instructional approaches.

5.6 Conclusion

The use of NLP tokenization and frequency analysis to evaluate student input provides significant insights into learner understanding and drive. This research article investigated the use of NLP techniques to analyze feedback from learners' data, especially tokenization and frequency analysis. The examination of student input results in critical recommendations for curriculum and instructional enhancement. It is critical to recognize the study's limitations, such as sample size, data representativeness, and the accuracy of NLP approaches. To confirm and build on the findings, future research should overcome these constraints and investigate more NLP approaches in diverse educational situations. Overall, employing NLP tokenization and frequency analysis to evaluate student input provides a data-driven method for analyzing student viewpoints. Educators may acquire greater insights into learner knowledge and motivation by employing computational resources, resulting in well-informed choices and ongoing enhancement in education.

References

1. González, A.: Turning a traditional teaching setting into a feedback-rich environment. Int. J. Educ. Technol. High. Educ. **15**(1), 1–21 (2018)
2. Hassan, A., Abiddin, N.Z., Yew, S.K.: The philosophy of learning and listening in traditional classroom and online learning approaches. High. Educ. Stud. **4.2**, 19–28 (2014)
3. Alghazo, A.M.: Comparing effectiveness of online and traditional teaching using students' final grades. Online J. Workforce Educ. Dev. **1.3**, 6 (2005)
4. McGuire, C.: Transforming traditional teaching practices with 21st century skills in K-12 classrooms (2018)
5. Bozkurt, A., et al.: The impact of COVID-19 on education: a meta-narrative review. TechTrends **66.5,** 883–896 (2022)
6. Singh, V., Thurman, A.: How many ways can we define online learning? A systematic literature review of definitions of online learning (1988–2018). Am. J. Distance Educ. **33**(4), 289–306 (2019)
7. Gowda, R.S., Suma, V.: A comparative analysis of traditional education system versus e-learning. In: 2017 International Conference on Innovative Mechanisms for Industry Applications (ICIMIA). IEEE (2017)
8. Abuhassna, H., et al.: Development of a new model on utilizing online learning platforms to improve students' academic achievements and satisfaction. Int. J. Educ. Technol. Higher Educ. **17**, 1–23 (2020)
9. Liu, Z.Y., Lomovtseva, N., Korobeynikova, E.: Online learning platforms: reconstructing modern higher education. Int. J. Emerg. Technol. Learn. (iJET) **15**(13), 4–21 (2020)

10. Ahea, M.M.A.B., Ahea, M.R.K., Rahman, I.: The value and effectiveness of feedback in improving students' learning and professionalizing teaching in higher education. J. Educ. Pract. **7**(16), 38–41 (2016)
11. Plante, S., LeSage, A., Kay, R.: Examining online course evaluations and the quality of student feedback: a review of the literature. J. Educ. Inform. **3**(1), 21–31 (2022)
12. Sun, J., Yan, L.: Use topic modeling to understand comments in student evaluations of teaching (2023)
13. Katz, A., Norris, M., Alsharif, A.M., Klopfer, M.D., Knight, D.B., Grohs, J.R.: Using natural language processing to facilitate student feedback analysis. In: 2021 ASEE Virtual Annual Conference Content Access (2021)
14. Yamano, H., et al.: Understanding students' perception of sustainability: educational NLP in the analysis of free answers. Sustainability **14.21**, 13970 (2022)
15. Sengar, N.S., et al.: Student feedback prediction using machine learning. Int. J. Adv. Res. Ideas Innovations Technol. **5.2**, 818–821 (2019)
16. Kolchinski, Y.A., Ruan, S., Schwartz, D., Brunskill, E.: Adaptive natural-language targeting for student feedback. In: Proceedings of the Fifth Annual ACM Conference on Learning at Scale, pp. 1–4 (2018)
17. Kang, H., Zhang, J., Kang, J.: Analysis of online education reviews of universities using NLP techniques and statistical methods. Wireless Commun. Mob. Comput. (2022)
18. Toçoğlu, M.A., Onan, A.: Sentiment analysis on students' evaluation of higher educational institutions. In: Intelligent and Fuzzy Techniques: Smart and Innovative Solutions: Proceedings of the INFUS 2020 Conference, July 21–23, 2020. Springer International Publishing, Istanbul, Turkey (2021)
19. Webster, J.J., Kit, C. Tokenization as the initial phase in NLP. In: COLING 1992 Volume 4: The 14th International Conference on Computational Linguistics (1992)
20. Baydogan, C., Alatas, B.: Detection of customer satisfaction on unbalanced and multi-class data using machine learning algorithms. In: 2019 1st International Informatics and Software Engineering Conference (UBMYK). IEEE (2019)
21. Badgujar, A., Shaikh, A., Kamble, T., Makhija, G.: Text summarization using natural language processing
22. Balakrishnan, V., Lloyd-Yemoh, E.: Stemming and lemmatization: a comparison of retrieval performances (2014)
23. Shehab, A., Elhoseny, M., Hassanien, A.E.: A hybrid scheme for automated essay grading based on LVQ and NLP techniques. In: 2016 12th International Computer Engineering Conference (ICENCO). IEEE (2016)
24. Khensous, G., Labed, K., Labed, Z.: Exploring the evolution and applications of natural language processing in education. Rom. J. Inf. Technol. Autom. Control **33**(2), 61–74 (2023)
25. Chary, M., Parikh, S., Manini, A.F., Boyer, E.W., Radeos, M.: A review of natural language processing in medical education. West. J. Emerg. Med. **20**(1), 78 (2019)
26. Younis, H.A., Ruhaiyem, N.I.R., Ghaban, W., Gazem, N.A., Nasser, M.: A systematic literature review on the applications of robots and natural language processing in education. Electronics **12**(13), 2864 (2023)

Chapter 6
Enabling Technology for Smart and Intelligent Transportation: Analyse the Uses and Impact of Blockchain Technology and Artificial Intelligent on Trucking Industry for Better Results

Mahendra Parihar and Akshaya

Abstract As a backbone of our economy, playing a very crucial role in the movement of our economy, the trucking industry still has not come out of the persistent problems it has been facing over the last few decades such as high cost of operation, low efficiency in operation, government issues, regulatory and legislative mechanism, information issues and many more. However, adopting these latest available technologies and their effective use can solve many of these problems significantly. One such is the adaptation of digitalization properly for Digital Governance which further results in Good Governance and another is the adoption of blockchain technology for data and information management to enhance operation efficiency of the trucking industry. Today in most cases especially for small truck operations, the operation is not viable due to poor technology adoption. Hence with this paper, an attempt is being made to initiate these improvements. The methodology adopted for the study is also relevant to the requirement of the work undertaken in the form of a structured approach. It consists of a relevant literature review, primary and secondary sources of information with available data, and a model framework about digitalization, digital governance, and possible application of blocking technology to bring improvement in operations of the trucking segment in our economy. The study found that the adoption of the latest technology in trucking operations in the form of digitization and blockchain reduces the problem of asymmetric information and brings

M. Parihar
Mukesh Patel School of Technology Management and Engineering, SVKM's NMIMS University, Mumbai, India

Akshaya (✉)
Department of Information Technology, Mukesh Patel School of Technology Management and Engineering, SVKM's NMIMS University, Mumbai, India
e-mail: akshaya.subramania@gmail.com

P. Nanda et al. (eds.), *Intelligent System and Data Analysis*, Smart Innovation, Systems and Technologies 398, https://doi.org/10.1007/978-981-97-5200-3_6

good governance to enhance operation efficiency. So, the study concludes by stating that if the government supports the trucking industry, provides consultation, and coordinates with other stakeholders of the industry regarding technology adaptation and other aspects, the trucking industry can yield better results and solve many of its problems independently.

6.1 Introduction

Governance is characterized by accountability, transparency, and participation. Cooperatives balance relationships between members, communities, and traders, and enable them to achieve economies of scale and devolve decision-making to the community level. Good governance principles apply to cooperatives and their success depends on proper practices and participation. Digital Governance is essential for accountability, decision-making, and change management in organizations' digital presence. It improves communication processes and contributes to employee well-being. The widespread use of technology has transformed businesses and created new roles, improving internal processes and competitiveness. Digital governance initiatives enhance organizational efficiency. Companies worldwide prioritize digital governance projects to stay competitive. Currently, amid the fourth industrial revolution digital transformation has permeated various aspects of society, economy, and industries, including transportation. Its introduction in transportation was driven by the need for automation and increased accelerated performance. Along with the creation of new opportunities, it has also brought forth challenges and threats, commonly referred to as digital disruption. Technology plays a crucial role in the transport sector in general and trucking industry in particular, and performs tasks like online monitoring, driving assistance, and optimization. With the advent of technological advancements, the trucking industry (although limited only so far) has seen positive growth which has showed that with the technology usage, technology has potential to transform this industry significantly. Although the trucking industry was not given required importance in the past for a longer period due to various reasons and therefore it has not come out completely from its inherent problems so far. However, the recent technologies like blockchain application and artificial intelligence have shown some good hope. Currently given the nature of operation and ownership profile, from the past, the trucking industry has always been highly competitive, focused on cost, and driven to meet customer demands. Moreover, it has become more technologically savvy due to technological changes and hence, its transformation is underlined by the recent trends of adopting IoT technology, emission-free electric vehicles, automating logistics, receiving real-time data, and making data-driven decisions. The conceptualization of blockchain technology started with bitcoins followed by blockchain, smart contracts, proof of stake, and blockchain scaling. Blockchain is a decentralized mode of payment. It is also known as a shared ledger where without the consent of all the people on the network changes cannot be made on the document. The study reviews how good governance is achieved through the effective implementation of technology. It analyses the impact of digital governance with blockchain and focuses on the digitalization trends in the fast-paced transportation sector in general and trucking industry in particular. It discusses how the dynamics of supply chain management have significantly changed and has led to sustainable development. Further in this paper, an attempt is being made to study the applications, challenges, complete working and architecture, integration with IoT and AI, and use in transportation sectors (trucking industry) of blockchain technology have all been discussed.

6.2 Literature Review

The authors discuss the concepts of blockchain, its architecture, the challenges faced, and the recent trends in the blockchain industry giving a comprehensive overview. It explores the architecture of blocks of data connecting multiple transactions which acts as a public ledger, encryption through public and private keys, and various consensus algorithms along with its pros and cons have been discussed. Moreover, the paper discusses the application of blockchain in potential areas of blockchain testing, stops the tendency to centralization, big data analytics, and blockchain application, and provides future prospects [1]. This paper focuses on explaining the basic working of blockchain technology and the issues concerned with security and privacy, consensus mechanisms, and applications. Major issues of the double spending problem which occurs when reusing the same currency in two transactions and the Byzantine Generals Problem which refers to change in communication contents because of nodes attacked are identified. In conclusion, it has been suggested that the government should provide consistent laws for this technology and build consensus mechanisms for different purposes for future work [2]. In this paper, the working and concepts of blockchain technology, the major problems associated with it, and the applications and challenges of this technology have been explained. Various applications such as the patent system, e-voting, network operations, and operational monitoring verify, identity, and define consumption of resources within the network along with challenges in regulation and standards, cost and efficiency, security and privacy, validation, and organization that have been highlighted [3]. This paper introduces its architecture and consensus algorithms along with several potential applications, opportunities, and technical challenges faced. Detailed mechanisms, strengths and limitations, comparisons of various consensus algorithms and applications in finance, IoT, public and social services, green energy, reputation system for academics, and web communities have been discussed. For privacy and scalability issues, proposed solutions include mixing services, improving anonymity, stubborn mining techniques, and selecting fresh blocks [4]. Key features of blockchain, benefits and consensus protocols, and applications across multiple domains have been presented. Applications, particularly in finance, accounting, supply chain management, energy, advertising and media, legal, real estate, health care, and IoT, have been discussed. The paper concludes by highlighting the need to address the limitations of scalability, security, and regulation [5]. The paper discusses the application of blockchain technology in e-government services. Areas such as authentication, e-service framework, e-voting, land property systems, e-delivery systems, HRM systems, and government contracting can be enhanced with blockchain. It concludes that most of the proposed solutions are still in the early stages and future research is necessary to make it sustainable [6]. The paper aims to study the latest advancement in blockchain technology and its application in the Internet of Vehicles, an emerging concept of intelligent transport systems. Applications include sharing spare computational and spectrum resources, smart parking management, content broadcasting, anonymous payment systems, traffic accident analysis, road safety applications, and vehicular big data auctions. Future research must focus to develop more security and privacy [7]. The paper discusses the concepts, applications, and risks of blockchain in AI, IoT, and smart logistics and supply chain management.AI provides concise instructions and reduces repetitive tasks by automating, while IoT helps monitor various parameters for quality control and also enforces contract control during shipping along with various projects are discussed. Security concerns, limited scalability, expensive to implement, and energy concerns are a few of the challenges. In conclusion, there is relative immaturity in some areas, whereas the effect of blockchain can be amplified using IoT and AI [8]. Digitalization is a significant technology trend impacting society but there is a gap in digitization and digital transformation. Findings reflect that digitalization is closely linked to advancements in digital technologies, including AI and automation, which are enabling new service delivery systems and manufacturing innovations. Future research should aim to develop a broader comprehension of digitalization, including related concepts like digital disruption, convergence, and the digital divide [9]. The digitalization concept has been introduced and its importance in various fields is discussed. Digital India programme and digital fields of culture, photography, computer, education, economy, and media are highlighted. The paper concludes by proposing a framework to achieve profitable digitalization [10]. The paper discusses the significance of digitization. Digitalization of infrastructure and supply chains, robotization of production processes, automation of management processes, and the introduction of autopilot systems are key focus areas. The paper concludes by stating how ignoring the trends in digitization might risk companies leaving the market [11]. Innovative technologies help IT effectively succeed in competitive environments, and their changing role and leveraging opportunities in digitalization are discussed. The paper then highlights that despite the dynamically changing industry, the core tasks of IT remain relevant in the digital age. It then suggests the emerging need to integrate IT into the organization [12]. The transformative effects of digital technologies on the labour market, with a focus on the transport industry, are highlighted. Findings indicate a clear relationship between high unemployment rates and the increased risk of job automation. It concludes by stating the ultimate effects on the labour market will depend on the skill requirements of jobs, with a potential decline in less skilled positions in fully autonomous scenarios [13]. The paper examines the sustainability impact of digitization in logistics. Findings suggest that the sustainability potential of digitization in logistics is still emerging and has not reached its full potential. The social implications of digitization in logistics have shown mixed outcomes. Digitization is expected to generate greater societal value along with economic impact [14]. The trucking industry has always been highly competitive and Various aspects of the industry, including operations, human resources, industrial relations, business strategy, marketing, accounting, safety, information technology, legal changes and deregulation, and its impact on the trucking industry are discussed. It concludes that labour productivity from larger and heavier trucks may face huge public resistance. Industrial relations solutions also are of high importance [15]. It aims to explore the relationship between organizational commitment and driver safety and develop a model illustrating its impact. It seeks to explain variations in retention, crash rates, and related costs. The theoretical framework provides practical benefits and has the potential to assist HRD practitioners [16]. It focuses on transforming the trucking industry through ZET adoption in India, for it to become a global leader by achieving significant benefits in the economy, energy, and emissions. Findings include reduced logistics costs and India becoming a battery manufacturing hub. Active leadership and steps are suggested for progress [17]. The adoption of IoT technology in trucking logistics in Indonesia is explored. Findings include the highest level of IoT adoption by industry being observed in the cement industry followed by chemical manufacturing and by location in Java and Sumatra islands. The implementation of IoT in the supply chain for both technological and management efficiency is suggested [18]. TRANSCOM and the ATA Foundation conducted a demonstration to assess the benefits and market value of providing real-time traffic information to trucking companies. Major benefits found were minimizing time spent, confident decision-making, improved scheduling, reduced repair costs, customer service, and increased productivity. Concluding, other benefits and confirmation by TRANSCOM's OIC are stated [19]. The adoption of (IT) in the trucking industry, specifically in freight mobility systems, is examined. Findings include areas of growth such as Route Planning and Scheduling, Weather and Construction Information, Map Databases, Tracking Power Units, Containers, and Trailers, Mobile Traffic Probes, Private Fleets, and challenges like Privacy of Information, Government Oversight, Monitoring the Driver, and Maintenance. Concerns need to be addressed in conclusion [20]. The role of information and technology in trucking industry is examined. TL, LTL, and Package Express industries and their contractual relationships with shippers and recipients are focused. It concludes by stating increased availability of information is impacting drivers, dispatchers, firms, and the industry [21]. Issues regarding the international transport system are being studied and a structure is proposed while establishing sustainable development. The 3 main components of ITS are found to be communications infrastructure, transportation systems, and institutional structures involving organizations, policies, funding mechanisms, and business processes. The use of intelligent technology to help improve strategic management decisions is suggested [22]. The impact of ITS on the overall performance of commercial road transport organizations in Zimbabwe and the challenges is discussed. Findings include coordination, technological factors, government policies, funding, and leadership to be significant hindrances. It concludes that the commercial roads in Zimbabwe have not fully embraced intelligent transport systems due to many issues [23]. Aspects of digital transformation in the supply chain are highlighted. Findings include contributions of digital transformation to customers, suppliers, investors, and employees. Risk management studies performance measures are suggested in future research [24]. It aims to determine which sectors utilize real-time data and leverage the value of data-driven analytics in the trucking industry. Findings include the need for stronger connections between supply chain and logistics platforms and the disparity between business practices in the trucking industry and technological advancements. Conducting research in underrepresented areas is suggested [25]. A multivariate discrete choice model (MPM) is used to analyse the adoption of seven technologies simultaneously, considering their relationships. Operating characteristics are found to affect technology adoption. Ongoing study of IT adoption by commercial vehicle operators is crucial [26]. Companies worldwide prioritize digital governance projects to stay competitive. Challenges of simultaneous triage, localization and bespoke solutions, growing adversities, and segmentation are highlighted. It is crucial to explore innovative ways to leverage technology, especially due to slow economic growth post-COVID-19 [27]. The focus is on blockchain governance in the public sector. Through thematic analysis using grounded theory, 3 categories of blockchain governance are infrastructure, operational processes, and governance as a technology. The future of blockchain technology in the public sector depends on comprehensive solutions that consider all aspects [28]. AI in government sectors is discussed. Applications of AI in fields of public health care, environment, transportation, government, education, economic, and financial sectors are stated. Concludes by stating gaps in literature and future research directions [29]. E-governance and the promotion of good governance through effective public service delivery are explored. A hypothesis stating that effective e-government service is positively related to e-service centres is supported. Framework to achieve e-governance to good governance is proposed [30]. The concept of "good governance" and its relevance to development policy is studied. Findings indicate, there is a correlation between good governance and economic development. Future research should prioritize bridging disciplines and formulating precise hypotheses to improve governance outcomes [31]. Good governance to achieve development programmes is defined. The focus is on governance and good governance in the context of regional development, challenges, and their indicators are discussed. Equitable approaches and inclusive views promote a sustainable and prosperous future [32]. Governance practice and its impact on the performance of Ethiopian-based community organizations are assessed. The conceptual framework of good governance was developed. Insufficient knowledge of cooperative values, prevalence of corruption, lack of honesty, transparency, responsiveness, and weak relationships were factors affecting performance. To address these issues, recommendations are proposed [33]. It aims to compare the disclosure levels of governmental financial statements using the International Public Sector Accounting Standards (IPSAS) as a benchmark. Findings indicate that both Indonesia and Malaysia have low levels of financial statement disclosure. Strengthening the infrastructure of financial reporting, including human resources and government support, is crucial to improve their financial statement disclosure [34]. Factors affecting the implementation of good government governance (GGG) and accountability performance are examined. The hypothesis that Competence and professionalism influence the implementation of GGG and its implementation influences performance accountability which is supported by findings. Study is limited to Riau province and further research is needed in other provinces [35]. How good governance is manifested in the sustainable development strategies of EU countries is examined. Findings include all investigated strategies demonstrating long-term focus, with specific objectives and indicators outlined for each area. Policies are found to be coherent. Suggestions include enhancing vertical coordination mechanisms and promoting bottom-up approaches to engage local communities in sustainable development efforts [36]. The effects of good management of local governments on individual subjective well-being (SWB) are studied. Findings include social connections which influence life satisfaction, age has a U-shaped relationship with life satisfaction, accountability and government efficiency have positive effects, presence of corruption has a long-term negative impact. Individual factors and good municipal governance are important determinants of individual SWB [37].

6.3 Methodology

The research in this paper is exploratory and descriptive. It aims to examine the uses of technology especially blockchain and artificial intelligence and its impacts on the trucking industry. The methodology includes the conduction of an exhaustive survey on the available research, the use of secondary data, the proposition of a framework model, findings, and recommendations in the light of the study performed. The secondary sources of information involving government reports, industry publications, literature, and websites have been used in this study. A comprehensive literature review is undertaken to acquaint with the existing research available and explore different ideas and perspectives from the authors. It involves understanding blockchain technology and other emerging technologies on the rise. An extensive review of the trucking industry and its evolution over the years has been studied. The possibility of achieving good governance through digitalization and e-governance has also been explored. Based on the insights gained, the formation of a blockchain and artificial model has been explained in detail. A blockchain framework model has also been proposed highlighting the use of technology in the trucking industry for digital governance. After an in-depth assessment, the paper provides an analysis on the adoption of blockchain and artificial intelligence into the trucking industry and suggests recommendations to promote the same.

6.4 Objectives

The objectives of the paper are stated as follows:

1. To understand the current situation of the trucking industry in India and identify issues related to it.
2. To explore the possibility of digitization, e-governance and blockchain technology used in trucking operation for its betterment.
3. To analyse the impact of digital governance and introduction of blockchain technology in the trucking industry.
4. To suggest policy recommendations on the basis of the study.

6.5 Analysis of Blockchain and Artificial Intelligence in Trucking Industry and its Impacts

The increasing use of technology and bringing the operation on the digital platform via digitalization in various perspectives has changed the function pattern of industry and sector too. Along with this, enabling technologies in the form of blockchain and artificial intelligence further benefit the industry, and the tracking industry in India is also no exception to it. The only problem is that the technology penetration level is very low when we look at trucking operations and various components associated with it.

The trucking industry in India is one of the most important segments and plays a very important role in sustaining our economy and economic activities. It is also considered as the backbone of our economy. The scenario of the trucking industry concerning its growth varies in pre- and post-independence as well as in pre- and post-economic reform eras in our economy. Today, there are more than 12 million trucks (all categories) operating on Indian roads to make sure that the goods are delivered on time and in an efficient manner. However, despite the significant number of trucks, there hasn't been much improvement in the nature and pattern of operations. This has led to issues such as time wastage, high costs, various barriers, regulatory and legislative problems, norm violations, governance issues, and viability problems. As a result, the trucking industry has not been able to fulfil its potential. There may be many reasons behind it one of the possible reasons is its very fragile ownership profile. That is a large number of truck operators who are unorganized as well more than 80% of the operators are small operators with up to five trucks with no market power or reach properly. This leads to exploitation and results in low levels of performance. Another possible reason is the government's long-term negligence in the past. However, in today's situation, given the availability and utilization of technologies such as the field of blockchain, information technology, artificial intelligence, and more, we are attempting in this paper to bring some significant positive improvements in trucking operations for better results. It is important to note that truck operations are based on specific areas such as local, interstate, intrastate, and international. During this, they come across numerous challenges, but those challenges can be ruled out with the implementation of enabling technologies. Moreover, with the increasing number of trucks on Indian roads over some time from merely 87,000 in 1948 to more than 60 lakhs during early 2010 to more than 10 million now as per the Motor Vehicle Statistics yearbook of the Government of India, to maintain an optimum level of tracking operations, the use of appropriate technology is crucial. Thus, in this regard, digital technology via digitalization, blockchain, artificial intelligence and various types of technology may play a very important role in enhancing trucking activities. The emergence of technologies in the trucking industry marks a major point in digitalization in India. Ever since its introduction, there has been an increased improvement in the effectiveness of repetitive work and productivity. The trucking industry has flourished in terms of increased efficiency, safety, reduced costs, and improved customer satisfaction and profitability. This has been possible through the immense adoption of digital technology. Below we are discussing two such evolving technologies of blockchain and artificial intelligence which can help evolve the trucking industry greatly.

6.5.1 Blockchain

Blockchain is known to be a distributed ledger technology without the need for an intermediary to perform transactions. A blockchain is nothing but a sequence of blocks, each block representing a transaction. It is transmitted to every node in the network and once it has been verified it gets added to the chain. The first block of the blockchain is known as Genesis. Each blockchain consists of 2 parts: a header and a body. The header consists of various components like the block version, Merkel tree root hash, timestamp, Bits, Nonce, and parent block hash. Given in (Fig. 6.1) is the architecture and important components of a block in a blockchain.

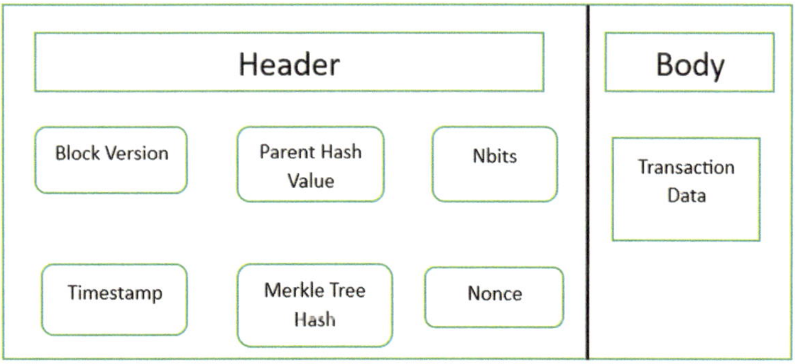

Fig. 6.1 Architecture of a block in a blockchain

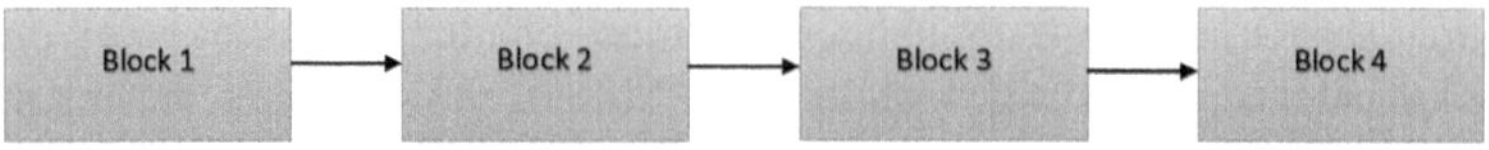

Fig. 6.2 Blockchain linkage (forming process)

Block version: It is very essential to know since there are different versions and implementations of blockchain. It helps to ensure that all blocks interpret the data well and use the same rules to process the transaction. *Parent Hash value*: The hash value of the preceding block is included. This aids in forming a link in the chronological order of the transaction and hence prevents any tampering of information. *Merkel Tree root hash*: It is a cryptographic hash function that is a single root hash formed by combining all the hashes of the transactions in a block. *Timestamp*: It consists of information on when the block is created which is crucial for consensus algorithms. *Nbits*: Also known as target difficulty is the level of difficulty of the problems in the consensus algorithms which is set while encoding the information. More computational power is required for solving problems of greater difficulty. *Nonce*: It is a random value assigned to the header, important for proof of work algorithms. It is short for the number used once.

The figure (Fig. 6.2) depicts the formation of a blockchain through Linkage.

A transaction once completed enters a block and subsequently every block gets added using the previous block's hash value thus forming a chain. Security in the blockchain is achieved through cryptography. Asymmetric cryptography uses a pair of keys namely a public key used for encrypting the data and verification of digital signatures. While a private key is used for decryption and creating digital signatures. Digital signatures are unique to each transaction which helps authenticate data. Consensus algorithms ensure an agreement regarding the state of the blockchain among the participants. There also exists a ledger that contains a copy of the record of who is participating in the network and the consensus protocols attached to it. These self-enforced agreements can be known as smart contracts. Smart contracts help ensure transparent interactions. There are various consensus algorithms as listed further. proof of work (*PoW*): Miners compete to solve the mathematical computations and the first one to solve is rewarded with cryptocurrency, i.e. right to add another block. It is energy intensive; hence, proof of stake (*PoS*) is a better alternative. Validators are chosen based on the amount of cryptocurrency they choose and are willing to lock up (stake) as collateral. Delegated proof of stake (*DPoS*) was introduced wherein the participants vote for delegates who produce blocks in a predefined order, enhancing scalability and transaction throughput. In *PBFT*, validators exchange messages to reach an agreement on block validity and transaction order providing fault tolerance and fast consensus. *Ripple* (RCPA) is designed for real-time gross settlement, currency exchanges, and remittances. A unique node is selected to create a new ledger every few seconds, using distributed agreement protocol. Another protocol is *Tendermint* which relies on a fixed set of validators providing high throughput and low latency for permissioned networks. The types of blockchains are private, public, and consortium based on accessibility restrictions. These elements of blockchain help build its unique characteristics such as immutability, decentralization, anonymity, fault tolerance, and transparency. Hence, the architecture, components, working, and characteristics are summarized.

6.5.2 *Artificial Intelligence*

Artificial intelligence is a machine technology that is a simulation of human cognition. It learns from experience via repetitive learning processes from which algorithms are built. The diagram below (Fig. 6.3) illustrates the entire process of building an artificial intelligence model or algorithm.

A typical AI model consists of a training library that contains datasets to build algorithms and models for a specific task. Neural Network Layers are one of the framework models which consist of input layers, hidden layers, and output layers. The input layers contain our input data which is processed in hidden layers through complex mathematical computations and the output layer consists of the results. Optimizers include optimization algorithms to minimize the loss function generated. Loss functions help the model learn and improve during training. It indicates the difference between the predicted output and the actual desired target output. Through this, performance is measured and evaluated, and necessary changes and adjustments are made to the algorithm and the models' architecture. Based on the evaluation results, the iterative training process continues until the desired outcome is achieved. Once the output is achieved, predictions or analysis is made and the inferences are generated. This is the brief working of an AI model. AI has penetrated various fields and has helped increase productivity through its competence. It has transformed the technology sector with various disciplines like Machine learning, Deep learning, Neural Networks, Natural language processing, and much more. Right from chatbots, facial and speech recognition, virtual assistants, Google Maps, self-driving cars, and open-AI like Chat-GPT AI has opened new directions for technology advancements. It has found its applications in several domains and a few are discussed below. *Health*

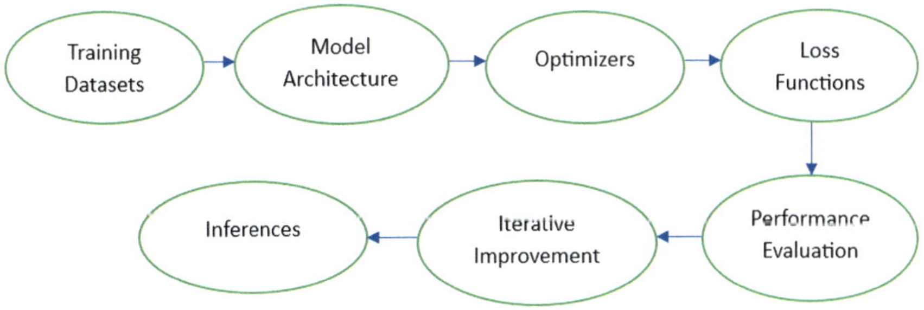

Fig. 6.3 Artificial intelligence model building process

care: It has helped doctors diagnose faster and more precisely by aiding in image processing. *Finance:* By detecting fraudulent transactions, automating transactions, and enabling algorithmic trading it has helped improve efficiency. *Social Media and entertainment:* It provides recommendations to users by analysing their choices and behaviour patterns. *Logistics:* It helps automate operation tasks like managing inventory, making schedules, and other mundane activities. *Transportation:* In this field, AI has helped provide real-time traffic and congestion information, help identify shortest routes, and self-driven cars are also emerging. Thus, AI helps automate repetitive tasks and makes unbiased, accurate, and strategic decisions, achieving high levels of accomplishment of the tasks.

6.5.3 Proposed Framework Model

The above-mentioned technologies can help in addressing various challenges faced by the trucking industry which ultimately hinders its growth. The challenges include overloading, vigilance issues, high cost and low efficiency of operation, en route barriers primarily concerned with regulatory measures, information constraints related to load availability, freight rates, route information, and so on. Primarily, this can be done by introducing digitalization into the system of trucking operations which further helps in e-governance and digital governance. Presently, limited digitalization is available in the form of certain portals providing information related to vehicles and drivers though with limited capacity. Thus, with digitalization, integrating blockchain technology where all digital information is stored in blocks and interconnected for additional information. Applying artificial intelligence to this interconnected system would greatly improve convenience, efficiency, and overall outcomes. The proposed blockchain (Fig. 6.4) defines the framework model with blocks inside of it indicating the storage of information. Different blocks within the blockchain will contain information about trucks, drivers, and other stakeholders. This information will include the number of times rules have been violated regarding traffic signals, accidents, overloading, issues faced en route, truckloads, capacity, and more.

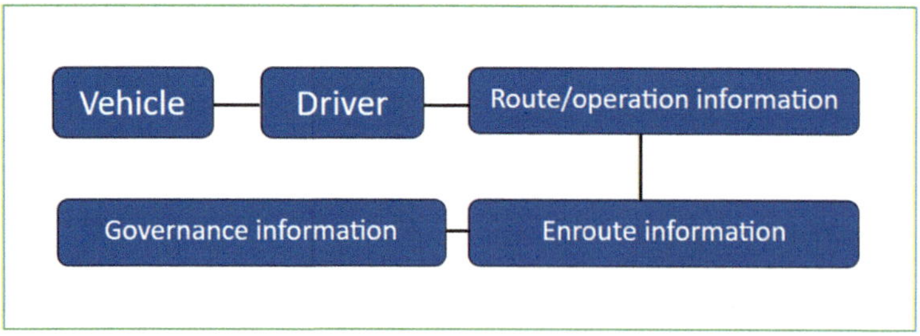

Fig. 6.4 Proposed blockchain model

To enhance the efficiency of digital governance, it can be integrated with blockchain technology. If all the relevant information is available digitally, it can be easily shared and accessed by all blocks. The Government of India, in this regard, has already initiated the concept of Digilocker, where individuals can upload authorized documents that can be utilized by concerned departments or blocks within the system. Artificial intelligence can also be employed to verify the authenticity of the documents, as Digilocker is linked to Aadhar numbers, PAN numbers, TAN numbers, and other identifying information. Fastag is another automotive system that is being used for not only the collection of toll tax but also for storing various information via QR codes. Scanning the QR code will store the received information in blocks, which can then be accessed by relevant authorities through the digital governance process. Given this, many en route issues related to regulatory measures can be resolved without causing delays or harassment to truck operators. This technology enhancement complements the existing e-way bill technique. Additionally, the received information can be accessed by financial institutions, as blockchain applications are integrated into the system. This facilitates efficient economic functioning as well.

6.6 Findings and Recommendations

The study analyses the benefits of using blockchain technology along with artificial intelligence to give optimum support for digital governance in the trucking industry. Digitalization or digital governance brings challenges such as overloading, norm violations, vehicle conditions (POC), vehicle and cargo information, and financial and operational data, among others. By employing the concept of blockchain, all captured information can be stored in blocks using blockchain technology. Later, with the application of artificial intelligence, concerned authorities can utilize this information for decision-making and planning to facilitate the development of the trucking industry. This data-driven decision-making will help the trucking industry bloom in terms of efficiency and competitiveness. The study also analyses the positive relationship between economic growth and the trucking industry. As national income increases, the truck population is expected to grow. However, without optimal utilization of available resources, it may not lead to desired growth and positive impacts such as efficiency, transparency, smoother governance, income growth, and reduced violations. Furthermore, the study suggests that technological enhancements in the trucking industry can lead to a reduction in trip time through symmetric information on load availability, minimized trouble with authorities (as all data is in Digilocker), improved operational efficiency, good governance, and more.

Based on these findings, the study recommends the following actions:

1. The government should promote the use of blockchain technology and AI in the trucking industry and provide necessary support and assistance to concerned stakeholders.

2. The government should establish training centres to enable stakeholders to effectively utilize these technologies.
3. Truck operators and trucking companies should actively participate in this mission for better results.
4. Financial assistance and affordable technology, along with the required infrastructure, should be provided to the trucking industry for technological enhancement.
5. The trucking industry should be provided with infrastructure sectors' component statuses.
6. The trucking companies should try integrating blockchain with other technologies like artificial intelligence into existing built systems might prove to be more beneficial.
7. Government should promote collaboration among technology and sustainable development companies which strive to automate processes to reduce carbon emissions.
8. For government contracts and verification of documents, the identity of the truck driver should be recorded by blockchain technology.
9. AI models and algorithms should be developed for better maintenance, optimization of resources, reduced costs, and enhanced safety.
10. Using the proposed framework model as the base, an entire tracking and monitoring system should be built for improved traceability and logistics.

By implementing these recommendations, the trucking industry can leverage technology to overcome challenges, improve efficiency, and contribute to overall growth and development.

6.7 Conclusion

In this study, we explored the persistent challenges faced by the crucial trucking industry and proposed solutions through digitalization and blockchain technology. We have examined blockchain technology, AI, and digitalization. It also proposes a framework model for integrating blockchain into the trucking industry to improve efficiency. Emphasizing good governance and operational efficiency, we have integrated pioneering technologies. The study recommends promoting blockchain technology and AI adoption in the trucking industry with government support, financial aid, and coordination from stakeholders among other suggestions. Introducing skill development will incentivize more companies to adopt the technology. Keeping in mind the rising environmental concerns, collaborations with organizations working towards sustainable development goals will prove to be beneficial. By implementing these actions, the trucking industry can overcome challenges, improve efficiency, and foster overall growth and development. Limitations of the paper include that we have primarily focused on the Indian trucking industry, this cannot be generalized to other countries. The research has not delved deeply into potential barriers to technology adoption among small truck operators. Possible research areas in future can be to explore innovative strategies to incentivize small truck operators to embrace advanced technologies. It can also extend to integrate other advanced and emerging technologies for optimizing the effectiveness of the transportation sector. Technology advancements are moving at a fast pace and leveraging this technology to improve regulation will be advantageous to all organizations. Improving India's trucking industry will lay the foundations for more digitized governance and change the growth trajectory of this sector. Adopting emerging technologies will help propel the trucking industry to new heights and produce multiplier effects in the economy with fruitful returns.

References

1. Zheng, Z., et al.: An overview of blockchain technology: architecture, consensus, and future trends. ResearchGate (2017)
2. Sri Aruna, S.G., Lalitha, B.: A study on blockchain technology. ResearchGate (2018)
3. Singh, S., et al.: A detailed study of blockchain: changing the World. Int. J. Appl. Eng. Res. **13**(14) (2018), ISSN 0973-4562
4. Zheng, Z., et al.: Blockchain challenges and opportunities: a survey. Inderscience Enterprises Ltd. (2018)
5. Baiod, W., et al.: Blockchain technology and its applications across multiple domains: a survey. J. Int. Technol. Inf. Manag. (2021)
6. Lykidis, I., et al.: The use of blockchain technology in e-government services. MPSDI-Comput. (2021)
7. Mollah, M., et al.: Blockchain for the internet of vehicles towards intelligent transportation systems: a survey. IEEE Internet Things J. **8**(6) (2021)
8. Idrissi, Z., et al.: Blockchain, IoT and AI revolution within transport and logistics. In: 14th International Conference of Logistics and Supply Chain Management. LOGISTIQUA (2022)
9. Belzer Michael, H.: Technological innovation and the trucking industry: information revolution and the effect on the work process. J. Labor Res. (2001)
10. Graham, C., et al.: Work climate, organisational commitment, and highway safety in the trucking industry: toward causal modelling of large truck crashes (2008)
11. Niti Aayog: Transforming Trucking in India (2022)
12. Wedha, B., et al.: Analysis of IoT adoption on trucking logistics in various industry in Indonesia. IEEE-ICIC (2021)
13. Batz Thomas, M.: The utilization of real-time traffic information by the industry. IEEE Trans. Veh. Technol. **40**(1) (1991)
14. Bander, J., et al.: IEEE strategic management of intelligent transport at ion systems: the case of freight mobility systems in the trucking industry (1998)
15. Bander, J., et al.: Information and the trucking industry. IEEE (1997)
16. Tagiltseva J., et al.: The intelligent transportation system in multimodal freight transportation. IEEE (2021)
17. Muchaendepi, W., et al.: Intelligent transport systems and its impact on performance of road freight transport in Zimbabwe. IEEE (2018)
18. Özkanlısoy, Ö., Akkartal, E.: Digital transformation in supply chains: current applications, contributions and challenges. ResearchGate (2021)
19. Keepers, M., Wuest, T.: Smart Trucking—Status of Digital Transformation of the Trucking Industry: A Bibliometric Analysis. Elsevier, Amsterdam (2019)

20. Golob Thomas, F., Regan Amelia, C.: Trucking Industry Adoption of Information Technology: A Multivariate Discrete Choice Model. Elsevier, Amsterdam (2001)
21. Abdul Salam, M.: E-governance for good governance through public service delivery. Institute of Governance Studies, BRAC University (2013)
22. Gisselquis Rachel, M.: Good governance as a concept, and why this matters for development policy. ResearchGate (2012)
23. Rahim, A.: Governance and good governance a conceptual perspective. J. Public Adm. Gov. (2019)
24. Dayanandan, R.: Good governance practice for better performance of community organizations—myths and realities!!. J. Power Polit. Gov. **1**(1) (2013)
25. Sukmadilagaa, C., et al.: Good Governance Implementation in Public Sector: Exploratory Analysis of Government Financial Statements Disclosures Across ASEAN Countries. Elsevier, Amsterdam (2015)
26. Nofiantia, L., Suseno, N.S.: Factors Affecting Implementation of Good Government Governance (GGG) and Their Implications Towards Performance Accountability. Elsevier, Amsterdam (2014)
27. Kardosa, M.: The reflection of good governance in sustainable development strategies. Procedia. Soc. Behav. Sci. (2012)
28. Carcaba, A., et al.: Does Good Local Governance Improve Subjective Well-Being? European Research on Management and Business Economics. Elsevier, Amsterdam (2022)
29. Tiwari, S.P.: Organizational competitiveness and digital governance challenges. ResearchGate (2022)
30. Tan, E., et al.: Blockchain Governance in the Public Sector: A Conceptual Framework For Public Management. Elsevier, Amsterdam (2021)
31. Sharma, G., et al.: Artificial Intelligence and Effective Governance: A Review, Critique and Research Agenda. Elsevier, Amsterdam (2019)
32. Reis, J., et al.: Digitalization: a literature review and research Agenda. ResearchGate (2020)
33. Narula, T.K., Dr. Rana, S.: Digitalization in India. Int. J. Emerg. Technol. (2017)
34. Mashkina, N., et al.: Digitalization of the Transport Industry. Department of Finance and Credit, SWSU, Kursk, Russia, EDP Sciences (2020–21)
35. Urbach, N., et al.: The impact of digitalization on the IT department. Springer, Berlin (2018)
36. Chinoracký, R., Čorejová T.: Impact of Digital Technologies on Labor Market and the Transport Sector. Elsevier, Amsterdam (2019)
37. Yasanur, K.: Sustainability Impact of Digitization in Logistics. Elsevier, Amsterdam (2018)

Chapter 7
Detecting Parkinson Disease Using Various AI Enabled Methods: Recent Advances, Requirements, and Open Challenges

Sourabarna Roy, Tannistha Pal, and Swapan Debbarma

Abstract Parkinson's disease is a widespread and complex neurological condition characterized by a range of motor symptoms, including trembling, rigidity, and impaired coordination. Additionally, Parkinson Disease affects non-motor functions, such as facial expressions, speech, and mobility, progressively worsening as the disease advances. While commonly associated with older age, Parkinson Disease can manifest in individuals as young as 20, underscoring its diverse demographic impact. This paper aims to provide a comprehensive survey of techniques employed for the detection and analysis of Parkinson Disease. Despite significant advancements in managing Parkinson Disease symptoms through medication and deep brain stimulation, the precise etiology of Parkinson Disease remains elusive, particularly regarding the potential involvement of immunological factors. This ambiguity underscores the intricate nature of neurodegenerative diseases, necessitating continued research to unravel the enigma surrounding Parkinson Disease origins. Notably, this paper highlights the promise of deep learning algorithms, particularly convolutional neural networks and recurrent neural networks, in medical image processing tasks. Deep learning models can discern subtle patterns in medical images, such as brain scans or muscle activity recordings, aiding in Parkinson Disease diagnosis. Finally, this paper provides a comprehensive overview of current methods for diagnosing and analyzing Parkinson's disease. It emphasizes the challenges posed by the subtle early symptoms and the uncertainty surrounding its etiology. Furthermore, it underscores the potential of deep learning to advance our understanding and treatment of this intricate neurodegenerative condition, paving the way for more precise and tailored medical interventions in the future.

S. Roy · T. Pal (✉) · S. Debbarma
Department of Computer Science and Engineering, National Institute of Technology Agartala (NITA), Agartala, India
e-mail: tannisthapaul@gmail.com

P. Nanda et al. (eds.), *Intelligent System and Data Analysis*, Smart Innovation, Systems and Technologies 398, https://doi.org/10.1007/978-981-97-5200-3_7

7.1 Introduction

The degeneration of nerve cells in the substantia nigra, a part of the brain that controls movement, causes Parkinson's disease (PD), a degenerative disorder. The production of the essential neurotransmitter dopamine is reduced as a result of these nerve cells dying or being damaged. PD is a neurological condition that cannot be cured and is brought on by a decline in dopamine in the human brain. Dopamine is a neurotransmitter that aids in communicating with the basal ganglia, the area of the brain that controls movement and coordination. Due to the variability of the disease, PD symptoms might progress differently from one person to the next. Recently, deep learning (DL) techniques have dominated categorization issues. DL methodologies excel at automatically deriving intricate representations and patterns from huge datasets. Neuroimaging methods such as positron emission tomography (PET) and magnetic resonance imaging might help us better understand the anatomical and functional changes occurring in the brains of persons with PD. They are well-suited for analyzing a variety of medical data, from wearable sensor data to neuroimaging scans and speech recordings. This enables the creation of precise and MRI. The ability to identify between PD patients and healthy people using deep learning models applied to neuroimaging data has shown encouraging results. These models are able to spot small changes in brain morphology, pinpoint particular PD-related biomarkers, and even forecast how the illness will develop. Although PD cannot be cured, there are numerous therapies and treatments that help minimize the symptoms. Early diagnosis of PD is necessary for effective treatment with drugs, surgery, and physical therapy. MRIs can help in the early diagnosis of PD before any symptoms appear. MRIs can help determine the severity of impending symptoms by identifying damaged brain neurons that may be a marker of PD. An MRI can provide data that enables doctors to identify PD early and begin treating individuals to slow the progression of symptoms. PD is treated with medication, dietary changes, exercise, speech therapy, and deep brain stimulation (Figs. 7.1 and 7.2).

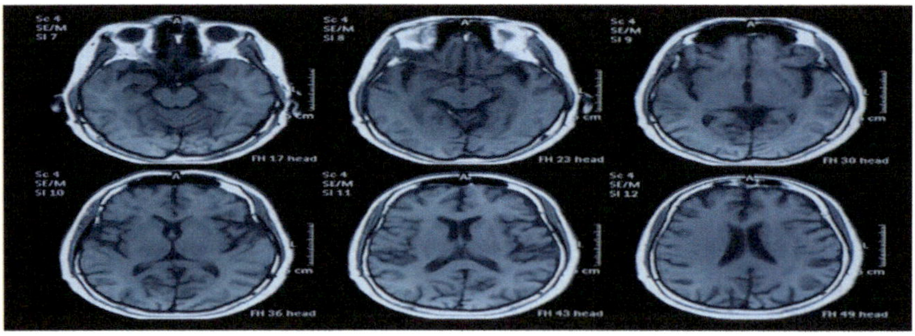

Fig. 7.1 Stages of PD on MRI scan [1]

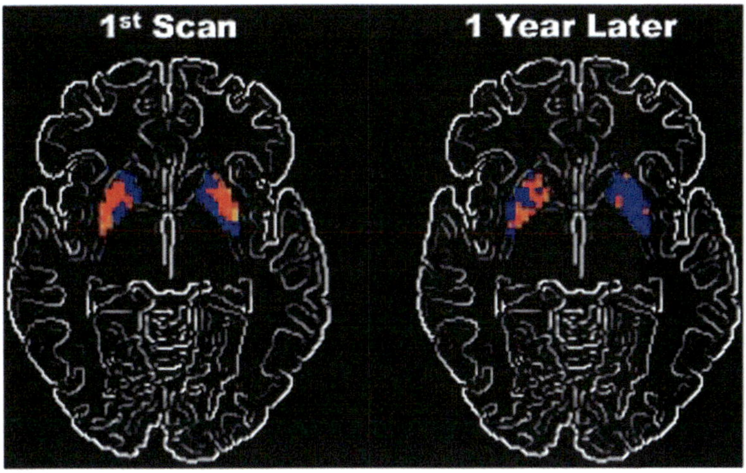

Fig. 7.2 PD increment level within 1 year [2]

7.2 Background

In recent years, the identification and monitoring of critical diseases has made extensive use of health informatics systems. Assessing the severity of symptoms using a number of methods is the aim of PD detection systems. PD is a neurological ailment that gets worse with time and exhibits symptoms that are both motor and non-motor. Early on in the course of the disease, voice difficulties are a regular occurrence in PD patients [3]. Consequently, current investigations on the detection of PD have placed a focus on diagnosis methods based on voice problems. Parkinson is an extremely complicated condition, and there is no reliable scale to determine how severe it will be. Because of the loss in dopamine levels in the brain, it is a neurodegenerative condition that disrupts motor functions and has physical implications on the body. Dopamine, a chemical fluid that is only produced by neurons, is what causes the body to move and how neurons communicate with one another. The several brain communication modes influence the neurological condition's slowing down when dopamine levels start to decline with age. A novel deep learning method is presented to quickly determine if a person has PD or not based on promoter traits. Parkinson's disease (PD) is a degenerative neurological condition marked by the gradual loss of nerve cells in the substantia nigra, which reduces the amount of dopamine produced in the brain, which is necessary for maintaining balance and motor function. The vast range of symptoms that PD presents with, which might differ from person to person, makes early diagnosis and treatment essential. Recent advancements in deep learning (DL) techniques have made them effective classification and analysis tools that can automatically uncover complex patterns from large datasets. Effective PD treatment, which may include medication, dietary modifications, exercise, speech therapy, and deep brain stimulation, depends on early diagnosis. MRIs are crucial in detecting early-stage PD and determining the severity of the symptoms, allowing for timely interventions to halt the progression of the disease (Fig. 7.3).

7.3 Literature Review

Over the past two decades, AI techniques have advanced dramatically, and as a result, these technologies are now fairly well-known across a range of industries. The detection of PD has also been studied using a number of symptoms, such as olfactory loss, walking patterns, speech patterns, handwriting tests, and other motor skill assessments. Abdullah et al. [4] used handwritten records from a common NewHandPD dataset, a novel framework was presented for the accurate diagnosis of PD. The suggested architecture is built on transfer learning models like ResNet, VGG19, and InceptionV3 to reduce the load of training time. Wang et al. [5] proposed to early diagnosis of PD is essential in order to better understand its origins, begin therapy approaches, and enable the creation of powerful drugs. This study provided a DL strategy to automatically distinguish between healthy people and PD patients based on promoter signs (such as rapid eye movement (REM), sleep behavior disorder (RBD), and olfactory loss). Qin et al. [6] also proposed the sole use of sMEG signals for PD tremor severity quantification. We make it possible for them to be utilized to automatically evaluate PD patients by using the widely accepted MDS-UPDRS evaluation criteria. Vasquez Correa et al. [7] worked on a study of the multimodal motor abilities of PD patients using DL architectures based on TFRs and convolution neural networks (CNNs) to incorporate data from speech, handwriting, and gait signals. The suggested strategy mimics how difficult it is for patients to begin and stop verbal and physical movements of their upper and lower limbs. Wahid et al. [8] compared utilizing MR normalization, compared the spatial–temporal gait features of PD patients and controls, and assessed the capacity of machine learning (ML) algorithms to classify PD gait after data normalization.

Joshi et al. [9] presented the most significant risk variables to suggest a new classification scheme for Alzheimer's and PD. In order to assess the relationship between risk factors, this model has utilized the attribute assessment technique concept. It has been found that the classification accuracy changes depending on how one risk factor is modified by another. Noor et al. [10] using the most popular DL techniques, the three most prevalent neurological diseases have been identified from the MRI scan data in this study. DL approaches have been proposed for the categorization of neurological diseases that have been described in the literature. The benefits, drawbacks, and effectiveness of DL methods. According to Martinez-Eguiluz et al. [11], non-motor PD symptoms also appear early and are easily measurable in routine clinical practice. The effectiveness of nine different machine learning algorithms for differentiating idiopathic Parkinson's disease (PD) patients from controls using a variety of non-motor clinical symptoms. Govindu et al. [12] approaches employing ML approaches, audio recordings can be used to detect PWP early. Our research compares and contrasts a range of ML models for disease categorization that are

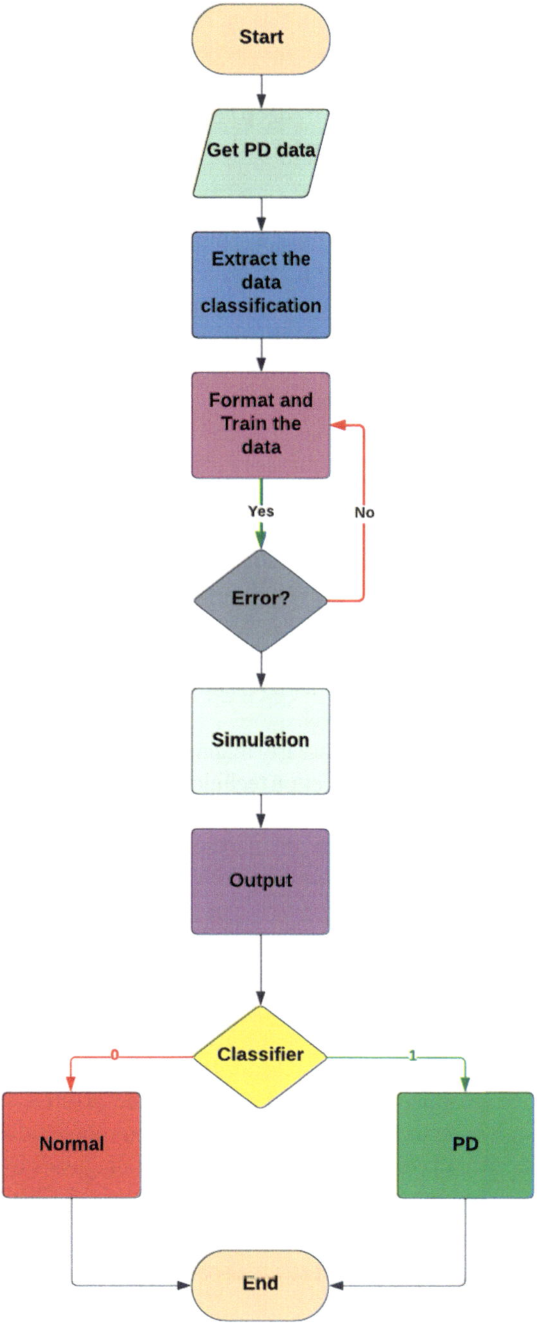

Fig. 7.3 Workflow of PD detection

not only quicker but also more memory efficient than deep neural network learning models. Our promising results should encourage advancements in PD telemedicine. Ogawa et al. [13] in their research, tested classical neural networks vanishing gradient issue has been solved by deep neural networks, notably CNNs. The vanishing gradient problem, which has returned with more layers, was addressed with the idea of a residual network.

Shivangi et al. [14] in their different research articles on PD were examined in order to develop an effective method for identifying the disease. It classified patients into the categories of normal and suspect based on a number of symptoms using a variety of ML and DL Algorithms. Wodzinski et al. [15] presented when the reader considers that the speech recordings were only classified using frequency-based features from spectrograms, the results are even more impressive. Alharthi et al. [16] also presented builds, tests, and shows the ability of DL models to classify gait disability in PD patients. The models overcome the heterogeneity of human gait from person to person and surpass earlier manual feature extraction approaches. They also show robustness to perturbation noise. Quan et al. [17] proposed vocal cues in a DL-based method for PD identification is suggested. The suggested strategy for capturing the time-series characteristics of continuous speech signals creatively combines the dynamic articulation transition features and the bidirectional long short-term memory (LSTM) model. Gunduz [18] suggested a variety of vocal (speech) variables, deep CNN architectures were used in this work to categorize PD. To do this, we develop two frameworks that let us distinguish between healthy individuals and PD patients and are based on CNN.

In their research on early diagnosis utilizing voice signals, Lamba et al. [19] looked at several feature selection techniques and classifiers to see which one performed best. The system can be a useful second opinion for a Parkinson's diagnosis, but it is not intended to replace medical experts. By putting the suggested technique to the test on sizable speech and voice datasets and investigating a diagnostic system based on handwritten drawings, taking into account the influence of slowness and tremors on patients' handwriting, which can be an early indicator of Parkinson's, the authors want to further their study. Senturk [1] built a decision support system for early Parkinson's diagnosis using speech cues from PD patients and healthy participants. By applying a range of feature selection (FS) and classification techniques, they increased efficiency and accuracy while reducing computational costs. The findings demonstrated that FS had a pronounced positive impact, especially when dealing with speech signals that contained various phonetic features. The developed approach showed excellent accuracy in the early diagnosis of PD, perhaps delaying the onset of severe symptoms. According to Pahuja et al. [20], on datasets for audio, Wisconsin breast cancer, and Pima Indian diabetes, the article assesses ANN, KNN, and SVM classifiers. On the audio dataset, ANN and Levenberg–Marquardt obtain an accuracy of 95.89, assisting doctors in the diagnosis of Parkinson's disease. The handwritten data used in [2] discussion of PD detection was quite uneven. Training on unbalanced data led to biased machine learning models, which had high sensitivity and low specificity for PD diagnosis. In order to eliminate bias and provide unbiased models, random subsampling was performed. To anticipate Parkinson's disease [3], employed

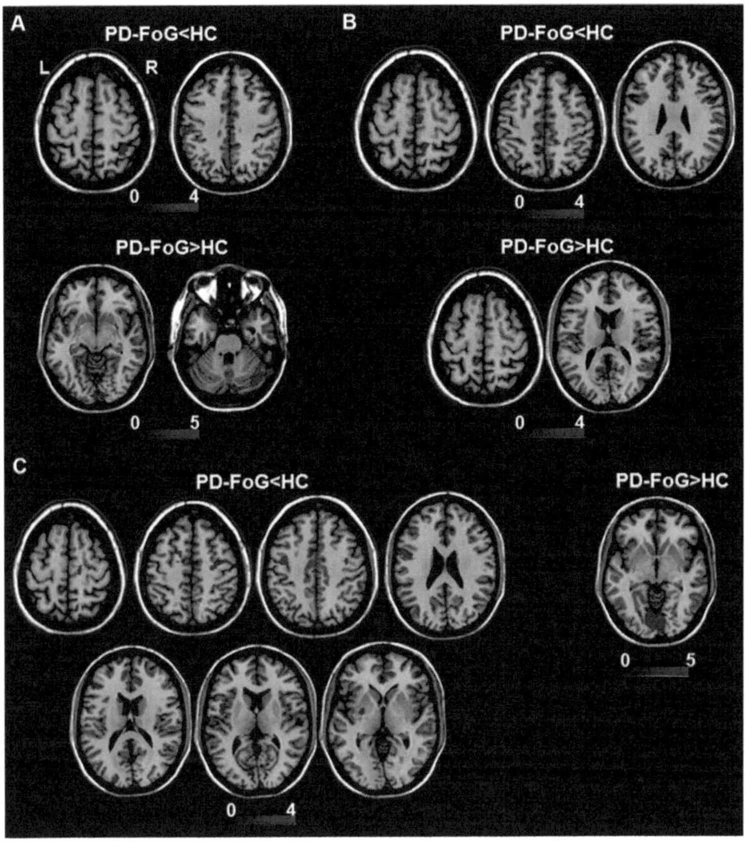

Fig. 7.4 Detecting stages of brain having PD in different existing techniques [20]

several machine learning classification techniques. Feature expansion techniques based on correlation analysis of fundamental characteristics, including PCA, IG, and original features, are presented to increase the classification accuracy (Fig. 7.4).

7.4 Comparative Analysis

A machine learning system is used to identify PD. For the detection of PD, CNN and recurrent neural networks and other DL techniques have received a lot of interest. To identify irregularities in the brain related to PD, CNNs have been used to assess neuroimaging data, including MRI images. Transfer learning, a technique that uses previously trained models as a jumping off point for new tasks, has been used to study PD detection.

The field of Parkinson's disease (PD) detection and diagnosis has been dramatically impacted during the past 20 years by the quick development of AI approaches. Numerous researches have investigated various methods for diagnosing PD using a variety of symptoms and data sources. Using transfer learning models, Sura Mahmood Abdullah et al. developed a unique approach for precisely diagnosing PD from handwritten records. Deep learning (DL) was used by Wu Wang et al. to identify between healthy people and PD patients based on premotor symptoms, emphasizing the value of early PD diagnosis. Using the MDS-UPDRS evaluation criteria, Zengyi Qin et al. concentrated on sMEG signals for tremor severity estimation. J. C. Vasquez Correa et al. evaluated the multimodal motor abilities of PD patients using DL architectures incorporating data from speech, handwriting, and gait signals. In order to evaluate ML algorithms for PD gait classification, Ferdous Wahid et al. investigated MR normalization and spatial–temporal gait variables. According to risk variables, Sandhya Joshi et al. developed a classification system for Alzheimer's and Parkinson's disease, emphasizing the value of determining how the two diseases interact. Using DL methods, Manan Binth Taj Noor et al. diagnosed neurological disorders from MRI scan data. Early non-motor PD symptoms were the subject of Maitane Martinez-Eguiluz et al.'s comparison of various ML algorithms for discrimination. Aditi Govindu et al. investigated the use of memory-efficient ML models on audio recordings for the early PD identification. Deep neural network vanishing gradient issues were addressed by Mitsuhiro Ogawa. Based on a variety of symptoms, Shivangi et al. categorized patients using ML and DL algorithms. With the use of speech recordings categorized using frequency-based characteristics, Marek Wodzinski produced outstanding results. Abdullah S. Alharthi showed how DL models can accurately detect gait impairment. Hakan Gunduz used vocal variables and deep CNN architectures, while Changqin Quan suggested a DL-based technique for capturing voice cues for PD recognition. This research highlights the complexity of PD diagnosis, using a variety of data sources, symptomology, and AI techniques to improve early detection and comprehension of this crippling condition (Table 7.1).

7.5 Qualitative Evaluation

The accuracy of any model is the percentage of accurate forecasts it produces compared to all other forecasts. This means that we split the total number of projections that were successfully recognized as positive or correctly identified as negative by the total number of accurate and inaccurate forecasts.

Accuracy is formulated as:

$$\mathrm{ACC} = \frac{\mathrm{TP} + \mathrm{TN}}{\mathrm{TP} + \mathrm{TN} + \mathrm{FP} + \mathrm{FN}} \tag{7.1}$$

Where,

Table 7.1 Summary of various Parkinson Disease detection method

Method	Pros	Cons
Transfer learning [10]	While enhancing neural network performance and needing less data, transfer learning can shorten training times	This only happens if the source and target are too dissimilar, which causes the first training to be inaccurate
Machine learning [4]	Machine learning is able to analyze large amounts of data and spot particular trends and patterns that people might overlook. To display them with suitable advertisements, it takes advantage of the findings. Spam detection using ML is effective	Large, thorough, objective, and high-quality datasets are necessary for machine learning's training. On rare occasions, they might have to wait until fresh data is produced
Deep learning [5]	Deep learning algorithms don't need manually designed features because they can automatically learn them from the data. State-of-the-art performance on a variety of issues, such as image and speech recognition, natural language processing, and computer vision, has been proven using deep learning methods	Strong GPUs and a lot of RAMS are needed for deep learning model training, which consumes a lot of computing power. A model is considered to be overfit when it performs poorly on new, untrained data and overperforms on training data. Deep learning algorithms must be taught on high-quality data, which is essential
Spatial–temporal gait features [13]	The GAIT Rite device can quickly measure up to 70 gait parameters, and this data is easily retained as a performance log. To show changes during a care episode, performance can be tracked over time	Metrics that may be used include step length, stride length, stride length to leg length ratio, stride or step width, longitudinal dimension, horizontal dimension, foot dimensions, foot line, direction line, foot angle, gait line, etc.
Neural networks [9]	Complex problems can be solved with neural networks since they have the capacity to learn from data. Additionally capable of generalization, neural networks can identify patterns in data that conventional algorithms might not be able to	Due of their complexity, neural networks need a lot of data to be trained. Additionally, neural networks are prone to overfitting, which is when they recognize patterns in the data that aren't really there. This might produce unreliable outcomes
Residual networks [12]	Remaining networks can combat degeneration by short-circuiting superficial levels to deep layers by taking shortcuts or skipping connections	ResNet has proven effective in a variety of applications, but one key drawback is that a deeper network often requires weeks of training, making it practically unusable in real-world scenarios

TP True Positives
TN True Negatives
FP False Positives
FN False Negatives

One of the most well-liked performance metrics is accuracy. It measures how many observations out of all the observations could correctly be anticipated. However, claiming that a model is the best based just on accuracy is incorrect. Accuracy is a crucial factor to take into account when there are almost as many FPs as FNs and the dataset is symmetrical. In the case of asymmetric datasets, we must resort to other performance measures because the number of improperly classified positive and negative predictions is also a cause for concern.

Recall is expressed as:

$$\mathrm{Recall} = \frac{\mathrm{TP}}{\mathrm{TP} + \mathrm{FN}} \tag{7.2}$$

The proportion of true positives that were correctly identified is measured by recall. The total samples that were correctly forecast as positive or incorrectly projected as negative (TP, FN) are divided by the total samples that were correctly forecast as positive (TP).

And, the precision is accounted as:

$$\mathrm{Precision} = \frac{\mathrm{TP}}{\mathrm{TP} + \mathrm{FP}} \tag{7.3}$$

One measure of a ML model's effectiveness is precision, or the standard of a successful prediction the model makes. The proportion of overall true positives to overall positive projections is known as precision.

There have been several novel strategies investigated, each with distinct advantages. To accurately diagnose PD from handwritten data, Sura Mahmood Abdullah et al. suggested a novel approach utilizing transfer learning techniques. The need of early PD diagnosis was emphasized by Wu Wang et al., who provided a DL technique that uses premotor symptoms to automatically distinguish between healthy people and PD patients. In accordance with the widely accepted MDS-UPDRS criteria, Zengyi Qin et al. developed a novel approach to measure the severity of PD tremors by focusing on sMEG signals. J. C. Vasquez Correa et al. used a multimodal method to demonstrate the complexity of motor abilities in PD patients by merging speech, handwriting, and gait data. In his comparison of various methods for classifying PD gaits, Ferdous Wahid emphasized the need of data normalization. From MRI scans, Manan Binth Taj Noor was able to identify common neurological illnesses, highlighting the potential of Deep Learning in this field. Maitane Martinez-Eguiluz examined various ML algorithms and gave insights into their effectiveness by concentrating on non-motor PD symptoms. Aditi Govindu used memory-effective ML models to investigate audio recordings for the early PD identification, showing promise for telemedicine applications. In order to solve disappearing gradients, Mitsuhiro Ogawa introduced the idea of residual networks to address neural network difficulties. Shivangi used a variety of ML and DL algorithms to do substantial study on PD categorization and provide a thorough methodology. With the help of frequency-based features and speech recordings, Marek Wodzinski produced impressive results. Abdullah S. Alharthi proved the reliability of DL models for categorizing gait impairment. Changqin Quan cleverly combined dynamic articulation transition characteristics and LSTM models to identify PD using vocal cues. Hakan Gunduz introduced two frameworks for patient differentiation by proposing voice variables and deep CNN architectures for PD categorization. Collectively, these investigations demonstrate the varied and dynamic landscape of AI in PD diagnosis, presenting potential opportunities for early detection and enhanced knowledge of the illness (Table 7.2).

7.6 Result and Discussion

After analyzing all the existing researches, we can say that neural networks have a number of benefits over conventional methods. They are able to learn from data and be applied to the resolution of challenging issues. Additionally capable of generalization, neural networks can identify patterns in data that conventional algorithms might not be able to. Thus, we conclude that Joshi et al.'s [9] model has the best performance in detecting PD Disease. The final results of the algorithm demonstrate accuracy of 99.25 percent and [5, 7, 10, 16] have moderate accuracy in performance, we can also use that techniques and try to improve the accuracy level. However, as PD worsens, symptoms will become much more. PD treatment can assist you in controlling your symptoms. People with extremely early Parkinson's frequently don't initially have any symptoms. PD cannot currently be cured or slowed down in its course. Giving the algorithm symptoms as an attribute enables ML-based diagnosis of this elusive illness. Young individuals are rarely affected by this condition; instead, it usually shows up in middle or later life. A range of AI techniques, including ML, DL, feature selection, feature fusion, and transfer learning, have shown promise for the diagnosis of PD. These techniques could aid in the early diagnosis of illnesses, the monitoring of disease progression, and the creation of customized treatment programs for PD patients. More research and validation on larger and more diversified datasets are needed in order to increase the precision and generalizability of these AI-based strategies in actual clinical situations. When differentiating PD patients from healthy individuals using ML algorithms like SVM, RF, and artificial neural networks (ANNs), studies have shown positive results. These algorithms may extract relevant information from clinical data, such as demographic information, medical history, and motor complaints, in order to develop prediction models. Although the accuracy of these models varies, several studies have found that they have high accuracy rates of between 80% and over 95%. Achieving high accuracy in PD detection

Table 7.2 Performance evaluation of different existing models

Model	Accuracy	Dataset used
K-nearest neighbor classifier [4]	0.9529	NewHandPD dataset has been taken into consideration
Feed-forward neural network classification [5]	0.9645	Parkinson's Progression Markers Initiative (PPMI) database
S-Net and computational efficient CNN [6]	0.9055	The IRB of Tsinghua University Yuquan Hospital
Convolution neural network [7]	0.973	The data contain recordings of speech, handwriting, and gait collected
Kernel Fisher Discriminant (KFD), Naive Bayes, k-Nearest Neighbor (kNN), Support Vector Machine (SVM) and Random Forest (RF) [8]	0.926	Using raw the only significant differences in spatial–temporal gait features
Neural networks and ML [9]	0.9925	487 patient records totaled from ADRC, USA, make up a dataset
CNN, DNN, RNN, AE, Deep Belief Network (DBN), and Probabilistic Neural Network (PNN) [10]	0.9686	AIBL Australian Imaging Biomarkers and Lifestyle Study of Aging (AIBL) dataset contains MRI, PiB PET images
Supervised algorithm [11]	0.844	This database included data from the Biscay PD Association (ASPARBI) and the Cruces University Hospital's Department of Neurology
Logistic regression for classification, random forest SVM, KNN [12]	0.9183	31 individuals biomedical voice measurements have been compiled.
CNN and novel residual network-type 1-d [13]	0.888	The Department of Neurology at Istanbul University's Cerrahpasa Faculty of Medicine provided the dataset
DL, neural network [14]	0.8915	The debilitating symptoms of PD is gait and another one is based on symptom
Neural networks, decision trees, regression and DM neural algorithm [15]	0.9	The PC-GITA dataset was used
Deep CNN, Layer-Wise Relevance Propagation (LRP) [16]	0.95	Open access benchmark from PhysioNet.org
LSTM [17]	0.7556	The GYENNO Science PD Research Centre
CNN, SVM [18]	0.869	The Department of Neurology in Cerrahpasa Faculty of Medicine, Istanbul University

tasks, DL algorithms have demonstrated promising outcomes, frequently outperforming conventional ML approaches. Large datasets are needed for these models to be trained, which might be a drawback in some circumstances. PD detection has been investigated using transfer learning, a method in which previously trained models are employed as a starting point for a new task. To extract pertinent characteristics and

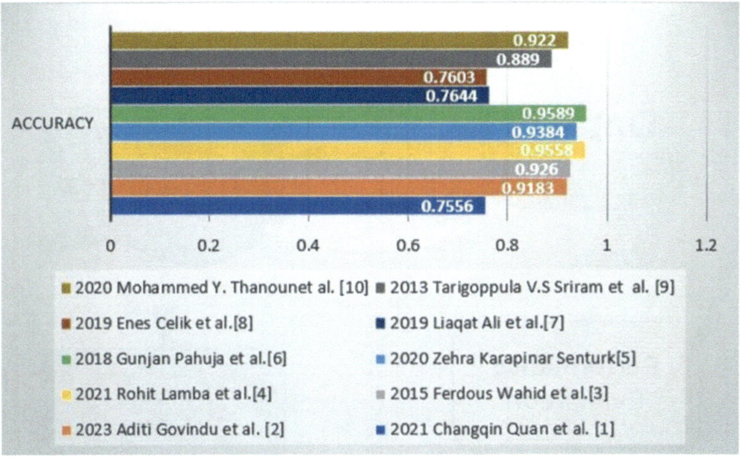

Fig. 7.5 Analysis of the accuracy of existing models [4–13]

achieve reliable identification, pre-trained models, such as those built on ImageNet or other large-scale datasets, can be fine-tuned using data unique to PD (Fig. 7.5).

7.7 Conclusion

This paper attempted to provide a survey to compare the various techniques for Parkinson Disease detection and analysis. A neurological ailment that affects millions of individuals worldwide is Parkinson's disease (PD). The subtle and even invisible early indicators of PD are one of the most difficult parts to manage. There have been major advancements in the treatment of PD symptoms, including the use of drugs and deep brain stimulation. However, the actual origin of PD is still unknown, and it is still unknown how immunological elements contribute to its emergence. With an emphasis on the potential contribution of deep learning features to enhancing classification accuracy, this work intends to give a thorough assessment of various methodologies utilized for the diagnosis and analysis of Parkinson's disease. Although PD is now more understood than ever before, its specific cause is still unknown. Although there is some data pointing to a possible connection between immunological factors and PD, immune factors have not yet been proven to be the primary cause. The use of deep learning algorithms is one of the potential areas of research in PD identification and analysis. Convolutional neural networks (CNNs) and recurrent neural networks (RNNs) in particular have demonstrated outstanding performance in a variety of medical image processing tasks. Deep learning models can be trained to find minute variations in medical pictures, like brain scans or recordings of muscle activity, that might be suggestive of PD. Deep learning models can recognize complex patterns that conventional machine learning techniques would miss by developing hierarchical

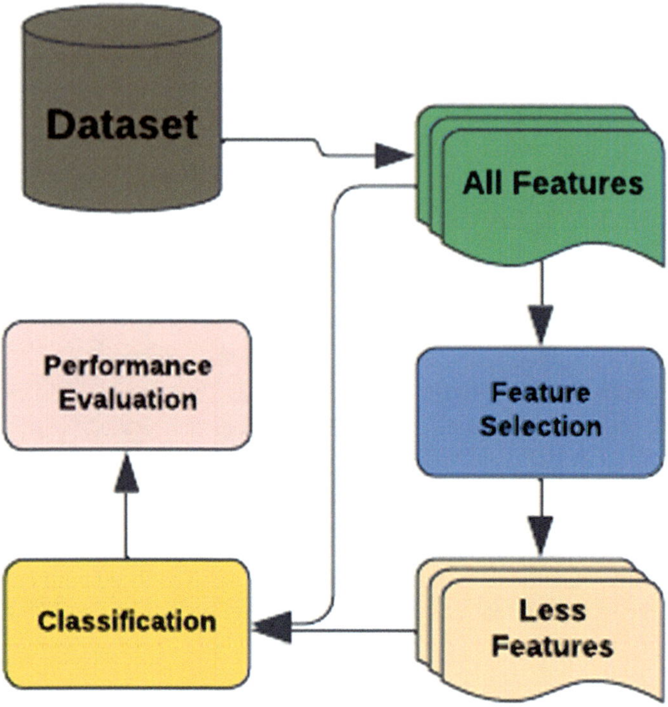

Fig. 7.6 Architecture of feature selection for predicting PD [15]

representations of the input. The importance of this increase in accuracy cannot be overstated because early and precise diagnosis is vital for prompt action and better patient outcomes. Future study avenues are suggested in the paper's conclusion. Investigating the suitability of the suggested deep learning approach for categorizing various stages of PD is one such direction. Parkinson's disease develops in a number of stages, each with its own set of symptoms and medical needs. Creating a trustworthy classification system can help medical practitioners customize treatment plans for specific patients. The research also discusses the possibility of utilizing more intricate deep learning network topologies to improve performance. Researchers can experiment with innovative architectures and methodologies as deep learning develops to extract even more insightful characteristics for PD diagnosis and analysis. This paper concludes by giving a thorough summary of current methods for diagnosing and analyzing Parkinson's disease. It draws attention to the difficulties presented by the disease's mild early symptoms and the ambiguity around its source. Additionally, it highlights how deep learning features could enhance categorization accuracy. Deep learning is anticipated to become more crucial as PD research proceeds since it will help us better understand and treat this complicated neurodegenerative condition (Fig. 7.6).

References

1. Senturk, Z.K.: Early diagnosis of Parkinson's disease using machine learning algorithms. Med. Hypotheses **138**, 109603 (2020). 0306-9877/ © 2020 Elsevier Ltd.
2. Fadzal, C.C.W., Mansor, W., Khuan, L.Y., Zabidi, A.: Short-Time Fourier Transform Analysis of EEG Signal from Writing. 2012 IEEE 8th International Colloquium on Signal Processing and Its Applications, pp. 525–527 (2012)
3. Parvez, Z., Paul, M.: Epileptic seizure detection by analyzing EEG signals using different transformation techniques. Neurocomputing **145**, 190–200 (2014)
4. Abdullah, S.M., Abbas, T., Bashir, M.H., Khajja, I.A., Ahmad, M., Soliman, N.F., El Shafai, W.: Deep transfer learning based Parkinson's disease detection using optimized feature selection. https://doi.org/10.1109/ACCESS.2023.3233969
5. Wang, W., Lee, J., Harrou, F., Sun, Y.: Early detection of Parkinson's disease using deep learning and machine learning. https://doi.org/10.1109/ACCESS.2020.3016062
6. Qin, Z., Jiang, Z., Chen, J., Hu, C., Ma, Y.: sEMG-based tremor severity evaluation for Parkinson's disease using a light-weight CNN. IEEE Signal Proc. Lett. **26**(4) (2019)
7. Vasquez-Correa, J.C., Arias-Vergara, T., Orozco-Arroyave, J.R., Eskofier, B., Klucken, J., Noth, E.: Multimodal assessment of Parkinson's disease: a deep learning approach. IEEE J. Biomed. Health Informatics
8. Wahid, F., Begg, R., Hass, C.J., Halgamuge, S., Ackland, D.C.: Classification of Parkinson's disease gait using spatial-temporal gait features. JBHI-00745-2014.R1
9. Joshi, S., Shenoy, D., Simha G.G.V., Rrashmi, P.L., Venugopal, K.R.: Classification of Alzheimer's disease and Parkinson's disease by using machine learning and neural network methods. IEEE (2010). ISBN: 978-0-7695-3977-5
10. Noor, M.B.T., Zenia, N.Z., Kaiser, M.S., Mamun, S.A., Mahmud, M.: Application of deep learning in detecting neurological disorders from magnetic resonance images: a survey on the detection of Alzheimer's disease, Parkinson's disease and schizophrenia. Brain Inf. **7**, 11 (2020)
11. Martinez-Eguiluz, M., Arbelaitz, O., Gurrutxaga, I., Muguerza, J., Perona, I., Murueta-Goyena, A., Acera, M., Del Pino, R., Tijero, B., Gomez-Esteban, J.C., Gabilondo, I.: Diagnostic classification of Parkinson's disease based on non-motor manifestations and machine learning strategies. Neural Comput. Appl. **35**, 5603–5617 (2023)
12. Govindua, A., Palweb, S.: Early detection of Parkinson's disease using machine learning. Procedia Comput. Sci. **218**, 249–261 (2023). 1877-0509©2023
13. Ogawa, M., Yang, Y.: Residual-network-based deep learning for Parkinson's disease classification using vocal datasets. IEEE (2021). ISBN: 978-1-6654-1875-1
14. Shivangi, Johri A., Tripathi, A.: Parkinson disease detection using deep neural networks. IEEE (2019). ISBN: 978-1-7281-3591-5
15. Wodzinski, M., Skalski, A., Hemmerling, D., Orozco-Arroyave, J.R., Noth, E.: Deep learning approach to Parkinson's disease detection using voice recordings and convolutional neural network dedicated to image classification. IEEE (2019). ISBN: 978-1-5386-1311-5
16. Alharthi, A.S., Casson, A.J., Ozanyan, K.B.: Gait spatiotemporal signal analysis for Parkinson's disease detection and severity rating. IEEE (2020). 1530-437X (c)
17. Quan, C., Ren, K., Luo, Z.: A deep learning based method for Parkinson's disease detection using dynamic features of speech. https://doi.org/10.1109/ACCESS.2021.3051432
18. Gunduz, H.: Deep learning-based Parkinson's disease classification using vocal feature sets. https://doi.org/10.1109/ACCESS.2019.2936564
19. Lamba, R., Gulati, T., Alharbi, H.F., Jain, A.: A hybrid system for Parkinson's disease diagnosis using machine learning techniques. Int. J. Speech Technol. https://doi.org/10.1007/s10772-021-09837-9
20. Pahuja, G., Nagabhushan, T.N.: A comparative study of existing machine learning approaches for Parkinson's disease detection. IETE J. Res. https://doi.org/10.1080/03772063.2018.1531730

Chapter 8
Synergistic Fusion of Steganography and Cryptography in Medical Imaging: A Comparative Analysis to Enhance Image Security

Sharmila Ghosh, Ashim Saha, and Tannistha Pal

Abstract Image steganography is a technique for encrypting data, allowing covert communication, and enhancing data security inside image files. This study provides an in-depth analysis of image steganography, covering its core ideas, methods, uses, difficulties, and developments. Numerous embedding algorithms and methodologies, including transform domain methods and LSB-based strategies, are examined in this study. The paper highlights the need for effective defenses against steganalysis attacks, while also discussing the challenges and potential weaknesses in steganography. It examines common steganography techniques and discusses countermeasures such as adaptive embedding, key-based embedding, and algorithmic resistance. Additionally, the paper also highlights the significance of assessing the efficacy and caliber of image steganography techniques. Objective metrics such as PSNR and MSE are discussed, providing insights into the trade-offs between security and image fidelity. For those interested in understanding image steganography, its applications, and its challenges, this study offers a complete resource.

8.1 Introduction

Digital technology has brought about a revolutionary shift in the healthcare industry, particularly in the way medical images are captured, stored, and shared. With the aid of advanced imaging technologies like X-rays, MRI scans, CT scans, and ultrasound images, healthcare professionals now have access to crucial insights that facilitate accurate diagnoses and treatment planning, referring to Fig. 8.1 [1]. However, alongside these advancements comes the imperative need for robust cybersecurity measures to protect the privacy, reliability, and accessibility of sensitive patient data. In this rapidly evolving landscape, the reliance on picture archiving and communication

S. Ghosh · A. Saha · T. Pal (✉)
Department of Computer Science and Engineering, National Institute of Technology Agartala, Agartala, India
e-mail: tannisthapaul@gmail.com

A. Saha
e-mail: ashim.cse@nita.ac.in

P. Nanda et al. (eds.), *Intelligent System and Data Analysis*, Smart Innovation, Systems and Technologies 398, https://doi.org/10.1007/978-981-97-5200-3_8

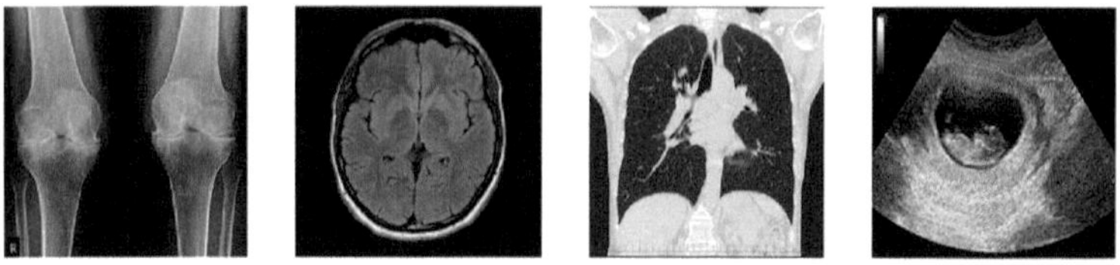

(a) X-Ray (b) MRI (c) CT (d) USG

Fig. 8.1 Different types of medical images [3–6]

systems (PACS) and the transmission of medical images across networks necessitate proactive cybersecurity protocols to mitigate potential risks and safeguard patient information [2].

Ensuring the integrity of medical images is paramount, as any unauthorized alterations or modifications can have severe consequences for patient care, leading to incorrect diagnoses and potentially harmful treatment choices. The region of interest (ROI) and the region of no interest (RONI) must be designated as two separate areas within medical imaging in order to handle this problem. The ROI encompasses delicate and highly sensitive information, while the RONI represents areas where crucial information may be absent. In the domain of image security, forensic techniques play a vital role in detecting modifications and manipulation applied to medical images [7]. By employing digital signatures, cryptographic algorithms, and data validation methods, the authenticity of medical images can be verified, ensuring their integrity remains intact throughout their lifecycle. This survey paper focuses specifically on investigating the combined features of steganography and cryptography within healthcare facilities. By exploring the integration of these techniques, we aim to enhance the security and confidentiality of medical images, thereby bolstering patient privacy and reinforcing the reliability of diagnostic processes.

8.2 Background

Cryptography is the practice of generating encrypted text through the encoding and decoding of messages. Medical records are encrypted using cryptographic algorithms, which assume that only authorized persons have access to the password. A public-key cipher and hashing techniques are used to achieve this. The majority of encryption techniques, including RSA, DES, Blowfish, and AES, are hybridized as well and ensure the highest level of obscurity [1].

Steganography describes techniques for blending a cover message with a secret message so that its existence is totally concealed. A hidden message can be encoded using plaintext, an image, or cipher text, that is recognizable as bits [8]. Digital steganography was created to prevent snoopers from monitoring securely sent data. Occasionally, a stego-key (secret key) may be used to control the embedding process.

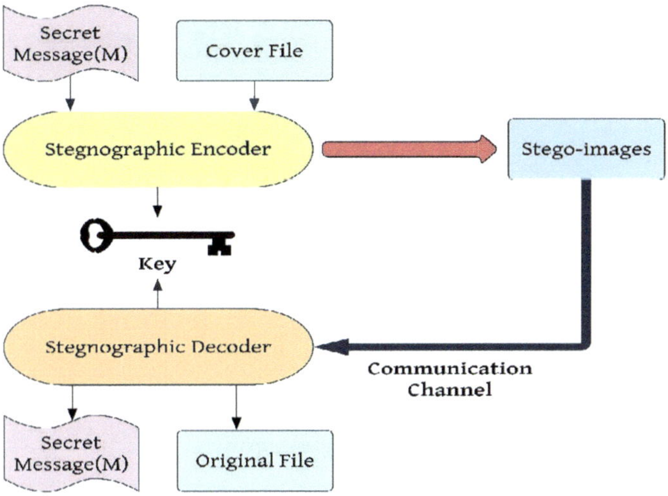

Fig. 8.2 Workflow of steganography framework: concealing information in images

This key must be known in order to identify and recover the concealed message. Once a communication has been concealed by a cover message, it is known as a stego-object. Figure 8.2 shows a general framework of the steganography model.

Safety, quantity, and optical imperceptibility are three important parameters that can be used to gauge the effectiveness of the steganography process. Image steganography's goal is to conceal secret information in nondescript portions of the carrier picture so that the alterations made to the image are undetectable and the hidden data itself can only be retrieved by permitted, smart individuals. There are two categories of image steganography techniques exist: image domain and transform domain. In the case of the image domain technique, the message is directly incorporated in the pixel intensity, whereas in the case of the transform domain, images are first transformed before the message is added. Image domain techniques employ bit-wise methods that make use of bit infusion and noise elimination methods. The picture format determines how lossless image domain techniques work [9]. The transform domain includes both the transformation of images and the modification of algorithms [10] (Fig. 8.3).

Over the past ten years, numerous steganographic methods for images have been developed in both the spatial and frequency domains. The various methods that can be utilized for secure communication are discussed in Table 8.1.

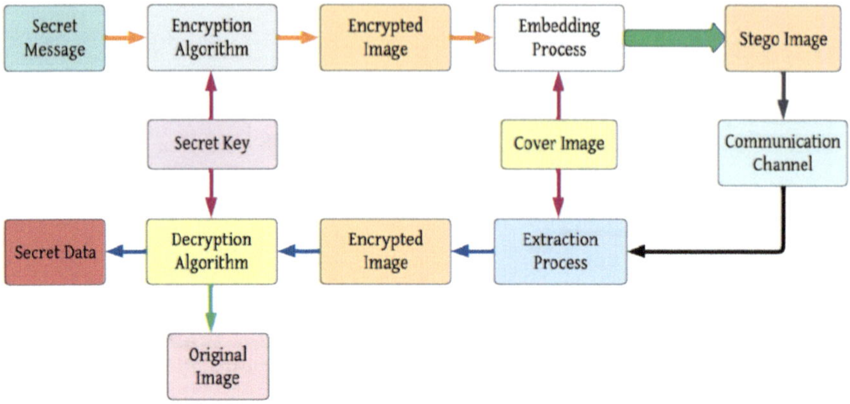

Fig. 8.3 Combining cryptography and image steganography

8.2.1 Spatial Domain Technique

The spatial domain in image steganography refers to the direct alteration of pixel values in a covering image to conceal secret information. To incorporate concealed data in this domain, the approach directly modifies the pixel values of the cover image [11].

(a) **Least Significant Bit (LSB)**—LSB is a renowned and established substitution-based steganography method. Despite its intrinsic simplicity, it has the ability to successfully obfuscate large hidden messages. This technique directly substitutes secret message bits for part or all of the cover picture's least significant bits (LSBs). A little degree of contrast is typically produced when the LSBs are changed, thus guaranteeing the stego-image closely replicates the host image. Each pixel in a color picture has 24 bits, with 8 bits representing every RGB color section (Red, Green, and Blue), as opposed to a conventional grayscale image's 8 bits per pixel. The range of the 8-bit-plane's bit values is 1–255 and can be written as $2n - 1$, where n is one of 1–8. Adjustments to the eighth bit-plane only cause one alteration in pixel value, in contrast to more significant alterations that reduce the image clarity when made to the first bit-plane. The LSB embedding process can be shown using a variety of image forms, depending on the bit-plane chosen. Assume the bit-plane is [12, 14].

(b) **Pixel-Value Differencing**—The Pixel-Value Differencing (PVD) method conceals sensitive information by making use of the variations between neighboring pixels. Each of the ranges that these variants fall within has a clearly defined bottom and upper bound. The areas with a less abrupt shift and lower range metrics are those where a greater capability for data hiding can be identified. On the other hand, larger metrics display regions with stark contrasts and offer a greater chance for data reduction. The process starts by dividing the frame or cover image into adjacent pairs of non-overlapping pixels [15]. After that, the

Table 8.1 Various steganographic methods for images

Method	Description	Pros	Cons
Least significant bit (LSB) [11]	This technique is used to hide data within the images. The eighth bit of each byte of the carrier file is replaced by one bit of the secret information	Without considerably changing the total appearance of the image, the LSB of each pixel can be gently changed	This technique can be subject to bit flipping, lossy compression, and limited storage capabilities
Masking and filtering [12]	These techniques include altering the pixel values using specific masks or filters to obscure the hidden information	It offers the benefit of precisely altering or extracting particular data aspects.	If not used appropriately, it may cause data loss or distortion.
Transform domain techniques [11]	These techniques utilize mathematical techniques like DFT or DCT to convert the image into the frequency domain, which is then secretly data-encoded	The fact that these modifications are less obvious is advantageous to the human eyes	It could introduce artifacts or distortions in the restored data
Spread spectrum techniques [13]	Spread spectrum is a technique used to describe a method of obscuring information by spreading it out over a wider bandwidth, making it more elusive to detect and extract	It helps in maintaining resilience against interference	It is challenging to find the secret data that is modulated and encoded in specific frequency bands
Pixel-value differencing (PVD) [13]	The PVD approach, which calculates the difference between neighboring pixel values, is used to implant the hidden bits	It has the ability to encrypt data in the least important bits of pixel values	It might be detectable if the changed pixels show observable patterns or differences from the original image
Texture synthesis-based methods [12]	It incorporates the hidden information and creates textures with visually identical looks	It has the ability to make realistic-looking, high-quality textures data	This technique can be computationally expensive
Distortion minimization technique [11]	These methods aim to lessen the distortion brought on by concealed data embedding	It includes reduced observed distortion and high image quality during data embedding	Its restricted embedding capacity, limits how much information may be efficiently buried in an image

differential value and its associated range are calculated. The number of bits that must be buried is then determined using the range index. The required secret bits are finally taken from the private message to create a new version and aid in steganographic concealing [16].

(c) **Texture Synthesis-Based Method**—In the spatial realm of image steganography, texture synthesis-based techniques are employed to hide data. To ensure that the hidden information mixes in smoothly, these techniques focus on simulating the rich textures and patterns of the original image [1]. By analyzing and replicating the local aspects of the image, they attempt to maintain visual coherence while successfully hiding the hidden information. Because they are particularly effective at concealing data in a form that is invisible to the human eye, texture synthesis techniques are useful tools in image steganography for covert communication [17].

(d) **Distortion Minimization Technique**—In order to conceal sensitive information while reducing visible distortion, steganography of images depends on distortion minimization approaches that act in the spatial domain. These solutions prioritize preserving the cover image's artistic integrity by carefully arranging information in parts of the carrier image where human perception is less sensitive. Select embedding locations and intensities in an effort to render the delivered alterations unnoticeable to the human eye. In distortion minimization tactics, sophisticated algorithms are typically utilized to optimize the trade-off between data concealing ability and image fidelity. This technique finds a balance, making them ideal for steganographic applications like secure picture transfer and watermarking when maintaining visual quality is important [12] (Figs. 8.4 and 8.5).

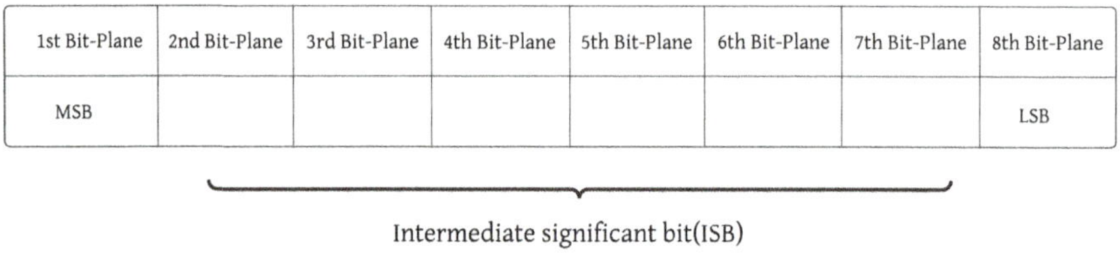

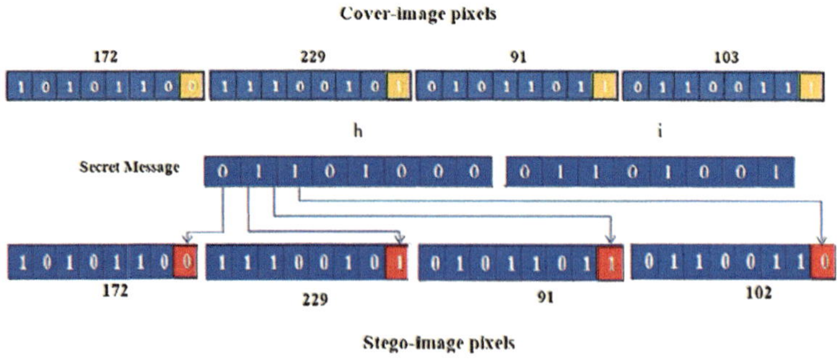

Fig. 8.4 Visualization of encoded message in grayscale image [14]

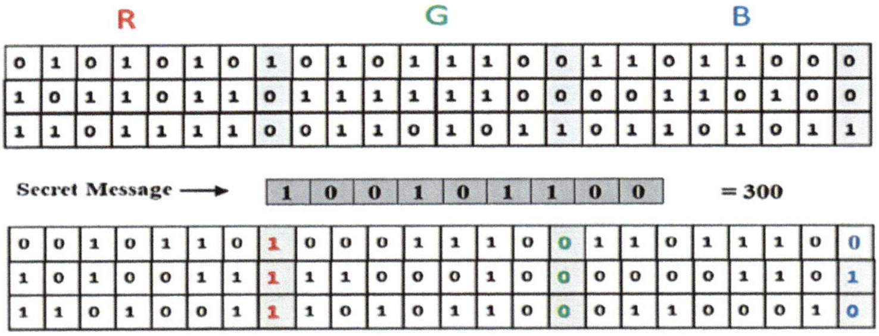

Fig. 8.5 Elaborating the encoded message in a color image [14]

8.2.2 *Masking and Filtering*

Masking and filtering are crucial components of picture steganography because they allow secret data to be hidden within digital photos while maintaining visual quality. Masking, which frequently makes use of spatial or frequency domain masks, entails selectively changing particular portions of the image in order to insert the hidden message [1]. This technique is limited to grayscale images or images with up to 24 bits. To ensure that changes are invisible to the human eye, filtering techniques are utilized to limit the visual impact of the embedded message. These techniques are key in image steganography for covert communication because they are essential for striking a balance between message concealing and keeping the appearance of the cover image [8].

8.2.3 *Transform Domain Technique*

The robustness and perceptual transparency of transform domain steganography are higher despite its complexity. These processes frequently involve numerous important steps, including First, a frequency domain conversion of the cover image is performed. A portion or all of the adjusted coefficients are then quietly updated to incorporate the covert message [14]. The layout of the old carrier photo is then applied to the revised coefficients. There are several other transformation options, including the discrete wavelet transform (DWT), discrete cosine transform (DCT), and discrete Fourier transform (DFT). DFT has applications in steganography while being less popular. DCT is well-recognized and used extensively, especially in image and video compression.

The initial operations of the DCT-based picture steganography technique are the loading of a carrier(i.e., cover) image and the binary encoding of a secret message. The cover image is divided into non-overlapping 8×8 pixel blocks before the discrete cosine transform (DCT) is applied to each block. The program determines the number of bits to be concealed within each DCT coefficient and softly modifies

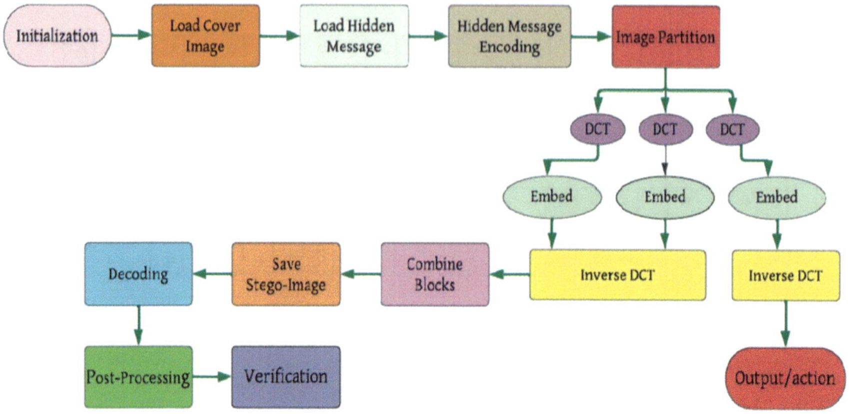

Fig. 8.6 Block diagram: DCT-based image steganography algorithm

these coefficients in order to insert the data while reducing visual distortion [17]. After that, inverse DCT transformation is used to rebuild the stego-image. To extract the hidden message, the stego-image is once more treated to DCT. Post-processing and verification come next. Then, based on the message that was recovered, the pertinent activities of maintaining the stego-image are carried out [18]. Figure 8.6 demonstrates the flow of information and procedures from startup to output or action for the DCT-based steganography algorithm. DWT has received recognition for its role in signal processing and compression thanks to its multi-resolution analysis for different frequencies and resolutions. The main benefit of DWT is that it can effectively capture both frequency and position data due to its high temporal resolution. The figure shows a simplified flowchart illustrating the major processes of the DCT-based steganography technique, including the embedding and extraction procedures. The arrows represent the data flow among each block's various activities, such as message extraction, embedding, and DCT transformation.

8.3 Attacks in Image Steganography

Attacks on secured images are defined differently from attacks on other types of transmitted data. Image steganography attacks are quite easy to carry out, especially with the help of the many free applications that are currently available and the DIY toolkits that are easily accessible online. Anyone who uses image steganography maliciously has the potential to abuse it and severely harm their target [11]. In this case, attackers attempt to train data rather than crack ciphers and damage encrypted media. When fraudsters start an attack using image steganography, they cloak their malicious code inside innocent-looking image files. By altering a few pixels in a typical JPEG image, which contains several megabytes of pixel data, an attacker

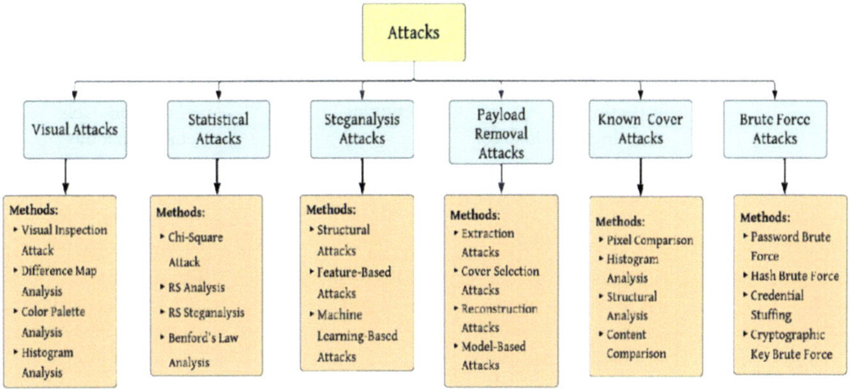

Fig. 8.7 Attacks on image steganography

may inject malicious code [7]. Figure 8.7 illustrates how several attack types can affect the transmission of medical images through health networks. Let's look at how simple it is to conceal text in an image

- **Determining the Attack Target and Payload for Image Steganography**—To construct an image using the steganography approach, an attacker must follow a number of steps. They must decide which healthcare network to target, which target to choose inside of it, which credentials that target has, and how precisely to execute the attack. Additionally, the payload—or what they anticipate the steganography technique to accomplish—must be established. Do they want to seize the target's computer or steal information in secret [1]?
- **Make use of steganography**—To modify an image Steganography allows for the modification of an image's data by integrating hidden information [14]. By subtly altering the pixels or other elements of the image, sensitive information can be concealed while maintaining the aesthetic look, providing an additional layer of protection. By altering a handful of an image's pixels, one can use steganography to hide files or messages within a common, innocent-looking image.
- **Getting the malicious Image Code Out**—Once the file has been extracted, attackers can run the payload to harm the victim [2]. By hiding the payload within an image file, image steganography techniques can easily get beyond standard anti-malware and APT technologies because these tools are not designed to detect this form of malware. Fortunately, there are sophisticated security technologies available that can thwart image steganography attacks [10, 19].

The following are some typical steganography attacks discussed in Table 8.2.

Table 8.2 Demonstration on various steganography attacks

Attacks	Description	Algorithm	Mitigation techniques
Visual attack [15]	Focus on visually inspecting the stego-image to look for any patterns or irregularities	LSB (least significant bit) substitution	Random pixel-value perturbation—this method randomly modifies or perturbs the image's pixel values in order to disturb the hidden information
Statistical attack [11]	Attempts to exploit cover image or stego-image statistics. They examine the distribution of pixel values, color channels, or higher-order statistics to identify hidden data	RS (Reed–Solomon) code	Adaptive data hiding—this technique changes the amount of concealed information that is buried in particular areas
Steganalysis attacks [16]	It's examining and identifying steganography in transmitted data or images	Feature extraction algorithms (e.g., SIFT, DCT)	Data whitening: it alters the statistical properties of the concealed information to make it appear like random noise, lowering the likelihood that an attack will detect it
Payload remova l attacks [16]	It tries to revert the embedding process in order to retrieve the concealed data and assume knowledge of the steganography algorithm utilized	Digital watermark embedding	Content-aware steganalysis: these methods concentrate on discovering steganography payload alterations and can help detect payload removal attempts
Known cover attack [20]	This attack occurs when an attacker has the original cover image used for steganography. It compares the original cover image with the stego-image	Spatial domain analysis	Cover generation technique—it creates customized cover images, each with a slight variation from the original to make sure that it is difficult to match or reverse-engineer the known cover image
Brute force attack [9]	To find the concealed information, brute-force assaults test every key pairing and password combination	Advanced encryption standard (AES)	Serpent—a symmetric encryption technique noted for its high security. It uses 128-bit blocks as its o.s and accepts keys with sizes 128–256 bits, offering a strong defense against these attacks

8.4 Literature Survey

Images are one of the most efficient file types for steganography. Digital image steganography is a method for concealing sensitive data without raising any red flags. This section referring to Table 8.3 of the paper gives a thorough review of the literature survey on image steganography with the goal of giving readers an understanding of the fundamental ideas, approaches, and developments in this area. Lin et al. [21] in his paper presents a clever symmetric cryptography solution for info security protection of medical images. The method combines a quantum-based key generator and a chaotic map for encryption. The solution ensures the privacy and integrity of sensitive medical imaging data. Hamza et al. [13] provides a reliable way to conceal information within images while enhancing the security of image steganography through the use of pattern matching. Spread spectrum steganography is a technique proposed as a unique method for integrity verification in medical image retrieval systems, suggested by Eze et al. [15]. Integrity verification codes are embedded inside the frequency domain of the images.

Advanced image steganography techniques have revolutionized the field by incorporating cutting-edge approaches to boost security and durability. Raja [10] suggests a safe combined strategy for quickly shrinking and encoding medical images. It uses multiscale transform, symmetric key encryption, and EBCOT model for improved compression and secure encoding. Konyar et al. [19] proposed a method by including hidden data with error-correcting codes into the image. The method ensures reliable data recovery regardless of the face of disturbance. This technique enhances the safety and reliability of medical image transmission and storage spaces and safeguarding confidential information. BOOST medical image steganography is a method suggested by Stoyanov et al. [18] utilizing nuclear spin generators. The method makes excellent use of the erratic behavior of nuclear spin and ensures accurate diagnosis while providing secure medical image distribution and preservation.

8.5 Comparative Analysis

In the modern world, digital communication is now a necessary part of communication. There are a lot of Internet-based applications, and in some cases, it's better if the communication is secret. To do this, steganography and cryptography are used [8]. In contrast to the original image, a number of digital steganography techniques are applied in this study to create hidden images that are undetectable to human sight. A comparison analysis is performed to demonstrate the effectiveness of the indicated techniques using metrics like PSNR and MSE. The visual quality or VQ is the condition of the stego-image and is quantified by the peak signal-to-noise ratio (PSNR) of the image [22].

Table 8.3 Literature survey

S. No.	Author and proposed year	Method used	Dataset used	Results	Observation
01	Hussein et al. [20]	RSA, SHA-256	12 medical image sizes from 700 KB–19 MB	The experimental result shows a finding of the highest compression ratio **90%**	Creative medical image compression and encryption is produced utilizing DWT, quantization, and adaptive arithmetic coding technique
02	Karar et al. [22]	CNN, ResNet	306 X-ray images	The investigation showed achieved an excellent accuracy of **99.90%**	A cascaded deep learning classifier has been created to improve the accuracy of CAD models for severely detected COVID-19 in X-ray photos
03	Benssalah et al. [12]	ECC, Arnold's cat map, and hyperchaotic Lorenz generator	DICOM, grayscale image	The experimental result shows a finding of **100 and 33.46** for the NPCR and UACI	This paper suggests a more sophisticated, effective TMIS medical image encryption approach to circumvent the weaknesses of different attacks
04	Hizukuri et al. [23]	DCNN with optimized Bayes' theory	DCE-MRI	The classification accuracy, sensitivity, and specificity are **92.9, 93.3, and 92.3%**	A flexible CAD model was presented to discriminate between benign and malignant tumors on breast DCE-MRI images
05	Priyadharshini et al. [2]	One-time pad encryption, LSB	TIFF, BMP, JPEG, and DICOM	Not disclosed	A novel steganography and cryptography mechanism is offered to safeguard the veracity and security of a healthcare image
06	Guesmi et al. [16]	DNA encoding, SHA-2, and a hybrid chaotic map	BITE, USC-SIPI	The average NPCR and UACI are **99.600124 and 33.459415**	By employing the SHA-2, chaotic maps, and DNA masking, a hybrid cryptography paradigm for medical imaging security was presented
07	Liao et al. [17]	Two-stream CNN	UCID	The proposed model provides an excellent accuracy of **96.23%**	An image operator chain detection order forensics system based on CNN is presented
08	Ozcan et al. [7]	LSB and SVM	DICOM	The analysis yields a **99.28%** accuracy rate	The aim was to determine whether a particular medical image includes concealed data or not using a steganalysis classifier approach

Peak signal-to-noise ratio (PSNR)
The term "PSNR" refers to the relationship amid the highest power of the signal and the power of noise that affects the signal's quality. When evaluating the effectiveness of image compression, denoising, or any other operation that modifies the original image, PSNR is frequently utilized. It offers a quantified numerical value [14]. The PSNR equation is given by

$$\text{PSNR} = 10\log_{10}\left(\frac{\text{MAX}^2}{\text{MSE}}\right)\text{dB} \tag{8.1}$$

- The maximum pixel value that can be obtained for the image is 255 (for 8-bit images)
- PSNR is measured in decibels (dB), which is a logarithmic scale used to express the ratio between two values.

Mean Square Error (MSE)
The term "MSE" determines the mean squared difference or similarity amid the expected image and the real image [1]. It calculates the average squared deviation among the real and predicted values of the target variable. The mentioned parameter is calculated using the subsequent formula

$$\text{MSE} = \frac{1}{n}\sum_{i=1}^{n}(x_i - y_i)^2 \tag{8.2}$$

- $\sum$: Summation sign, indicating that all related pairs of variables (x_i, y_i), should add up the squared differences.
- x_i: The i-th value from the starting set of information.
- y_i: The i-th value extracted from the compressed or reconstructed data.
- n: Denotes the total number of data points, which must be the same for both the original and the reconstructed data.

Table 8.4 shows an analysis of various image encryption and steganography techniques in terms of peak signal-to-noise ratio (PSNR) and mean squared error (MSE), one methodology stood out above the rest. With a stunning PSNR of 76.87 dB, the "Ebcot Encoding technique and Symmetric Key Cryptography" showed the strongest encryption capabilities and minimal data loss during the process. This method also maintains image quality the best, as seen by the fact that it has the lowest MSE. On the other hand, the "Hybrid Image Steganography" method showed the lowest PSNR values, such as 33.82 dB for "Baboon" and 33.36 dB for "Lena," which indicates a significant deterioration in image quality. The high MSE values serve as additional evidence of the considerable alterations made to the photos throughout the stegano graphic process. It is crucial to look into advanced steganographic algorithms, modify encoding variables, and take error correction techniques into account in order to get over the challenge of low PSNR and high MSE in steganography approaches like "Hybrid Image Steganography." A more equitable trade-off between visual quality

Table 8.4 Quantitative evaluation of image security methods using PSNR, MSE

Refs.	Method	Images	PSNR	MSE
[2]	Encrypting medical image	Lena	51.141	0.4999
	using one-time pad encryption	Baboon	51.181	0.495
[13]	Hybrid image steganography	Baboon	33.82	26.98
	on pattern	Lena	33.36	29.99
	matching	Jet	33.84	29.99
		Boat	33.52	28.91
		Couple	33.25	30.76
		Peppers	33.44	29.44
[1]	2D chaotic map based PRNG	Image a1	67.7270	0.0110
	model	Image a2	64.8506	0.0213
		Image a3	70.0143	0.0065
		Image a4	73.5965	0.0028
[19]	Reed–Solomon codes based data	CT	50.80	0.519
	hiding	MRI	52.17	0.375
		US	50.81	0.464
		X-ray	51.73	0.326
[18]	Image stego hiding using	Brain	113.5319	0.0192
	BOOST technique	Knee	120.1618	0.0043
		Liver	116.4383	0.0103
[10]	Ebcot encoding technique and symmetric key cryptography	6 input medical images	76.869	0.0009
[11]	Gaussian based cuckoo	Lena search	47.0963	0.0171
	algorithm	Pepper	46.163	
		Baboon	45.128	
		Barbara	46.086	

and data hiding capacity may also be offered by hybrid systems that combine a variety of steganographic approaches. These results emphasize the need to select the optimum strategy based on specific application requirements, where strong encryption, little image distortion, and low MSE are all essential considerations (Figs. 8.8 and 8.9).

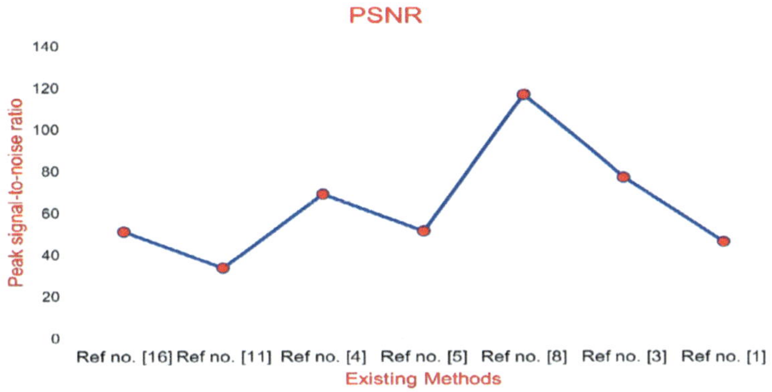

Fig. 8.8 Peak signal-to-noise ratio (PSNR) visualization using graphical illustration

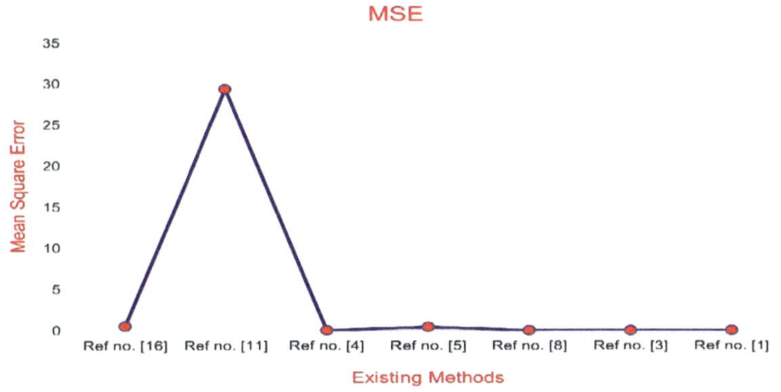

Fig. 8.9 Mean square error (MSE) visualization using graphical illustration

8.6 Conclusion

This review paper has provided a complete overview of image steganography and emphasized its significance in the field of data privacy and security in healthcare networks. By concealing information within digital images, image steganography offers a surreptitious means of communication. The essay examined the fundamental principles and techniques of visual steganography, including several anchoring strategies and algorithms in medical image security. By discussing numerous steganography attacks, including visual attacks, statistical attacks, etc., the potential flaws that exist in the systems were brought to light. The preventative measures aim to make it more difficult for intruders to locate or extract secret data. A comparative analysis of image security techniques was also done using objective metrics like PSNR, MSE, and SSIM. These evaluations allowed us to learn how effectively different image security methods preserved image quality and maintained imperceptibility. Future work should concentrate on evaluating steganography's practicality and efficiency in the context of medical imaging security.

References

1. Hamza, R., et al.: A privacy-preserving cryptosystem for IoT E-healthcare. Inf. Sci. **527**, 493–510 (2020)
2. Priyadharshini, A.: Securing medical images using encryption and LSB steganography In: International Conference on Advances in Electrical, Computing, Communication and Sustainable Technologies (ICAECT), 1–5. IEEE (2021)
3. Fradiopaedia. MRI. Accessed 15 June 2023 (2022). https://www.Fradiopaedia.org/normal-brain-mri
4. GaneshDiagonostics. X-Ray. Accessed 15 June 2023 (2022). www.ganeshdiagnostic.com/x-ray-right-knee-joint-lat
5. jewelwc. USG. Accessed 15 June 2023 (2022). https://www.jewelwc.org/Fearly-pregnancy-ultrasound
6. melbourneradiology. CT. Accessed 15 June 2023 (2022). https://www.melbourneradiology.com/diagnostic-imaging
7. Özcan, F.B., Karakiş, R., Güler, İ.: Steganalysis on medical images with support vector machine. In: 28th Signal Processing and Communications Applications Conference (SIU), 1–4. IEEE (2020)
8. Mohammed, Z.F., Abdulla, A.A.: An efficient CAD system for ALL cell identification from microscopic blood images. Multimedia Tools Appl. **80**(4), 6355–6368 (2021)
9. Shen, Y., et al.: Optical selective encryption based on the FRFCM algorithm and face biometric for the medical image. Opt. Laser Technol. **138**, 106911 (2021)
10. Raja, S.P.: Multiscale transform-based secured joint efficient medical image compression-encryption using symmetric key cryptography and EBCOT encoding technique. Int. J. Wavelets Multiresolution Inf. Process. **17**(05), 1950034 (2019)
11. Alif Siddiqua Begum, A., Nirmala, S.: Secure visual cryptography for medical image using modified cuckoo search. Multimedia Tools Appl. **77**, 27041–27060 (2018)
12. Benssalah, M., Rhaskali, Y., Drouiche, K.: An efficient image encryption scheme for TMIS based on elliptic curve integrated encryption and linear cryptography. Multimedia Tools Appl. **80**(2), 2081–2107 (2021)
13. Hamza, A., et al.: Novel secure hybrid image steganography technique based on pattern matching. KSII Trans. Internet Inf. Syst. **15**(3) (2021)
14. Shtayt, B.A., Zakaria, N.H., Harun, N.H.: A comprehensive review on medical image steganography based on LSB technique and potential challenges. Baghdad Sci. J. **18**(2 Suppl.), 0957 (2021)
15. Eze, P., et al.: Integrity verification in medical image retrieval systems using spread spectrum steganography. In: Proceedings of the 2019 on International Conference on Multimedia Retrieval, pp. 53–57 (2019)
16. Guesmi, R., Ben, M.A., Farah: A new efficient medical image cipher based on hybrid chaotic map and DNA code. Multimedia Tools Appl. **80**, 1925–1944 (2021)
17. Liao, X., et al.: Robust detection of image operator chain with two-stream convolutional neural network. IEEE J. Sel. Top. Signal Process. **14**(5), 955–968 (2020)
18. Stoyanov, B., Stoyanov, B.: BOOST: medical image steganography using nuclear spin generator. Entropy **22**(5), 501 (2020)
19. Konyar , M.Z., Öztürk, S.: Reed Solomon coding-based medical image data hiding method against salt and pepper noise. Symmetry **12**(6), 899 (2020)

20. Hussein, N.H., Ali, M.A.: Medical image compression and encryption using adaptive arithmetic coding, quantization technique and RSA in DWT domain. Iraqi J. Sci., 2279–2296 (2022)
21. Lin, C.-H., et al.: Intelligent symmetric cryptography with chaotic map and quantum based key generator for medical images Infosecurity. IEEE Access **9**, 118624–118639 (2021)
22. Karar, M.E., Hemdan, E.E.D., Shouman, M.A.: Cascaded deep learning classifiers for computer-aided diagnosis of COVID 19 and pneumonia diseases in X-ray scans. Complex Intell. Syst. **7**, 235–247 (2021)
23. Hizukuri, A., et al.: Computer-aided diagnosis scheme for distinguishing between benign and malignant masses on breast DCE-MRI images using deep convolutional neural network with Bayesian optimization. J. Digital Imaging **34**, 116–123 (2021)

Chapter 9
Detecting Inattentive and Aggressive Driver Behavior Using Deep Learning: Recent Advances, Challenges with Performance Evaluation

Tina Debbarma, Tannistha Pal, and Nikhil Debbarma

Abstract Road safety faces a grave threat from the prevalence of aggressive and inattentive driving behaviors, contributing significantly to the global rise in traffic accidents and fatalities. Identifying, tracking, and proactively addressing instances of aggressive and inattentive driving are pivotal steps toward mitigating accidents and enhancing road safety. This research project seeks to explore the potential of deep learning methods for analyzing human driving behavior, with a specific focus on identifying inattentive and aggressive driving behaviors. The ultimate goal is to develop a robust and reliable system capable of accurately recognizing and categorizing such behaviors. To achieve this objective, we harness the power of deep learning algorithm. This paper also discusses the identification and categorization of aggressive and distracted driving behaviors among human drivers, using deep learning methodologies which offer substantial promise for elevating traffic safety. The literature review uncovered several noteworthy techniques, including CNN, MTCNN, transformer-based action detection systems, and fusion models inspired by densely connected convolution networks. Comparative analysis revealed improvements in accuracy and precision through the utilization of these techniques. However, the pursuit of larger and more diverse datasets, real-time implementation capabilities, and robustness in various environmental conditions remain essential requirements. Despite these challenges, deep learning-based systems offer the potential for automatic and precise analysis of intricate driving patterns, pivotal in accident prevention and overall road safety enhancement. Future research should focus on surmounting current limitations and advancing the development of more intricate deep learning models for accurate recognition and categorization of human driving behaviors.

T. Debbarma · T. Pal (✉) · N. Debbarma
Department of Computer Science and Engineering, National Institute of Technology Agartala, Agartala, India
e-mail: tannisthapaul@gmail.com

P. Nanda et al. (eds.), *Intelligent System and Data Analysis*, Smart Innovation, Systems and Technologies 398, https://doi.org/10.1007/978-981-97-5200-3_9

9.1 Introduction

Human drivers exhibit a variety of driving behaviors and habits as a result of their varied driving abilities, experiences, and emotions. Numerous studies have been conducted in an effort to recognize distinct human driving behavior through the collection and analysis of data. Inattentive and aggressive driving are serious challenges in the field of road safety, resulting in innumerable accidents, injuries, and fatalities around the world. Aggressive driving refers to activities such as excessive speeding, tailgating, and sudden lane changes, whereas inattentive driving refers to situations in which drivers are unable to focus on the road due to distractions, fatigue, or other concerns. A deep learning-based system for driver activity detection is described in order to continuously track and understand driving activities. To improve road safety and reduce risks, it is critical to understand and effectively manage these habits (Figs. 9.1 and 9.2).

Traditional methods of evaluating driver conduct typically relied on subjective surveys, manual observations, or limited sensor data. However, technological advancements, particularly in the field of deep learning, are opening up opportunities for more precise and automated analysis of complex driving habits. Convolutional neural networks (CNNs), a type of deep learning approach, have shown promise in a variety of computer vision applications such as object detection and recognition. As a result, they are ideal for predicting driving behaviors (Fig. 9.3).

Fig. 9.1 Inattentive human driver [4]

Fig. 9.2 Accident due to inattentive driving [11]

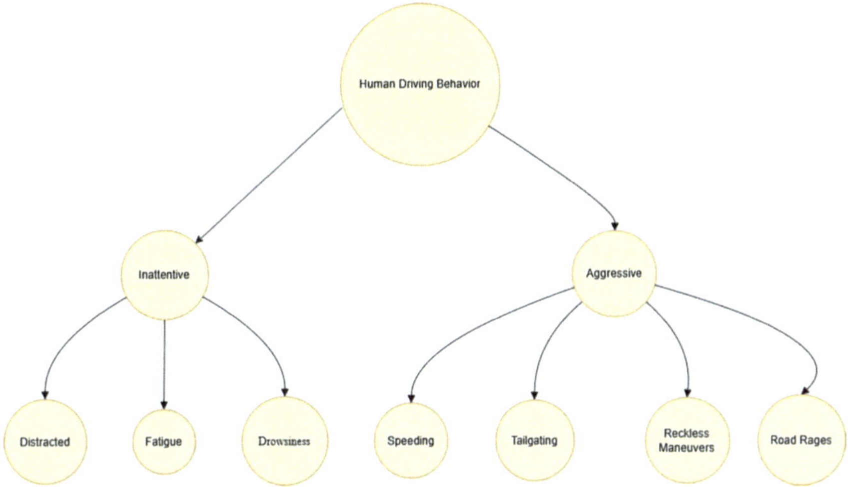

Fig. 9.3 Different unusual behaviors during driving

9.2 Literature Review

Alkinani et al. [1], in their research work, they provide background information on the deep learning algorithms. For identifying anger, weariness, and inactivity in human drivers, they also assessed the most recent deep learning-based systems and techniques. To perform well, this approach needs substantial knowledge. Chen et al. [2], proposed a multi-task cascaded neural network (MTCNN) method to overcome the difficult problem of fatigue driving detection. They used the three courts and five eyes method to determine the eye positions. When compared to convolutional neural networks (CNNs), the suggested technique improved detection accuracy and reduced detection time, providing reliable technical support for fatigue driving detection study. Liang et al. [3] provide the multi-scale vision transformers in an improved version as a component of the transformer-based stargazer action detection system. The proposed system comes in second place in the Naturalistic Driving Action Recognition category of the AI City Challenge 2022. He et al. [4], in this study, deep learning-based multi-feature fusion detection approaches are proposed. It can tell whether a driver is talking on the phone by analyzing their voice patterns and phone call behaviors. In a complex environment, this approach offers a better detection accuracy. Huang et al. [5], they focused on deep learning-based fusion techniques and proposed three new deep learning-based fusion models that were inspired by closely connected convolution networks. This approach is quite useful for detecting irregular driving behavior based on video. This techniques can be used only in the autonomous cars.

Cengil et al. [6], in order to complete the task of categorizing driving behaviors, they used the pre-trained CNN architecture GoogleNet. For picture preprocessing, a Laplacian Gaussian function was employed. The method aids in the development of autonomous vehicles and accident avoidance. Xing et al. [7], in this study, they suggested using deep learning to identify driving behavior. Transfer learning is used to make improvements to the pre-trained CNN models. A segmentation technique based on GMM is used to process the raw pictures. The applications in real life are examined at in this study. Yan et al. [8], in their research, a vision-based approach based on a convolution neural network was used to recognize he driver's behaviors. To create action labels, they use R*CNN. This approach is good for recognizing driving actions. Jegham et al. [9], their proposed method uses the depth modality to extract informative sections of RGB data and is created to discriminate between driver actions in a realistic environment. The frame system, which is based on depth information, delivers superior accuracy when compared to commonly used to identification systems since it focuses on the silhouette and motion of divers. Bruno et al. [10], they propose a system that is used to support the environment-based, semi-autonomous control of robotic vehicles and to identify human errors. The Multi-Layer Perceptron with Recurrence is then used to achieve the goal of learning the patterns of failures. The proposed system has five modules: "State vector", "Offline Traffic Signs Map", "Online Traffic Signs Detection", "NeuroFSM", and "Security Routines". They developed a method which is capable of classify and detect human failures while driving using internal data from vehicle.

Islam et al. [11], the method employed a Sigmoid Activation Function for the last FC layer and an ADAM optimizer for optimization. For training and testing, they employed real-world drowsiness detection video dataset, which were then translated into frames. Their model demonstrated that the optimal approach is a deep learning-based fatigue detection system. Bruno et al. [12] use deep learning to create a novel model for anomalous driving behaviors. For generic driving behavior, stacked sparse auto-encoders are employed. Dropout technology is introduced to prevent overfitting. The suggested deep network recognizes various driving behaviors with success. Camlica et al. [13], Parallel Convolution Neural Networks were developed for classifying aggressive driving conduct using maneuver categorization. In real-world datasets, two CNNs are trained and tested. This model can recognize high-level aspects of many aggressive driving behaviors. Bichicchi et al. [14], to develop output layers in RGB colors, an upgraded deep learning model with an unsupervised Denoising Stacked Auto-encoder was applied. The approach detects simple driving actions and the complexity of the road environment efficiently. Noor et al. [15], in this study, the proposed method is based on the DenseNet 3D backbone. The proposed model is feasible if adequate sample rates are chosen.

Shirabayashi et al. [16], in order to categorize photos reflecting driving behavior, they analyzed different pre-trained convolutional Neural Network designs that have been published in the literature. They were able to determine that changing specific parameters, such as using the optimizer in conjunction with a sufficient learning rate, can significantly improve the performance of the models. Hossain et al. [17], to recognize driver distraction patterns and their causes, they developed a distraction detection architecture based on convolutional neural networks. Systems for real-time driver attention detection can be developed using the findings of the proposed model. Kashevnik et al. [18] provided an examination of the research on distracted driving. The suggested structure offers a thorough identification of distractions throughout the sensor data collection chain to data processing, inference of behavior, and type of distraction inference. As a result, this page provides an overview for all academics interested in driver distraction research and distraction detection technology, making the text useful for those who work in the transportation industry. Aljohani et al. [19] propose a hybrid GA and CNN combination as a remedy for the problem of driver attention. The choice of feature extractors and classifier architectures is guided by GA. Therefore, the suggested procedures can also be utilized with the new dataset. Each module additionally showed real-time data. The suggested paradigm would still benefit from some data refinement. Khosravi et al. [20], they suggests leveraging real-time smartphone sensor data to discover safe driving practices and driving mishaps. So that they may focus on creating quick and low-cost systems for smartphone driver monitoring, and they give guidance to auto-mobile developers.

The investigations that have already been published in the literature were conducted utilizing a wide range of deep learning methods. Depending on the problem and the used dataset, they function differently.

9.3 Comparative Analysis

Inattentive driving is characterized as driving with a lack of focus and attention. Distracted driving (texting or talking on the phone), eating, grooming, or utilizing in-car gadgets are all examples of this conduct. Drivers who are distracted usually fail to detect potential hazards, respond poorly to traffic situations, and violate traffic signals, resulting in accidents and near misses [1]. Aggressive driving, on the other hand, comprises aggressive and deliberate road behavior. Excessive speeding, tailgating, frequent lane changes without indicating, running red lights, and irrational driving behavior are a few examples. Aggressive drivers frequently demonstrate impatience, contempt for traffic laws, and disregard for other road users, creating a hazardous driving environment. It has been demonstrated that deep learning approaches are very helpful and successful in correcting human inattentive and aggressive driving behavior. These methods employ deep neural networks strengths to analyze vast volumes of data and identify interesting patterns and traits. Deep learning models can be trained using labeled data that include instances of aggressive and careless driving, allowing researchers to teach the algorithms how to spot these behavior. Deep learning algorithms are able to properly identify and categorize current aggressive and careless driving tendencies. Using the information provided, advanced driver assistance systems (ADAS) that provide warning signals or take preventive action to avoid collisions can be developed. The deep CNN model and transfer learning method are used to develop a driving-related activity identification system. To improve the accuracy of the identification, the raw RGB photos are first processed with a GMM-based segmentation algorithm, which rapidly removes irrelevant components and identifies the driving location from the background context. According on the classification results, segmentation contributes to a far more exact detection result than the model trained with raw photos [7]. For instance, the system might warn the driver if they show signs of aggression or inattention, causing them to adjust their behavior or refocus on the road (Table 9.1).

9.4 Performance Evaluation

Many classification strategies are offered in the literature for recognizing driver conduct. Deep learning techniques are frequently employed nowadays to achieve great accuracy, and several useful systems have been proposed to detect human inattentive and aggressive driving behaviors. To illustrate the effectiveness of the offered strategies, performance evaluation was done using measures such as precision, recall, and accuracy.

Precision

It is a performance metric used in binary classification problem. It determines the proportion of accurately predicted positive cases out of each case that is projected to be positive. A precision equation is:

$$\text{Precision} = \frac{\text{TP}}{\text{TP} + \text{FP}} \tag{9.1}$$

Table 9.1 Summary of different existing methods in detecting unusual human driving

Method used	Description	Pros	Cons
Convolutional neural network (CNN)	This technique is particularly useful for finding patterns in images to recognize objects, classes, and categories	Can handle large amounts of data and achieve high accuracy, no need for manual feature extraction	Can be prone to overfitting if not enough data or proper regularization is used
Multi-task cascaded convolutional neural network (MTCNN)	This method is used to detect human faces and face landmarks on images	Acceptable false detection and acceptable accuracy	Slow detection process, complex methodology
Recurrent neural network (RNN)	This method is used for temporal problems such as speech recognition, image recognition	RNN can recognize each and every piece of information through time, it is used in CNN to extend the pixel	Training an RNN is difficult, and it cannot process very long sequences as an activation function
ResNet model	Residual network is used for computer vision applications	It improves the efficiency of deep neural networks with more neural layers while minimizing the percentage of errors	Time consuming
Hidden Markov model	This model enables us to speak about visible events and hidden events in probabilistic model	HMMs are able to represent the variance of	The time spent in a given state is not captured explicitly
Laplacian Gaussian model	This method is useful for detecting edges that appear at various image scales	It is used to find the correct places of edges, testing and emphasizes the pixel	Malfunctioning at the corner, curves. and the gray-level intensity varies
LSTM	This technique is used to process single data points like image and keep the hidden state	It is efficient to model complex sequential data, less susceptible to the vanishing gradient	More complicated then RNN and require more training data. Slow to train datasets.

Recall

Recall is a parameter used to evaluate the efficiency of categorization computation. It is sometimes referred to as sensitivity or true positive rate. The percentage of all positive occurrences that were correctly predicted to be positive cases is known as recall. The calculation looks like this:

$$\text{Recall} = \frac{\text{TP}}{\text{TP} + \text{FN}} \tag{9.2}$$

Accuracy

The performance of the model is often described by the metric of accuracy. It determines the percentage of predictions that were typically accurate (Table 9.2).

$$\text{Accuracy} = \frac{\text{TP} + \text{TN}}{\text{TP} + \text{TN} + \text{FP} + \text{FN}} \tag{9.3}$$

Table 9.2 Performance evaluation of different existing models

Author	Method used	Dataset	Accuracy (%)
Alkinani et al. [1]	CNN, RNN, DBN, LSTM, generative model, discriminative model	UAH-drive set	71
Chen et al. [2]	MTCNN	WIDER FACE data of the Chinese University	97.58
Liang et al. [3]	CNN and vision transformers, action recognition, temporal action localization	AI city challenge datasets	74.08
He et al. [4]	PFLD_CPN key point detection model, retina for face recognition, YOLOV5	State farm distracted driver detection	96.03
Huang et al. [5]	ResNet, DenseNet	State farm distracted driver detection	
Cengil et al. [6]	Laplasian of Gaussian, CNN	State farm distracted driver detection	97.79
Xing et al. [7]	DeepCNN, TransferLearning, unsupervised GMM-based segmentation method, Alex-Netmodel, GoogleNet, ResNet model	RGB images are collected using the kinetic camera	91.4
Yan et al. [8]	Gaussian mixture model, R*CNN	SEU datasets	97.76
Hu et al. [9]	Novel deep learning-based ADD system, SVM, and BPNN	Driver behavioral data	98.33
Jegham et al. [10]	3DCNN or space time networks, multi-stream CNN, hybrid network	MDAD dataset	75
Islam et al. [11]	CNN, CNN-R	Real-life drowsiness detection video dataset (UTA)	93.57
Bruno et al. [12]	ANN, CNN, GoogleNet, DCNN		92
Camlica et al. [13]	ANN, SVM, dynamic Bayesian networks (DBN), Random Forest (RF), CNN, and hidden Markov models (HMM)	Driving behavioral dataset	86
Bichicchi et al. [14]	SDAE	V-BOX GPS/camera data logger	
Noor et al. [15]	DriftNet, ConvLSTM, C3D	New car drifting datasets	77.5

where

TP stands for True Positive,
TN stands for True Negative,
FP stands for False Positive,
FN stands for False Negative.

Improving road safety is a critical goal, and larger datasets, real-time implementation, and robust models are essential components in achieving this objective. A larger dataset allows for a more comprehensive representation of various driving scenarios, road conditions, and driver behaviors. The diversity is crucial because road safety issues can manifest differently under different circumstances. Aggressive and inattentive driving behaviors may be relatively rare compared to normal driving. A larger dataset ensures that there are enough instances of these behaviors for the model to learn effectively. With a larger dataset, deep learning models can generalize better, making them more capable of recognizing subtle and nuanced instances of dangerous driving behaviors. Drivers who receive real-time feedback on their driving behavior are more likely to modify their actions and adopt safer habits, contributing to long-term road safety improvements. Real-time systems can continuously collect data, which can be invaluable for analyzing trends, identifying high-risk areas, and informing policy decisions for road safety. Road conditions and driver behaviors can change rapidly. Robust models should be adaptable and capable of handling novel situations or evolving driving patterns. The UAH-Drive set, WIDER FACE data, AI City Challenge datasets, State Farm's Distracted Driver Detection, RGB images obtained from the Kinetic camera, SEU datasets, Driver Behavioral data, MDAD dataset, Real-life Drowsiness Detection Video Dataset (UTA), V-BOX GPS/camera data logger, and New Car Drifting Datasets are among the datasets used in the studies. These mentioned datasets are essential for developing and validating deep learning models that can recognize and categorize driving behaviors. Having access to vast and varied datasets that accurately represent actual driving situations is crucial for effectively enhancing road safety. Additionally, creating reliable and real-time models is crucial for use in practical applications in the field of recognizing driving behavior (Fig. 9.4).

9.5 Discussion

We analyze the accuracy of different existing models to evaluate their performance. Both a table describing the results and a graphic summary of the models performance are given. Comparatively Hu et al. [9] in this study, their proposed method achieved 98.33% the high accuracy. Cengil et al. [6] in their research the model achieved the accuracy 97.79%, which is also a good performance. Alkinani et al. [1] they achieved the low accuracy 71% and to perform well, their approach needs substantial knowledge. Islam et al. [11], their proposed method performed better than the Lenet-5 architecture in terms of accuracy and f-1 score. They showed that their model

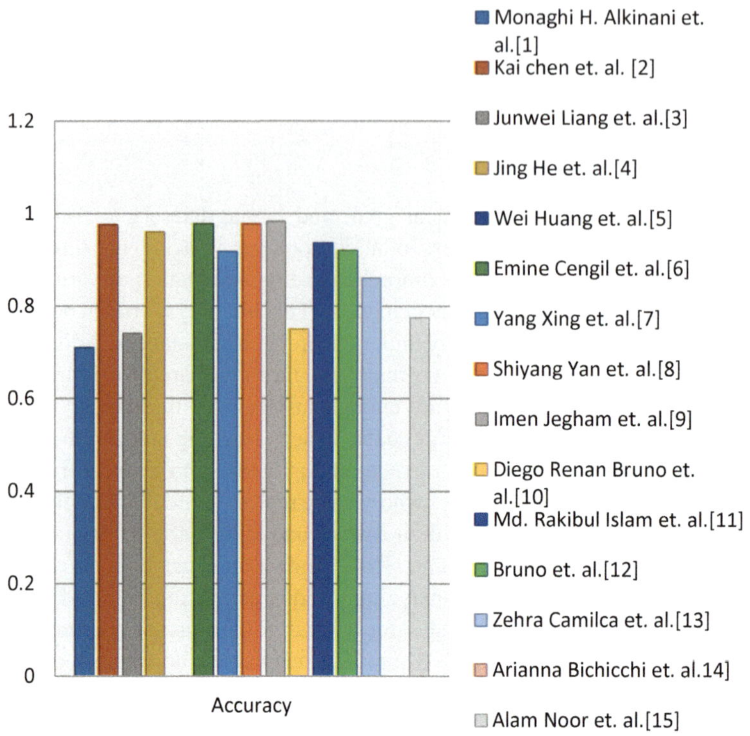

Fig. 9.4 Graphical analysis of different existing model [1–15] in terms of accuracy

is the most effective approach to deep learning-based fatigue detection. Yan et al. [8], they evaluated the proposed approaches on the Southeast University Driving-posture Dataset and achieved mean Average Precision on the dataset, illustrating the proposed method is effective in recognizing drivers actions.

9.6 Conclusion

In conclusion, recognizing and categorizing aggressive and distracted driving behaviors in human drivers using deep learning approaches holds great promise for improving traffic safety. Some of the techniques that the literature review discovered were convolutional neural networks (CNNs), multi-task cascaded neural networks (MTCNNs), transformer-based action detection systems, and fusion models impacted by densely connected convolution networks. The comparative analysis shows the accuracy and precision were all improved by using these techniques. The demand for larger and more varied datasets, real-time implementation, and robustness to various environmental conditions are a few needs. Despite the difficulties, deep learning-based systems provide automatic and accurate analysis of complicated driving patterns, which can help prevent accidents and improve overall road safety. Future studies should concentrate on overcoming the current constraints and creating more intricate deep learning models to correctly recognize and categorize driving actions in humans.

References

1. Alkinani, M.H., Khan W.Z., Arshad, Q.: Detecting human driver inattentive and aggressive driving behavior using deep learning: recent advances, requirements and open challenges (2020). https://doi.org/10.1109/ACCESS.2020.2999829
2. Chen, K., Wang, X., Yang, C., Liu, Z.: Fatigue driving detection method based on multi-task cascaded convolutional network. 978-1-6654-4854-3/21/$31.00 ©2021 IEEE
3. Liang, J., Zhu, H., Zhang, E., Zhang, J.: Stargazer: a transformer-based driver action detection system for intelligent transportation. 978-1-6654-8739-9/22/$31.00 ©2022 IEEE
4. He, J., Wu, Y., Xiao, J.: Calling behavior detection of port truck drivers based on deep learning. 978-1-6654-5462-9/22/$31.00 ©2022 IEEE
5. Huang, W., Liu, X., Luo, M., Zhang, P., Wang, W., Wang, J.: Video-based abnormal driving behavior detection via deep learning fusions. https://doi.org/10.1109/ACCESS.2019.2917213
6. Cengil, E., Cinar, A.: Classification of human driving behaviour images using convolution neural network architecture. ©Springer Nature Switzerland AG 2019
7. Xing, Y., Lv, C., Wang, H., Cao, D., Velenis, E., Wang, F.Y.: Driver activity recognition for intelligent vehicles: a deep learning approach. IEEE Trans. Veh. Technol. **68**(6), 5379–5390 (2019)
8. Yan, S., Teng, Y., Smith, J.S., Zhang, B.: Driver behavior recognition based on deep convolutional neural networks. 978-1-5090-4093-3/16/$31.00 ©2016 IEEE
9. Hu, J., Zhang, X., Maybank, S.: Abnormal driving detection with normalized driving behavior data: a deep learning approach. IEEE Trans. Veh. Technol. **69**(7), 6943–6951
10. Jegham, I., Khalifa, A.B., Alouani, I., Mahjoub, M.A.: Soft spatial attention-based multimodal driver action recognition using deep learning. IEEE Sens. J. **XX**(XX), XXXX (2020)
11. Islam, Md.R., Matin, A., Kamruzzaman, T.: Automatic identification of driver inattentiveness using convolutional neural networks. 978-1-6654-1917-8/20/$31.00 ©2020 IEEE
12. Bruno, D.R., Gomes, I.P., Osorio, F.S., Wolf, D.F.: Advanced driver assistance system based on NeuroFSM applied in the detection of autonomous human faults and support to semi-autonomous control for robotic vehicles. 978-1-7281-4268-5/19/$31.00 ©2019 IEEE
13. Camlica, Z., Quesenberry, J., Carballo, D., Crowley, M.: Aggressive driver behavior detection using parallel convolutional neural networks on simulated and real driving data. 979-8-3503-2045-9/22/$31.00 ©2022 IEEE
14. Bichicchi, A., Belaroussi, R., Simone, A., Vignali, V., Lantieri, C., Li, X.: Analysis of road-user interaction by extraction of driver behavior features using deep learning. https://doi.org/10.1109/ACCESS.2020.2965940
15. Noor, A., Benjdira, B., Ammar, A., Koubaa, A.: DriftNet: aggressive driving behavior detection using 3D convolutional neural network. IEEE (2020)
16. Shirabayashi, J.V., Braga, A.S.M., Da Silva, J.: Comparative approach to different convolutional neural network (CNN) architectures applied to human behavior detection. Neural Comput. Appl. **35**, 12915–12925 (2023)
17. Hossain, Md.U., Rahman, Md.A., Islam, Md.M., Akhter, A., Uddin, Md.A., Paul, B.K.: Automatic driver distraction detection using deep convolutional neural networks. Intell. Syst. Appl. **14,** 2667–3053, 200075 (2022)

18. Kashevnik, A., Shchedrin, R., Kaiser, C., Stocker, A.: Driver distraction detection methods: a literature review and framework. IEEE Access **9** (2021). https://doi.org/10.1109/ACCESS.2021.3073599
19. Abeer, A., Aljohani, F.: Real-time driver distraction recognition: a hybrid genetic deep network based approach. Alex. Eng. J. **66**, 377–389 (2023)
20. Khosravi, E., Hemmatyar, A.M.A., Siavoshani, M.J., Moshir, B.: Safe deep driving behavior detection (S3D). IEEE Syst. Man Cybern. Soc. Sect. **10** (2022). https://doi.org/10.1109/ACCESS.2022.3217644

Chapter 10
Improved Brain Tumor Classification Accuracy Using CNN Architecture with Efficient Net on Magnetic Resonance Imaging

Poonam Jyoti and Makhan Singh

Abstract Brain tumor classification from magnetic resonance imaging (MRI) scans performs an essential task in early detection as well as planning of treatment. Convolutional Neural Networks (CNNs) have exhibited outstanding performance in task of image classification, as evidenced by numerous research studies. In this research, we propose a CNN architecture utilizing Efficient Net as the backbone for accurate and efficient brain tumor MRI classification. Efficient Net, a state-of-the-art CNN architecture, has gained significant attention in the field of research for its remarkable combination of superior performance and efficiency. It leverages a compound scaling technique that optimizes model parameters by balancing depth, width, and resolution. This is what makes Efficient Net highly suitable for resource-constrained scenarios, such as medical imaging. This pooling operation summarizes the learned information, enabling more efficient learning. Fully connected layers are then employed to transform the pooled features into a classification decision. Total number of units at final fully connected layer represents number of target classes, in which case would represent different tumor types. The SoftMax activation function applied to final layer, providing possibility scores for each class, indicating the likelihood of the input belonging to a specific tumor category. Experiments show that our proposed method achieves a considerable improvement of 5% in performance metrics, surpassing existing methods in brain tumor classification accuracy.

P. Jyoti (✉) · M. Singh
Computer Science and Engineering, University Institute of Engineering and Technology, Panjab University, Sector-25, Chandigarh 160014, India
e-mail: Poonamchoudhary355@gmail.com

M. Singh
e-mail: singhmakhan@pu.ac.in

P. Nanda et al. (eds.), *Intelligent System and Data Analysis*, Smart Innovation, Systems and Technologies 398, https://doi.org/10.1007/978-981-97-5200-3_10

10.1 Introduction

Nowadays, brain tumors are among the most frightening illnesses. An abnormal clump of cells that develops uncontrollably in brain or spinal canal is called a brain tumor [1]. The tumor can only spread so far within skull, increasing pressure inside skull and perhaps endangering brain. Brain tumors can develop due to several circumstances, including genetic abnormalities and exposure to specific substances.

Now diagnosis of brain tumors needed to raising survival rates and boosting the likelihood of successful treatment.

Glioma tumors, meningioma tumors, pituitary tumors, and no tumors are four forms of brain tumors that are typically recognized [2]. There are several locations in the brain where glioma tumors can develop, including the brainstem and spinal cord. There are multiple symptoms associated with different forms of gliomas, such as headaches, nausea, vomiting, and irritability. Tumors called meningiomas are frequently seen in brain and developed from the meninges, both brain and main nervous system are enclosed by it [2, 3]. On other hand, pituitary tumors are abnormal growths that may grow in pituitary gland, which oversees generating hormones that control variety of physiological processes. These tumors can result in hormonal imbalances and produce variety of symptoms, such as fatigue headaches and eye issues [3].

Using imaging data to make medical diagnoses requires a variety of methods, including X-ray, CT scan, and MRI. Out of all methods used, MRI is especially interesting because it involves measuring magnetic fields and using strong radio waves and magnetic fields to stimulate the hydrogen atoms in water molecules inside patient's body [4]. Previously, tumor detection in MR images required laborious human inspection, which took time. Now, MRI images are very important in the field of automatic medical analysis because they help us see different parts of brain clearly. This allows us to get detailed information about brain's structure [5].

In this research, different techniques for learning were examined such as machine learning and deep learning to automatically analyze tissues and identify tumors. Brain imaging analysis is a field that benefits from application of ML techniques, which entails creation of models and algorithms that allow machines to learn and make decisions or predictions without requiring specific programming for each task [6]. These techniques leverage mathematical models that consider image qualities such as texture, brightness, and contrast for feature extraction [7]. In medical image analysis, various machine learning algorithms such as *K*-Nearest Neighbors (KNN), Support Vector Machines (SVM), Decision Trees (DT), Naive Bayes (NB), expectation maximization (EM), random forest (RF), and others are commonly used [8–10]. These techniques play a significant role in understanding and analyzing medical pictures, enabling accurate diagnosis and decision-making.

Deep learning focuses on training artificial neural networks to analyze large datasets and make predictions [11]. Deep learning (DL) has developed as a prominent and popularly utilized approach in field of brain image analysis, with applications ranging from classification of normal and abnormal brain tumors to segmentation of different tumor regions (such as enhancing and non-enhancing regions), segmentation, diagnosis, and others [12]. In this paper, features are being extracted using CNN to capture essential patterns related to tumor characteristics. A widely used deep learning model CNN employs a series of feed-forward layers and is commonly utilized in image processing. It includes layers for instance input, convolution, ReLU, pooling, and fully connected layers. The convolutional layer performs mathematical linear operations on input. The dimension of image is progressively reduced in each layer, preserving essential information [13].

10.2 Literature Survey

Various approaches require to introduce in literature to detect and classify brain tumors. ML and DL techniques earned significant popularity in current times for the recognition and classification of brain cancers, particularly utilizing MRI brain scans as the primary imaging modality. Bahadure et al. [14] suggested approach which combines (SVM) for classification with Berkeley wavelet transformation (BWT) for segmentation. Moreover, by extracting appropriate features from segmented tissues, the accuracy of the SVM classifier is improved. Ayad et al. [15] developed a method that classifies MRI brain imaging into two categories: normal tissue and abnormal tissue. Methods used in this approach are noise subtraction, gray-level co-occurrence matrix (GLCM) feature elimination and wavelet-based region growing segmentation. Then, the feature vectors acquired by each technique are processed by *k*-Nearest Neighbors (KNN) classifier. To enhance precision of classification, dissimilarity test values produced through classifiers are merged using fusion operator. This fusion operation aims to combine outputs of several classifiers and enhance overall classification accuracy. Arunachalam and Royappan Savarimuthu et al. [16] present a highly effective and automatic approach for the detection of glioma brain tumors. Such approach leverages the shift-invariant shearlet transforms in combination with neural networks to achieve accuracy.

Tumor detection. The authors also highlight the significance of feature selection in reducing the dimensionality of feature spaces and improving classification performance. Vasantha et al. [7] explore numerous feature selection procedures such as filter, wrapper, and embedded methods, their applications in the field of medical image analysis. Authors developed different classification algorithms commonly used in medical image analysis, e.g., decision trees, SVM, k-NN, and ANN. Munira and Hafiza along with Islam et al. [17] proposed a novel method that relates (CNNs) with long short-term memory (LSTM) networks. This hybrid architecture enables the integration of both spatial and temporal information, leading to improved accuracy in classification tasks. According to Sharif et al. [18] method involves fine-tuning Densenet201 pre-trained DL model and employing deep transfer learning techniques during training process. Moreover, features are obtained from average pool layer and two feature selection techniques such as EKbHFV and MGA, which are applied and using multi-class SVM cubic as a classifier. Hence, experimental results on the BRATS2018 and BRATS2019 datasets demonstrate an accuracy of over 95%.

According to Deepak et al. [19], classification of brain tumors holds significant importance in computer-aided diagnosis (CAD). Researchers proposed methodology classifying brain tumors utilizing deep Convolutional Neural Network (CNN) features through transfer learning. Their methodology involves employing pre-trained Google Net to obtain features that begin with a brain MRI image.

Another researcher Ramdas Vankdothu, Mohd Abdul Hameed et al. [20] present a novel approach for brain tumor identify and categorization that use deep learning techniques. Authors proposed the CNN-LSTM method that combines the power of CNNs and LSTM networks to capture both spatial and temporal information from brain tumor images. Hence, experimental findings demonstrate the efficiency of the recommended method accurately on identifying and classifying brain cancers, achieving high accuracy rates compared to traditional methods.

10.3 Methodology

The proposed methodology presents an approach for brain tumor classification that includes clustering, preprocessing, data splitting, CNN training model, CNN with Efficient Net training, prediction, and testing (Fig. 10.1).

10.3.1 Dataset Collection and Preprocessing

Dataset that used in this research was obtained from various publicly available online resources [21]. The dataset consisted of brain tumor images categorized in different categories (labels): glioma tumor, no tumor, meningioma tumor, and pituitary tumor.

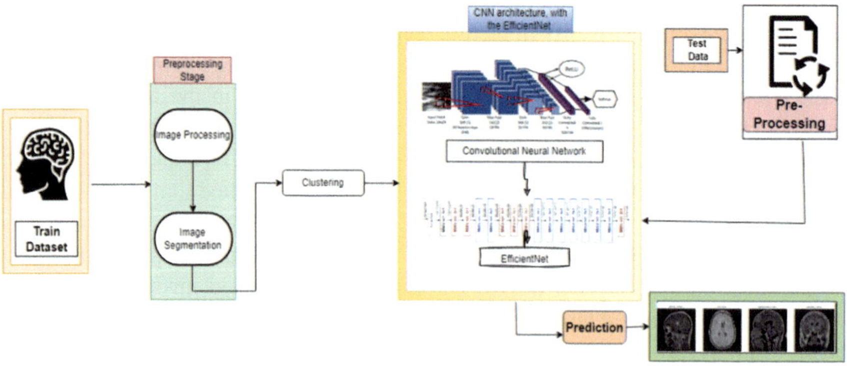

Fig. 10.1 Basic structure of the methodology proposed

There is total 3264 imaginings besides these images have resized to a standardized size, typically 150 × 150 pixels, ensuring uniformity across the dataset. Secondly, pixel value normalization is performed, scaling intensity values to a standardized assortment, typically between 0 and 1. After that additional preprocessing techniques are applied in the next step of clustering.

10.3.2 Clustering-Based Segmentation

In our study, we applied K-means clustering to the preprocessed MRI data to identify distinct clusters or groups within the dataset. This technique allowed us to partition the images into clusters based on pixel intensities. By doing so, we can isolate the tumor regions of interest from surrounding healthy tissue. K-means segmentation helps in extracting tumor regions accurately, which is crucial for subsequent analysis and classification.

Following the preprocessing and segmentation of MRI data, we proceeded to split it into training and testing sets. Each MRI sample was assigned respective labels during this process. During training phase, our model learned from labeled data, enabling it to make accurate predictions on new, unseen MRI images based on their features. This step was essential for developing reliable and effective classification model. By employing K-means clustering and subsequent training on labeled data, our study aimed toward improving accurateness and effectiveness of brain tumor categorization from MRI scans.

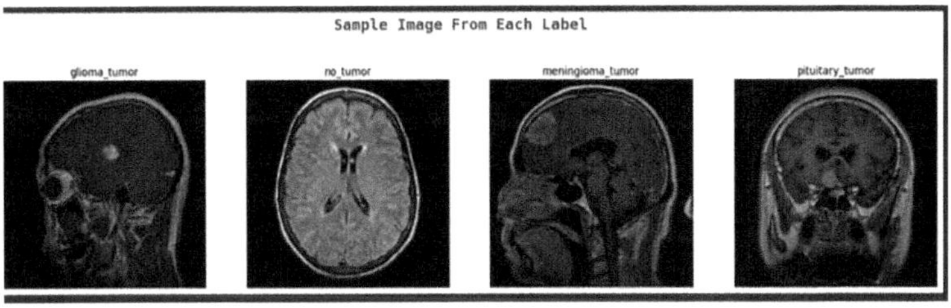

10.3.3 CNN Model Architecture and Training

CNN model working involves passing input images through the convolutional blocks, which extract features using filters and ReLU activation. This CNN architecture typically consists of multiple convolutional layers. First in the convolutional, a layer contains 32 filters and one kernel size of 5 × 5. The subsequent convolutional layers have increasing filter sizes: 64, 128, 128, 128, 128, and 64. Final layers include global average pooling, dropout to prevent overfitting and Global Average Pooling reduces the spatial dimensions to a fixed-length vector and a dense layer with SoftMax

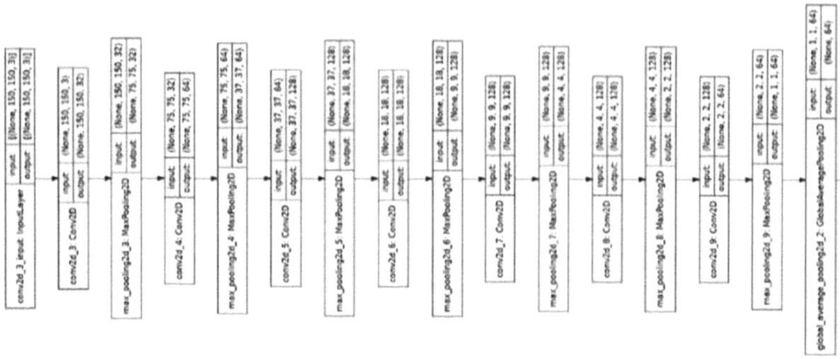

Fig. 10.2 CNN architecture

activation for classification with the output size set to 4 (assuming it is a multi-class classification problem with 4 classes). Different hyperparameter are used learning rate, batch size, number of epochs. Here, training is carried out for 30 epochs using batch size 32. The model stands therefore assembled with Adam optimizer, definite cross-entropy loss, and accuracy metric.

Additionally, better models based on validation accuracy is saved during training (Fig. 10.2).

10.3.4 CNN with Efficient Net

The segmented tumor regions obtained from the *K*-means segmentation are then fed into a CNN architecture. We leverage the Efficient Net model as the backbone of our CNN.

Efficient Net is state-of-the-art, whereas CNN architecture notorious representing its efficiency and effectiveness. It scales on model depth, width, and resolution in balanced manner, allowing it to capture intricate features while optimizing computational resources. The CNN architecture with Efficient Net backbone is responsible for extracting relevant features from segmented tumor regions. It comprises convolutional layers that learn hierarchical representations of the tumor regions, capturing important visual patterns indicative of different tumor types.

Efficient Net model combines convolutional layers for feature extraction, a GlobalAveragePooling2D layer for dimensionality reduction, a Dropout layer for regularization, and dense layer for classification. This combination enables models to effectively study and represent complex patterns in images, leading to improved accuracy and generalization.

Firstly, load the pre-trained Efficient Net-B0 model: Efficient Net is a powerful CNN which pre-trained on huge dataset (e.g., ImageNet). By loading a pre-trained model, we utilize learned features and weights, which can be helpful in improving the accuracy and performance of our model.

Freeze pre-trained layers: Obstruction pre-trained layers means that we can prevent their weights from updating during training and only newly added layers will be updated. Since the pre-trained layers have already learned useful features, freezing them helps to retain this knowledge and prevents them from being altered.

Add custom layers for brain tumor detection: To adapt the pre-trained Efficient Net for brain tumor detection, we add custom layers on top of the base model. The output of our existing CNN model acts as the input to the modified Efficient Net. These custom layers utilize the learned features from the base model to make the final predictions. This integration combines strengths of Efficient Net with the insights from our CNN model, resulting in improved accuracy for brain tumor detection.

- Global Average Pooling Layer:
  After pre-trained layers, we normally add a Global Average Pooling layer. Minimizing spatial dimensions of the feature maps effectively summarizes learned features. It calculates average value for each feature map, resulting in fixed-length vector representation of the features.
- Dense Layers:
  After global Average Pooling layer, we add some Dense layers. These fully connected layers responsible for mapping split up features to desired output, which in this case is brain tumor detection. The Dense layers perform nonlinear transformations and capture complex relationships between the features.
- Output layers: The finale layer of network and it generates predicted output.
- Compiling and Training the Model: Compile models by identifying loss function, optimizer and evaluation metrics. Train model with prepared training dataset, which specifies the number of epochs, batch size, and any desired callbacks (e.g., Tensor Board, Model Checkpoint) for monitoring and saving the best model.
- Evaluate the model's performance on the validation: Calculate: By compiling the modified Efficient Net model, we prepare it for training and evaluation. Leveraging its optimized architecture and pre-trained weights, the Efficient Net model outperforms traditional CNNs in terms of performance and efficiency, making it ideal for image classification tasks.

10.4 Implementation and Results

10.4.1 Performance Evaluation of CNN Model

The performance of our trained CNN model is evaluated by monitoring for training and validation accuracy and loss throughout the training process. General accuracy of our model on the test set is 91%. Additionally, model achieved high precision and recall values for each class, with f_1-scores ranging from 0.88 to 0.97 (Fig. 10.3; Table 10.1).

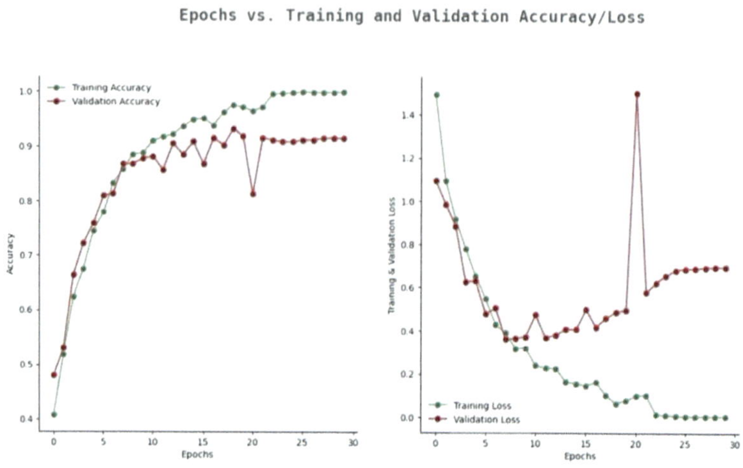

Fig. 10.3 Training and validation accuracy/loss curves

Table 10.1 Classification metrics

Class	Precision	Recall	F_1 score	Support
0	0.93	0.85	0.89	93
1	0.87	0.90	0.88	51
2	0.90	0.91	0.90	96
3	0.95	1.00	0.97	87
Accuracy				0.91
Macro avg.	0.91	0.91	0.91	327
Weighted avg.	0.91	0.91	0.91	327

10.4.2 Performance Evaluation CNN with Efficient Net

In the second part of the code, combining the Efficient Net model with the existing CNN model improve the accuracy up to 96%. These experimental results demonstrated a significant improvement of 5% in performance metrics compared to existing methods in brain tumor classification accuracy (Table 10.2).

Such research highlights the potential of deep learning and CNNs, specifically using Efficient Net, in accurately and efficiently classifying brain tumor MRI scans. The proposed approach has implications for early diagnosis, treatment planning, and improving survival rates for brain tumor patients.

By applying heat map visualization to brain tumor classification, it becomes possible to identify the specific regions within an MRI scan (Fig. 10.4).

Table 10.2 Classification metrics

Class	Precision	Recall	F_1-score	Support
0	0.97	0.92	0.95	93
1	0.96	0.98	0.97	51
2	0.94	0.96	0.95	96
3	0.99	1.00	0.99	87
Accuracy				0.96
Macro avg.	0.96	0.97	0.96	327
Weighted avg.	0.96	0.96	0.96	327

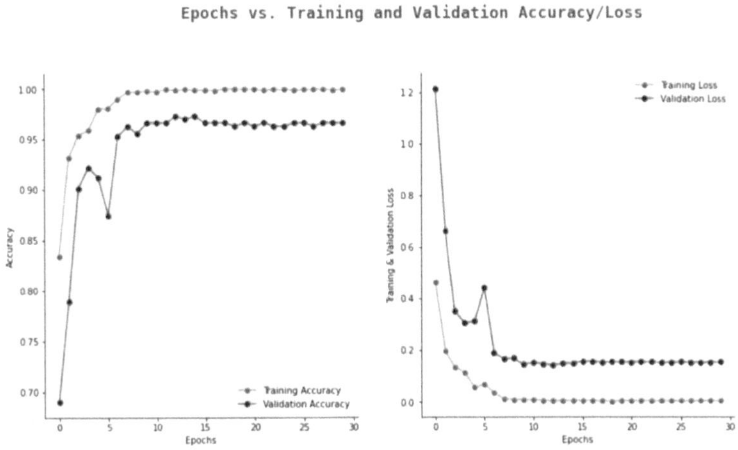

Fig. 10.4 Training and validation accuracy/loss curves

10.4.3 Heatmap Confusion Matrix

By applying heat map visualization to brain tumor classification, it becomes possible to identify the specific regions within an MRI scan (Figs. 10.5 and 10.6).

This visualization helps to recognize the brain areas that strongly indicate the range of presence or type of a tumor whether it is glioma tumors, meningioma tumors, pituitary tumor and no tumors. Heatmap analysis illustrates that the CNN architecture utilizing the Efficient Net model outperforms the existing CNN model in brain tumor classification.

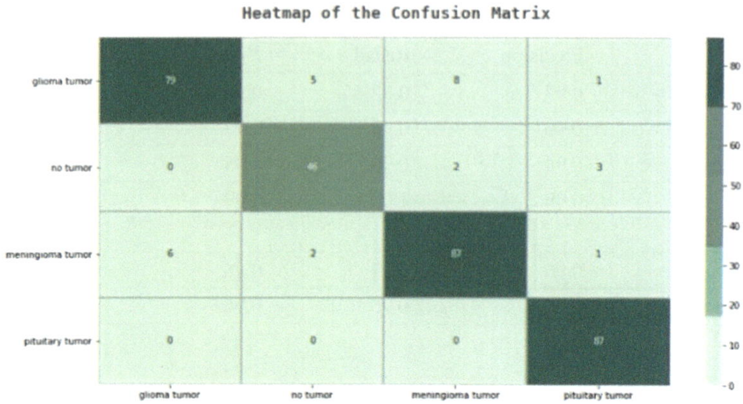

Fig. 10.5 Heatmap of confusion matrix for the CNN model

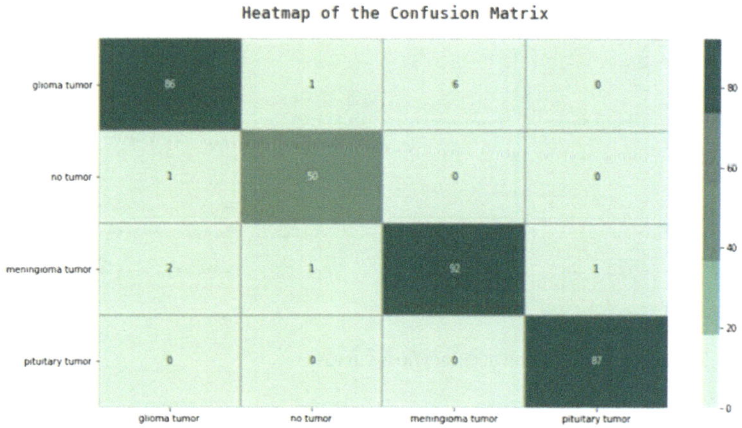

Fig. 10.6 Demonstrates heatmap of the confusion matrix for the CNN architecture using Efficient Net

10.5 Result

Experiments evaluating tumor detection on brain MRI dataset were conducted using 11th Gen Intel(R) Core (TM) i3-1115G4@ CPU @3.00 GHz with 16 GB RAM and NVIDIA GPU. The CNN architecture, Efficient Net was implemented in Python using TensorFlow and the Keras library. Training was performed for 30 epochs including batch size 32. All networks are being trained for 30 epochs, batch size 32. Adam optimizer applied using learning rate 0.001. Outcomes demonstrate high accuracy 96% in classifying different tumor types, showcasing the potential of our approach in assisting medical professionals in accurate diagnosis and treatment planning. The proposed approach has implications for early diagnosis, improving survival rates for brain tumor patients.

10.6 Conclusion

CNN architecture with Efficient Net as the backbone classifying of brain tumors in MRI scans. By combining the powerful features of Efficient Net with the capabilities of CNNs, we have achieved high accuracy and robust performance in differentiating between different brain tumor types. The utilization of Efficient Net, known for its exceptional performance and efficiency, ensures the practical applicability of our architecture in real-world scenarios with limited computational resources. This makes it a valuable tool for accurate and efficient diagnosis, enabling timely treatment decisions for patients with brain tumors. The experimental evaluation on a public available MRI dataset confirms effectiveness of our recommended CNN architecture. The results highlight its potential in improving the diagnostic process by accurately categorizing brain tumors formed on their MRI scans. By harnessing capabilities of deep learning and leveraging state-of-the-art architectures like Efficient Net, we contribute domain of medical imaging and offer promising approach for automated brain tumor classification. Further research and validation on larger, separate datasets are warranted to consolidate the performance afterward generalizability of our proposed architecture. Ultimately, our work aims assist health care consultants generating reported results and providing better care to patients with brain tumors.

References

1. Swamy, G.J., Anusha, L., Shree, S.M.: Brain tumor detection using convolutional neural network
2. Cole, S.: Brain cancer types (2017). https://www.cancercenter.com/cancer-types/brain-cancer/types//www.vice.com/enus/article/gydydm/gal-gadot-fake-ai-porn//
3. Louis, D.N., Perry, A., Reifenberger, G., Von Deimling, A., Figarella-Branger, D., Cavenee, W.K., Ohgaki, H., Wiestler, O.D., Kleihues, P., Ellison, D.W.: The 2016 World Health Organization classification of tumors of the central nervous system: a summary. Acta Neuropathol. **131**, 803–820 (2016)
4. Litjens, G., Kooi, T., Bejnordi, B.E., Setio, A.A.A., Ciompi, F., Ghafoorian, M., Van Der Laak, J.A., Van Ginneken, B., Sánchez, C.I.: A survey on deep learning in medical image analysis. Med. Image Anal. **42**, 60–88 (2017)
5. Mabray, M.C., Cha, S.: Advanced MR imaging techniques in daily practice. Neuroimaging Clin. **26**(4), 647–666 (2016)
6. Sharma, K., Kaur, A., Gujral, S.: Brain tumor detection based on machine learning algorithms. Int. J. Comput. Appl. **103**(1) (2014)
7. Vasantha, M., Bharathi, V.S., Dhamodharan, S.: Medical image feature, extraction, selection and classification. Int. J. Eng. Sci. Technol **2** (2010)

8. Altman, N.S.: An introduction to kernel and nearest-neighbor nonparametric regression. Am. Stat. **46**(3), 175–185 (1992)
9. Bokulich, N.A., Dillon, M.R., Bolyen, E., Kaehler, B.D., Huttley, G.A., Caporaso, J.G.: q2-sample-classifier: machine-learning tools for microbiome classification and regression. J. Open Res. Softw. **3**(30) (2018)
10. Thayumanavan, M., Ramasamy, A.: An efficient approach for brain tumor detection and segmentation in MR brain images using random forest classifier. Concurr. Eng. **29**(3), 266–274 (2021)
11. Saleh, A., Sukaik, R., Abu-Naser, S.S.: Brain tumor classification using deep learning. In: 2020 International Conference on Assistive and Rehabilitation Technologies (iCareTech), pp. 131–136. IEEE (2020)
12. Ravi, D., Wong, C., Deligianni, F., Berthelot, M., Andreu-Perez, J., Lo, B., Yang, G.-Z.: Deep learning for health informatics. IEEE J. Biomed. Health Inform. **21**(1), 4–21 (2016)
13. Hashemzehi, R., Mahdavi, S.J.S., Kheirabadi, M., Kamel, S.R.: Detection of brain tumors from MRI images base on deep learning using hybrid model CNN and NADE. Biocybern. Biomed. Eng. **40**(3), 1225–1232 (2020). https://doi.org/10.1016/j.bbe.2020.06.001
14. Bahadure, N.B., Ray, A.K., Thethi, H.P., et al.: Image analysis for MRI based brain tumor detection and feature extraction using biologically inspired BWT and SVM. Int. J. Biomed. Imaging **2017** (2017)
15. Khalil, M., Ayad, H., Adib, A.: Performance evaluation of feature extraction techniques in MR-brain image classification system. Procedia Comput. Sci. **127**, 218–225 (2018)
16. Arunachalam, M., Royappan Savarimuthu, S.: An efficient and automatic glioblastoma brain tumor detection using shift-invariant shearlet transform and neural networks. Int. J. Imaging Syst. Technol. **27**(3), 216–226 (2017)
17. Munira, H., Islam, M.: Hybrid deep learning models for multi-classification of tumour from brain MRI. J. Inf. Syst. Eng. Bus. Intell. **8**, 162–174 (2022). https://doi.org/10.20473/jisebi.8.2.162-174
18. Sharif, M., Khan, M., Alhussein, M., Khursheed, K., Raza, M.: A decision support system for multimodal brain tumor classification using deep learning. Complex Intell. Syst. **8** (2021). https://doi.org/10.1007/s40747-021-00321-0
19. Deepak, S., Ameer, P.M.: Brain tumor classification using deep CNN features via transfer learning. Comput. Biol. Med. **111**, 103345 (2019). https://doi.org/10.1016/j.compbiomed.2019.103345
20. Oksüz, C., Urhan, O., Güllü, M.K.: Brain tumor classification using the fused features extracted from expanded tumor region. Biomed. Sig. Process. Control **72**, 103356 (2022). https://doi.org/10.1016/j.bspc.2021.103356
21. Dataset (2014). https://www.kaggle.com/datasets/masoudnickparvar/brain-tumor-mri-dataset

Chapter 11
Noise Prediction: A Case Study of Gjilan City

Buen Bajrami and Kostandina Veljanovska

Abstract Recently, we have started to hear more often about the smart city. This is a new term, but recently there is a growing trend, which uses computer sciences to create an integrated, intelligent system, which is easily used and managed by people. Such smart grid technologies can directly impact everything from operations to infrastructure maintenance and planning. A key part of this research will be noise monitoring in specific areas of the city depending on population, traffic density, number of cars, and wind speed. We will make real-time measurements in four areas per day and hourly. The data will be placed in a database, where its data will be processed through two algorithms on our platform designed in PHP. The two forecasting algorithms we used are Linear Regression and the ARIMA Forecasting Model. We have also included the correlation coefficient to see the report of the parameters. At the end, we have presented the factors that influence the increase in noise as well as the recommendations that we believe will help prevent noise in the city.

11.1 Introduction

Our topic that has chosen regarding noise pollution is become a significant issue in many cities around the world, impacting the health and well-being of residents and wildlife, and disrupting the environment. In response, our project aims to develop a noise prediction platform for the city of Gjilan (Kosova) that can accurately predict noise levels in different areas at different times. This platform will use algorithms and data analysis techniques to take into account factors such as traffic flow, construction sites, and weather patterns that can contribute to noise pollution. For residents, the

B. Bajrami (✉) · K. Veljanovska
Faculty of Information and Communication Technologies, University St. Kliment Ohridski - Bitola, 1 Maj bb., 7000 Bitola, North Macedonia
e-mail: bajrami.buen@uklo.edu.mk

K. Veljanovska
e-mail: kostandina.veljanovska@uklo.edu.mk

P. Nanda et al. (eds.), *Intelligent System and Data Analysis*, Smart Innovation, Systems and Technologies 398, https://doi.org/10.1007/978-981-97-5200-3_11

noise predictions will provide valuable information to help them make informed decisions about their daily lives, such as choosing quieter residential areas or avoiding noisy commercial areas during peak hours. For city authorities, the platform will provide a valuable tool for developing and implementing effective strategies for reducing noise pollution and improving sustainability. In addition, the noise prediction platform will help to obtain information regarding noise pollution more easily and also the consequences that can appear in people or the environment. By providing accurate predictions and encouraging a greater understanding of the issue, we hope to promote a more sustainable, healthier, and quieter urban environment.

11.2 Literature Review

The forecasting process is important in every institution. The authors of this paper have assessed that it is important to use one or two prediction models in this process [1]. Based on the paper [2], noise prediction is also quite necessary in the design of highways. The main goal is to predict the sound pressure level. There are already several types of software that measure the noise emission in urban areas. We have IMMI software, which offers noisy areas in urban areas by creating and integrating digital maps, with the surrounding topography, roads, green spaces, etc. Observing a visualization of noise emission in 3D dimension [3]. We have also other software CadnA [4] which provides analysis of the noise map in urban spaces in the USA. Based on specific directives related to noise measurements and management.

11.2.1 Different Noise Prediction Methods

There are various methods and algorithms that are used to predict noise in urban areas. Therefore, to arrive at a more accurate noise prediction, many factors must be carefully analyzed [5]. The authors estimate that the duration of collected the data obtained help a lot in predicting noise values. Also, the authors paid attention to the duration of high and low noise, since we have differences in prediction.

11.2.2 Forecast Algorithm—ARIMA Model

The Forecast ARIMA model algorithm is one of the two algorithms we use in our noise forecasting platform. According to [6], this algorithm creates different patterns based on time series. And, this criterion deals with the evaluation of the models by comparing the ratio of the average of the real values and the average of the stochastic errors during the forecast time. Also, the authors think that this algorithm is more accurate if the forecast is applied to the time series in large value, then it would save efforts and also time.

In [7] this project, the authors have used the forecast algorithm to predict the weather, based on four conditions, minimum temperature, maximum temperature, humidity, and precipitation. These data were taken every day and the Matlab2010a program was used for forecasting. While in the case of our project, we used three conditions which are mentioned above and the measurements were made every hour for every day.

Authors [8] have assessed that it is important that in the selection of a subset of the most suitable models, the peculiarities of the time series should be taken into account, and also during the online phase, there is no need to test all combinations of individual models. Therefore, this can have a positive impact on the inclusion of judgment or expertise by practitioners to arrive at a decision-making and forecast, and model ranking offers an alternative solution.

According to this paper, the authors [9] show a new model that provides the short-term forecast of the new high-tech procedure. After collecting real data to build stock market data, an ARIMA model is applied to the dataset to improve short-term forecasting. About 200 observations were collected to implement the predictions. The best model based on the most popular MSE criterion was selected ARIMA.

11.2.3 Linear Regression Algorithm

In addition to the ARIMA model algorithm, we have also used the linear regression algorithm, which is widely used in forecasting systems. Regression [10] can be either a simple linear regression or multiple regression.

Usually, the use of linear regression [11] is a process, which is carried out through mathematical analysis, measuring the predicted values and having new models in the input data. Therefore, we would have a new modeling of the analysis reports, the learning process and up to the final results. According to [12], it is estimated that within the two variables, we can have a linear relationship through linear regression.

A linear regression [13] shows also the use of analysis which helps to assess how strongly the training and its effect are related within a dependent variable.

11.3 Methodology

The decibel (dB) values for an urban city can vary greatly depending on various factors such as time of day, location, and the presence of different sources of noise, for example roads with cars, buildings under construction, premises near factories, etc. On average, daytime noise levels in an urban city can range from 60 to 70 dB, while nighttime levels can be lower, ranging from 40 to 60 dB. It's important to note that decibel levels are logarithmic and not linear, which means that an increase

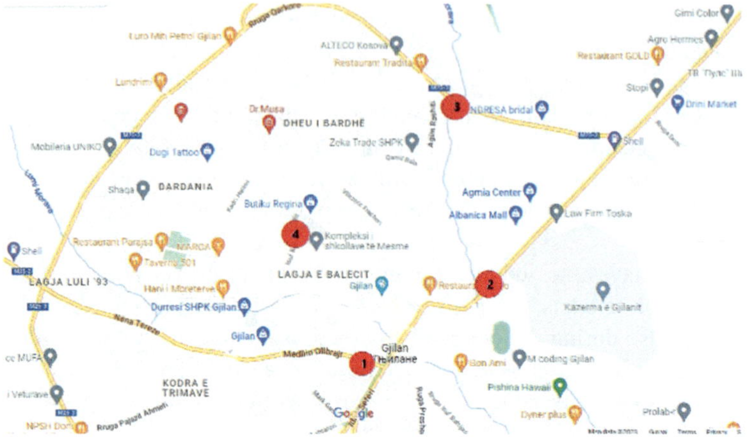

Fig. 11.1 Map of the Gjilan city with measurements points

of 3 dB indicates that we have a double increase in sound energy, and an increase of 20 dB represents a 20-fold increase in sound energy. As a conclusion, even a small increase in decibel levels can have a high effect on the environment and human health. We have decided to measure in four different points. Starting from 08:00 to 22:00 o'clock, in every hour.

11.3.1 Selection of Measurement Points

Below we will present the graphic view of the city map of Gjilan obtained from Google Map (Fig. 11.1). We also have four red dots which illustrate the points from which the measurements were made.

- Point 1—Center Zone 1.
- Point 2—Center Zone 2.
- Point 3—Transit road.
- Point 4—Public University and High schools.

11.3.2 Correlation Coefficient of Measurement Parameters

In this project, we have analyzed a total of three parameters.

- Prediction time.
- Number of cars.
- Wind speed.

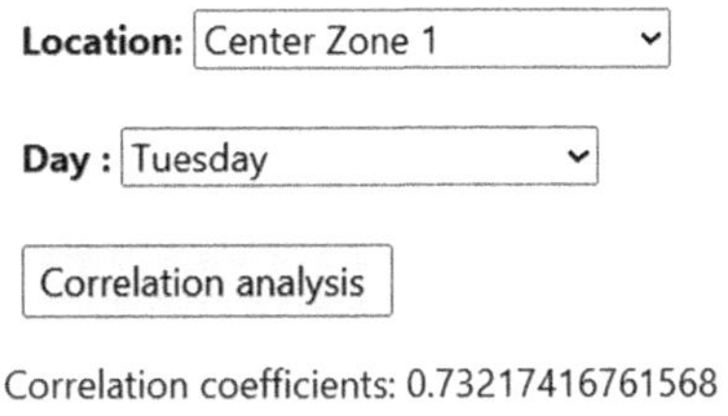

Fig. 11.2 Correlation coefficient for center zone 1 parameters

In order to have a strong correlation, the value must be as close as possible to the number 1; on the other hand, as close as possible to the value 0, we will have a weak correlation between the parameters. Below (Fig. 11.2) we will show the correlations obtained for the four measuring points. We first display the correlation results for point 1, Center Zone 1.

Here (Fig. 11.3) we present the correlation results for point 2, Center Zone 2.

In Fig. 11.4 we present the correlation results for point 3, Transit Road.

Finally in Fig. 11.5 we have the results of the four measuring point, Public University and High schools.

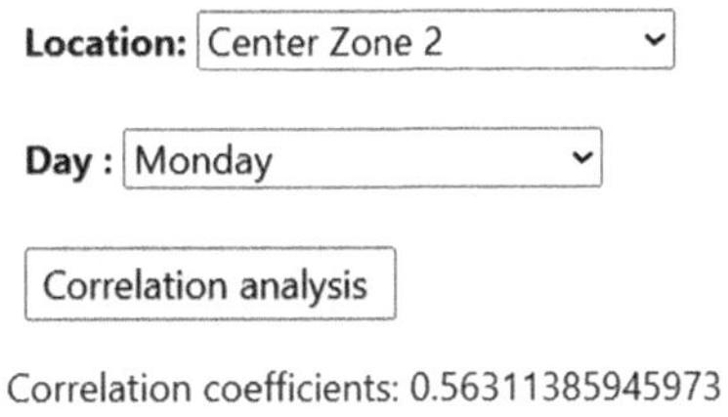

Fig. 11.3 Correlation coefficient for center zone 2 parameters

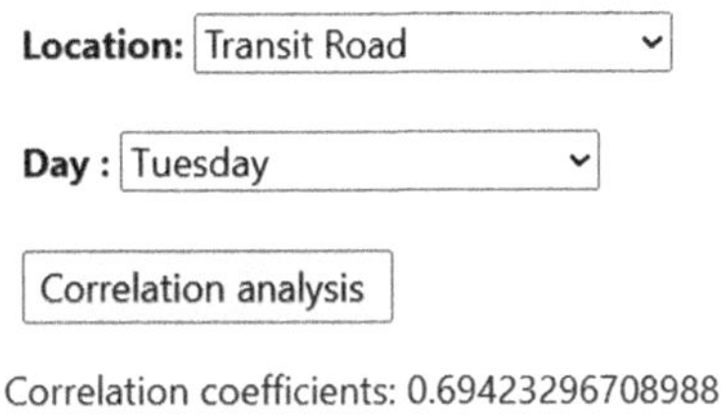

Fig. 11.4 Correlation coefficient for transit road parameters

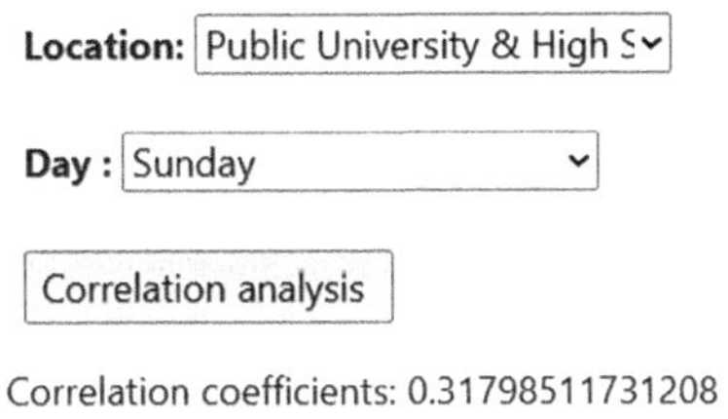

Fig. 11.5 Correlation coefficient for public university and high schools' parameters

Through the results obtained from the correlation, we come to the conclusion that from three points, we have a strong correlation of the parameters, except for the fourth point where we see that we have a weaker correlation.

11.3.3 Sound Level Meter

To take measurements for our requirements, we used the Unit-T 353 instrument (Fig. 11.6). The UT353 instrument enables the measurement of sound and the conversion of measured values into electrical signals. It offers a visual image on the LCD as well as the possibility of constant monitoring [14].

Fig. 11.6 Sound level meter

11.3.4 Description of the Calculation Method

In this research, we are based on the temporary method for calculating traffic noise [15] which is called "NMPB-Routes-96(SETRA-CETRU-LCPC_CSTB)". The method provides details in the measurement of noise near the road as well as taking into account meteorological side effects.

11.3.5 Measurement Procedure

XPS-31-133 is taken as the complete emission model for the calculation of road traffic noise [16].

11.3.6 Type of Vehicles

There are two types of vehicle categories that have been used in our case:

- Category 1—(vehicles less than 3.5 t of net load).
- Category 2—(vehicles with a higher load or the same 3.5 t net load).

Table 11.1 shows allowed values of noise emissions in traffic.

11.4 Algorithms Used for Prediction

In our platform, we have used two algorithms for noise prediction. Below we will give a more detailed explanation about the two algorithms.

Table 11.1 Allowed values of noise emissions in traffic

Environment	Noise source limit values (dB)	
	Day	Night
Industrial areas	69	59
Villages	64	54
Habitable areas	59	49
Hospitals, schools, and nursing homes	57	47

11.4.1 Prediction of Noise Level with Linear Regression Algorithm

Linear regression offers calculation of a dependent parameter, based on one or several independent parameters [17]. This method was used in our case as an algorithm that could have predicted noise level in a city based on the movement of vehicles from 08:00 to 22:00 during the day, as well as the number of cars and wind speed. In this case, the dependent variable would be traffic level (y) and the independent variable would be time (x). To calculate the linear regression, we will need to use the data collected about the traffic level at different hours of the day and use a mathematical formula to find a line of fit that describes the relationship between time and traffic level. Once the regression parameters are found, we can use linear regression to predict the level of traffic flow at certain hours in future by entering the desired time into the linear regression formula. The formulas to find the parameters a and b are as follows (Eqs. 11.1and 11.2):

$$b = \sum ((x(i) - x\_mean) * (y(i) - y\_mean)) / \sum (x(i) - x\_mean)^2, \qquad (11.1)$$

$$a = y\_mean - b * x\_mean, \qquad (11.2)$$

where $x(i)$ and $y(i)$ are the values of x and y for a given point, x_mean and y_mean are the means of x and y, and Σ is the summary symbol. The formula for predicting the value of y is as follows (11.3):

$$y = a + b * x, \qquad (11.3)$$

where a and b are the parameters found in linear regression and x is the value of x for which you want to predict y.

11.4.2 Prediction of Noise Level with ARIMA Model

There are many models and techniques that can be used for forecasting in statistics and time series analysis. Here are some of the main formulas and concepts used in one of the most well-known forecasting method:

ARIMA model: This model offers mathematical analysis based on time segments in order to provide more accurate predictions. This model can be autoregressive if it manages to provide future predictions based on the values obtained in the past. For example, we have the AR model, which is based on its own delays. That is, Y_t is a function of the "lags of Y_t" [18] as shown in Eq. (11.4).

$$Y_t = \alpha + \beta_1 Y_{t-1} + \beta_2 Y_{t-2} + \cdots + \beta_p Y_{t-p} + \epsilon_t, \qquad (11.4)$$

where Y_{t-1} is the lag1 of the series, β_1 is the coefficient of lag1 that the model estimates, and α is the intercept term, also estimated by the model. An ARIMA model is one where the time series was differenced at least once to make it stationary and you combine the AR and the MA terms. So Eq. (11.5) becomes:

$$Y_t = \alpha + \beta_1 Y_{t-1} + \beta_2 Y_{t-2} + \cdots + \beta_p Y_{t-p} \in_t + \emptyset_1 \in_{t-1} + \emptyset_2 \in_{t-2} + \cdots + \emptyset_q \in_{t-q} . \tag{11.5}$$

11.5 Platform Development

We have analyzed the data extracted from the measurements and created a platform using PHP and MySQL database.

11.5.1 PHP Code of ARIMA Model

Below is the code for forecast ARIMA algorithm in php.

11.6 PHP Code of Linear Regression

Below is the code for linear regression in php.

```
echo "Prediction noise level with Forecast ARIMA algorithm </br>";
echo "________________________________________</br>";
echo "Prediction noise level at time" . $time . ":" . round($noise_forecast, 2) .
"dB</br>";
echo "Prediction noise level with vehicles = " . (round($noise_forecast, 2)
+ round(10 * log10($cars), 2)). "dB" . "</br>"; //ku $cars is the number of
vehicles
echo "Prediction noise level with vehicles & Wind = " . (round($noise_
forecast, 2) + round(10 * log10($cars), 2) + abs(round(20 * log10(1.2 * $v *
0.28/20), 2))) . "dB" . "</br>";}
echo "Prediction noise level with Linear Regression algorithm </br>";
echo "________________________________________</br>";
echo "Prediction noise level at time" . $x . "o'clock" . round($y, 2) . "dB" . "</
br>";
```

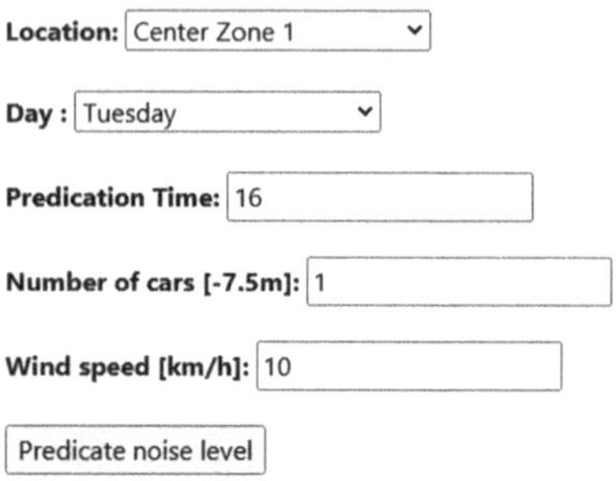

Fig. 11.7 Prediction form for the first example

```
echo "Prediction noise level with vehicles = " . round($y, 2) + round(10 *
log10($cars), 2) . "dB" . "</br>"; // $cars is the number of the vehicles
echo "Prediction noise level with vehicles & Wind = " . round($y, 2) +
round(10 * log10($cars), 2) + abs(round(20 * log10(1.2 * $v1 * 0.28/20),
2)) . "dB" . "</br>";}
```

We are going to take two examples for comparison.

First example (Fig. 11.7):

Noise level in location "**Center Zone 1**":

Selected location = Center Zone 1.
Selected in day = Tuesday.
The number of noise level measurements is: 15.
The number of cars is: 1 and generated noise is: 0 dB
The wind speed is: 10 km/h or 2.8 m/s and generated noise is: 15.49 dB.
Prediction noise level with Linear Regression algorithm is:
Prediction noise level at time 16 o'clock 75.18 dB.
Prediction noise level with vehicles = 75.18 dB.
Prediction noise level with vehicles and Wind = 90.67 dB.
Prediction noise level with Forecast ARIMA algorithm is:
Prediction noise level at time 16 o'clock 55.56 dB.
Prediction noise level with vehicles = 55.56 dB.
Prediction noise level with vehicles and Wind = 71.05 dB.

With its graphical representation shown in Fig. 11.8:
Below is the table of algorithms' performance.

Second example is shown in Fig. 11.9 with its graph in Fig. 11.10:

Selected location = Center Zone 2.
Selected day = Wednesday.
The number of noise level measurements is: 15.
The number of cars is: 3 and generated noise is: 4.77 dB.

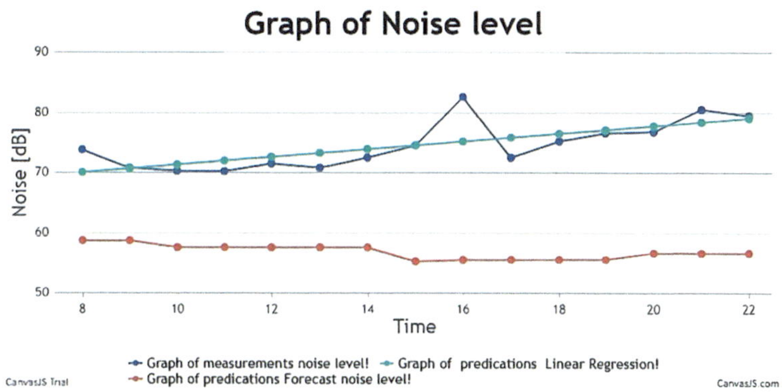

Fig. 11.8 Graph view for the first prediction example

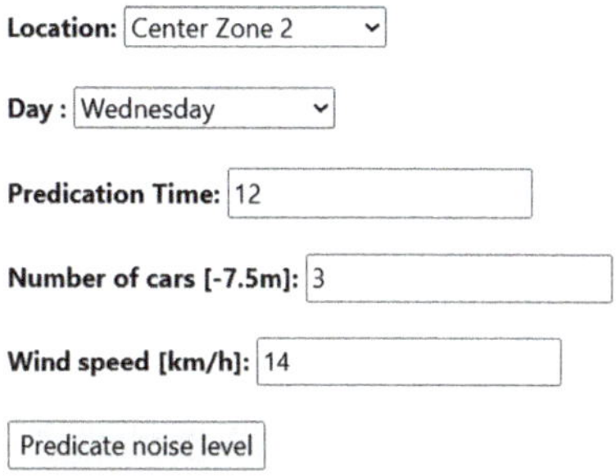

Fig. 11.9 Prediction form for the second example

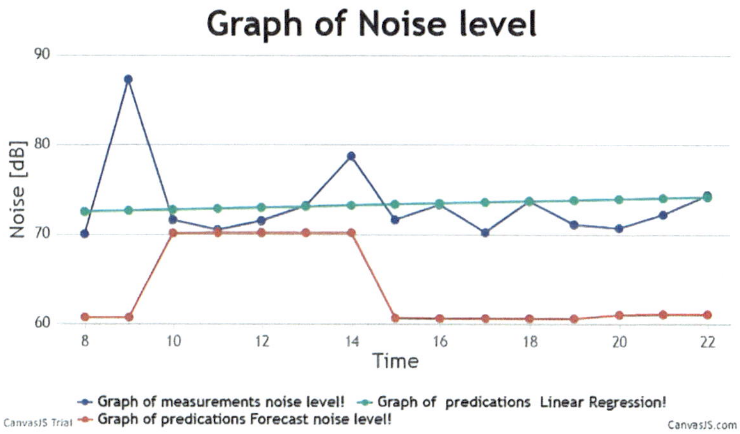

Fig. 11.10 Graph view for the second prediction example

The wind speed is: 14 km/h or 3.92 m/s and generated noise is: 12.57 dB.
Prediction noise level with Linear Regression algorithm.
Prediction noise level at time 12 o'clock 72.97 dB.
Prediction noise level with vehicles = 77.74 dB.
Prediction noise level with vehicles and Wind = 90.31 dB.

We have shown in Table 11.3, the algorithms performance for our second case.

11.7 Results and Discussion

In this research, we have presented two models of algorithms through which we have made noise prediction based on several factors, which we have mentioned above. We took a total of 4 measuring points in the city. From these points, we extracted a total of 412 measurements, which were placed on the platform and then it was possible to predict the noise at every hour, for every day of the week at each measuring point. In addition to these input data, we have placed the correlation coefficient on the platform to show the relationship between the variables. Also, we have presented the result in numerical and graphic form. In order to evaluate which algorithm gives more accurate results, we have integrated four performance evaluation factors into the platform, i.e., MAE, MSE, RMSE, and R^2. Based on Tables 11.2 and 11.3, we estimate that the ARIMA model provides more accurate noise prediction.

Table 11.2 Algorithms' performance for the first prediction example

Id	Performance	Linear regression algorithm	Forecast algorithm
1	Mean absolute error (MAE)	17.31	8.33
2	Mean squared error (MSE)	311.14	83.03
3	Root mean squared error (RMSE)	17.64	9.11
4	*R*-squared (R^2)	1.29	0.56

Table 11.3 Algorithms' performance for the second prediction example

Id	Performance	Linear regression algorithm	Forecast algorithm
1	Mean absolute error (MAE)	17.47	8.33
2	Mean squared error (MSE)	328.63	83.59
3	Root mean squared error (RMSE)	18.13	9.14
4	*R*-squared (R^2)	0.12	0.11

11.8 Conclusion

In conclusion, we estimate that we can reach some points after sampling, web application development, and comparison of prediction algorithms.

1. The online platform developed in PHP for noise prediction has been suitable and fulfilled our expectations in the integration of prediction algorithms.
2. Based on the noise measurements we have made, we estimate that in three measuring points, except in the yard of the university and high schools, the noise exceeds the standards both during the day and at night in comparison of XPS 31-133.
3. We conclude that the Arima model is more accurate in predicting noise based on the performance criteria of the algorithms.
4. We estimate that the integration of noise (dB) measurements, number of vehicles, and wind has increased the reliability of the prediction.

We also have some recommendations, which we have listed below.

1. We recommend that in addition to PHP, other programming languages can be tried which can develop online and offline applications for prediction.
2. To try other algorithms which help to improve the prediction, always comparing to their performances.
3. To add the other factors in correlation along with noise dB, number of vehicles and wind.
4. To take measures from the municipal or central government level to reduce traffic, to bring the city within the standard noise norms. Which would also have side effects in lower carbon dioxide emissions.

References

1. Tashman, L.J., Leach, M.L.: Automatic forecasting software: a survey and evaluation. Int. J. Forecast. **7**(2), 209–230 (1991). http://www.sciencedirect.com/science/article/pii/0169-2070(91)90055-Z
2. Steele, C.: A critical review of some traffic noise prediction models. ScienceDirect **62**(3), Mar 2001, 271–287 (2000). https://www.sciencedirect.com/science/article/abs/pii/S0003682X000030X
3. Panainte, M., Lazar, G., Nedeff, F., Nedeff, V., Tomozei, C.: Reducerea poluării fonice în mediul industrial, pp. 132–137. Alma Mater Bacau (2014)
4. Duncan, E., Cowan, J., Kaliski, K.: Community and regional noise mapping in the United States. Sound Vib. 14–17 (2007)
5. Bharadwaj, S., Dubey, R., Biswas, S., Zafar, M.I.: Different scales of urban traffic noise prediction. The International Archives of the Photogrammetry, vol. XLIII-B2-2020, pp. 1181–1188 (2020). https://isprs-archives.copernicus.org/articles/XLIII-B2-2020/1181/2020/isprs-archives-XLIII-B2-2020-1181-2020.pdf
6. Onkov, K., Tegos, G.: Forecasting algorithm adaptive automatically to time series length. In: IFIP International Conference on Artificial Intelligence Applications and Innovations, pp. 537–545 (2014). https://doi.org/10.1007/978-3-662-44654-6_53

7. Kapoor, P., Bedi, S.S.: Weather forecasting using sliding window algorithm. ISRN Signal Process. **2013**, 1–5, Article ID 156540 (2013). https://doi.org/10.1155/2013/156540
8. Talagala, T.S., Li, F., Kang, Y.: FFORMPP: feature-based forecast model performance prediction. Int. J. Forecast. **38**(3), 920–943 (2022)
9. Almasarweh, M., Wadi, S.A.: ARIMA model in predicting banking stock market data. Mod. Appl. Sci. **12**(11) (2018). https://doi.org/10.5539/mas.v12n11p309
10. Kavitha, S., Varuna, S., Ramya, R.: A comparative analysis on linear regression and support vector regression. In: Online International Conference on Green Engineering and Technologies (IC-GET), pp. 1–5 (2016)
11. Lim, H.-I.: A linear regression approach to modeling software. In: IEEE 43rd Annual Computer Software and Applications Conference, pp. 942–943 (2019)
12. Abdulazeez, A.M., Maulud, D.: A review on linear regression comprehensive in machine learning. J. Appl. Sci. Technol. Trends **1**(4), 140–147 (2020)
13. Wei, D., Xing, M., Zhang, J., Zhang, C., Cao, H.: Applied research of multiple linear regression in the information quantification of Chinese medicine bone-setting manipulation. In: IEEE International Conference on Bioinformatics and Biomedicine (BIBM), pp. 1912–1916 (2018)
14. Uni-T: https://meters.uni-trend.com, 23 Apr 2023 [Online]. Available: https://meters.uni-trend.com/product/ut353-ut353bt/#Specifications
15. France: eur-lex.europa, 2003 [Online]. Available: Official Journal L 212, 0049–0064, 22 Aug 2003. https://eur-lex.europa.eu/LexUriServ/LexUriServ.do?uri=CELEX:32003H0613:EN:HTML
16. Alan Stimac, S.S.R.N.: The equivalence of road traffic noise assessments using XPS 31-133 Interim, RLS90, CRTN, RMW2006 and Harmonoise. In: Euronoise (2009)
17. Kumari, S.Y.K.: Linear regression analysis study. J. Pract. Cardiovasc. Sci. **4**(1), 33 (2018)
18. Nau, R.: The Mathematical Structure of ARIMA Models. Fuqua School of Business, Duke University, Durham (2014)

Chapter 12
Economic Crimes in Current Era: World and Indian Context

Arjun Choudhary, Ashish Tripathi, Meenakshi Punia, Chetanya Kunndra, and Vikas Kumar Kharbas

Abstract With the advancement of technology in every aspect of life, its adaptability by various businesses has increased manifolds. Economic crime has also taken the shape of a techno-cyber-related crime in current times. This advancement in technological usage by various finance-related companies and the access to the internet by majority of the population without any guided training or manual has given attackers a niche point over the victims. It has been evident since the inception of technology that the first ones to commercialize the technology are criminals. In this research, it has tried to showcase the new variants of cyber economic crimes and their modus operandi. This technology is only the medium through which the miscreants have been duping victims with financial-related frauds. At the end of this research, its context with the Indian scenario is also discussed in detail.

12.1 Introduction

Economy of any country is termed as the backbone or strength of that nation itself. Since, with this the nation actually shows its production, consumption, or purchasing power and the term economy as defined in ancient Greek language which means management of a household, administration. Economy not only tells us about the management but also about the culture, history, societal organizations, political status,

A. Choudhary · C. Kunndra
Centre for Cyber Security, Sardar Patel University of Police, Security and Criminal Justice, Jodhpur, India

A. Tripathi
School of Computing Science and Engineering, Galgotias University, Greater Noida, India

M. Punia (✉)
Centre for Economic Crime and Counter-Corruption Studies Sardar, Patel University of Police, Security and Criminal Justice, Jodhpur, India
e-mail: meenakshilaw@policeuniversity.ac.in

V. K. Kharbas
Vivekanand Global University, Jaipur, India

P. Nanda et al. (eds.), *Intelligent System and Data Analysis*, Smart Innovation, Systems and Technologies 398, https://doi.org/10.1007/978-981-97-5200-3_12

legal structure, educational background, as well about the technological advancement of an entity. All the above-mentioned factors actually inform us about the content, context and set the limitations under which the economy of an individual, organization or of a nation shall govern or thrive. Alternatively, economy is not a stand-alone term or an activity but is an inter-related activity based on several parameters of an entity to make a transactional ecosystem and the main objective is to make this whole system either sustainable or thriving with various means. As the history became elongated and the society tried to become more laureate by utilizing the earlier mistakes as examples or case studies, economy shaped in various types. Currently the economy of a nation is compared with the help of Gross Domestic Product (GDP). GDP is an estimate of total produced goods and services in a country during a specified period. Through this means, the size of the economy of a nation is well defined in current times. The USA is currently heading the group of all nations with the highest GDP across the world. Currently India is termed as the fastest growing economy and fifth largest economy across the world as per the statistics made available by world bank prior to COVID-19. In other words, as we put the cloud computing, it is a technology which is efficiently utilizing the underlying resources to create internet-based micro-service, which in turn are turning into larger and much complex services. Sometimes, also working dynamically as and when user needs it across the globe.

Every sector has pros and cons; similarly, an economy which involves money also attracts lots of people for doing unlawful activities. In this direction after the First World War Charles Ponzi created a luring offer to Bostonian people for creating quick money. In this scheme, an investor can make \$1500 from \$1000 in just 45 days of time by purchasing the promissory note from his firm. He proclaimed to provide such huge profit in such a short time using arbitrage between the postal reply coupons in international rate differences. However, Ponzi was just repaying the earlier investor with the newer investor, which was then coined as "robbing Peter to pay Paul." This was initially being used by Banco Zarossi, but the scheme got famous due to Ponzi only, as he made millions out of it and remained under the radar for a long time. This is the single example which has been quoted here, but in today's scenario, a variety of economical related crime has been found across the globe. Be it against the firm, organization or a nation, there is no dearth of limitations or type of economic crimes (Fig. 12.1).

12.2 Cyber Economic Crime

Economic and financial crime are the activities done in digital world which can inflict losses to the any governmental, public funds. This also can be done under the social security schemes, research and scholarship schemes or any defense related schemes. Emergence of global economy and that too intertwined with the help of digitalization of all the finance-related activities. One cannot even imagine that in this era there might be an organization which is doing financial transaction in an offline or bookkeeping manner. This exponential growth in this direction has increased the

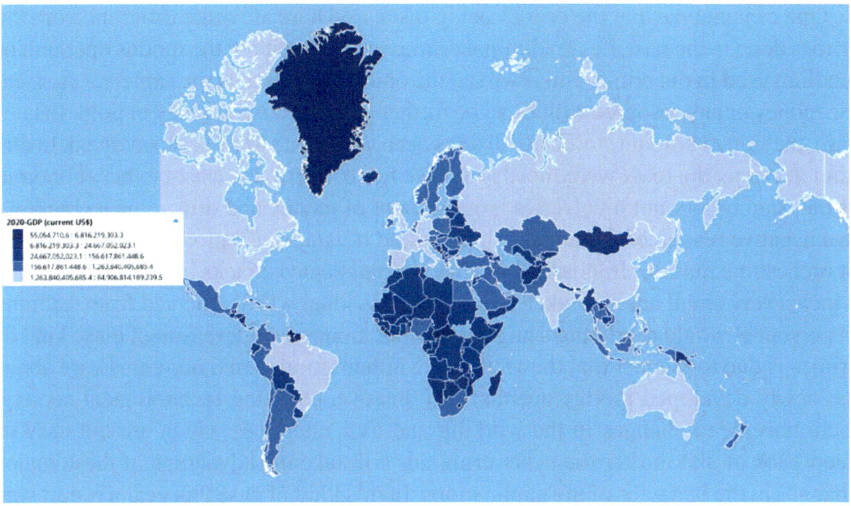

Fig. 12.1 GDP based on share of 2020 world total in percentage [1]

scope of and scale of the risk which an organization can face. There have been unprecedented manners through which cyber perpetrators have been siphoning the money from the organizations.

12.3 Characteristics of Cyber Economic Crime (CEC)

To understand the basic nature of cyber economic crime one has to understand the root cause of the crime, i.e., very basic idea of doing this kind of crime. This crime has been also termed as white-collar crime. As in the meaningful manner that this kind of crime is generally done and managed by a person of higher status or knowledge who understands then nitty gritty of the economic system. Earlier it was only the economic crime but when everything digitalized then the new coined term is cyber economic crime. Thus, it can be termed as white-collar forgery, since the involvement of person or a conglomerate ripping off the money covertly from top brass officials or through banking and financial organizations. When it comes to about laws related to these crimes, then it is surprisingly quite low. Since these CEC-related crimes are somehow being booked under normal civil or criminal procedures across the world and the countries who does have some Act or regulations related to IT or cyber are also not regularly updated as per the current trend. Since in current time, cryptocurrency related crime, stock manipulations, evading tax using tax heaven states, credit or debit card related frauds, intellectual property infringements or patent violations etc. are some of the crimes along with the very newly development of ransomwares [2].

One can analyze that the crimes being discussed here are quite different from the crimes done in the past, i.e., traditional crimes in terms of both the modus operandi or medium used in the crime. But if we see the objectives remains the same, i.e., to steal the money or any resource which can be further converted into money or gold. In later case, the sum or amount stolen had always been small with lots of risk involved. In this kind of crime, the tools were mostly of brute force nature. In case of cyber economic crime, sum or amount involved is large amount of money and utilize the technology in a negative resourceful manner [3]. This kind of attack also modifies the databases or orchestrate massive frauds by doing very sophisticated attacks like salami attack in which a very small amount taken from each individual when collected from millions of personnel would result into a huge amount of money. Major reason of these kind of crimes is due to the fact that the victims are either ignorant and quite unaware about the newly developed modus operandi. In this ever-evolving technological society, there have been changes in the working and user interfaces which are not easy to keep track of and mostly the cyber criminals will take the advantage of these major changes in the behavior of the applications. In this kind of case this year a report was published in one of the leading national daily newspapers in India, that the people in Chandigarh, India alone lost a huge amount of around Rs. 4 crores in a duration of 13 months [4].

12.3.1 Low Risk, High Profit

One of the characteristics of economic crimes that is extremely lucrative for criminals is the low risk associated with the crime and the associated high profits. Criminals have low chances of being prosecuted due to the sheer complexity of the crime, the rules associated with it, making it difficult for investigators to investigate the crime.

With the introduction of computers, even economic offenses have crossed international boundaries, a criminal can commit offenses in another country, again making it difficult for investigators to investigate and prosecute criminals due to jurisdictional clashes among various investigating agencies. In order to investigate such a crime, a collaborative international effort must be made. Owing to such complications while investigations, the chances of being caught reduce significantly.

Owing to such low risk and high profits, enable organized criminals to commit large-scale economic offenses.

12.3.2 Untraceable

In the wake of this digital era, when everything is online including the money which is kept with banking institutions. Embezzlement, money laundering and financial frauds are becoming easier to do on behalf of some other personnel. In India alone every year financial frauds are reaching in thousands of crores, data related to various banking institutions is mentioned in Table 12.1.

In Table 12.1, RBI has showcased that the financial frauds were on rise from 2018 to 2019, but in the year 2020 due to COVID-19 issue fraud amount decreased by 25% from 2019, whereas the numbers of frauds got decreased by 15% only. This somehow proves the hypothesis that most financial frauds are being done through physical means rather than online mode only. This also showcases that the banking sector in India is still to grow and attain maturity. Since in other nations, mostly the financial frauds are increasing through online means. Although there is a misconception among the researchers that using e-governance [5] accountability and transparency increase in the social system. In reality, fraudsters using low-income group personnel or proletarians are being used as money mules to create several accounts and identity proofs. These are now being used for financial or economic frauds. Thus, making this an untraceable crime, since the actual perpetrator will be doing the crime sitting somewhere other part of the world and the law enforcement agencies based on the findings will be trying to find out the mule which never understood the greater game. This way crime is on the rise in India currently.

12.4 Types of Economic Crime

Economic crimes are being shaped in various types in today's time. Based on research and findings most of them have been chalked out and discussed in the following sections.

12.4.1 Asset Misappropriation (Embezzlement)

Cressey [7] mentioned in year 1953 about the psychological evaluation of people at higher position having any issues related to the financial stability of their own and they do believe that they can get away with this kind of situation by secretly utilizing the money which has been entrusted with them by common public, trust or by an organization. This hypothesis stands out correct in today's scenario as well. In Indian context usually the public perception is that the politicians, public workers/officers be it from government or contractual staff, are the people who habitually indulge in this kind of activity [8].

- *Remedies*
  - Regular or random audits.
  - Inclusion of professional ethics in trainings.
  - Reward for the whistleblowers and privacy maintenance of those.

Table 12.1 Fraud cases—bank group-wise [6]

Bank group/Institution	2018–19		2019–20		2020–21	
	Number of frauds	Amount involved	Number of frauds	Amount involved	Number of frauds	Amount involved
1	2	3	4	5	6	7
Public sector banks	3704(54.5)	64,207(89.8)	4410(50.7)	148,224(79.9)	2903(39.4)	81,901(59.2)
Private sector banks	2149(31.6)	5809(8.1)	3065(35.2)	34,211(18.4)	3710(50.4)	46,335(33.5)
Foreign banks	762(11.2)	955(1.3)	1026(11.8)	972(0.5)	521(7.1)	3315(2.4)
Financial institutions	28(0.4)	553(0.8)	15(0.2)	2048(1.1)	25(0.3)	6839(4.9)
Small finance banks	115(1.7)	8(0.0)	147(1.7)	11(0.0)	114(1.6)	30(0.0)
Payment banks	39(0.6)	2(0.0)	38(0.4)	2(0.0)	88(1.2)	2(0.0)
Local area banks	1(0.0)	0.02(0.0)	2(0.0)	0.43(0.0)	2(0.0)	0(0.0)
Total	6798(100.0)	71,534(100.0)	8703(100.0)	185,468(100.0)	7363(100.0)	138,422(100.0)

  - Use of Artificial Intelligence and Machine Learning in payment and decision-making systems as well in the audits.
  - Strong punishable laws shall be made in these kinds of crimes.

12.4.2 Investment Fraud

In this kind of fraud, the individual is fed with falsely provided information about a bogus or low value share and after the investment is being done at higher prices, company shares are being bought out by the same firm or individual who were actually forcing the individual to invest. A prime example of this fraud is the 1992 Securities scam, perpetrated by the infamous Harshad Mehta [9], it is estimated that this scam amounted to approximately INR 5000 crores, making it one of the biggest investment fraud scams.

- *Myths*
  - Guaranteed returns.
  - It is sounds too good to be true.
- *Reasons*
  - Everyone is buying it.
  - Pressure to send the money Right Away.
  - Obligation from peer pressure.
- *Remedies*
  - Ask questions.
  - Before investing do the research related to the stock or firm.
  - Know the salesperson history or previous record.
  - Be wary of unsolicited offers.
  - Beware of "halo" effects.

12.4.3 Payment-Order Fraud

In this form of fraud attackers impersonate themselves as a high-ranking official such as the CEO and try to fool the accountings or HR departments into making payments to the attacker, this form of fraud is also called the CEO Fraud or more commonly known as Business email compromise (BEC). In most of the cases, the attacker targets an international organization, again using jurisdictional differences to get away with the fraud.

- *Reasons*
  - Lack of knowledge among employee to identify phishing emails.
  - Insufficient contact among upper management and lower management.
- *Remedies*
  - Implementation of strict guidelines, requiring approval from upper management while issuing finances.
  - Confirm with the executive whether the money is to be transferred.
  - Train employees to identify phishing emails and implement a security division that will handle such cases.

12.4.4 Subsidy Fraud

In this type of fraud, criminals submit false applications for tenders, subsidies, schemes. Typically, criminals forge false documents, declarations, reports and even invoices to get the amount associated with the subsidy. Prime example of subsidy fraud is the PM Awas Yojana and Dewan Housing Finance Limited scam [10], which amounted to a fraud of INR 1880 crores in interest subsidy under the PMYA scheme.

- *Reasons*
  - Improper checks to validate records.
  - Unwillingness of employees to initiate proper checks.
  - Collusion between perpetrators and employees.
- *Remedies*
  - Implementing a strict application validation framework.
  - Implementing a multi-level committee to oversee the application process from start to end.

12.4.5 Insurance Fraud

In this form of fraud, the primary victims are organizations that provide insurance related services. Typically, a criminal will try to claim an insurance amount by falsifying insurance records, such as a death certificate, health bills etc.

- *Reasons*
  - Collusion between multiple parties such as certifying authorities, employees, perpetrators.
  - Improper verification of documents before issuing insurance.

- *Remedies*
  - Implementation of multi-level verification of documents before issuing of insurance.

12.4.6 Banking Fraud

As opposed to bank robbery banking fraud is considered a white-collar crime. In this form of economic fraud perpetrators use fraudulent means, such as forging documents to defraud financial institutions of money. In some cases, the criminals abscond with the money as seen in the case of PNB Scam [11] and SBI Scam [12].

- *Reasons*
  - Low risk and high score philosophy.
  - Long legal remedial process.
  - Insider personnel become greedy for smaller bribes.
- *Remedies*
  - Random and regular auditing and background checks for all personnel in banking institutions.
  - Increase the use of Artificial Intelligence, Machine Learning and Business Intelligence.

12.4.7 Procurement Fraud

In this very fraud, mainly the information plays the important role which is bought by either bribing the employees of counter company or try to do industrial espionage using much sophisticated manners like BEC mentioned earlier. This information will eventually will help the competitors in winning bid in order to get orders for any public works. This kind of information manipulation is very much evident in the field of public works department, construction ministry, energy sectors or information technology fields [13] (Fig. 12.2).

12.4.8 Marketing Fraud

In this kind of fraud mostly variety of communication channels like SMS, mobile calls, WhatsApp messaging, internet advertisements in social media platforms, mails using subscription mechanisms are utilized by the criminals. The main objective is to showcase the victim with attractive offers which cannot be turned down especially if the attacker does have any pre-identified information related to any shopping done

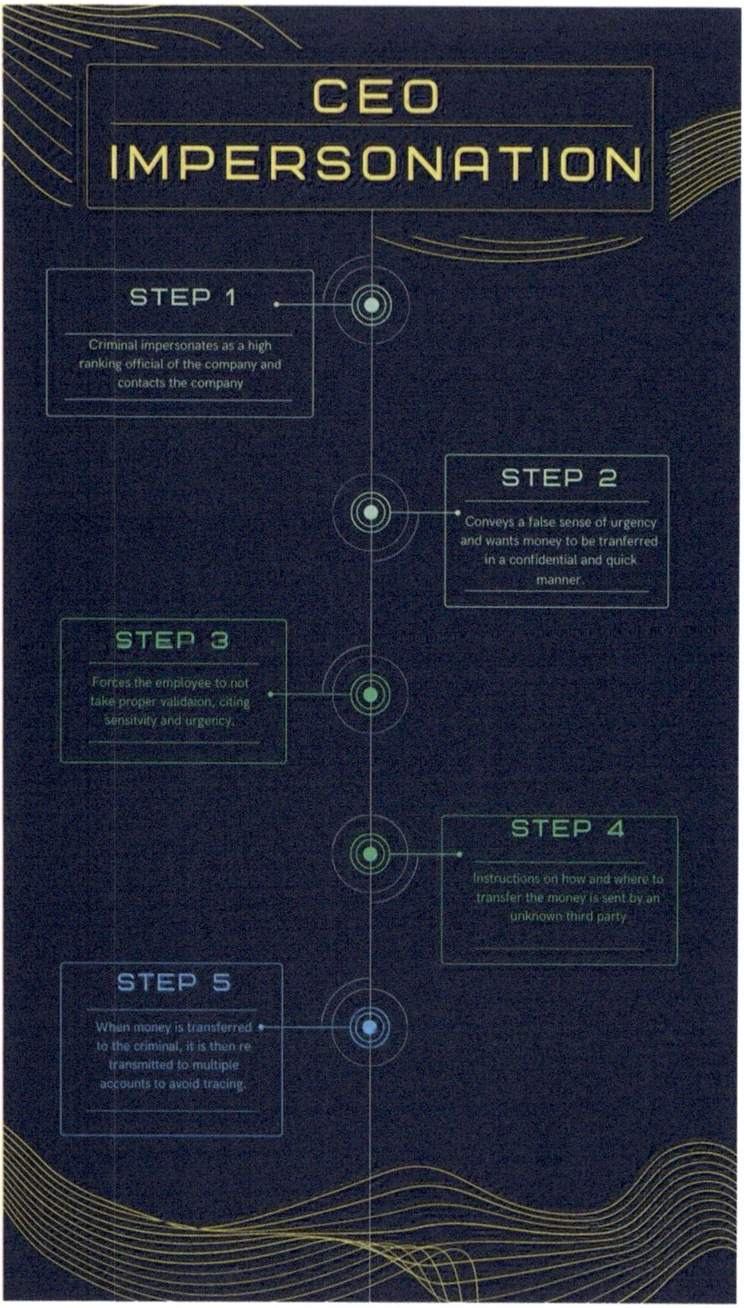

Fig. 12.2 Cyber fraud showcasing how a person can impersonate a CEO of one company can dupe other people with money

by the victim will be a niche point over the common attacks. Offers will always very from the educational scholarships, share market tips, very low price of a product of a popular brand or a Ponzi kind of scheme with much lucrative returns. In one of a kind of this case in 2016 many urban personnel were duped with mobile tower installation at their premises. Since rent of a mobile tower is a persistent money without any serious efforts, thus giving criminals a bait for the victims. Recently government of India has posted that in 2022–23 it will be investing more than 25K crore INR for mobile tower installation in order to equip country with 5G technology. In this kind of crime, people were contacted with TRAI looking SMS and were asked with minimal documental requirements initially and further would ask a token money of first month of rent. In this kind of attack, now TRAI had to publish an advisory for common people using regular SMS or through news channels or newspapers.

12.4.9 Customer Care Fraud

The modus operandi of a customer care fraud is a bit different, criminals use SEO (search engine optimization) and other techniques to display fraudulent customer care numbers for legitimate service, victims thinking that they are calling a legitimate service are led to the criminal's fraudulent customer care center, where agents dupe the victim of their money.

- *Reason*
  - Trust in the first result that shows up on search engines.
  - Less knowledge about these kings of scams.
- *Remedies*
  - Always check the customer care number before calling.
  - Use the numbers provided on the official website only.
  - Never share personal details such as card numbers, CVV to a customer care representative.

12.4.10 Benefit Fraud

In this kind of fraud mostly the personnel attached with any social security scheme or any scholarship scheme are targeted and duped with money. Any students from low-income family who want to go for higher study in other country are also soft targets for the attackers. Basic mitigation in this kind of case is to cross verify from other means about any offer being provided by the attackers.

12.4.11 Forgery of Money and Means of Payment

In this technological era cash soon will be forgone but will not be properly forgotten though usage of electronic means of payment. In recent technological advancements, criminals can copy the bank notes with much precision and accuracy. Since the raw material being used for bank notes printing is available easily with the help of dark web using sophisticated software like TOR or I2P.

For combating against any counterfeit money minting cartels, worldwide agencies have been working tirelessly and doing research to detect this kind of forgery in bank notes.

In fight against any payment-related fraud recently Ministry of Home Affairs (MHA) under Government of India has established a new helpline for any bank or NBFC related financial frauds. Initially, the scheme was launched in beta stage in some of the states, and after successful testing and results, it has now been implemented pan India with 1930 as helpline number. Similarly, Europol's European Cybercrime Centre (EC3), using its Joint Cybercrime Action Taskforce (J-CAT), several high-profile cybercrime operations and investigations, such as Operation Imperium, has been handled in a much-targeted manner, which targeted an organized crime network active in payment fraud.

In 2016 Internet Organized Crime Threat Assessment (IOCTA), Europol recom mended that EU law enforcement shall focus on the following areas:

- ATM malwares and skimming devices usage.
- E-commerce frauds with special emphasis on transport (airlines), retail and housing sectors.
- The acquisition and trading of compromised financial users' data and credentials.

On similar grounds, Europol has also worked-up its activity for *CEO fraud* cases in Europe as described in Fig. 12.1. In this kind of case, fraudsters will pose as the CEO or a member of the senior management team in a company and trick the employee at the company into wiring funds to them or to some other agency on the behest of this senior person.

12.4.12 Money Laundering

This kind of economic crime has been existing for a much longer time but with the advancement of technology and the cost associated with each transaction getting lower. The cost of Money laundering has been lowered due to high-speed transactions and above-mentioned reasons. Due to sheer complex nature of technology which provides a platform for any kind of financial transaction provides an easier medium to do the placement, layering and integration. This kind of frauds can be helpful for various organized crimes like drugs selling, funding terrorism, human trafficking or illegal weapon trading in darknet. Money laundering is not only being used by basic financial transactions but also from various alternate delivery channels. With the introduction of crypto currencies and other internet-based currencies like being used in various games where there has been intricate system working like a barter system, money laundering kind acts are difficult to identify. It is something like finding a needle in a haystack (Fig. 12.3).

While accessing the scale of money being washed with this kind of act, it is envisaged that it is next to impossible to successfully identify the act and measure its sheer strength. Recently, United Nations Office on Drugs and Crime (UNODC) predicted an estimation that it has been growing between 2 and 5% of global GDP for each year which is coming between EUR 715 billion or 1.87 trillion each year. As an investigator if we try to find that most of the organized crime shares a common denominator i.e., the financial motive. The Organized crime cartels eventually boost their assets and then inject the assets into the legal economy through various money laundering schemes possible. If with any possible trick or using insider mechanism this whole process is traced, then the whole network can be traced back.

12.4.13 Excise Fraud

The European Union levies excise duty on production or on import of commodities such as alcohol, cigarettes and fuel.

Organized crime groups employ various illegitimate tactics to avoid such excise duties. These groups are fueled by the high profit margins incurred on selling both genuine and counterfeit excise goods. Most of the time such goods are sold at a lower price as compared to their licit market value.

Excise tax fraud is fueled by legislative differences and varying excise tax rates applied by different states.

The main sectors of excise fraud in Europe include but are not limited to

- The illegal importation of excised goods via the means of smuggling.
- The illegitimate manufacturing of excised goods.
- 'Diversion', the act of moving excised goods without paying the associated excise duty.

Serious and Organized Crime Threat Assessment (SOCTA) 2017 identified frauds associated with fuels as a growing phenomenon. These categories of frauds typically involve base-oil fraud, also known as "designer fuel" fraud and fuel laundering. Unlike other frauds, these frauds require significant expertise. This required amount of expertise can only be expected from trained chemists or people having similar professions.

Excise fraud causes deprivation of significant revenues. This revenue otherwise could be used to fund critical public services such as healthcare, education and public welfare. Excise fraud can also act as a crime enabler by acting as a finance source for other organized crime groups. This poses a serious threat to the nation's internal security. In 2010 a UN Security Council investigative body linked excise fraud with

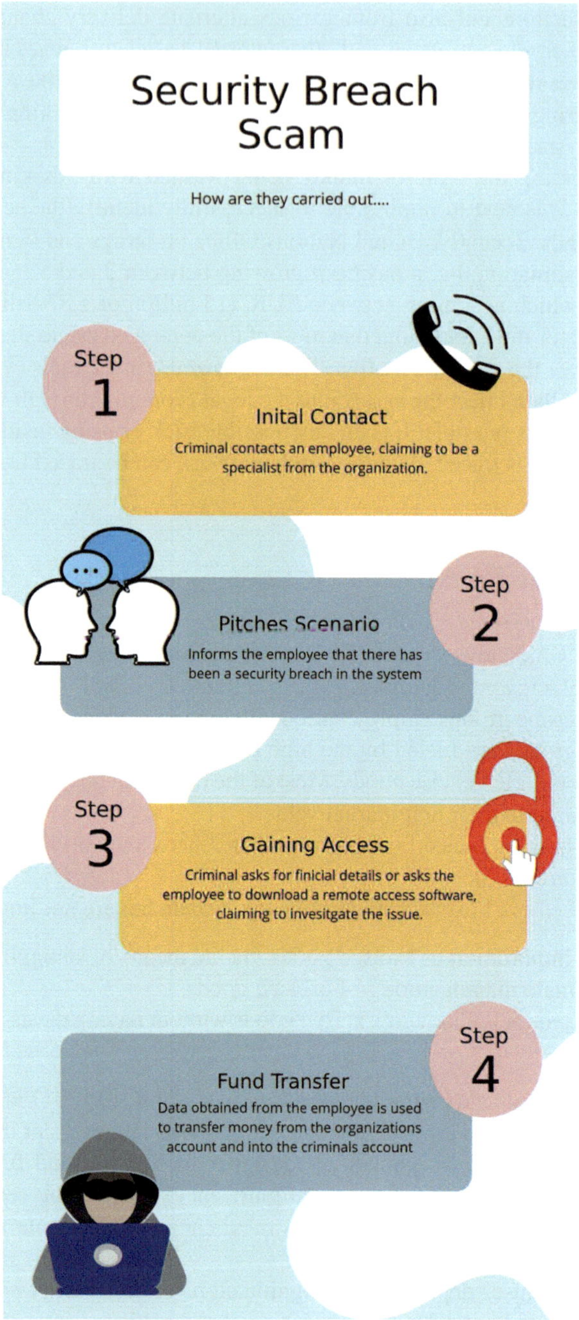

Fig. 12.3 Showcasing the modus operandi of a cybercriminal to create a scene of security breach

terror funding. The investigative body reported that millions of dollars were being made in revenue by perpetrators by trading in illicit tobacco. The said revenue was then being used to fund terror activities of radical groups such as the Al-Qaeda, Taliban or ISIS. The revenue from such illicit trades was also being used to fund other terrorist organizations and organized crime groups.

According to a report by KPMG, the EU consumed a total of 48 billion illicit cigarettes in 2016. These 48 billion illicit cigarettes deprived EU governments of a total of 10.2 billion euros in taxes. KPMG, in their report also estimated that illicit cigarettes comprise nearly 9% of all the cigarettes consumed in the EU. Nine% may seem like a small number, but in retrospect, it amounts to a huge amount of evaded taxes. Another research conducted by the Trans crime research center in 2015 estimated that the illegal cigarette market in the EU yielded nearly 7.8–10.5 billion euros in criminal proceeds. This gross amount is comparable to the revenues generated by cocaine and heroin markets. This makes these illicit trades a huge problem for any nation state.

12.5 Means by Which Economic Crime Was Detected Originally

- Internal tipoff: Employees working inside an organization.
- External tipoff: Tip received either from the third party auditing team or third party vendor who works in some of the projects.
- By Accident: Many a time the crime gets detected by accidental means like it happened in the case of Vyapam scam [14], when some of the dummy candidates were arrested and then after further investigation a huge scam was unearthed.
- Internal Audit: Internal audits, give indications about discrepancies in the system, leading to full-blown investigations that often end up uncovering economic crimes.
- External Audit: Audits carried out by third party agencies, either on a regular basis or on demand lead to uncovering of crimes.
- Corporate Security: Policies and strategies implemented to protect an organization from any form of harm are defined under corporate security. These policies flag any irregularities in the system that can bring harm to the organization.
- Law Enforcement Agencies: LEA sometimes uncover economic offenses, while investigating other cases or on complaints from individuals.
- Whistle Blower System: Whistle blowing is an act to expose information about activities that can be deemed illegal, fraud or against the community, committed by organization, public, private and even government agencies. Whistle blower systems are such programs that enable people to report malicious deeds performed by their organizations without the fear of being reprimanded and pushbacks.

- Change of personnel duties: In several organizations, various personnel do the duties assigned to them throughout their working tenure, which is considered a bad practice in the industry. Eventually when a newly appointed employee joins the organization in the vacant position, sometimes the economic fraud is uncovered.
- Regulatory authority: Authorities that provide regulatory oversight require an organization to work within certain codes of conduct, any organization violating the said codes face investigations and penalties. Many times such regulatory bodies uncover malpractice within organizations, leading to investigations, which even uncover economic crimes.
- Changes in working process.

12.6 Economic Crime in Indian Context

India is a Union of States and is governed by a written Constitution, which came into force on November 26, 1949. India consists of 28 states and 7 union territories. As per constitution of India, states or center can create law as per the union list, state list or concurrent list. In Union list, issues pertaining to the finance or banking can be taken and required laws can be created. Similarly, any policy or law related agricultural, education and especially Police and Public Order comes under state list. In concurrent list criminal laws and criminal procedure has been put, due to which the situation is quite tricky in terms of nabbing and putting the cyber economic criminals behind the bars. These loopholes have always been milked by the criminals keeping banking regulatory body, law enforcement agencies and legislation bodies into the abyss. Since the criminal will do the activity in one state or a group of states while sitting in some other location, in response to investigate and solve this economic criminal activity the police of one state will leave for other state. This will eventually be dealt by the law in the other state in case of any eventuality.

With this government of India has also implemented central investigation agencies like Enforcement Directorate, Central Bureau of Investigation etc. These organizations will take up certain cases which are of national level and required any central level interventions to work up on any pressing issues or cases.

12.7 Conclusion

Economic fraud arises out of the greed of individual(s), who feel the necessity to hoard money over morals. One of the most enabling factors of economic frauds are the low chances of being prosecuted, this is due to the use of technology, jurisdictional differences and complexity. In the post COVID-19 era, numerous new economic crimes are taking shape every day. Authors like Liu et al. [15] have tried to uncover these using technological analysis by implementing Intention-aware Heterogeneous Graph Attention Networks for the identification of fraudulent transactions. Similarly,

Zhu et al. [16] have also tried to showcase their intentional financial fraud detection research work that ponders over the use of the new extremely sophisticated deep learning-based algorithms and its comparisons with similar techniques.

References

1. The World Bank: World Development Indicators. The World Bank DataBank (2021, Oct 28). Retrieved 21 Nov 2021, from https://data.worldbank.org/indicator/NY.GDP.MKTP.CD
2. Meland, P.H., Bayoumy, Y.F.F., Sindre, G.: The ransomware-as-a-service economy within the darknet. Comput. Secur. **92**, 101762 (2020). Miyachi, C.: What is "Cloud"? It is time to update the NIST definition? IEEE Cloud Comput. **5**(03), 6–11 (2018)
3. Karim, Y., Hasan, R.: Taming the digital bandits: an analysis of digital bank heists and a system for detecting fake messages in electronic funds transfer. In: National Cyber Summit, pp. 193–210. Springer, Cham (2020, June)
4. The Times of India: Cyber cheats took Rs 4 crore in 13 months, reckon cops: Chandigarh News—Times of India. The Times of India (2021, Aug 23). Retrieved 20 Nov 2021. Source: https://timesofindia.indiatimes.com/city/chandigarh/cyber-cheats-took-rs-4-crore-in-13-months-reckon-cops/articleshow/85550882.cms
5. Pina, V., Torres, L., Acerete, B.: Are ICTs promoting government accountability? A comparative analysis of e-governance developments in 19 OECD countries. Crit. Perspect. Account. **18**(5), 583–602 (2007)
6. Reserve Bank of India Annual Report 2020–21. Reserve Bank of India (2021, May 24). Retrieved 25 Nov 2021. Source: https://rbidocs.rbi.org.in/rdocs/AnnualReport/PDFs/0RBIAR202021_F49F9833694E84C16AAD01BE48F53F6A2.PDF
7. Cressey, D.R.: Other People's Money; A Study of the Social Psychology of Embezzlement (1953)
8. Sukhtankar, S., Vaishnav, M.: Corruption in India: bridging research evidence and policy options. In: India Policy Forum, vol. 11, no. 1, pp. 193–276. National Council of Applied Economic Research, Delhi, India (2015, July)
9. Das, K.: Story of Harshad Mehta, financial frauds and the lessons that were never learned (2020, Dec 03). Retrieved from https://www.indiatoday.in/business/story/harshad-mehta-securities-scam-india-legacy-of-bank-fraud-1733374-2020-10-20
10. Service, T.N.: PM Awas Yojana: DHFL faked a/cs to get Rs 1,880-cr subsidy (n.d.). Source: https://www.tribuneindia.com/news/business/pm-awas-yojana-dhfl-faked-a-cs-to-get-rs-1-880-cr-subsidy-229995. Last accessed 22 Dec 2022
11. PNB Scam: Properties worth Rs 500 cr of Nirav Modi firms restored to bank (2021, Aug 27). Retrieved 25 Nov 2021. Source: https://economictimes.indiatimes.com/news/india/pnb-scam-properties-worth-rs-500-cr-of-nirav-modi-firms-restored-to-bank/articleshow/85691573.cms
12. India, P.T.: SBI-led consortium gets over Rs 5,800 crore in Vijay Mallya case (2021, June 25). Retrieved 25 Nov 2021, from https://www.ndtv.com/india-news/sbi-led-consortium-gets-over-rs-5-800-crore-in-vijay-mallya-case-2472693
13. PricewaterhouseCoopers: PwC's Global economic crime and fraud survey 2020 (n.d.). Retrieved 25 Nov 2021. Source: https://www.pwc.com/gx/en/services/forensics/economic-crime-survey.html
14. Mullick, J.: The Vyapam mess, and why it matters. Curr. Med. Res. Pract. **5**(5), 245 247 (2015)
15. Liu, C., Sun, L., Ao, X., Feng, J., He, Q., Yang, H.: Intention-aware heterogeneous graph attention networks for fraud transactions detection. In: Proceedings of the 27th ACM SIGKDD Conference on Knowledge Discovery & Data Mining, pp. 3280–3288 (2021, Aug)
16. Zhu, X., Ao, X., Qin, Z., Chang, Y., Liu, Y., He, Q., Li, J.: Intelligent financial fraud detection practices in post-pandemic era: a survey. Innovation 100176 (2021)

Chapter 13
Analyzing the Citation Networks Using Community Detection Approaches: A Review

R. Kiruthika and N. Radha

Abstract Community detection (CD) in a citation network is necessary for detecting the relationships and patterns between scientific articles which results to discover the knowledge and a better understanding of significant research works. Citation network analysis is a field of research in which it demonstrates how the academic articles and their citations are interrelated, most influenced and illustrating the trends of research. CD is an important element for studying citation networks and it provides a way to understand the relationships between different elements of the network. The primary goal of CD in citation networks is to identify groups of closely related publications that have common topics or subjects. This may lead to a better understanding of the topic. The goal of this study is to find the communities based on their citation patterns of linked sources in citation networks to understand the implicit structure and connections between the academic papers using different CD algorithms.

13.1 Introduction

Community detection (CD) is an essential approach in network research that involves discovering groups of nodes that have strong connections inside the group but have less connection with nodes outside the community. This domain has its origins which has been derived from graph theory and has been used to a wide range of fields such as social network analyses, bibliometrics, and computational sciences, etc.,. This also helps us for understanding the complex structure of connected systems. It allows researchers and analysts to uncovered deeper network structures and patterns, which can provide important insights into the organization and dynamics of complex networks systems [1] like citation networks.

R. Kiruthika (✉)
Department of Computer Science, PSGR Krishnammal College for Women, Coimbatore, India
e-mail: kirthikamole@gmail.com

N. Radha
Department of Data Analytics (PG), PSGR Krishnammal College for Women, Coimbatore, India
e-mail: radha@psgrkcw.ac.in

P. Nanda et al. (eds.), *Intelligent System and Data Analysis*, Smart Innovation, Systems and Technologies 398, https://doi.org/10.1007/978-981-97-5200-3_13

A key challenge in real-time network analysis is to extract and evaluate the data from citation networks. Citation Network Analysis (CNA) provides the best solution for this dilemma. CNA is a collection of methods and methodologies for analyzing exactly how the research publications are linked through references, enabling us to increase a better perception of how information is exchanged and impacted within the world of the academic community. The citation network is partitioned into clusters, which results in a formation of communities that retrieve similar research articles, disclosing the major links and interactions across the Citation network [2].

This review article on identifying communities using citation networks undergoes in-depth research which focuses on the main findings, methodologies [3], performance evaluation metrics, and applications CD and also how it may be used to identify research trends and recognize significant works. This enhances the organization and discovery of information in academic literature and provides valuable knowledge into the various disciplines. This study provides a better perceptive of the importance CNA in the current academic landscape and also for future research.

13.2 Literature Review

Multiple studies have been carried out on the problem of CD [4] working with complex network systems. This exhaustive review of the literature seeks to give an in-depth study of studies on citation networks applying to CD. Community identification across reference networks is critical for identifying relevant trends and patterns in research articles. This study examines the methodologies, algorithms, evaluation criteria, and applications used in the subject of citation networks and community detection by evaluating a variety of publications. This provides useful analyses on the significance of the various disciplines, identifies gaps, and proposes future study possibilities. Most of the preceding work has been concentrated on enhancing CD in ambiguous undirected networks; current research is mostly focused on directed networks. The following review provides a complete summary of the most important research findings on this dilemma.

Gao et al. [3] conducted investigation on a D-SCORE metric used for identifying groups in statistical citation networks. The leading four statistics publications citation networks from 2001 to 2018 were used as datasets in this paper. Fluid Communities (FluidC), the Clauset–Newman–Moore algorithm, and SCORE are three important algorithms applied to this study. This research got an outcomes that group includes a total of four communities and 11 sub-communities, with tags such as "High-dimensional Model," "Variable Selection," and "Covariance Matrix Analysis."

Chen et al. [5] undertaken a research on a certain issue considering the complexity of real-world networks and its challenging to use standard community detection algorithms to reliably and quickly identify community structure in large-scale networks. This work used the compact sample of 272 the edges was retrieved from DBLP for vector representation. DeepWalk network embedding method and the variational

Bayesian Gaussian mixture model (VBGMM) were selected as a clustering algorithm as implementation in this paper. This results that detection of 11 communities with respect to the title information for the community vertex.

Rahul et al. [6] carried out a study on an innovative methodology for identifying neighborhoods in social networks that makes use of overlapping communities in a social network graph. This study developed a method for converting a social network graph into a more compact collection of characteristics. The system is adaptable when it comes to examining a node's surroundings and can detect overlapping communities. It distinguishes between two types of node neighborhoods depending on distance and purpose, and it contains both local and global graph representations. The technique was validated by running it over huge social network datasets and seeing the resulting clusters. These clusters represent the many parameters into which the communities were classified. The more clusters there are, the more complex the communities separated within the Blog Catalog dataset.

Rajita et al. [7] investigates on the two frequently used strategies for detecting communities in large-scale social networks. The main aim of this study was to understand the concepts better by underpinning CD approaches and picking the best one for their practical applications. The efficiency of the algorithm was assessed using both real-world and simulated data like Zachary Karate club and Enron email datasets. The obtained metrics give insight into the suitability and scalability of each technique for datasets of varied sizes. The study utilizes comparative analysis, graphical modeling, and the production of random graphs. The Louvain method takes around 0.14 s on average, while the LP algorithm takes 0.37 s. Community Count, Modularity, and Clustering Coefficient are evaluation measures used in this paper. The prediction has been observed that Label Propagation works well with compact datasets and the Louvain method seems more effective when dealing with large datasets.

Fiallos et al. [8] conducted an experiment on the article that makes use of data from the bibliographic database SCOPUS, which contains 4552 scholarly papers from 29 distinct study fields in Ecuador. The data gathered comprises the "Scopus identifier," "Abstract," "Title," "Authors," "Research area," "University name," and "Number of citations." This data was retrieved by querying the Scopus site using the Elsevier developer API. A novel technique has been proposed to identify the key research topics in Ecuador. The technique is based on Social Network Analysis (SNA) and Natural Language Processing (NLP), which were implemented to a shape file including 4552 scientific publications relevant to study fields of 29 in Ecuador. According to the findings, the most collaborative research disciplines in Ecuador are environmental sciences, engineering, and medicine. The most popular themes in these fields are biodiversity, water resources, and public health. The study suggests that this technique can assist governments in ensuring that scientific investigation and study institutions are associated with the priority research topic with respect to country.

Yudhoatmojo et al. [9] accomplished a challenge using LinkRank technique in which used to find communities within the citation network of a sample batch of DBLP data to detect 10,442 communities from 188,514 nodes in the citation network.

Wang et al. [10] commenced a procedure for CD Based on Graph Neural Network on heterogeneous networks consists of Enzymes, Github, Proteins Cora, Citeseer, and PubMed datasets. This paper applied the Heterogeneous Graph Neural Network (HetGNN) algorithm for this model. HetGNN achieves higher modularity scores and lower running times compared to other methods. This experiment results that Cora, Citeseer, and PubMed datasets achieved 86.5%, 75.6%, and 82.6%, respectively.

With respect to the preceding reviewed literature, several community discovering strategies were suggested for all different kinds of networks. The inclusive approaches for these studies were reviewed and the observed general methodologies were discussed below. The above literature survey comprises identified the problem of CD in a citation network. The General approach consists of eliminating unnecessary or redundant documents, building a graph representation using citation links, collecting and preprocessing the relevant academic paper dataset. The next step was to split the data into three sections for model training as train, validation, and test. The main goal of this experimental study is to employ a best CD algorithm like modularity-based algorithms, Leiden, Infomap, Louvain Algorithm and Graph Convolutional Networks (GCN), etc., as a model to find communities inside the citation network. By applying the best suitable CD algorithm to find the community structure and developing the model for detecting communities. The model was detected the communities from the networks that are visualized based on their interconnections. The results are evaluated, interpreted, and analyzed using appropriate metrics. The final step is to discuss the findings and results are presented and visualized. Overviews of the experimental comparison study for reviewed articles are compared in Table 13.1.

13.3 Community Detection in Citation Network

Community identification is categorizing vertices in a complex system into coherent communities or clusters based on their strong connections inside the group and weaker interconnections outside the group. This fundamental idea in graph theory and network analysis tries to expose underlying patterns and structures in complex networks, providing important insights in domains such as biology, computer science, and social network research. Various techniques are used to detect these communities, allowing researchers and analysts to gain a better knowledge of the structure and dynamics of linked systems. CD in citation networks is a significant area of investigation due to its practical applications which has obtained a lot of attention in recent years. As the volume of research in various fields continues to increase, community detection can help researchers make sense of the complex network of relationships between authors, papers, and fields of study. Citation networks are depictions of the relation between academic papers based on references. CD encompasses detecting cluster of papers within these networks that share similar characteristics, such as topic, subject area, or methodological approaches. An outline for the process of CD [4] is shown in Fig. 13.1.

Table 13.1 Summary of experimental comparison for existing works

Author	Objective	Datasets	Algorithm	Results
Gao et al. [3]	D-SCORE is a metric used for identifying groups in statistical citation networks	The leading four statistics journals citation network from 2001 to 2018	FluidC, Clauset–Newman–Moore and SCORE algorithms	Four communities and 11 sub-communities
Chen et al. [5]	DeepWalk approaches for recognizing community structure in large-scale networks	DBLP of 272 the edges in vector representation	DeepWalk network embedding method and VBGMM	Based on the title information for the community vertex, to identify 11 communities
Rahul et al. [6]	A technique for locating neighborhoods in social networks that makes use of intersecting groups in an extensive system the network	Blog catalog dataset with 10,312 nodes and 333,983 edges	BFS and DFS using random walks on the network	The combined effort of overlap node community and clustering is scalable and resistant to disturbances
Rajita et al. [7]	This research compares the effectiveness of CD methods using actual data and artificially generated data	Real-world and synthetic data like Zachary Karate club and Enron email	Label propagation (LP) and Louvain algorithm	Louvain method took 0.14 and LP algorithm took 0.37 s
Fiallos et al. [8]	An approach based on scientific collaboration evidenced by publications, a suggested technique utilizing SNA and NLP	Bibliographic database SCOPUS which contains 4552 scholarly papers from 29 distinct study fields in Ecuador	Social network analysis and natural language processing	The modularity of Ecuadorian scientific publications revealed a network is 0.85

(continued)

Table 13.1 (continued)

Author	Objective	Datasets	Algorithm	Results
Yudhoatmojo et al. [9]	The LinkRank technique was used to find communities within the citation network of a sample batch of DBLP data	DBLP data sample Set	LinkRank algorithm	Detect 10,442 communities
Wang et al. [10]	Community detection based on complex networks specifically in heterogeneous networks	Enzymes Github Proteins cora, Citeseer, and PubMed datasets	HetGNN method	Cora, Citeseer, and PubMed datasets achieved 86.5%, 75.6%, and 82.6%, respectively

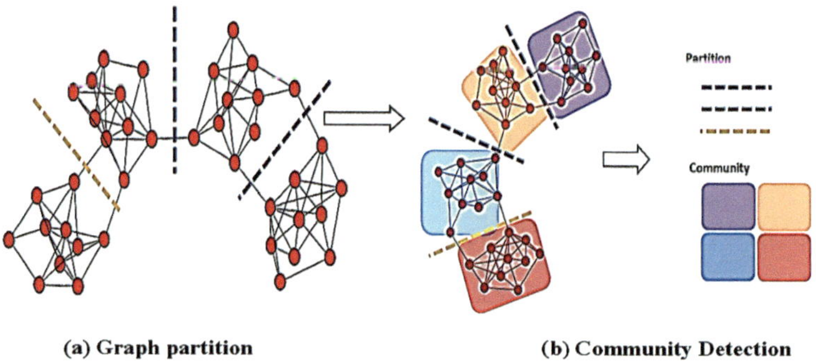

Fig. 13.1 Outline of community detection process

One of the significant benefits of CD in citation networks is the ability to detect dominant authors and papers. By appreciative the structure and association of a citation network, scholars can identify the most highly cited authors and papers within a network, providing perceptions into significant exploration directions of research and potential collaborators. This information helps the scholars, publishers, etc., to make experienced decisions about future research investments.

13.3.1 Citation Network

Citation networks for CD can also be used to recommend relevant papers to scholars based on their study of interests and citation patterns. This can help students to stay up-to-date with the latest research in their arena and recognize potential collaborators. Overall, the current research CD in citation networks is an important area of study with several practical applications. The insights gained from CD help to guide future research investments and collaborations, predict future research directions, and recommend relevant papers to researchers. As the volume of research continues to increase, CD can help researchers navigate the complex network of relationships between authors, papers, and fields of study, making it a valuable tool for researchers, publishers, and funding agencies equally. The basic network datasets' repository which contains network datasets is shown in Table 13.2.

Citation Network Datasets. Citation network datasets [11] are made up of interlinked research papers that are visualized as graphs in which it contains nodes and edges. Nodes represent articles of research in citation networks, while edges define the links connecting them based on their citation patterns. This methodology facilitates the identification of connected groups of publications and builds as communities from the networks is known as community detection. The goal of community detection is to uncover distinct groups or communities within a network by analyzing its structural patterns and connections. CD in citation networks has numerous applications such as recognizing patterns of study, finding prominent articles, promoting collaborations, enhancing recommendation systems, and organizing and summarizing scholarly literature, etc. Such applications facilitate the growth of knowledge and assist scholars in navigating the broad field of studies in academia. Some of the benchmark datasets [11] for citation related networks are listed in Table 13.3.

Citation Network Databases. Network of citations Databases are enormous collections of scholarly articles and the reference connections that correlate with them. Such interconnections produce a complex structure facilitating investigation into the influence on academic excellence, the dissemination of information, and the assessment of impact in diverse disciplines of study. The important distinct types of databases are bibliographic databases and database of citation.

Bibliographic databases and citation databases are inextricably linked. Citation databases frequently take bibliographic data from bibliographic databases and supplement along with citation information. Analysts may use both forms of data to create an accurate representation of the intellectual community. This enables them

Table 13.2 Network datasets' repository

S. no.	Network datasets' repository	Link for repository
1	Network repository	http://networkrepository.com/
2	Stanford large network dataset collection	http://snap.stanford.edu/data/
3	UCI network data repository	http://networkdata.ics.uci.edu/
4	KONECT—Koblenz network collection	http://konect.cc/networks/
5	Citation network dataset	https://www.aminer.org/citation
6	Kaggle—datasets	https://www.kaggle.com/datasets

Table 13.3 Citation network datasets' collections

S. no.	Dataset	Description of dataset
1	DBLP (digital bibliography and library project)	Bibliography details for key computer science conferences and publications are available in a computerized library
2	CORA	Dataset of machine learning research papers
3	CiteSeerX	A large citation network dataset containing scientific articles from various domains
4	Microsoft academic graph (MAG)	A large-scale academic graph that includes citation data among research papers, authors, institutions, and fields of study

to conduct in-depth analyses of research communities and find their links. Bibliographic databases serve as the foundation for building networks in CD research. Citation databases provide extra information on citation relationships that may be used to assess the effect and influence of research. Combining these types of databases allows academics to undertake advanced community discovery analyses that take into account both scholarly ties and citation trends. Some of the databases are Scopus [12], Dimensions, WOS (Web of Science), Crossref, ACM Digital Library, PubMed, IEEE Explore, Google Scholar, Microsoft Academic Search, etc.

Different Types of Analysis. The article represents a comprehensive examination of the fascinating topic of citation network analysis, which comprises a variety of analyses that are applied to investigate the citation patterns of scientific publications. Bibliometrics approaches [13] such as co-citation and bibliographic coupling are used in the study to reveal hidden patterns that influence academic communication. This review study underlines the crucial role interconnection plays in influencing the development of scientific knowledge by evaluating it through co-authorship, co-citation, and collaborative networks.

13.3.2 Bibliometrics Analysis

Bibliometrics Analysis is the quantitative examination of academic and intellectual publications. This is used to measure the effect, visibility, and relationships of publications and authors within a certain topic. Bibliometrics tools aid to measure the impact of research papers, identifying patterns of collaboration, and revealing the relationships between publications and authors.

Bibliographic coupling method evaluates the similarity between two publications based on the references. Two publications are regarded as related and may have a similar research focus if they have a large number of references.

Co-citation analysis looks at how frequently two publications are mentioned in tandem in other papers. When two articles are widely mentioned together, they are seen as connected and are probably covering the same or related issues.

Co-author analysis examines how writers in a certain field collaborate. This is also beneficial for identify the notable researchers, authors and measure the effectiveness of their teamwork.

Co-word analysis investigates the frequency with which specific words or phrases appear together in research paper titles, abstracts, or keywords. This is also very advantageous for performing the identification of study themes and subjects that are commonly mentioned in conjunction.

Collaboration analysis focuses on the patterns and dynamics of researcher cooperation. It may entail examining co-authorship networks, institutional connections, and research partnerships across multiple regions or nations.

Keyword co-occurrence analysis is the approach for comprises examining the frequency and arrangement of terms that appear together in academic works in order to uncover basic research topics and interconnections across subjects.

Citation analysis is the process of assessing the citations obtained by a publication in order to determine its effect and influence within the scientific community. The number of citations received by a work might reflect its significance and relevance in the area.

13.3.3 Network Analysis Measures

Network analysis is the examination of linkages and interactions among diverse entities, such as authors, articles, institutions, or keywords, utilizing network theory and graph theory. This is also effective for identifying major participants, clusters, and primary nodes within a scholarly network. Some of the measures are Nodes, Edges, and Ground-Truth Communities [4], Triadic Closure, *D*-Score [3], Average Clustering, Density, Coefficient, Modularity, Average Degree, Maximum Degree, Radius, Minimum Degree, Diameter. The main aim of the observation is to increase accuracy of the outcomes. Modularity scores, *F*-measure, clustering coefficient, NMI similarity assessments, precision, recall, and runtime analyses are utilized to assess the correctness and effectiveness and evaluate of these methods. The modularity score is analyzed, and discrepancies between the found communities are noted, to compare the performance of the algorithm.

13.4 Computational Tools and Methods

Community detection in citation networks has a significant challenge in the study of citations that helps researchers understand the structure and patterns of scientific collaboration and data exchange. Several computational tools and methodologies [14]

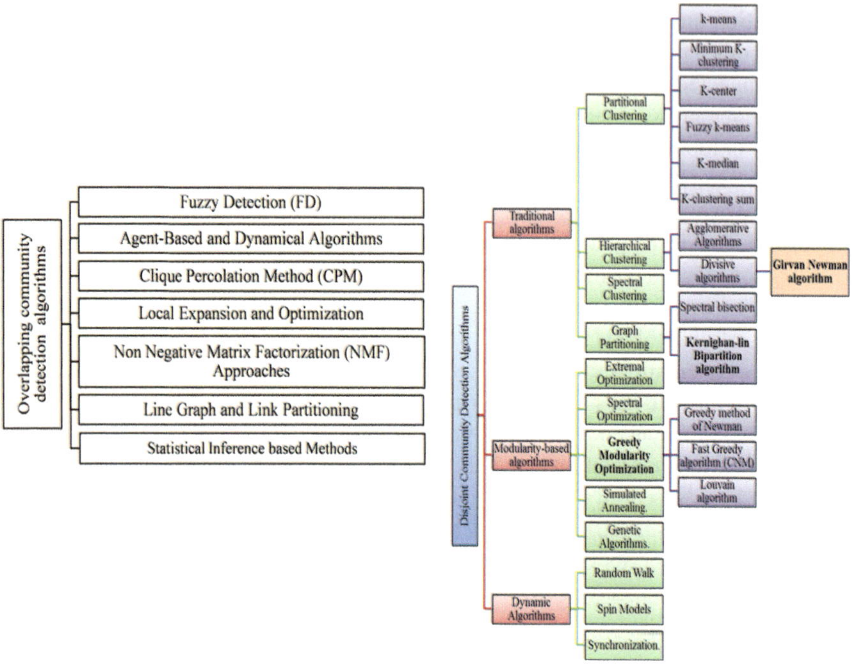

Fig. 13.2 Community detection algorithms

were developed to solve this issue. The main classifications of CD algorithms are overlapping community and disjoint CD algorithms which are classified in Fig. 13.2. Some of the important notable techniques and resources for detecting communities in citation networks are listed below.

Louvain Method. The Louvain method is notable for its speed and ability to handle large-scale networks well. It frequently provides high-quality community partitions with a decent combination of computational efficiency and accuracy.

Infomap. Infomap is extensively used and well-regarded for its ability to discover hierarchical community structures and for its strong theoretical foundation based on information theory.

Walktrap. This approach is very beneficial for discovering communities in networks with high modularity. It has demonstrated strong performance in numerous sorts of networks, including citation networks.

Fastgreedy. Fastgreedy is well-known for its speed and scalability. It uses hierarchical agglomerative clustering to effectively uncover community structures in huge networks.

Girvan–Newman. This edge-betweenness-based technique which detects communities by progressively eliminating edges with the highest betweenness until communities is discovered.

Label Propagation. Label propagation techniques find communities by iteratively assigning node labels depending on their neighbors' labels.

13.4.1 Computational Tools

Wide ranges of software are available as solutions to implement the bibliometrics analysis [15]. Also will be utilizing tools [15] as a Python packages like NetworkX, Scikit-learn, Matplotlib, NetworkX, Visualization, pyBibX [16], Pandas, litstudy [17], Louvain Method, pybliometrics [12], etc., and Jupyter Notebooks as tools to preprocess data, apply community detection algorithms, conventional and deep learning algorithms, and visualize the detected communities for in-depth analysis in the experiment on CD in citation networks. Some of the tools are listed below in Table 13.4.

13.5 Conclusion

Community detection is a critical challenge in complicated network systems. Although prior research focused mostly on directed networks, subsequent work has turned its focus to the community discovery problem on citation networks. Several ways have been presented, and these methods are also applicable for demonstrating significantly in detecting communities in various similar types of networks. Larger research is required to increase the accuracy and efficiency of these algorithms, as well as to expand their application to larger complicated networks. According to the results of the experiment, modularity-based algorithms and deep learning are efficient in spotting community structures in citation networks. The communities that have been found shed important light on the fundamental research questions and suggest possible cross-disciplinary research partnerships across academic articles. The future direction and scope for community detection in citation networks resides with the establishing the extensively effective, suitable and reliable model using deep learning algorithms. These models are highly capable for handling the gradually growing massive and dynamic datasets with respect to the multidimensional aspects of node relationships to uncover meaningful knowledge patterns and research trends.

Table 13.4 Citation network analysis—software and tools' collections list

S. no.	Bibliometrics/CAN-software/tools	Description of software
1	Python	A popular computing framework for bibliometrics and citation network analysis applications. It is a flexible programming language with a wide range of libraries that is frequently used in bibliometrics and CNA
2	R	A prominent statistical program used in bibliometrics and CNA. This is a robust statistical program that is commonly used in bibliometrics investigations for its data manipulation and display capabilities
3	VOSviewer	A visualization tool for generating and examining bibliometrics networks. This is a popular tool for creating and displaying bibliometrics, co-authorship, and co-citation networks
4	CitNetExplorer	An interactive tool for examining and analyzing citation networks that supports scholars in learning more about the relationships between pieces of scientific literature
5	Bibexcel	An application for bibliometrics analysis that allows academics to extract and manipulate data from bibliographic files
6	Gephi	An open-source program for network visualization, analysis, and exploration that is frequently used in the study of citation and cooperation networks
7	CiteSpace II	Specialized software for recognizing and displaying developing trends and patterns in scientific literature using citation data, supporting knowledge discovery
8	Pajek	A software package for large-scale network analysis and visualization, including citation networks, to help in the research of complicated citation structures
9	Publish or Perish	A software program that obtains and analyzes academic citations and bibliographic data from Google Scholar, supporting academics in determining scholarly impact
10	Bibliometrix/Biblioshiny	R packages for bibliometrics analysis and interactive visualization, allowing academics to investigate publication and citation trends

(continued)

Table 13.4 (continued)

S. no.	Bibliometrics/CAN-software/tools	Description of software
11	ScientoPy	A Python toolkit for bibliometrics network analysis and visualization, providing academics with tools to examine scientific collaboration and citation trends
12	BiblioTools/BibioMaps	Tools for constructing bibliometrics maps and analyzing citation data, allowing for the visual depiction and investigation of intellectual linkages
13	CitedReferences Explorer (CRExplorer)	A tool for displaying and studying citation data and related information, which assists researchers in comprehending citation linkages
14	SciMAT	A software application for evaluating and mapping science, technology, and innovation networks, which gained knowledge in dissemination and multidisciplinary cooperation research
15	Science of Science (Sci2) Tool	An analytical tool for studying and comprehending scientific research trends and patterns, hence offering insights into the dynamics of scholarly communication

References

1. Zhou, X., Su, L., Li, X., Zhao, Z., Li, C.: Community detection based on unsupervised attributed network embedding. Expert Syst. Appl. **213**, 118937 (2023). https://doi.org/10.1016/j.eswa.2022.118937
2. Hentona, A., Sakumoto, T., España, H.A.M., Nonaka, H., Kataoka, S., Hiraoka, T., Nakai, K., Carreón, E.C.A., Hirota, M.: Community detection and growth potential prediction from patent citation networks (2019). https://doi.org/10.48550/ARXIV.1904.12040
3. Gao, T., Pan, R., Wang, S., Yang, Y., Zhang, Y.: Community detection for statistical citation network by D-SCORE. Stat. Interface **14**, 279–294 (2021). https://doi.org/10.4310/20-SII636
4. Kiruthika, R., Vijaya, M.S.: Community detection using Girvan–Newman and Kernighan–Lin bipartition algorithms. In: Jacob, I.J., Kolandapalayam Shanmugam, S., Bestak, R. (eds.) Data Intelligence and Cognitive Informatics, pp. 217–231. Springer Nature, Singapore (2022). https://doi.org/10.1007/978-981-16-6460-1_16
5. Chen, Y., Wang, L., Qi, D., Ma, T., Zhang, W.: Community detection based on deepwalk model in large-scale networks. Secur. Commun. Networks **2020**, 1–13 (2020). https://doi.org/10.1155/2020/8845942
6. Rahul, P.B., Goel, P., Nayak, P.: Community detection using graphical relationships. In: Ranganathan, G., Chen, J., Rocha, Á. (eds.) Inventive Communication and Computational Technologies, pp. 985–996. Springer, Singapore (2021). https://doi.org/10.1007/978-981-15-7345-3_84
7. Rajita, B.S.A.S., Kumari, D., Panda, S.: A comparative analysis of community detection methods in massive datasets. In: Goel, N., Hasan, S., Kalaichelvi, V. (eds.) Modeling, Simulation and Intelligent Computing, pp. 174–183. Springer, Singapore (2020). https://doi.org/10.1007/978-981-15-4775-1_19

8. Fiallos, A., Jimenes, K., Vaca, C., Ochoa, X.: Scientific communities detection and analysis in the bibliographic database: SCOPUS. In: 2017 Fourth International Conference on eDemocracy & eGovernment (ICEDEG), pp. 118–124. IEEE, Quito, Ecuador (2017). https://doi.org/10.1109/ICEDEG.2017.7962521
9. Yudhoatmojo, S.B., Samuar, M.A.: Community detection on citation network of DBLP data sample set using LinkRank algorithm. Procedia Comput. Sci. **124**, 29–37 (2017). https://doi.org/10.1016/j.procs.2017.12.126
10. Wang, P., Kong, B., Bao, C.M., Zhou, L.H., Wang, C.Y.: Community detection based on graph neural network. In: 2021 6th International Conference on Intelligent Computing and Signal Processing (ICSP), pp. 89–93. IEEE, Xi'an, China (2021). https://doi.org/10.1109/ICSP51882.2021.9408828
11. Velasquez, J.D.: TechMiner: analysis of bibliographic datasets using python. SoftwareX **23**, 101457 (2023). https://doi.org/10.1016/j.softx.2023.101457
12. Rose, M.E., Kitchin, J.R.: Pybliometrics: scriptable bibliometrics using a python interface to Scopus. SoftwareX **10**, 100263 (2019). https://doi.org/10.1016/j.softx.2019.100263
13. Moral-Muñoz, J.A., Herrera-Viedma, E., Santisteban-Espejo, A., Cobo, M.J.: Software tools for conducting bibliometric analysis in science: an up-to-date review. EPI **29** (2020). https://doi.org/10.3145/epi.2020.ene.03
14. El-Moussaoui, M., Agouti, T., Tikniouine, A., Adnani, M.E.: A comprehensive literature review on community detection: approaches and applications. Procedia Comput. Sci. **151**, 295–302 (2019). https://doi.org/10.1016/j.procs.2019.04.042
15. Joshi, D., Patalia, T.: Community detection methods and tools for various complex network. Int. J. Eng. Res. Technol. **13**, 1386 (2020). https://doi.org/10.37624/IJERT/13.6.2020.1386-1390
16. Pereira, V., Basilio, M.P., Santos, C.H.T.: pyBibX—a python library for bibliometric and scientometric analysis powered with artificial intelligence tools (2023). https://doi.org/10.48550/ARXIV.2304.14516
17. Heldens, S., Sclocco, A., Dreuning, H., Van Werkhoven, B., Hijma, P., Maassen, J., Van Nieuwpoort, R.V.: Litstudy: a python package for literature reviews. SoftwareX **20**, 101207 (2022). https://doi.org/10.1016/j.softx.2022.101207

Chapter 14
A Hybrid Correlation Feature Selection Based Support Vector Machine (CFS-SVM) for Effective Software Defect Prediction

D. R. Medhunhashini and K. S. Jeen Marseline

Abstract Software Engineering is seamlessly developed with advancement in the development tools and framework as so the development of the defect also. Software defect prediction focus on ensuring and delivering quality software product. Machine learning algorithms plays a vital role in effectively predicting the software bugs early in the software life cycle. The main objective of this study is to identify a hybrid machine learning model incorporating Correlation feature selection. The radial basis function kernel of SVM maps the features in the input space. The CFS based SVM model encompasses a higher defect prediction when employed. The CM1 dataset from NASA spacecraft instrument were subjective for the evaluation of CFS based SVM model. Confusion matrix, Precision, Recall, and F_1 Score are the metrics for CFS based SVM. The results show a higher rate of prediction with 64.32% for precision, 65.13% for Recall, and 64.72% for F Score, respectively. This study has been able to show the defective classes have a big impact on software defect prediction and opting the correct parameters with correlation feature selection effectively performs predictions among the other classifiers.

14.1 Introduction

Software defect prediction is an essential aspect of software engineering aimed at identifying and mitigating potential defects or bugs in software systems. It involves the use of various techniques and approaches to analyze historical data and build models that can predict the likelihood of defects in future code. To identify and

D. R. Medhunhashini (✉) · K. S. Jeen Marseline
Sri Krishna Arts and Science College, Coimbatore, India
e-mail: medhun.hashini@gmail.com

K. S. Jeen Marseline
e-mail: jeenmarselineks@skasc.ac.in

P. Nanda et al. (eds.), *Intelligent System and Data Analysis*, Smart Innovation, Systems and Technologies 398, https://doi.org/10.1007/978-981-97-5200-3_14

address these defects early in the development lifecycle, software defect prediction helps improve software quality, reduce maintenance costs, and enhance overall productivity. Machine learning techniques are commonly employed in software defect prediction, leveraging historical data to train predictive models. These models utilize a range of features, including code complexity metrics, code change metrics, software process metrics, and code smells, to capture patterns and correlations associated with defects [1]. The selection and relevance of features depend on the specific context and availability of data.

Cross-project defect prediction is a significant area of research, where models are trained on data from one project and applied to predict defects in another project [2]. This approach allows for knowledge transfer across projects, identifying common characteristics and factors contributing to defect. The incorporation of cutting-edge methods like graph neural networks (GNNs) and natural language processing (NLP) has recently made it possible to forecast software defects. NLP techniques enable the analysis of textual data, such as bug reports and commit messages, to extract valuable information related to defects. GNNs, on the other hand, capture relationships and dependencies between code elements, enabling more nuanced predictions by considering the context and structure of the code [3, 4].

Evaluation of software defect prediction models typically involves metrics such as precision, recall, F_1 Score, and AUC-ROC. These metrics assess the model's ability to accurately identify defective modules while minimizing false positives and false negatives [5, 6]. Despite the progress made in software defect prediction, challenges persist. Imbalanced datasets, data quality issues, and the need for effective feature selection and preprocessing remain areas of concern [7]. The dynamic nature of software development and the evolving complexity of software systems pose ongoing challenges for accurate defect prediction.

14.1.1 Problem Statement

The problem of software defects continues to persist in software development, leading to increased maintenance costs, reduced software quality, and negative user experiences. Predicting and preventing these defects in advance is crucial for ensuring the reliability and stability of software systems. However, accurately identifying and predicting software defects remains a challenge due to the complex and evolving nature of software code, the interdependencies among code components, and the vast amount of data involved. Existing defect prediction approaches often struggle to achieve high accuracy and may not effectively capture the underlying patterns and factors contributing to defects.

14.1.2 Motivation

The motivation behind software defect prediction research is to proactively address the problem of defects in software development. By predicting and identifying potential defects early on, developers can allocate resources efficiently, prioritize testing efforts, and focus on improving the quality of critical code components. Effective defect prediction helps organizations save costs associated with bug fixing, reduce maintenance efforts, enhance customer satisfaction, and minimize the risks of software failures or security vulnerabilities. Accurate defect prediction facilitates the adoption of agile development practices by providing timely feedback and enabling iterative improvements during the development process. By leveraging advanced techniques such as machine learning, natural language processing, and graph analysis, researchers aim to enhance the accuracy and efficiency of defect prediction models, contributing to the overall improvement of software quality and reliability. The motivation behind software defect prediction is to create more robust, secure, and user-friendly software systems that meet the high standards of modern software development.

14.1.3 Objective

The objective of software defect prediction research is to develop accurate models and techniques that can identify and predict defects in software systems. This objective encompasses improving prediction accuracy, detecting defects early in the development lifecycle, enhancing software quality, optimizing resource allocation, facilitating decision-making, and enabling continuous improvement in software development processes. By achieving these objectives, defect prediction aims to enhance software reliability, reduce maintenance costs, and improve overall customer satisfaction.

14.2 Literature Review

"Interpretable defect prediction" [8] proposes an innovative technique that consider the inherent imbalance of data are not essential. These approaches encompass a range of strategies, including under sampling, oversampling, and hybrid methods, which aim to balance the class distribution and enhance the interpretability of defect prediction models. Dealing with imbalanced data poses a significant challenge in the realm of interpretable defect prediction. The presence of unequal class distribution hampers the effectiveness of traditional machine learning models and impedes the accurate identification of defects.

"Issue-driven software fault prediction" [9] incorporates features into fault prediction models, and it becomes possible to identify patterns and correlations between specific issues and subsequent faults, enabling proactive defect prevention and mitigation. Issues-driven features play a crucial role in software fault prediction by capturing key aspects related to the issues and defects encountered during software development and maintenance. These features encompass a wide range of attributes, such as the number of reported issues, the severity of the issues, the resolution time, and the frequency of code changes in response to issues.

"Precise software defect prediction" [10] employs several factors contributing to the challenge, including the inherent complexity of software systems, the dynamic nature of development processes, and the limitations of available data and prediction models. Predicting the precise number of software defects remains a challenging task, and it is an ongoing area of research and development. While significant advancements have been made in software defect prediction, achieving precise defect counts is still a complex endeavor. The nature of defects themselves can be elusive, with some latent or hard-to-detect issues surfacing only after deployment. Researchers continue to explore innovative techniques such as machine learning, statistical models, and data mining approaches to improve the accuracy of defect prediction. Collaborative efforts between academia, industry, and the open-source community, along with the accumulation of more comprehensive and diverse data sets, are essential to further advance the state of predicting precise software defect numbers.

"Cross-project defect prediction" [11] encompasses sufficient labeled defect data to be available for training, and selecting projects with diverse defect types ensures model robustness. A hybrid method can identify suitable projects for accurate and transferable defect prediction models. Candidate project selection in cross-project defect prediction using a hybrid method requires careful consideration of project similarity, data availability, and defect diversity. Metrics such as software metrics, project characteristics, and domain knowledge aid in assessing project similarity.

"Just-in-time defect prediction" [12] based on abstract syntax tree (AST) change embedding is a novel approach that enables the proactive identification of software defects during the development process. By analyzing and embedding the changes made to the AST, this method captures the semantic modifications in the codebase. It influences machine learning techniques to predict the likelihood of introducing defects with each code change, allowing developers to address potential issues early on. This approach enhances software quality and reduces debugging efforts by providing timely feedback and guidance for code improvements, ultimately resulting in more reliable and maintainable software systems.

"Salp Swarm Optimizer" [13] applies the SSO to software fault prediction. The algorithm can effectively search the solution space and find optimal model parameters, feature subsets, or classification thresholds. This approach helps improve the accuracy and efficiency of fault prediction models by dynamically adapting to the complexity and dynamics of software systems. The SSO algorithm shows potential for enhancing the modeling process in software fault prediction, contributing to more effective defect identification and prevention strategies.

"Three-stage transfer learning" [14] proposes a framework that enables knowledge transfer from source to target projects, improving prediction performance even with different project characteristics. A three-stage transfer learning framework is proposed for multi-source cross-project software defect prediction. It involves pre-training a deep learning model on source project data, fine-tuning it using limited target project data, and using the fine-tuned model for defect prediction. This framework enables knowledge transfer from source to target projects, improving prediction performance even with different project characteristics.

"Machine learning-based prediction" [15] investigates the various machine learning techniques employed, the types of software defects predicted, and the impact on business outcomes such as cost reduction, quality improvement, and resource allocation. By focusing on the business-driven aspects, this study provides valuable insights for organizations seeking to use machine learning in software defect prediction to enhance their software development processes and achieve tangible business benefits. A business-driven systematic mapping study explores the application of machine learning in software defect prediction from a business perspective.

"Kernel-based software defect prediction" [16, 17] evaluates various kernel functions, such as linear, polynomial, Gaussian (RBF), and sigmoid kernels. Researchers can assess their impact on the SVM's ability to accurately classify software defects. This analysis aids in understanding the strengths and limitations of SVM with different kernels, enabling practitioners to make informed decisions when selecting an appropriate kernel method for software defect prediction based on their specific dataset and prediction requirements. Examining the performance of kernel methods for software defect prediction, specifically based on the support vector machine (SVM), is a critical analysis that sheds light on the effectiveness of SVM with different kernel functions in predicting software defects.

"Graph neural networks-based prediction" [18] incorporates a semantic information, such as variable names, function calls, and structural information, like control flow and dependencies, and GNNs can effectively learn representations that capture the complex nature of code. This approach allows for more accurate identification and prediction of software defects. Software defect prediction using graph neural networks (GNNs) leverages semantic as well as structural information of code is an emerging approach.

14.3 Proposed Work

The proposed work in this section begins with compiling historical information on software development projects. Software modules, features, development metrics, and problem incidences are covered in this data. Version control systems, bug tracking systems, code repositories, and other pertinent source are used to gather the fault data. Figure 14.1 is the overall process of the proposed work CFS-SVM. Collected historical defect metric information is preprocessed. Correlation feature selection is applied to find the relation between features. Radial basic function kernel of support vector machine places the feature in high dimension space for the parameters. The hybrid CFS based SVM finally presents efficient prediction.

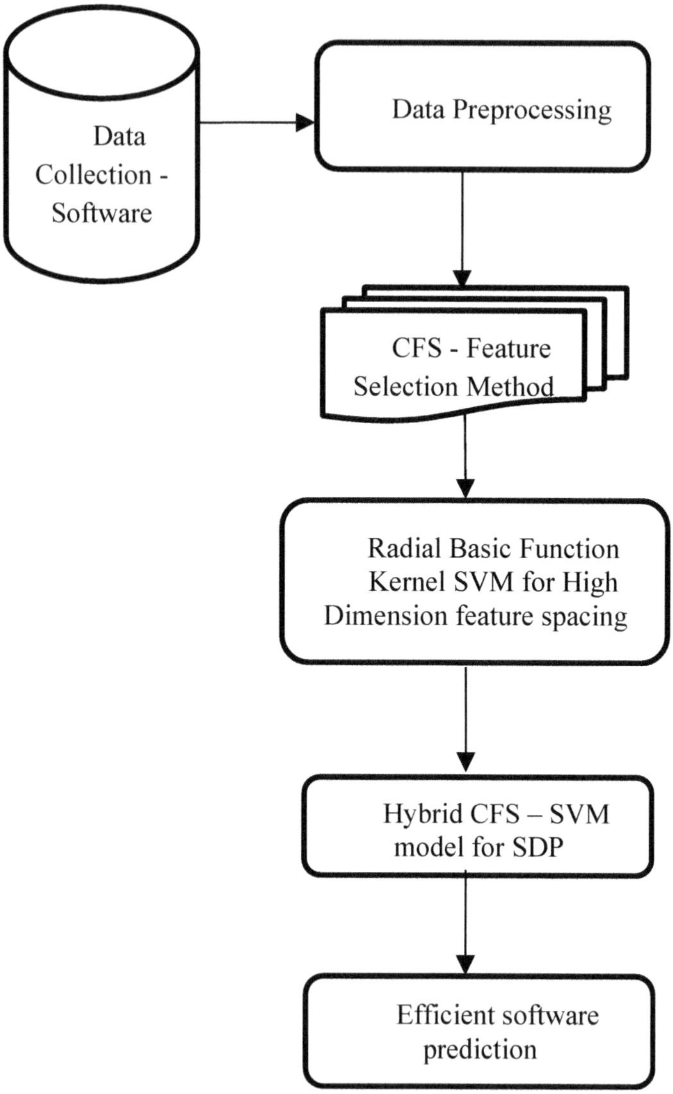

Fig. 14.1 A hybrid correlation feature selection based support vector machine (CFS-SVM) model

14.3.1 Correlation Feature Selection (CFS)

To ensure that the defect prediction model is accurate, careful feature selection is essential. Correlation-based feature selection selects a group of significant features to anticipate software flaws. Data preprocessing is done to ensure that the data is in a format that will be appropriate for analysis. This involves handling missing values, removing duplicates, normalizing data, and transforming variables.

Defective software modules are labeled as defective or non-defective based on historical defect data using a set of criteria. A well-defined machine learning technique support vector machine (SVM) is used for defect prediction in this model. SVMs are efficient in binary classification problems, making them useful for determining, from a collection of input features, whether a software module is broken or not. The proposed model for software defect prediction uses a hybrid correlation feature selection based support vector machine (CFS-SVM) model is to address the efficient classification with a benchmarked accuracy. The goal of CFS-SVM is to identify the best hyperplane for separating defective and non-defective cases.

14.3.2 Correlation Feature Selection (CFS)

Support vector machines (SVM) is a reliable algorithm for supervised learning for classification and regression tasks. SVM's primary objective is to locate the optimal hyperplane that maximally differentiates across classes of data points. In mathematical terms, when given a training dataset $\{(x_1, y_1), (x_2, y_2), \ldots, (x_n, y_n)\}$, where x_i represents the input feature vector and y_i represents the corresponding class label (-1 or 1), SVM strives to discover a hyperplane defined by the equation:

$$w \cdot x + b = 0 \tag{14.1}$$

In this context, w signifies the average vector pointing to the hyperplane, while b represents the bias term. The symbol $\cdot$ denotes the dot product operation between vectors. The hyperplane divides the feature space into two regions corresponding to distinct classes.

The primary goal of support vector machines (SVM) is to maximize the margin, defined as the angle between the hyperplane and the points closest to the distinct classes. These data points, known as support vectors, are important in determining the optimal hyperplane. To tackle nonlinear classification problems, SVM employs kernel methods. By employing these mathematical methods, a higher-dimensional space is implicitly mapped onto the input data. By transforming the input data into a higher-dimensional feature space, SVM can identify linear decision boundaries that correspond to nonlinear decision boundaries in the original input space.

The mapping function $\phi(x)$ represents the transformation of the input feature vector x from its originating space to a feature space with more dimensions. The kernel function, denoted as $K(x, y)$, calculates the inner product of the transformed feature vectors $\phi(x)$ and $\phi(y)$ in the higher-dimensional space. Importantly, this inner product can be computed without explicitly calculating the individual transformations $\phi(x)$ and $\phi(y)$. Using kernel functions eliminates the need to explicitly compute the transformed feature vectors in the higher-dimensional space, thereby avoiding the computational complexity associated with high dimensions. The kernel trick allows SVM to operate in the original input space while effectively utilizing the benefits of high-dimensional feature spaces.

14.3.3 Curse of Dimensionality

In high-dimensional spaces, the curse of dimensionality refers to the challenges arising from the increased number of features. As the dimensionality of the data increases, the available data becomes sparser, leading to difficulties in classification. The challenge known as the curse of dimensionality arises when there are many dimensions, leading to the risk of overfitting. When a model grows overly complicated and fits the training data closely enough, it is said to be overfitting, and it will have trouble generalizing to new data. With many features, the data points tend to be farther apart, making it harder to find meaningful patterns or boundaries between different classes. The curse of dimensionality highlights the need for effective techniques to handle high-dimensional data and mitigate its associated challenges.

14.3.4 RBF Kernel for High-Dimensional Data

The radial basis function (RBF) kernel is particularly well-suited for handling high-dimensional feature spacing in SVM due to its ability to capture complex relationships irrespective of the dimensionality. The input features are implicitly transformed into a feature space of infinite dimensions by utilizing the RBF kernel. This transformation enhances the possibility of effectively separating the classes within the data. The RBF kernel computes the similarity between two data points based on the Euclidean distance in the input feature space. The similarity is determined by the exponential function of the negative squared distance, controlled by the hyperparameter γ.

The RBF kernel can effectively capture intricate relationships between features and identify patterns that may not be apparent in the original input space. By implicitly mapping the input features to an infinite-dimensional feature space, the RBF kernel transforms the high-dimensional data into a space where the separation of classes can be achieved more quickly. The RBF kernel's ability to model complex relationships in high-dimensional spaces makes it a valuable tool for handling the curse of dimensionality. It allows SVM to find optimal decision boundaries that can effectively classify data points, even in scenarios with many features. Choosing an appropriate value for the γ hyperparameter is crucial, as it determines the smoothness or sharpness of the decision boundary.

14.3.5 RBF Kernel-SVM

Implicit Feature Mapping. The RBF kernel employed in SVM implicitly transforms the input features into a space of higher dimensions. This mapping allows SVM to find linear decision boundaries in the transformed feature space, even though the original feature space may be nonlinear. The RBF kernel uses a similarity measure between pairs of input samples based on the radial basis function:

$$K(x, y) = \exp\left(-\gamma \|x - y\|^2\right) \tag{14.2}$$

where x and y denote input feature vectors, $\|x - y\|^2$ indicates the squared Euclidean distance between them, and γ is a hyperparameter responsible for governing the bandwidth or extent of the kernel function.

The RBF kernel implicitly maps the input features x and y to an infinite-dimensional feature space, where the similarity between the mapped features determines the kernel value. This mapping allows for a more graphic representation of the data, potentially revealing nonlinear relationships that cannot be easily captured in the original feature space.

Nonlinear Separation. The RBF kernel enables SVM to find a nonlinear decision boundary using implicitly mapped features in the original feature space. Within the feature space after transformation, the decision boundary can form a hyperplane, representing a linear division. By implicitly mapping the input features to a higher-dimensional space, the RBF kernel expands the feature space to capture complex relationships. This expansion allows SVM to find linear decision boundaries corresponding to nonlinear ones in the original input space. The separation achieved in the transformed feature space leads to nonlinear separation in the original feature space, improving the classification performance of SVM.

Influence of γ. The hyperparameter γ in the RBF kernel plays a crucial role in determining the smoothness of the decision boundary. It controls the kernel function's bandwidth or spread and affects individual training samples' influence on the decision boundary. Higher values of γ result in a more localized decision boundary. This means that the decision boundary is highly influenced by the samples close to it, leading to a more intricate separation. On the other hand, lower values of γ produce a smoother decision boundary, where the influence of individual samples extends over a larger region. The choice of the γ hyperparameter is essential and should be carefully tuned. If γ is set too high, the decision boundary may become overly sensitive to individual samples, potentially leading to overfitting. Conversely, if γ is set too low, the decision boundary may become too smooth and fail to capture complex patterns in the data. Finding an appropriate value for γ involves balancing the trade-off between model complexity and generalization performance. It is typically determined through cross-validation or other optimization techniques to achieve the best classification results.

Algorithm CFS based SVM Model

Step 1: Pre-process the Data: Normalize or standardize the input features and split the dataset into training and test sets.
Step 2: Initialize SVM with RBF Kernel: Create an SVM model with the RBF kernel.
Step 3: Train the SVM Model: Train the SVM model using the training data. Set the hyperparameters, including the regularization parameter C and the RBF kernel's γ value.
Step 4: Hyperparameter Tuning: Perform hyperparameter tuning using techniques like grid search or cross-validation to find optimal values for C and γ.
Step 5: Test the SVM Model: Use the trained SVM model to predict class labels for the test set. Evaluate the model's performance using accuracy, precision, recall, or F_1 Score metrics.
Step 6: Feature Selection or Dimensionality Reduction: Apply CSA feature selection or dimensionality reduction techniques to reduce the dimensionality of the feature space.
Step 7: Fine-tuning: If the performance is unsatisfactory, adjust hyperparameters or modify the feature selection/dimensionality reduction process.
Step 8: Retrain and Test the SVM Model: If changes are made in steps 4–7, retrain the SVM model with updated settings. Test the retrained model on the test set to evaluate performance.
Step 9: Model Deployment: Deploy the trained CFS based RBF Kernel-SVM model to make predictions on new, unseen defect data.

14.4 About Dataset

This study conducted the experiment on the dataset CM1 NASA Metrics Data Program from PROMISE software engineering repository, which is available to the general public. This dataset includes details of number of lines in a software, Architectural Complexity, Decisional Statements (Cyclomatic Complexity), Structural Complexity (Essential Complexity), amount of effort involved, comments, and numerous more characteristics that determine whether a piece of software contains flaws or not. The background data needed for building prediction model is collected. The defect data collected is converted into product metrics that include the size, quality, complexity, and reliability of a software. Table 14.1 describes CM1 dataset features.

Table 14.1 CM1 dataset metrics

S.No.	Metrics type	Notation	Description
1.	McCabe	loc	No of lines in code
2.		v(g)	Cyclomatic complexity
3.		ev(g)	Essential complexity
4.		iv(g)	Design complexity
5.	Halstead	n	Total operators + operands
6.		v	Total capacity
7.		L	Length of the program
8.		d	Difficulty
9.		i	Intelligence
10.		e	Effort
11.		b	Delivered bugs
12.		t	Time estimator
13.		lOCode	Line count
14.		lOComment	No of comment lines
15.		lOBlank	Count of blank lines
16.		lOCodeAndComment	Numeric
17.		uniq_Op	Unique operators
18.		uniq_Opnd	Unique operands
19.		total_Op	Total operators
20.		total_Opnd	Total operands
21.		branchCount	Flow graph

14.5 Performance Metrics

The CFS based SVM model is applied on the CM1 dataset. The performance metrics like precision, recall, F_1 Score is calculated to find the prediction accuracy.

14.5.1 Correlation Heatmap

The correlation between features is inspected to find the best features for training the CFS-SVM model. Figure 14.2 shows heatmap for the model CFS-SVM and the features are highly correlated.

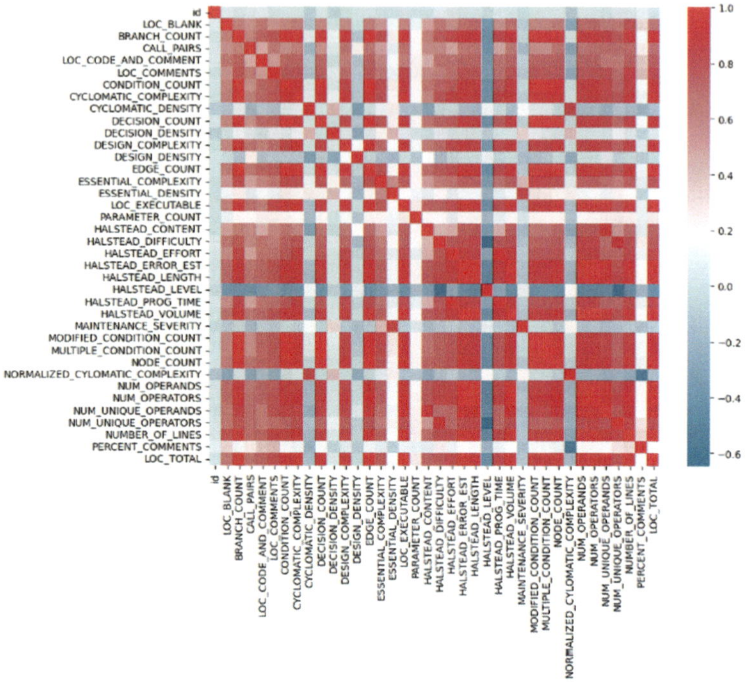

Fig. 14.2 Heatmap for correlation feature selection based support vector machine (CFS-SVM) model

14.5.2 Confusion Matrix

Confusion matrix identifies the incorrect areas of a model. This matrix contrasts the proportion of accurate and inaccurate predictions for each class.

True Positive. The number of positive observations that the model accurately predicted would be positive.

False Positive. The proportion of negative observations to the misinterpreted as positive.

True Negative. The number of negative observations that the model properly identified as negative is known as the true negative.

False Negative. The number of positive observations that the model falsely misinterpreted as negative.

14.5.3 *Precision*

The precision metrics of the model identifies the positive class with accuracy. Given by TP/(TP + FP).

14.5.4 *Recall*

The Recall metrics ensures the model's positive identifications in a dataset. Given by TP/(TP + FN).

14.5.5 *F_1 Score*

The mean of recall and precision is the F_1 score.

14.6 Results and Discussion

Table 14.2 represents the confusion matrix for the proposed CFS based SVM model for software defect prediction on CM1 dataset. The result shows the model correctly predicts the correct defect with a value 105. The performance is evaluated using the metrics precision for SVM and CFS based SVM with 64.32% accuracy for the latter. Recall has produced the result as 65.13 and the F-measure score is about 64.72% in Table 14.3. The final results show the correlation feature selection based support vector machine (CFS-SVM) model predicts the defects better to a plain SVM model (Figs. 14.3 and 14.4).

Table 14.2 Confusion matrix for CFS-SVM model

	Actual values		
Predicted values		Positive	Negative
	Positive	105	58
	Negative	56	108

Table 14.3 Evaluation results for CFS-SVM model

	SVM without CFS	CFS based SVM model
Precision	50.895	64.326
Recall	51.743	65.136
F-measure	51.315	64.728

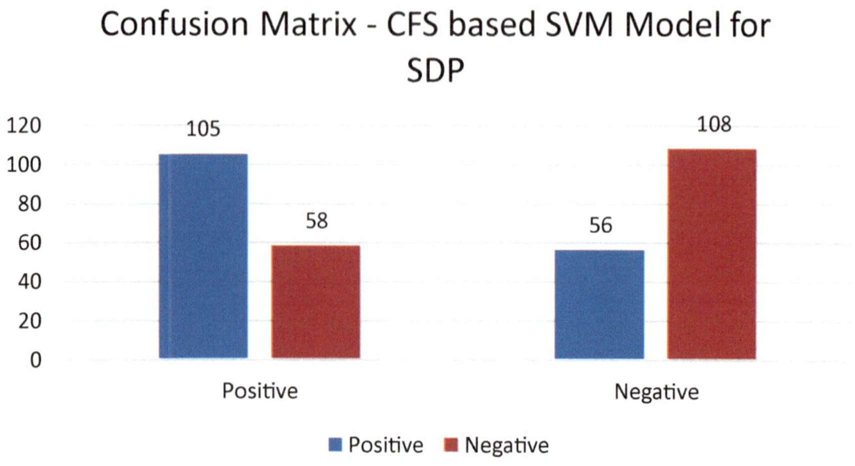

Fig. 14.3 Confusion matrix—correlation feature selection based support vector machine (CFS-SVM) model

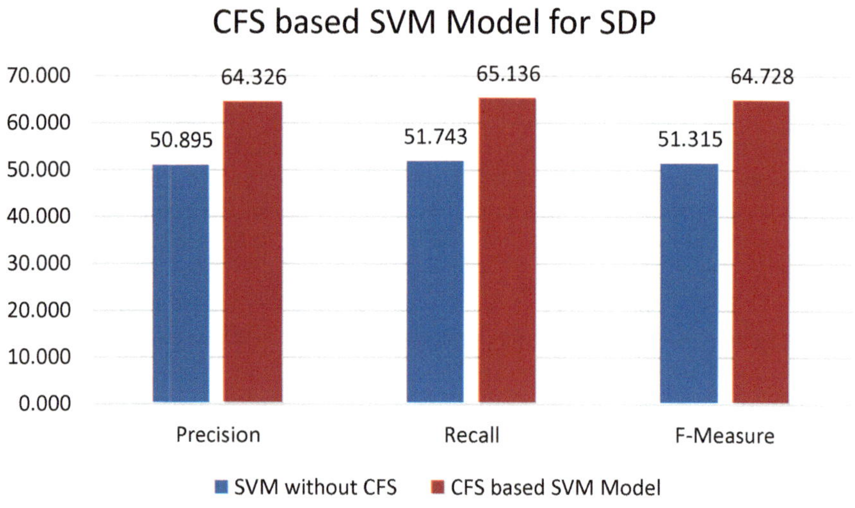

Fig. 14.4 Evaluation metrics—correlation feature selection based support vector machine (CFS-SVM) model

14.7 Conclusion

A critical component of software development is predicting defects. Software defect prediction seeks to locate and foresee possible flaws or problems in a software system. SDP uses of a variety of tools and techniques to analyze historical data, code metrics, and other aspects of software. Ensuring the quality data availability is a biggest challenge and measuring its trustworthiness also play a role. The consistency and accuracy of defect labeling is crucial for the efficacy of prediction models. To create reliable defect prediction models, machine learning algorithms and statistic methods are used. Still obtaining high accuracy seems difficult. The proposed CFS based SVM model enforces careful application of feature selection to improve prediction accuracy. Radial basis function is implemented optimally to tune the hyperparameter of the CFS based SVM Model. The model's efficiency is found to be 64.32% for precision, 65.13% for Recall, and 64.72% for F Score quite higher than the simple support vector machine. There are certain challenges pertain to the contextual dependency for SDP. Further a novel model can be developed that addresses the differential problem of context dependency including domain, architecture, programming languages, development methodologies, and software systems.

References

1. ElGhondakly, R., Moussa, S.M., Badr, N.: Service-oriented model-based fault prediction and localization for service compositions testing using deep learning techniques. Appl. Soft. Comput. **143**, 110430 (2023). https://doi.org/10.1016/j.asoc.2023.110430
2. Wu, F., et al.: Cross-project and within-project semisupervised software defect prediction: a unified approach. IEEE Trans. Reliab. **67**(2), 581–597 (2018). https://doi.org/10.1109/TR.2018.2804922
3. Psarommatis, F., Fraile, F., Ameri, F.: Zero defect manufacturing ontology: a preliminary version based on standardized terms. Comput. Ind. **145**, 103832 (2023). https://doi.org/10.1016/j.compind.2022.103832
4. Zheng, W., Tan, L., Liu C.: Software defect prediction method based on transformer model. In: 2021 IEEE International Conference on Artificial Intelligence and Computer Applications (ICAICA), pp. 670–674 (2021). https://doi.org/10.1109/ICAICA52286.2021.9498179
5. Nehi, M.M., Fakhrpoor, Z., Moosavi, M.R.: Defects in the next release; software defect prediction based on source code versions. In: Iranian Conference on Electrical Engineering (ICEE), pp. 1589–1594 (2018). https://doi.org/10.1109/ICEE.2018.8472535
6. Mohammed, M.Z., Saleh, I.A.: Predicted of software fault based on random forest and K-nearest neighbor. In: 2022 4th International Conference on Advanced Science and Engineering (ICOASE), pp. 43–48 (2022). https://doi.org/10.1109/ICOASE56293.2022.10075596
7. Li, W., Li, B., Niu, S., Wang, Z., Liu, B., Niu, T.: Selecting informative data for defect segmentation from imbalanced datasets via active learning. Adv. Eng. Inf. **56**, 101933 (2023). https://doi.org/10.1016/j.aei.2023.101933
8. Yin, Y., Dong, X., Xu, T.: Rapid and efficient bug assignment using ELM for IOT software. IEEE Access **6**, 52713–52724 (2018). https://doi.org/10.1109/ACCESS.2018.2869306
9. Qu, Y., Li, Z., Zhao, J., Li, H.: Unbalanced data processing for software defect prediction. In: 2022 4th International Conference on Data-Driven Optimization of Complex Systems (DOCS), pp. 1–6 (2022). https://doi.org/10.1109/DOCS55193.2022.9967755
10. Wang, J., Shen, B., Chen, Y.: Compressed C4.5 models for software defect prediction. In: 2012 12th International Conference on Quality Software, pp. 13–16 (2012). https://doi.org/10.1109/QSIC.2012.19
11. Pecorelli, F., Di Nucci, D.: Adaptive selection of classifiers for bug prediction: a large-scale empirical analysis of its performances and a benchmark study. Sci. Comput. Program. **205**, 102611 (2021). https://doi.org/10.1016/j.scico.2021.102611

12. Zheng, W., Shen, T., Chen, X.: Just-in-time defect prediction technology based on interpretability technology. In: 2021 8th International Conference on Dependable Systems and Their Applications (DSA), pp. 78–89 (2021). https://doi.org/10.1109/DSA52907.2021.00017
13. Daaji, M., et al.: On enhancing prediction abilities of vision-based metallic surface defect classification through adversarial training. Inf. Softw. Technol. Part B **142**(6), 106742 (2022, Oct). https://doi.org/10.1016/j.jss.2022.111537
14. Pandey, S.K., Tripathi, A.K.: DNNAttention: a deep neural network and attention based architecture for cross project defect number prediction. Knowl. Based Syst. **233**, 107541 (2021). https://doi.org/10.1016/j.knosys.2021.107541
15. Stradowski, S., Madeyski, L.: Machine learning in software defect prediction: a business-driven systematic mapping study. Inf. Softw. Technol. **155**, 107128 (2023). https://doi.org/10.1016/j.infsof.2022.107128
16. Sun, J., Jing, X., Dong, X.: Manifold learning for cross-project software defect prediction. In: 2018 5th IEEE International Conference on Cloud Computing and Intelligence Systems (CCIS), pp. 567–571 (2018). https://doi.org/10.1109/CCIS.2018.8691373
17. Zhou, Y., Shan, C., Sun, S., Wei, S., Zhang, S.: Software defect prediction model based on KPCA-SVM. In: 2019 IEEE SmartWorld, Ubiquitous Intelligence & Computing, Advanced & Trusted Computing, Scalable Computing & Communications, Cloud & Big Data Computing, Internet of People and Smart City Innovation (SmartWorld/SCALCOM/UIC/ATC/CBDCom/IOP/SCI), pp. 1326–1332 (2019). https://doi.org/10.1109/SmartWorld-UIC-ATC-SCALCOM-IOP-SCI.2019.00244
18. Zhao, Z., Yang, B., Li, G., Liu, H., Jin, Z.: Precise learning of source code contextual semantics via hierarchical dependence structure and graph attention networks. J. Syst. Softw. **184**, 111108 (2022). https://doi.org/10.1016/j.jss.2021.111108

Chapter 15
Mitigating Risks in Vehicle Fires: An Integrated Approach for Early Detection and Real-Time Alert System

Ranjeetsingh Suryawanshi, Harshita Bhagat, Aryan Karande, Parth More, and Vaishali More

Abstract Every year, there are about 170,000–220,000 car accidents each year that involve fire. The cause of the fire ranges from collision, engine malfunction, electric short circuits, and many more. Irrespective of the cause, the damage incurred is close to $2 billion in direct property damage and in severe accidents, even the passengers' lives. The rapid detection and timely notification of fire incidents are crucial for minimizing property damage and ensuring public safety. The current vehicle fire safety system relies on manual notification, leading to delayed responses and potential risks, especially in low-density areas and nighttime scenarios. Additionally, the lack of proactive fire detection technology hinders early identification, increasing the likelihood of extensive damage and harm to passengers. This paper proposes an integrated alert system automatically notifies fire stations, hospitals, and police upon fire detection, streamlining coordination and reducing response times. Rigorous evaluations are undertaken to assess the system's accuracy, sensitivity, and reliability under various conditions, solidifying its efficacy in addressing vehicle fire incidents.

R. Suryawanshi
Department of Computer Engineering, Vishwakarma Institute of Technology, Pune, India
e-mail: ranjeetsingh.suryawanshi@vit.edu

H. Bhagat (✉) · A. Karande · P. More · V. More
Department of Information Technology, Vishwakarma Institute of Technology, Pune, India
e-mail: harshita.bhagat21@vit.edu

A. Karande
e-mail: aryan.karande21@vit.edu

P. More
e-mail: parth.more21@vit.edu

V. More
e-mail: vaishali.more21@vit.edu

P. Nanda et al. (eds.), *Intelligent System and Data Analysis*, Smart Innovation, Systems and Technologies 398, https://doi.org/10.1007/978-981-97-5200-3_15

15.1 Introduction

Preventing vehicle fires is a huge challenge that will keep passengers safe and cause a lot of property damage. Current manual notification systems often cause delays in getting help, especially in sparsely populated areas or at night, putting travelers at risk. To solve this problem, we propose the development of a new comprehensive strategy to improve vehicle fire safety and emergency response. Our main goal in this project is to design and test a system that integrates advanced fire detection features to detect early fires. These features continuously monitor important parameters such as temperature and smoke levels, provide early detection and intervention to prevent injury to passengers and prevent fire damage.

Additionally, we integrate real-time location tracking to make it easier for emergency responders to reach the scene faster. To maximize the effectiveness of emergency response, our approach includes an immediate alert that automatically notifies fire departments, hospitals, police, and other relevant authorities. That vital information is this quick send makes communication easier and faster, ultimately reducing downtime.

Our system includes a built-in sound system and microphone for giving important audio instructions in fire detection. This rapid alert informs passengers of an emergency situation, allowing evacuation plans to be initiated quickly.

15.1.1 Objectives

1. To detect fire early: A system will be designed to detect any potential fire hazard in the vehicle as early as possible to avoid catastrophic damage. The system will be equipped with sensors that can detect smoke, heat, or any other signs of fire.
2. To detect location, the GPS component of the system should be able to accurately detect the location of the vehicle in real time. This information can be used to alert the relevant authorities.
3. To give an immediate alert, an alert system will be achieved through the use of a GSM model that sends notifications to connected smartphones.
4. To trigger automatic opening of doors for easy escape and to minimize injuries.

15.2 Literature Review

Saad Ur Rehman et al. suggest a model for an accident detection system for motorcycles. It uses a tilt sensor to calculate the inclination of the motorcycle. If the tilt angle exceeds the safe limit of $-40°$ to $40°$, it indicates an accident has occurred, and notifications are sent through GPRS and SMS using a GSM module. The system uses the SPI protocol for communication between the tilt sensor and the Arduino Nano [1]. Cheng-Yong Huang introduces a system that utilizes various sensors, like door position sensors and motorcycle proximity sensors, along with a fuzzy logic algorithm. Its purpose is to identify an approaching motorcycle and regulate the speed at which the vehicle's door opens to avoid collisions. The paper outlines the system's hardware and software setup, highlighting the utilization of a fuzzy inference system to make decisions based on the sensor data. Simulation tests demonstrated that the system successfully prevented accidents between motorcycles and vehicle doors, thereby lowering the risk of harm to motorcyclists [2]. Hamood Alqourabah et al. implemented a system to detect a fire and give alerts to property owners and emergency services. Singular sensors cannot detect amount of fire to address this problem, they have used detectors, such as smoke, flame, and heat, and the reading is sent to the Wi-Fi module for data analysis and trigger water sprinkle. The main feature is to minimize false alarms. Further systems can be integrated with ML to predict the possibility of fire [3]. J. Joy Mathavan et al. propose an automatic door unlock mechanism along with a local alarm for the passengers of the automobile in case of any automobile fire detection. The same is achieved by using servo motors interfaced with other fire detection sensors (temperature, CO_2, CO, etc.). To achieve maximum efficiency, each door is installed with its own system [4].

An ARM Cortex-M4 microprocessor-based system for accident prevention and vehicle speed control is suggested by Waladur Rahman et al. in their study. The system uses ultrasonic sensors of the vehicles to detect obstacles in the front and adjusts the speed of the vehicle accordingly to maintain a safe following distance. Processing of data is done using ARM M4 with the help of Arduino UNO which sends the signal to the wiper motor to apply brakes. The system also includes a collision avoidance feature that applies the brakes if the system detects an imminent collision [5]. Rohika Reddy Ala et al. found that lack of communication and delay in getting help increases the loss of property and life. So they have developed a fire detection system using a fire sensor. This system triggers an alarm when the fire is detected and simultaneously it sends a notification to our mobile, and a mail can be sent to the attached mail ID which will have information about the location and alert the fire station. It provides more safety than conventional fire alarm systems [6]. The system proposed by Asma Mahgoub et al. consists of several distributed nodes, each of which has an ESP8266 node MCU microcontroller connected to a variety of sensors that continuously scan for fire. The nodes communicate with a central node built with a Raspberry Pi microcontroller outfitted with a 4G module via their own Wi-Fi networks. A centralized node is alerted when a node detects a fire, causing it to send an SMS to the user and the fire department [7].

Tianxiang Li et al. introduced an intelligent fire-fighting system that incorporates the NB-IoT technology and IoT platform. The system utilizes a smoke-fire detection and alarm device based on the latest industry standard. By employing NB-IoT wireless smoke sensors, the authors addressed the challenges of complex wiring and effectively reduced installation expenses. Additionally, there are many connections in NB-IoT (> 10K connections that could interface smoke sensors). Additional benefits include extremely low power consumption and an extremely durable new overlay [8]. Rakshit Shah et al. have presented a fire and smoke detection system capable of identifying smoke presence and detecting temperature increases. The system is designed to activate a buzzer as an alert mechanism and send notifications to mobile applications using a Wi-Fi module (ESP8266). The hardware components include an Arduino Uno, a temperature sensor, a smoke sensor, a Wi-Fi module, and a buzzer [9]. Robert Sowah et al. proposed an automated fire detection and control system that integrates flame, smoke, and temperature sensors along with a customized mobile carbon dioxide (CO_2) air-conditioning unit. The system utilizes fuzzy logic control systems connected to an Arduino microcontroller for implementation. The system's fire detection capabilities cover four specific areas in the car: the engine compartment, passenger cabin, boot, and the vicinity around the fuel tank. The experimental results demonstrate that the system effectively handled false alarms and successfully detected and extinguished fires in less than twenty seconds, achieving an impressive 97.33% detection rate [10].

Yuepeng Cui et al. proposed a LiDAR data processing method that can acquire high-resolution (10HZ), high-accuracy and micro-level trajectories of all cars from a roadside LiDAR sensor. Among the techniques utilized in the process are background filtering, RGBC lane identification, RF classification, DBSCAN clustering, and GNN tracking. With the help of testing vehicles and field data, this was tested. The DSRC network is used to transmit and receive the extracted LiDAR data [11]. M. A. Desima et al. have made a fire detector and collision to avoid accidents due to leaky fuel drains and short circuit connections. The system is implemented using a smoke sensor (MQ-135), fire sensor module DFR0076 (range 16 cm) to detect fire and a push button to detect accident GSM SIM800L module for giving alerts and the system is controlled by Arduino. It is a fast method to register an accident by SMS and Google Maps [12]. In this paper, Azubuike Aniedu et al. have proposed a system that provides a real-time wildfire monitoring system and SMS alert. The environmental conditions are monitored using fire and smoke sensors. The signals regarding measured fire and smoke are given to the control unit the data is processed, then for abnormal conditions, fire alarm is triggered, and simultaneously, alert is sent to the mobile via SMS regarding environmental conditions [13]. Tianxiang Li et al. introduced an intelligent fire-fighting system that incorporates the NB-IoT technology and IoT platform. The system utilizes a smoke-fire detection and alarm device based on the latest industry standard. By employing NB-IoT wireless smoke sensors, the authors addressed the challenges of complex wiring and effectively reduced installation expenses. Additionally, there are many connections in NB-IoT (> 10K connections that could interface smoke sensors). Additional benefits include extremely low power consumption and an extremely durable new overlay [8]. Rakshit Shah et al. have presented a fire and smoke detection system capable of identifying smoke presence and detecting temperature increases. The system is
designed to activate a buzzer as an alert mechanism and send notifications to mobile applications using a Wi-Fi module (ESP8266). The hardware components include an Arduino Uno, a temperature sensor, a smoke sensor, a Wi-Fi module, and a buzzer [9]. Robert Sowah et al. proposed an automated fire detection and control system that integrates flame, smoke, and temperature sensors along with a customized mobile carbon dioxide (CO_2) air-conditioning unit. The system utilizes fuzzy logic control systems connected to an Arduino microcontroller for implementation. The system's

fire detection capabilities cover four specific areas in the car: the engine compartment, passenger cabin, boot, and the vicinity around the fuel tank. The experimental results demonstrate that the system effectively handled false alarms and successfully detected and extinguished fires in less than twenty seconds, achieving an impressive 97.33% detection rate [10].

According to a paper presented by Robert Sowah et al., an automatic system consisting of smoke sensors, temperature sensors, flame sensors, and a modified mobile carbon dioxide air conditioner is proposed. According to the findings, the auto fire detection and control system detects and extinguishes fire in less than 20 s without issuing any false alarms [14]. Charles B. et al. created a model that employs a temperature and smoke-sensing unit that detects temperature and smoke associated with fire and generates various levels of alarm based on the amount of time that the atmospheric changes associated with fire have been continuously increasing. The control panel stores previous data for each sensing unit and performs cross-correlations to detect fire accurately [15].

From this literature review, various innovative systems for accident detection, fire detection, and collision avoidance in different contexts are proposed. These studies demonstrate a range of innovative approaches to enhance safety and security in various scenarios, from motorcycles to vehicles and fire emergencies, highlighting the significance of advanced technology and intelligent systems in preventing accidents and mitigating potential risks. The existing system, however, cannot send the real-time location of the victim and alert message to the respective people and fire stations. To solve these problems, our system is designed with an A9G module which is integrated with GPS, GPRS, and a microphone.

15.3 Proposed Methodology

This system design mainly consists of five parts:

1. **Sensing Unit**
   The sensing unit detects the presence of fire using the REES52 IR Flame Sensor Module and MQ2 smoke sensor. The flame sensor uses the infrared flash flame method, which makes it easy to operate in case it gets coated with oil, dust, etc., which are prevalent conditions under the hood of any car. When a flame is detected, the flame sensor gives the signal as flame detected or not (0 or 1) or smoke above a certain threshold is detected then there may be fire detected.
2. **Software Unit**
   The software component of this project was implemented on Arduino IDE. The code for reading sensor data, comparing it with the threshold values defined for the ignition, and manipulating servo motors are written in C programming language.

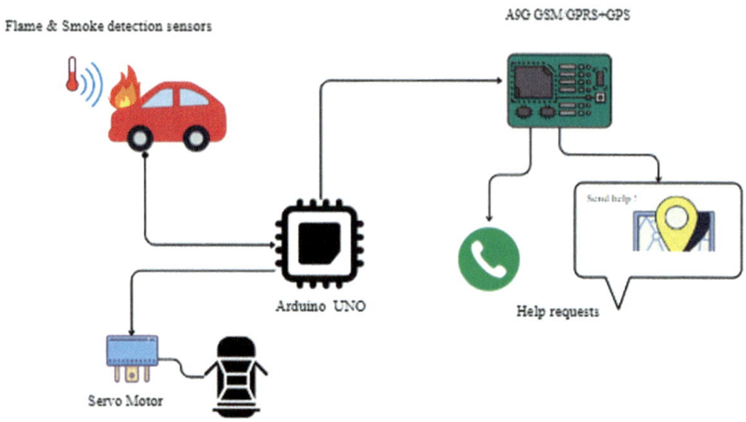

Fig. 15.1 System architecture of implemented system

3. **Processing Unit**
   The processing unit controls the system based on the signals from the sensing unit. ArduinoMega2560 board is used for the system control. ATMega328p microcontroller of the Arduino board takes the corresponding decisions based on code.
4. **Actuator Unit**
   The actuator unit takes the appropriate actions, such as opening the door of the vehicle and sending SMS messages. The servo motor is used to open the door of the vehicle automatically for the safety of the driver and other people in the vehicle and a buzzer starts ringing to alert the driver about suspicious conditions.
5. **Communication Unit**
   The communication unit sends SMS messages and calls to the fire station and related people on fire detection using the Ai Thinker A9G GSM/GPRS+GPS/BDS module. A9G module is an alternative for both GSM and GPS modules. As it is integrated with GSM and GPS technology, this A9G module sends the SMS with the location of the vehicle and calls to respective people, fire stations, hospitals, etc. We use AT commands to facilitate these features. The module also has a microphone integrated to facilitate audio conversations and give important information during times of emergency (Fig. 15.1).

15.3.1 Algorithm

1. Initialize the system:
2. Execution:

- 2.1. Read analog value from the smoke sensor and digital value from the flame sensor.
- 2.2. Check if the smoke sensor value exceeds a predefined threshold or if the flame sensor is triggered.
- 2.3. If a flame or smoke is detected:
  - 2.3.1. Activate the servo to unlatch the vehicle door.
  - 2.3.2. Activate the buzzer for an audible alert.
  - 2.3.3. Start monitoring GPS data.
    - 2.3.3.1. Wait for valid GPS data for a maximum of 2 s.
    - 2.3.3.2. If no valid data received, display a warning message.
    - 2.3.3.3. Send GPS data to designated phone number every 10 s.

3. GPS Data Monitoring and Sending:

- 3.1. Display latitude and longitude for debugging purposes.
- 3.2. If two sets of latitude and longitude data are collected:
  - 3.2.1. Reset the data counter.
  - 3.2.2. Prepare SMS with Google Maps link based on latitude and longitude.
  - 3.2.3. Set GSM module to text mode and configure notifications.
  - 3.2.4. Send the prepared SMS to the designated phone number.
  - 3.2.5. Wait for acknowledgement from the GSM module.
  - 3.2.6. Clear the message and reset the link for the next data collection.

4. Continue execution to monitor fire and GPS data.

15.4 Results and Discussion

The proposed methodology utilizes advanced fire detection sensors and real-time location tracking for early fire detection and swift emergency response. Upon fire detection, an integrated sound system activates, providing an audible alert within the vehicle, promptly notifying passengers of the emergency situation. Unlike the existing manual notification system, our approach automatically alerts fire stations, hospitals, and police, reducing response times significantly. Additionally, the integration of automatic door opening ensures quicker and safer evacuation, minimizing injuries. By combining these features, our comprehensive and proactive solution surpasses the limitations of current manual-based approaches, enhancing passenger safety and reducing potential damages caused by vehicle fires. The system Fig. 15.2 shows the successful integration of the components.

The system successfully detected vehicle fires at their early stages with a high level of accuracy, minimizing the risk of extensive damage and potential harm to passengers. Real-time location tracking using the GPS of the A9G module provided swift and precise information to emergency responders as shown in Fig. 15.4. Figure 15.3 demonstrates that immediate alerts are sent to fire stations, hospitals, and police facilitating prompt coordination and appropriate action using the GSM module which is

Fig. 15.2 Integration of the components

integrated with the A9G board. The A9G module's microphone feature helps to provide more information about the severity of the situation. The system started the fire alarm on detection of the fire and the integration of automatic door opening significantly improved evacuation procedures. Every component integrated into the system is very crucial to the overall success of the system. The implemented system provided an impressive solution using the module in response to network problems with 2G SIM.

Throughout this research, comprehensive evaluations are conducted to assess the system's performance under diverse scenarios encompassing various vehicle types and environmental conditions. By scrutinizing the system's accuracy, sensitivity, and reliability, we aim to substantiate its efficacy in meeting the challenges posed by vehicle fire incidents.

15.5 Conclusion and Future Scope

In conclusion, this project presents a comprehensive and innovative approach to enhance vehicle fire safety. The integration of advanced fire detection sensors, real-time location tracking, immediate alerts, an integrated sound system, and multi-sensor integration offers a proactive and efficient solution. By promptly detecting fires, optimizing emergency response, and facilitating safer evacuations, the proposed methodology aims to significantly reduce damages and enhance passenger safety during vehicle fire incidents. The results of this research highlight the efficacy of the system in mitigating risks and minimizing harm caused by vehicle fires. Emphasizing collaboration between technological advancements and regulatory adoption, the proposed methodology can pave the way for safer transportation and contribute to a safer road environment. Overall, this project underscores the importance of

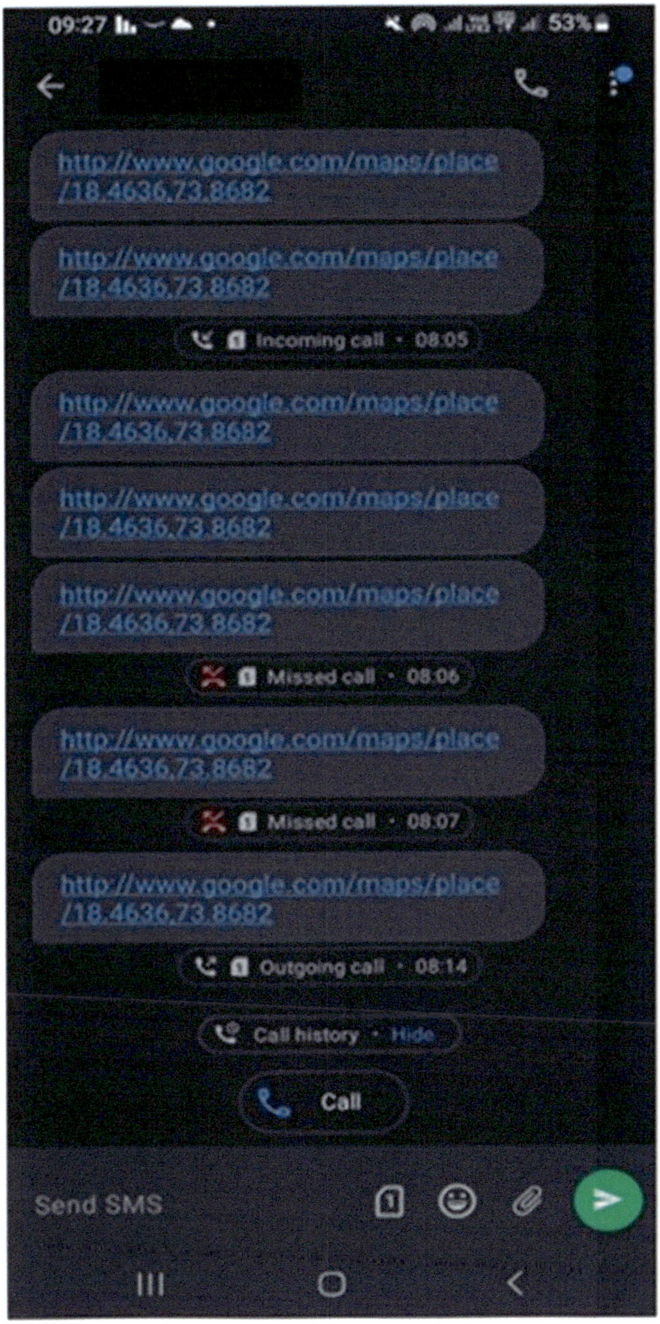

Fig. 15.3 Alert message with location details

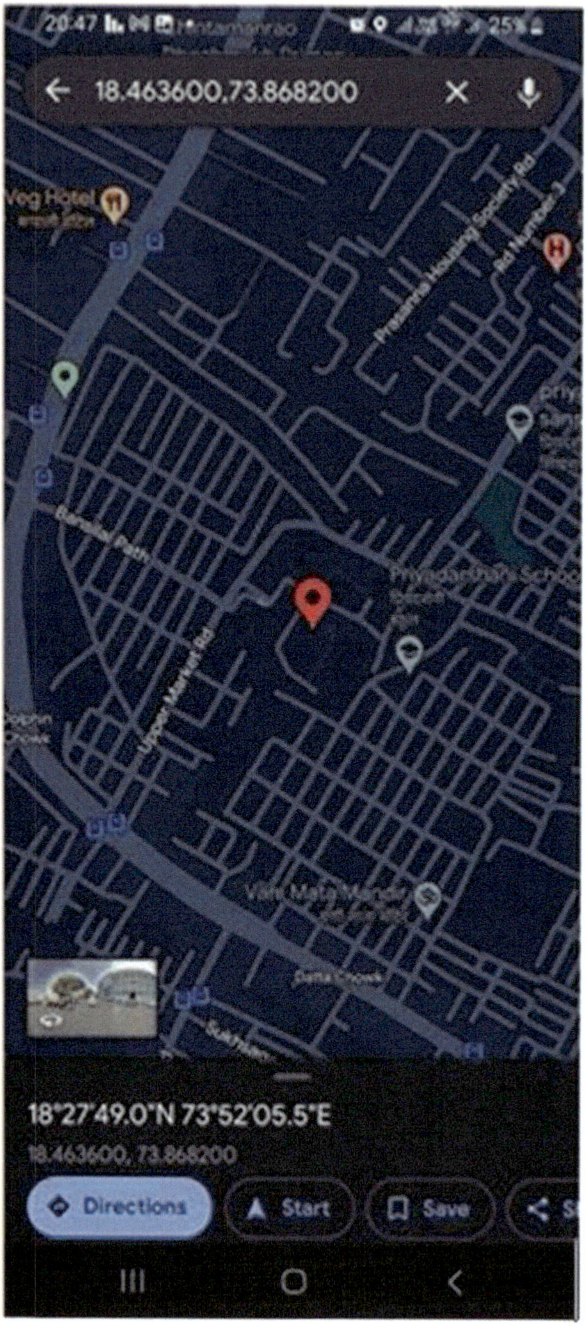

Fig. 15.4 Location of vehicle received with SMS

leveraging cutting-edge technologies to revolutionize emergency management and protect lives on road fire incidents and preserve human lives worldwide.

The future scope of the proposed vehicle fire safety system encompasses a range of transformative possibilities driven by technological advancements and collaborative efforts. The integration of AI and ML models holds promise for enhancing the system's fire detection capabilities through data pattern analysis and continuous learning, while predictive analytics can anticipate potential fire hazards, enabling proactive prevention. Cloud-based data sharing offers seamless communication with emergency responders and ensures the availability of critical information, even in remote areas, while accommodating vast datasets for improved efficiency. Vehicle-to-vehicle communication can create a safety network, providing real-time safety alerts to nearby drivers, thus enhancing overall road safety. Moreover, multi-sensor integration can refine the system's understanding of fire incidents, optimizing responses based on the specific nature of the fire. Regulatory adoption is pivotal in promoting widespread implementation of the system, making it a norm across the automotive industry, and safeguarding lives on a global scale. Embracing these future directions, the proposed system stands to revolutionize vehicle fire safety, preserving lives, and minimizing damages.

References

1. Ur Rehman, S., Khan, S.A., Arif, A., Khan, U.S.: IoT-based accident detection and emergency alert system for motorbikes. In: 2021 International Conference on Artificial Intelligence and Mechatronics Systems (AIMS) (2021)
2. Huang, C.-Y.: Vehicle door opening control model based on a fuzzy inference system to prevent motorcycle-vehicle door crashes. Sustainability **13**, 12558 (2021)
3. Alqourabah, H., Muneer, A., Fati, S.M.: A smart fire detection system using IoT technology with automatic water sprinkler. IJECE **11**, 2994 (2021)
4. Mathavan, J.J., Faslan, A.W., Basith, N.U.A., Wanigasinghe, W.V.S.D.: Hardware Implementation of fire detection, control and automatic door unlocking system for automobiles. In: 2020 4th International Conference on Trends in Electronics and Informatics (ICOEI) (48184) (2020)
5. Rahman, W., Ruman, Md.R., Jahan, K.R., et al.: Vehicle speed control and accident avoidance system based on arm M4 microprocessor. In: 2020 International Conference on Industry 4.0 Technology (I4Tech) (2020)
6. Reddy, A.R., Sai, M.V.R., Vamsi, J., Raju, C.S.K.: Automatic fire detection and alert system. J. Xi'an Univ. Archit. Technol. **12** (2020)
7. Mahgoub, A., Tarrad, N., Elsherif, R., et al.: IoT-based fire alarm system. In: 2019 Third World Conference on Smart Trends in Systems Security and Sustainability (WorldS4) (2019)
8. Li, T., Hou, P.: Application of NB-IoT in intelligent fire protection system. In: 2019 International Conference on Virtual Reality and Intelligent Systems (ICVRIS) (2019)
9. Shah, R., Satam, P., Sayyed, M.A., Salvi, P.: Wireless smoke detector and fire alarm system. Int. Res. J. Eng. Technol. (IRJET) **6**(1), 1407–1412 (2019)
10. Sowah, R., Ampadu, K.O., Ofoli, A.R., et al.: A fire-detection and control system in automobiles: implementing a design that uses fuzzy logic to anticipate and respond. IEEE Ind. Appl. Mag. **25**, 57–67 (2019)
11. Cui, Y., Xu, H., Wu, J., et al.: Automatic vehicle tracking with roadside LiDAR data for the connected-vehicles system. IEEE Intell. Syst. **34**, 44–51 (2019)

12. Desima, M.A., Ramli, P., Ramdani, D.F., Rahman, S.: Alarm system to detect the location of IOT-based public vehicle accidents. In: 2017 International Conference on Computing, Engineering, and Design (ICCED) (2017)
13. Aniedu, A.N., Chukwuneke, C.I., Asogwa, D.C., Nwokoye, C.S.: Real-time wildfire monitoring and alert system using GSM technology. IOSR **03**, 01–10 (2016)
14. Sowah, R., Ampadu, K.O., Ofoli, A., et al.: Design and implementation of a fire detection and control system for automobiles using fuzzy logic. In: 2016 IEEE Industry Applications Society Annual Meeting (2016)
15. Winterble, C.B., Seyouri, R.R.: Early fire detection using temperature and smoke sensing. U.S. Patent US6195011B1, filed 15 Apr 1998, and issued 27 Feb 2001

Chapter 16
Multimodal Data Fusion: Combining Big Data and Deep Learning for Enhanced Predictive Models

K. R. Sreelakshmi and Sajimon Abraham

Abstract In recent years, the integration of multimodal data and the application of advanced machine learning techniques have revolutionized the domain of predictive modeling. This research investigates the potential of multimodal data fusion, specifically combining textual and image-based information, using convolutional neural networks (CNNs) to predict house prices accurately. This research paper presents an advanced hybrid convolutional neural network (CNN) architecture designed to predict house prices based on house attributes, including numerical attributes, textual attributes, spatial information, and images. This defines the multimodal attribute of the data used in the study. The proposed model incorporates residual and skip connections to capture complicated and detailed spatial patterns in the input images and improve overall prediction performance. This helps to resolve the vanishing gradient problem that occurs with the deep neural networks. The study demonstrates the effectiveness of the proposed model on a real-world housing dataset, achieving accurate price predictions. The model is evaluated using mean squared error (MSE) as the loss function, and the Adam optimizer is utilized for efficient convergence during training. The research emphasizes the importance of integrating both numerical and visual information for better housing price estimation. Experimental results highlight the capability of advanced CNN to effectively learn features from images, leading to enhanced prediction accuracy in real estate valuation.

K. R. Sreelakshmi (✉)
School of Computer Sciences, Mahatma Gandhi University, Kottayam, Kerala 686560, India
e-mail: sreelakshmikr@mgu.ac.in

S. Abraham
School of Management and Business Studies, Mahatma Gandhi University, Kottayam, Kerala 686560, India
e-mail: sajimonabraham@mgu.ac.in

P. Nanda et al. (eds.), *Intelligent System and Data Analysis*, Smart Innovation, Systems and Technologies 398, https://doi.org/10.1007/978-981-97-5200-3_16

16.1 Introduction

As the title indicates the entire the scope of this research study is to utilize the advanced deep learning models for predictive analytics with multimodal data. The multimodal data indicates the data in various formats collected from different sources. This can include text, image, audio, video, time-series data, spatial data, or satellite data. This forms big data satisfying some of its characteristics like variety and volume. Processing and analyzing this data are a complicated task. The proposed system does take multimodal data as the input and processes it to predict the output. The following subsections give a brief idea on big data and multimodality of data, how the data used in this proposed model become a multimodal data and deep neural networks that are used for predictive analytics.

16.1.1 Big Data

The multiplication of data throughout the digital age has reached unprecedented proportions, giving rise to "big data." Big data refers to large and complicated datasets that are too large for typical data processing and analysis methods. But it is not just about how big the data is. It has many other characteristics that show its complexity. These properties are called v's of big data. Some of them are variety, velocity, value, veracity, viscosity, and more v's are being added to the list. These characteristics hinder the opportunities for gaining useful insights and knowledge. Deep neural networks have emerged as important tools for processing and analyzing big data in this context due to their capacity to learn sophisticated patterns and representations from massive datasets [1]. In their paper, they have given a wide view of how big data poses challenge to data processing and investigate some aspects of deep learning research that requires additional investigation to address unique big data analytics issues, such as streaming data, high-dimensional data, model scalability, and distributed computing. In our paper, we have taken a practical application and tried solving the task which shows considerably good results. Big data predictive analytics utilizing machine learning techniques is a hot topic. Deep learning has achieved great success in predicting numerous fatal diseases such as pneumonia as the number and complexity of medical data such as X-rays has increased. Jakhar et al. [2] Besides medical data, social media data, e-commerce data, maps and navigation systems, sensors data, satellite and surveillance data, business and real estate data contribute to big data. The architecture of big data analytics (see Fig. 16.1) is designed so as to make sure it can handle enormous data ingested into the system, regardless of streaming data or batch data, and then process and analyze data with utmost accuracy which is not possible with the traditional systems.

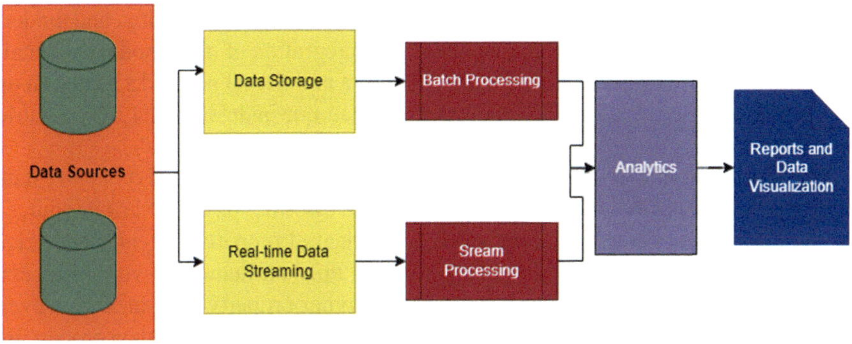

Fig. 16.1 Architecture of big data analytics

16.1.2 Multimodal Data Fusion

In today's data-driven world, the information we encounter is not limited to a single type or format. Instead, it often comprises multiple modalities, each conveying distinct aspects of knowledge. Multimodal data refers to datasets that contain information from diverse sources, including but not limited to text, images, audio, video, and sensor readings. By combining different modalities, multimodal data offers a richer and more comprehensive representation of real-world phenomena, enabling a deeper understanding of complex systems.

The widespread use of multimodal data has increased with the advancement of technology and the emergence of the Internet of Things (IoT), social media, and multimedia content. Computer vision, natural language processing, speech recognition, driverless vehicles, healthcare, and finance are just a few of the applications that use multimodal data. In healthcare, for example, patient records may contain textual medical reports, diagnostic imaging, and time-series data from wearable devices. Data streams from cameras, LiDAR, and GPS systems in autonomous cars form a multimodal input for navigation and decision-making.

Processing multimodal data poses novel challenges and opportunities. On the one hand, the fusion and integration of information from various modalities necessitates the use of advanced algorithms and approaches that accurately represent the links and dependencies between them. On the other hand, the mix of modalities opens up new pathways for data analysis, allowing academics and practitioners to tackle previously unsolvable complicated challenges.

The discipline of multimodal data analysis includes many research fields, such as multimodal machine learning, multimodal fusion, cross-modal retrieval, and multimodal interaction. Researchers hope to create novel approaches that take advantage of the complementary nature of many modalities in order to increase the accuracy, resilience, and interpretability of machine learning models.

As multimodal data becomes more prevalent, investigating its potential becomes increasingly important for generating comprehensive insights, making informed decisions, and developing intelligent systems capable of processing and understanding information from several sources. The integration of multimodal data analysis and emerging technologies like deep learning and neural networks can open up new horizons in artificial intelligence and advance our understanding of the world around us.

Multimodality in Real Estate Data. Multimodal data, in the context of real estate price prediction, refers to datasets that incorporate information from various sources or modalities to better estimate the value of residential properties. Real estate price prediction algorithms have traditionally focused only on numerical variables such as total area, number of bedrooms, location, and other property attributes. However, with the growing availability of multimedia data and developments in data processing techniques, adding multimodal data has shown to be a significant way for improving the accuracy and robustness of house price prediction models.

Multimodal data used in the proposed method. Multimodal data in the context of real estate price prediction can include:

Numeric Data: Traditional aspects such as the number of bedrooms, baths, square footage, lot size, and other property-specific qualities are included in numerical data. Numerical data gives useful information about the physical properties of the house.

Textual Data: Textual descriptions that provide more facts about the property, such as its distinctive features, amenities, and neighborhood characteristics, are frequently included in house listings. Analyzing and extracting information from textual data can provide insights that numerical features alone cannot provide.

Visual Data: Images and videos of the house and its environs provide a wealth of information for predicting house prices. Visual data allows the model to learn from the appearance, architecture, and curb appeal of the property, all of which are important elements.

Geographic Information: Location is an important aspect in deciding house values. Geographic data, such as maps and geospatial information, as well as accessibility to amenities, schools, and transit, can all have a substantial impact on the model's predictions.

Multimodal house price prediction models may take advantage of the value-added nature of the information and overcome limitations associated with depending simply on a single type of data by incorporating these different modalities. For example, while numerical features capture key property attributes, visual data can assist in determining the property's condition and aesthetic attractiveness. In contrast, textual data can provide additional context and insights about the property and its immediate surroundings.

16.1.3 Deep Neural Networks

Deep neural networks (DNNs) have transformed many fields by providing cutting-edge performance in complex tasks such as image identification, natural language processing, and speech recognition. However, the vanishing gradient problem can hinder the convergence and learning process, so training deep neural networks can be challenging.

The vanishing gradient problem occurs in deep neural networks during backpropagation, which changes the model's parameters based on the training error. As the gradient is carried backward via the network layers, it might rapidly diminish as it approaches the original levels. As a result, the weights in these early layers only receive minimal updates during training, resulting in weak convergence or stagnation. This is especially true in networks with several layers, such as deep convolutional neural networks or recurrent neural networks.

The vanishing gradient problem can have various negative implications on deep neural network performance and training:

Slow Training: When the gradients in the early layers become too small, the training process can become very slow, requiring an excessive number of iterations to produce satisfactory results.

Limited Depth: The vanishing gradient problem constrains the depth of the network that may be efficiently taught. The gradients become smaller as the network depth increases, eventually leading to the inability to learn meaningful representations.

Overfitting: Vanishing gradients can cause overfitting in some cases, where the model performs well on training data but fails to generalize to unseen data.

Several techniques have been developed to address the vanishing gradient problem and enable the training of deeper neural networks. One of the most effective methods is to use residual or skip connections.

The famous residual network (ResNet) architecture incorporated residual connections. Residual networks learn the residual mapping rather than the desired underlying mapping. This is accomplished by incorporating a shortcut or skip connection that bypasses one or more levels, allowing the network to learn the difference between input and output immediately. A residual block can be expressed mathematically as:

$$\text{Output} = F(\text{Input}) + \text{Input}. \tag{16.1}$$

Here $F(\text{Input})$ in Eq. (16.1) represents that the mapping to be learned by the layers within the block, and the addition of input preserves the original information. By using residual connections, the network can learn the identity mapping even when the layers learn to do nothing, mitigating the vanishing gradient problem.

Residual connections make it possible to train incredibly deep neural networks. They facilitate gradient flow during backpropagation by allowing data to flow straight across the network, reducing the danger of vanishing gradients. As a result, deeper models can be trained more successfully, resulting in greater performance and generalization.

16.2 Related Works

The use of skip connections in neural networks, particularly in the ResNet design, was a game changer in the field of deep learning. Skip connections, also known as residual connections, have had a significant impact on deep neural network training. Training incredibly deep networks with hundreds or even thousands of layers were extremely difficult prior to their emergence due to the vanishing gradient problem. Skip connections are used to eliminate singularities [3], image super-resolution [4], norm preservation [5], and more. Yang et al. [6] expand the benefits of residual networks for a common low-level vision task, such as single picture super-resolution, by using techniques like skip connection-induced quick training. Qin et al. [7] note that techniques based on deep convolutional neural networks (CNN) for single image super-resolution (SISR) have demonstrated remarkable performance in terms of quantitative metrics and visual impacts. Yamanaka et al. [8] propose a highly efficient and faster single image super-resolution (SISR) model with deep convolutional neural networks (Deep CNN). Jiang et al. [9] investigate deep distillation recursive networks for super-resolution in remote sensing imagery. For video satellite image SR, a straightforward yet effective CNN structure called the deep distillation recursive network (DDRN) is given. To this end [10] PECNN, a brand-new progressively upgraded network for satellite image SR that combines an enhanced dense connection network and a pretraining CNN-based network, is being proposed. Wen et al. [11] suggest a deep convolutional network called densely connected residual networks (DRNet). Pardede et al. [12] evaluate different deep learning models that are currently in use for the identification of plant diseases using low-resolution data. For learning a classification model from a large number of modulated signals, a novel deep convolutional neural network (DCNN) with several convolutional blocks specialized to simultaneously capture temporal intra-signal correlations and spatial intra-signal relations is proposed [13]. Qi et al. [14] explore a deep learning-based two-dimensional electromagnetic solution. Convolutional neural networks (CNNs) have been used to forecast the electromagnetic field (EM field) scattered by intricate geometries when illuminated by plane waves. Huang et al. [15] create a unique dual-path attention network that combines dual skip connections (DSCs) with dual attention groups (DPAGs) to combine the benefits of dense connections and residual connections for improved SR performance. Hybrid CNNs with two different CNN architectures are models which are used for segmentation [16]. Hybrid CNN is used for efficient for time-series analysis [17]. Gupta et al. [18] launch BRNet, a revolutionary deep regression neural network with branched residual learning that uses layer branching to maximize various features acquired from the preceding or input layer and skips connections after each layer to minimize information loss due to diminishing gradient. For efficient segmentation, multiple skip connections and an attention mechanism are used [19].

16.3 Methodology

This section primarily discusses the data used, the techniques involved in preprocessing the data, and the proposal of the hybrid deep CNN model to predict the house prices.

16.3.1 The Data

The dataset: The dataset for this research study has been taken from Kaggle, and it is freely available to use. Initially, this dataset was manually collected and labeled from publicly available information on real estate websites. There are no repeated data or missing one and made it publicly available. The dataset includes 536 samples and each sample has exactly four images, thus counting to a total of 2144 images in the dataset. All the numeric information is stored on a.txt file.

The data description: The dataset consists of numeric, spatial, and visual information. The number of rooms and bathrooms, the overall dimensions of the house or building, and the price of the property are all numerical facts. The spatial information includes zip codes that give the location or neighborhood details of the property. The visual information consists of four images of each of the properties in the dataset. These images are of the bedroom, bathroom, kitchen, and frontal view (see Fig. 16.2).

The four individual images are stacked and arranged as above using Python code so as to process it as a single image. Now it will be easy for feature extraction.

16.3.2 Data Preprocessing

The first and most important task in the data preprocessing step was to assign headers to the. txt file, as there were none. A function is defined to iterate over rows in text file and is assigned to the corresponding header. The function assigns column names to the data frames.

Removing Outliers: The program then loops through each unique zip code and its count, removing any rows (houses) that have fewer than 25 houses per zip code from the data frame. So that we get enough data for generalization.

Scaling: The data is scaled to bring them within the range [0,1] using min–max scaling technique. Min–Max normalization is a common method used to scale numerical data to a specific range, typically between 0 and 1. The formula for Min–Max normalization is as follows:

For a single data point x:

$$x_{\text{normalized}} = \frac{x - \min(X)}{\max(X) - \min(X)}, \tag{16.1}$$

Fig. 16.2 The sample images of one house. These images are stacked to form a single image as shown in the figure

where

x is the original value of the data point.

X is the set of all data points of the feature of interest in the dataset.

$\min(X)$ is the minimum value of the feature of interest in the dataset.

$\max(X)$ is the maximum value of the feature of interest in the dataset.

One-hot Encoding: Now the zip codes in the dataset are encoded using one-hot encoding technique and are then transformed into binary vectors.

Now another function is used to iterate through house IDs used in naming images and OpenCV is used to read and resize images. Then the processed images are concatenated into a single frame as in Fig. 16.2. Now this is kept in a list as NumPy arrays.

16.3.3 Hybrid Deep Convolutional Network

The basic idea is to use a hybrid deep convolutional network architecture that could handle both the numeric and image data and the output from both the input layers is concatenated and given to a fully connected CNN layer that predicts the final output which is the price of the property.

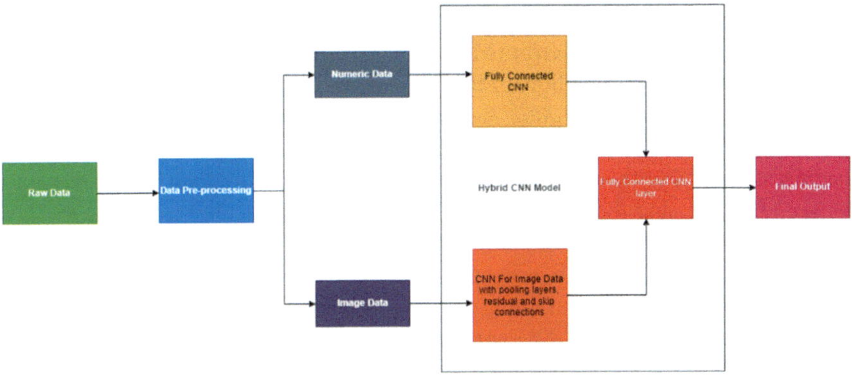

Fig. 16.3 The system process flow pipeline with hybrid CNN model

TensorFlow's Keras API is used to create an advanced convolutional neural network (CNN) model. The model contains some distinguishing features, such as the utilization of residual connections and dilated convolutions.

Hybrid CNN: The numeric data is given to a fully connected CNN layer and the image data is given to a convolutional network with pooling layers and residual or skip connections as shown in Fig. 16.3. This helps to keep track of gradients and information in the nodes of neural network. However, the complexity of architecture increases.

The model's Sequential() method creates an empty sequential model where layers can be added one after another. The model starts with two convolutional layers (Conv2D) with 32 filters each, followed by a MaxPooling2D layer to reduce spatial dimensions.

The output tensor of the first Conv2D layer is saved, which will be used later for a residual connection. Two more convolutional layers with 64 filters each and a MaxPooling2D layer are added to the model. A Conv2D layer with a stride of (2, 2) is used to adjust the spatial dimensions and number of channels in the output of first convolutional layer. This adjusted tensor is then added to the current output tensor using a residual connection. Two more convolutional layers with 128 filters each and a MaxPooling2D layer are added to the model.

Similar to the previous step, an adjustment to the spatial dimensions and number of channels in the second residual connection is made and then added to the current output tensor using another residual connection. Two more convolutional layers with 256 filters each are added, using dilated convolutions with a dilation rate of 2. This allows the model to have a larger receptive field without increasing the spatial dimensions.

The Flatten() layer is used to flatten the output tensor before passing it through the dense layers. The model includes a dense layer with 128 units and to make the computation easy for the network it uses ReLU activation function. Since this is a prediction task, the output layer consists of only a single neuron without any activation function, making it suitable for predicting the target variable, which is a regression task. The model is built using the Adam optimizer with a learning rate of 0.001 and the mean squared error (MSE) loss function for training.

The program flow: The flow of the program is defined below.

Load and preprocess the data. A user-defined method is used to load the house attributes data from a file called "HousesInfo. txt" and preprocesses the images by loading and resizing them to (32, 32) pixels. The images are then scaled to values between 0 and 1, and the target values (house prices) are scaled to the range [0, 1] by dividing them with the maximum price.

Train-Test Split. The train_test_split function from sklearn. Model_selection is used to divide the data into training and testing sets. 20% of the data are used for testing, while the remaining 80% are used for training.

Hybrid CNN model. The proposed hybrid architecture is used here. It consists of several convolutional layers, max-pooling layers, residual connections, and dense layers. The final output layer has a linear activation function, suitable for regression tasks. As stated, being a hybrid model, it consists of two input layers which then connect to two different layers of CNN and the output from each model is given to a fully connected layer.

Compile the Model. Adam optimizer and mean squared error (MSE) loss function are used in the model's construction. An initial learning rate of 0.001 is used by the optimizer.

Fit the Model to the Training Data. The model is trained on the training data using the fit method. It runs for 20 epochs with a batch size of 32. Initially, model was trained with 200 epochs which took long time as well as model tend to overfit with over training. Model converges somewhere between 15 to 20 epochs. During training, a validation split of 20% is used to monitor the validation loss and prevent overfitting.

Evaluate the Model on the Test Data. After training, the model is evaluated on the test data using the evaluate() method. The test loss (MSE) is computed and printed.

Predict house prices. The model. Predict() method is used to predict the house prices. The predicted house prices are in accordance with the mean and standard distribution of all values in the dataset taken together.

Plot Graphs. The loss vs epoch graph is plotted for 20 epochs and 200 epochs with batch size 32.

16.4 Result Analysis

The suggested hybrid CNN system outperforms the sequential CNN model because it has significantly low training and validation loss when compared to the sequential model on working with this dataset and it also tends to converge faster, between 15 and 20 epochs, than the sequential CNN model, which does not demonstrate substantial convergence and is quite unstable even after running 200 epochs.

When sequential CNN model was trained for 200 epochs, it is clear from the graph that the validation loss is quite large, and the model is prone to overfitting and fails to generalize (see Fig. 16.6) Then the same model was trained with 20 epochs but it only helped to avoid overfitting but the validation loss remained very high (see Fig. 16.4)

Now the other two graphs are showing the epoch vs. loss of hybrid CNN model. It is clear that (see Fig. 16.5) hybrid CNN model considerably reduces the test and validation loss. Besides that the model also tends to converge in very less time in comparison with the sequential CNN model and also the model seems to be much stable. However, when a number of epochs were raised to 200 the hybrid model also seemed to overfit the data and there is inconsistency in test and validation loss graph (see Fig. 16.7). It does not converge very well. This is why we go with hybrid model with 20 epochs. This experiment also highlights the importance of hyperparameter tuning.

The predicted house price houses prices are evaluated using MSE. In regression tasks, the mean squared error (MSE) is a regularly used metric to assess the average squared difference between predicted and actual values. It aids in determining how

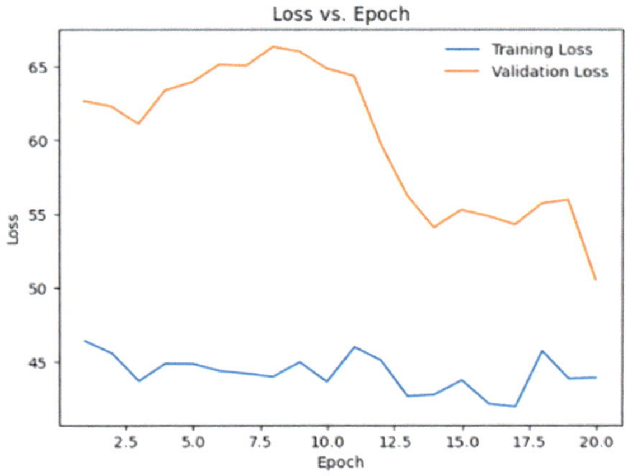

Fig. 16.4 Loss versus Epoch Graph showing the result of sequential CNN model for 20 Epochs

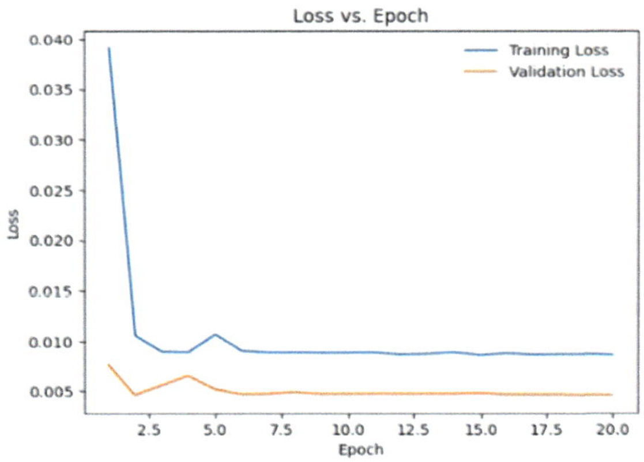

Fig. 16.5 Loss versus Epoch Graph showing result of Hybrid CNN model with residual or skip connections

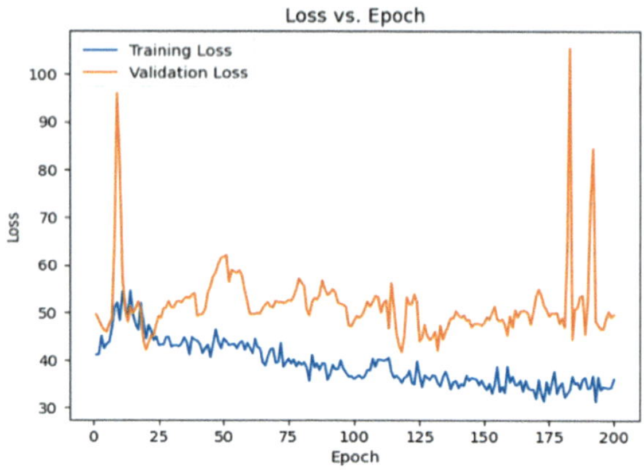

Fig. 16.6 Loss versus Epoch Graph of sequential CNN for 200 epochs

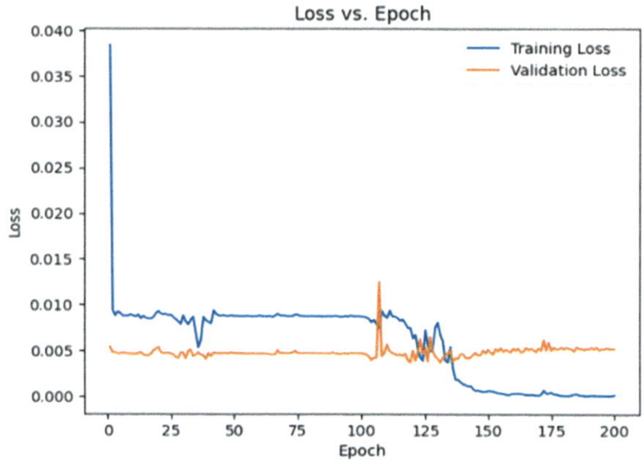

Fig. 16.7 Loss versus Epoch graph of Hybrid CNN model with residual or skip connections trained with 200 epochs

closely a model's predictions match the actual values. The MSE is determined as follows:

If we have N predictably valued data points $\hat{y}_i$ and actual values y_i for $i = 1, 2, \ldots, N$ the MSE is given by

$$\text{MSE} = \frac{1}{N}\sum_{i=1}^{N}(y_i - \hat{y}_i)^2, \tag{16.1}$$

where $(y_i - \hat{y}_i)$ gives the difference between the actual and anticipated values for each data point is represented in this equation. Squaring this difference guarantees that both positive and negative mistakes contribute to the overall error, and the MSE is calculated by taking the average of these squared differences.

The MSE measures the average prediction error, and a lower MSE implies that the model is performing better at predicting the target variable. It is commonly used in various regression tasks to assess the accuracy of predictions. The model gave considerably low MSE and significantly low training and validation loss compared to the sequential CNN model.

16.5 Conclusion

Integrating image and spatial data into big data analytics enables a thorough understanding of the data's underlying patterns and relationships. Using deep learning models, such as CNNs with residual and skip connections, improves the ability to handle large amounts of data with complicated features. These models can handle large-scale picture data and spatial information efficiently, extract significant features, and make accurate predictions, contributing to the efficient analysis and utilization of data in various fields.

16.6 Limitations

The residual or skip connections used in the architecture have both advantages and disadvantages. They may help in ensuring the retention of information and gradients, the combination of features, a cure of the vanishing gradient problem, and the formation of deeper networks. They can, however, raise complexity and memory requirements, introduce redundancy and noise, and necessitate careful design and tuning to fit the network architecture and data domain. The effects of different types and locations of skip connections on network performance can vary.

References

1. Najafabadi, M.M., Villanustre, F., Khoshgoftaar, T.M., Seliya, N., Wald, R., Muharemagic, E.: Deep learning applications and challenges in big data analytics. J. Big Data **2**(1), 1–21 (2015)
2. Jakhar, K., Hooda, N.: Big data deep learning framework using Keras: a case study of pneumonia prediction, In 4th International Conference on Computing Communication and Automation (ICCCA), Greater Noida, India, pp. 1–5, (2018) https://doi.org/10.1109/CCAA.2018.8777571
3. Yasrab, R.: SRNET: a shallow skip connection based convolutional neural network design for resolving singularities. J. Comput. Sci. Technol. **34**, 924–938 (2019)
4. Zhao, X., Song, Z.: Super-resolution reconstruction of deep residual network with multi-level skip connections. J. Electron. Inf. Technol. **41**(10), 2501–2508 (2019)
5. Zaeemzadeh, A., Rahnavard, N., Shah, M.: Norm-preservation: why residual networks can become extremely deep? IEEE Trans. Pattern Anal. Mach. Intell. **43**(11), 3980–3990 (2020)
6. Yang, Z., Zhang, K., Liang, Y., Wang, J.: Single image super-resolution with a parameter economic residual-like convolutional neural network. In: Amsaleg, L., Guðmundsson, G., Gurrin, C., Jónsson, B., Satoh, S. (eds.) MultiMedia Modeling. MMM 2017. Lecture Notes in Computer Science, Springer, Cham. Vol. 10132, (2017), https://doi.org/10.1007/978-3-319-51811-4_29

7. Qin, J., Zhang, R.: Lightweight single image super-resolution with attentive residual refinement network. Neurocomputing **500**, 846–855 (2022)
8. Yamanaka, J., Kuwashima, S., Kurita, T.: Fast and accurate image super resolution by deep CNN with skip connection and network in network. In Neural Information Processing: 24th International Conference, ICONIP 2017, Guangzhou, China, November 14–18, 2017, Proceedings, Part II 24 (pp. 217–225). Springer International Publishing. (2017).
9. Jiang, K., Wang, Z., Yi, P., Jiang, J., Xiao, J., Yao, Y.: Deep distillation recursive network for remote sensing imagery super-resolution. Remote Sensing **10**(11), 1700 (2018)
10. Jiang, K., Wang, Z., Yi, P., Jiang, J.: A progressively enhanced network for video satellite imagery superresolution. IEEE Signal Process. Lett. **25**(11), 1630–1634 (2018)
11. Wen, R., Fu, K., Sun, H., Sun, X., Wang, L.: Image superresolution using densely connected residual networks. IEEE Signal Process. Lett. **25**(10), 1565–1569 (2018)
12. Pardede, H.F., Suryawati, E., Zilvan, V., Ramdan, A., Kusumo, R.B.S., Heryana, A., Rahadi, V.P.: Plant diseases detection with low resolution data using nested skip connections. J. Big Data **7**, 1–21 (2020)
13. Huynh-The, T., Hua, C.H., Kim, J.W., Kim, S.H., Kim, D.S.: Exploiting a low-cost CNN with skip connection for robust automatic modulation classification. In 2020 IEEE Wireless Communications and Networking Conference (WCNC) (pp. 1–6). IEEE. (2020).
14. Qi, S., Wang, Y., Li, Y., Wu, X., Ren, Q., Ren, Y.: Two-dimensional electromagnetic solver based on deep learning technique. IEEE J. Multiscale Multiphys. Comput. Tech. **5**, 83–88 (2020)
15. Huang, Z., Li, W., Li, J., Zhou, D.: Dual-path attention network for single image super-resolution. Expert Syst. Appl. **169**, 114450 (2021)
16. Wang, Y.Z., Wu, W., Birch, D.G.: A hybrid model composed of two convolutional neural networks (CNNs) for automatic retinal layer segmentation of OCT images in retinitis pigmentosa (RP). Trans. Vis. Sci. Technol. **10**(13), 9–9 (2021)
17. Wibawa, A.P., Utama, A.B.P., Elmunsyah, H., Pujianto, U., Dwiyanto, F.A., Hernandez, L.: Time-series analysis with smoothed convolutional neural network. J. Big Data **9**(1), 44 (2022)
18. Gupta, V., Liao, W.K., Choudhary, A., Agrawal, A.: Brnet: branched residual network for fast and accurate predictive modeling of materials properties. In Proceedings of the 2022 SIAM International Conference on Data Mining (SDM) (pp. 343–351). Society for Industrial and Applied Mathematics. (2022).
19. Nodirov, J., Abdusalomov, A.B., Whangbo, T.K.: Attention 3D U-net with multiple skip connections for segmentation of brain tumor images. Sensors **22**(17), 6501 (2022)

Chapter 17
Analyzing Artificial Intelligence Based Intrusion Detection System in Detecting DDoS Attack

Sai Suchith Kurra, Likhitha Rayana, and Mohamed Sirajudeen Yoosuf

Abstract With the increasing complexity of the cyber-attacks, traditional intrusion systems have become inadequate in effectively detecting and preventing network attacks. Modern machine learning techniques have the capability of preventing sophisticated and aggressive cyber-attacks and have emerged as a promising approach to enhance the accuracy and efficiency of the intrusion detection. This research paper provides a comprehensive analysis and evaluation of the integration of machine learning and deep learning algorithms into intrusion detection system with respect to the Distributed Denial of Service (DDoS) attack. The study begins by presenting an overview of intrusion detection systems, highlighting their importance and significance in safeguarding network infrastructure. Subsequently, it explores supervised machine learning algorithms and deep learning models, such as Random Forest tree, Support Vector Machines (SVM), ResNet and TabNet. The paper presents the pros and cons of existing IDS models with respect to real-time processing, scalability, false positive and false negative results, and IDS vulnerabilities. The above-mentioned algorithms have been trained and evaluated with CICIDS-2017 dataset. The performance metrics used to evaluate the IDS models are accuracy, precision, F1 score, training time of each model, and adaptability to evolving attack patterns.

S. S. Kurra · L. Rayana
School of Computer Science and Engineering, VIT-AP University, Amaravati, Andhra Pradesh, India

M. S. Yoosuf (✉)
Department of Mathematics and Computer Science, Modern College of Business and Science, Muscat, Oman
e-mail: msyoosuf.research@gmail.com

P. Nanda et al. (eds.), *Intelligent System and Data Analysis*, Smart Innovation, Systems and Technologies 398, https://doi.org/10.1007/978-981-97-5200-3_17

17.1 Introduction

Network security has emerged as a paramount concern for individuals, companies, and organizations in today's digital landscape. The vulnerability of networks to various attacks has significantly escalated due to the increasing reliance on technology and the rapid advancement of the Internet of Things (IoT). Detecting cyber-attacks at an early stage is crucial for effectively safeguarding the data and IT infrastructure of any company. Modern artificial intelligence models, including machine learning and deep learning, aid in the development of efficient Intrusion Detection Systems (IDS) that can identify cyber-attacks at their initial level [1]. The AI models are capable of large-scale data analysis, pattern recognition, precise prediction, and decision-making.

Over the past few years, Distributed Denial of Service (DDoS) attacks have posed a significant challenge as they disrupt network operations and can cause severe damage to reputation and finances [2]. DDoS attacks involve overwhelming a target system or network with a substantial amount of malicious traffic originating from one or more sources [3]. The purpose of such attacks is to incapacitate the targeted system or network, rendering it inaccessible to authorized users. DDoS attacks involve a series of six phases.

- Phase 1 entails identifying the target and gathering information about their network infrastructure, vulnerabilities, and potential points of exploitation.
- In Phase 2, known as Botnet Recruitment, the attacker infects numerous devices, including PCs and IoT devices, with malware to transform them into "Zombie machines." These compromised workstations become part of the attacker's botnet, which can be remotely controlled.
- Phase 3, Command and Control (C&C), involves establishing connectivity with the infected devices through a command-and-control server. This server coordinates the attack by issuing instructions to the compromised devices.
- In Phase 4, Attack Execution, the attackers initiate the assault by directing the compromised devices to inundate the target devices or website with an overwhelming amount of traffic. This traffic can manifest in various forms, such as HTTP requests, UDP (User Datagram Protocol), TCP (Traffic Control Protocol), or other network protocols.
- Phase 5, Service Disruption, occurs when the target website or device becomes overwhelmed and incapable of handling the excessive influx of incoming requests. Consequently, legitimate users also face difficulties accessing the website, resulting in service disruption or complete inaccessibility.
- Phase 6, Attack Sustainment, involves the attackers maintaining the assault for an extended duration, often altering their attack strategies or intensity sporadically. This dynamic nature makes it challenging for defenders to restore services or mitigate the onslaught effectively.

DDoS attacks have emerged as a significant threat, disrupting network operations, and inflicting severe harm on reputation and finances. These attacks employ a series of phases to overwhelm target systems and render them inaccessible. The Mirai botnet attack, which happened in the year 2016, stands as a momentous event in the chronicles of DDoS attacks, renowned for its prominence and devastating consequences. Exploiting the vulnerabilities of unsecured Internet of Things (IoT) devices like cameras and routers, the Mirai botnet conducted massive-scale DDoS assaults, primarily targeting Dyn. This resulted in widespread disruption across well-known websites and services, including Twitter, Reddit, Netflix, and GitHub. The overwhelming volume of traffic directed at Dyn's servers led to a severe outage, affecting millions of customers [4]. The Mirai botnet attack serves as a crucial reminder of the vulnerabilities inherent in IoT devices and the potential for large-scale DDoS attacks. Consequently, it caused substantial disruption and financial losses for online businesses, with significant service interruptions, revenue decline, diminished user engagement, and reputational damage. The incident also incurred additional costs associated with response efforts, investigation, and the implementation of enhanced security measures. The cumulative impact of the Mirai botnet attack has prompted organizations to prioritize the strengthening of their defenses against similar future attacks.

In the face of complex and widespread modern DDoS attacks, conventional security measures such as firewalls and rule-based Intrusion Detection Systems (IDS) struggle to cope. To address this challenge, contemporary intrusion detection systems have turned to machine learning and deep learning techniques for enhanced early-stage detection [5, 6]. By implementing an effective AI-based IDS, organizations can significantly improve their ability to identify and mitigate such attacks, enabling proactive defense measures and minimizing potential damage [7].

The primary objective of this research article is to present a comprehensive review of artificial intelligence-based Intrusion Detection Systems. Specifically, the focus is on assessing the effectiveness of various machine learning and deep learning model-based IDS systems, including techniques such as anomaly detection and classification, in detecting DDoS attacks. To evaluate the performance of these models, extensive training was conducted using the CICIDS 2017 dataset, and their accuracy, precision, recall, and overall efficacy were carefully measured and compared.

17.2 Review

This chapter intends to showcase the recent research work in intrusion detection with respect to the DDoS attack.

Nazario et al. [2] The development of DDoS attacks. The document presents a succinct review of the evolution of distributed denial-of-service (DDoS) assaults. It demonstrates the evolving tactics and strategies employed by attackers throughout time, such as the change from simple flooding attacks to more complex and targeted approaches. Mahjabin et al. [3], Presents an overview of DDoS assaults and investigates several techniques for avoiding and mitigating these attacks. The authors address many forms of DDoS assaults, such as volumetric, application-layer, and protocol-based attacks, and propose solutions such as traffic filtering, rate limitation, and intrusion detection systems. The research emphasizes the necessity of collaborative defense mechanisms, early detection, and intelligent mitigation tactics in combating DDoS assaults in distributed sensor networks.

Antonakakis et al. [4] The document digs into a detailed investigation of the Mirai botnet, which targeted Internet of Things (IoT) devices. The study delves into its design, propagation methods, and assault capabilities, revealing insight on one of the most major botnets in recent history. Das et al. [8] This article provides a comprehensive overview of machine learning. It covers the core ideas, several machine learning methods, and a variety of applications across numerous industries. The readership of the text is intended to gain a comprehensive understanding of machine learning.

Haripriya et al. [5] The paper presents the issues that standard intrusion detection systems face and how machine learning techniques may improve the accuracy and efficacy of detecting network intrusions. Wason et al. [9] The article explores the evolution and expansion of deep learning. It covers the background of deep learning techniques, significant developments, and growing applications. The article discusses the development and significance of deep learning in the area of cognitive systems.

Liu et al. [6] The survey article provides a comprehensive overview of the application of ML and DL techniques in the field of intrusion detection systems. It examines several techniques, algorithms, and strategies used for identifying and preventing security breaches. The paper provides insightful information on the developments and difficulties associated with applying ML and DL to improve cybersecurity.

Verma et al. [10] The paper presents a thorough examination of the use of machine learning techniques for detecting Denial of Service (DoS) and Distributed Denial of Service (DDoS) assaults. The authors give a thorough examination of the various machine learning techniques used in the detection of these assaults. They analyze and assess the essential characteristics and datasets employed in these methods. The study emphasizes the efficacy of machine learning in improving DoS/DDoS attack detection skills, as well as future research goals in this subject. Ahmad et al. [7] This study examines the use of ML and DL techniques for network intrusion detection, offering analyses and insights into their efficiency and efficacy.

Cohen et al. [11] The author of this essay discusses the historical progression, present condition, and potential futures of machine learning. The chapter offers information on the developments, difficulties, and possible applications of machine learning in the study of pathology. Mathew et al. [12] This paper offers a thorough introduction to deep learning. It provides insights into deep learning's applications and breakthroughs while covering the underlying ideas, frameworks, and methods employed in the field. The article is an excellent resource for learning about deep learning. Halbouni et al. [1] This review paper investigates the use of ML and DL approaches in cybersecurity, offering a thorough overview of their uses, difficulties, and developments in enhancing cyber defense system.

17.3 Approaches for DDoS Defense

DDoS attacks can have a variety of purposes. They can be started by those who want to wreak havoc, interfere with services, or get even. DDoS-for-hire businesses are becoming more prevalent on the dark web, enabling even those with little technical experience to carry out destructive attacks. DDoS attacks can take several different shapes, such as volumetric attacks that bombard the target with a tone of traffic, protocol attacks that prey on flaws in network protocols, and application-layer attacks that target particular programs or services [2]. Organizations must establish effective defense tactics to reduce the impact of such attacks as attack methodologies develop and grow more sophisticated. A thorough understanding of network architecture, traffic patterns, and attack vectors is necessary for the implementation of effective DDoS defense strategies [3]. It is essential to have a layered defense strategy that combines preventative measures, like rate limitation and traffic filtering, with remedial ones, like traffic scrubbing and behavioral analysis, to lessen the impact of ongoing attacks.

The size, resources, complexity of the infrastructure, and risk tolerance of the organization are only a few of the variables that influence the DDoS defense strategy decision. Small businesses might rely on fundamental firewall setups and traffic filtering strategies, while larger businesses might use cutting-edge intrusion prevention systems (IPS) or invest in specialized DDoS mitigation services. A strong DDoS defense strategy must include ongoing monitoring, incident response planning, and frequent security reviews. In order to effectively detect and counteract DDoS assaults, organizations should proactively identify vulnerabilities, harden their network infrastructure, and train their staff on best practices. The Fig. 17.1 shows the methods in Artificial Intelligence that are discussed in this paper that could potentially help in defending the DDoS Attack.

17.3.1 Dataset Information and Feature Selection

All these methods were developed using the fixed dataset Friday-WorkingHours-Afternoon-DDoS.pcap_ISCX.csv from CICIDS-2017. The data set mostly consists of 225,745 rows and 79 columns, of which 128,023 records were records of DDoS attacks and 97,734 records were records of benign activity. In other words, the dataset originally contained 56.71% records of DDoS attacks and 43.28% records of benign activities. The data set has a significant imbalance. The random module is used to randomly select the DDoS records and equalize them to the benign records in order to balance the dataset. With the removal of all the extra DDoS records, the dataset is now balanced with 91,735 DDoS and benign entries.

This dataset is divided into two parts, the training dataset and the testing dataset in a ratio of 70:30 respectively. All the models are trained on the training dataset. The Machine learning models are trained on specific features of the dataset, in order

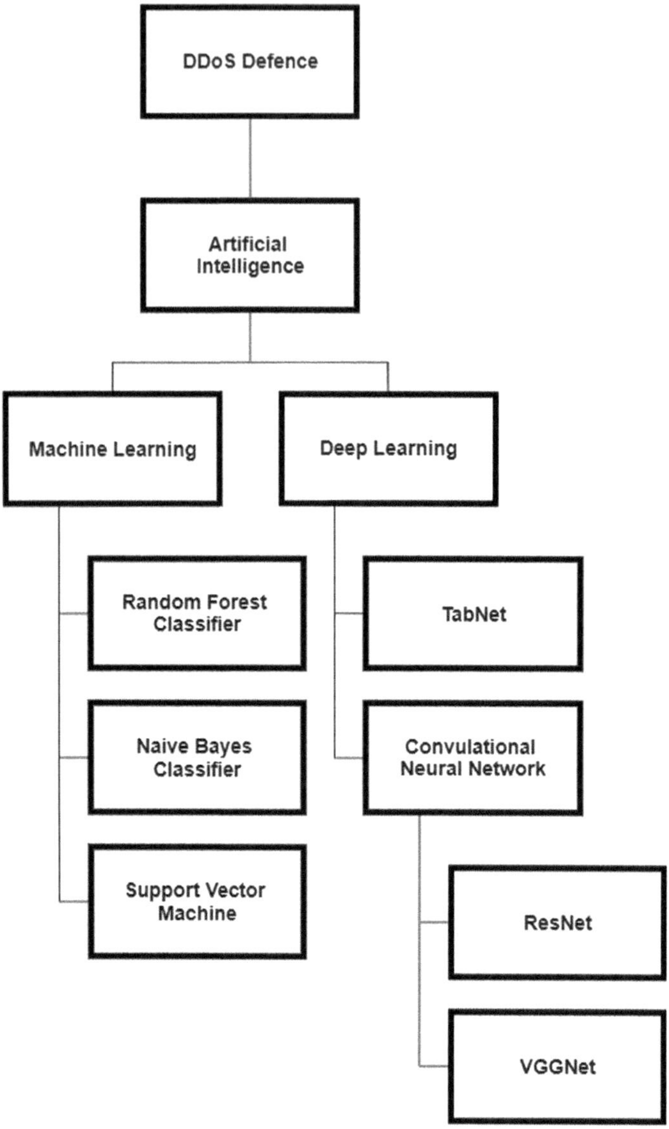

Fig. 17.1 Classification of methods

to extract these features, the correlation feature selection algorithm is used. All the models are trained on the features given by the correlation matrix. The correlation matrix helps in matching the relations between the existing features and the target variable on a scale of 0–1. The highly correlating features (> 0.8) are eliminated, thereby leaving the important features behind. Figure 17.2 shows the heatmap of the features that were used for training all the models discussed in this paper.

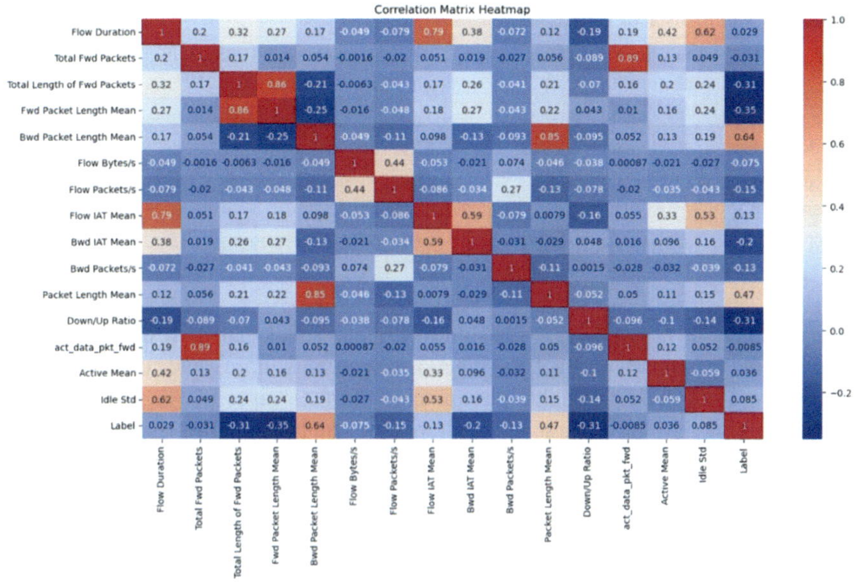

Fig. 17.2 The Correlation heatmap

17.3.2 Machine Learning Approach

The increasing complexity and sophistication of attack strategies provide the need for the machine learning approach for IDS to detect DDoS attacks [5]. As attack methods evolve, traditional rule-based IDS systems sometimes find it difficult to stay up since attackers can simply circumvent or overcome specified rules [10]. On the other hand, machine learning algorithms may recognize anomalies that static rules would miss, adapt to new attack routes, and learn from patterns [11]. The machine learning method uses past network traffic data, including both valid and malicious data, to train algorithms and create models that can distinguish between regular and anomalous traffic patterns [8]. Then, in real-time, these models may be used to examine incoming traffic and spot trends that point to a DDoS attack. The Machine Learning Approach uses algorithms such as Random Forest Classifier, Support Vector Machine, Naïve Bayes Classifier and the K-Nearest Classifier. Here, in order to get the best accuracy possible, the GridSearchCV is used for all the models. GridSearchCV trains on all the existing hyperparameter combinations of the specific model and extracts the best parameters which have the capability of the producing the best accuracy for that specific model. The results produced by all the machine learning models are shown in the Table 17.1.

Random Forest Classifier. A machine learning algorithm called Random Forest Classifier can be used with Intrusion Detection Systems (IDS) to identify and categorize DDoS Attacks. Random Forest Classifier can successfully handle the complexity and variety of cyber threats by utilizing the ensemble of decision trees. It has the

Table 17.1 Comparison of various ML Bases IDS System

Model	Feature selection method	Accuracy (in %)	Precision (in %)	F1-Score (in %)	Time taken for training (in sec)
Random forest	Correlation matrix	99.96	99.98	99.97	1479.60
Naïve Bayes classifier		94.6	94.2	94.95	173.6
SVM		99.80	99.80	99.80	2693.58

ability to handle noisy data, understand intricate attack patterns, and generalize successfully to unknown attackers. The GridSearchCV trains and tests the whole model based on all the available parameters, and then gives out the best parameters which give the best accuracy for that model. The best parameters were, {'max_depth': None, 'min_samples_split': 4, 'n_estimators': 50}.

Naïve Bayes Classifier. For classification problems, the Naive Bayes classifier is a straightforward but efficient machine learning technique. Based on the Bayes theorem, it makes the assumption that the features are conditionally independent of one another in light of the class label. The Naive Bayes classifier is a useful tool in intrusion detection systems for the detection of DDoS attacks due to its simplicity, efficiency, and effectiveness in some situations. It can help with the early detection and mitigation of harmful activity when used in conjunction with other defense strategies, which can lessen the effects of DDoS attacks on network infrastructure and services. The model has been trained with the best parameters that were given by GridSearchCV. The best parameter was {'var_smoothing': 1e-08}.

Support Vector Machine. SVM, or Support Vector Machine, is a commonly used abbreviation in machine learning. It is based on the idea of locating an ideal hyperplane in a high-dimensional feature space that divides data points into various classes. Finding a decision boundary that maximizes the margin between the classes is the basic goal of SVM. The margin is the separation between the decision boundary and the closest support vectors, or data points, from each class. SVMs are very good at classifying data, which makes them appropriate for identifying and categorizing various cyberattacks. SVMs can be used to create models that divide network traffic or system logs into groups, such as regular traffic or certain attack types such as DDoS attacks. In IDS applications, SVMs have shown to be excellent at identifying and categorizing cyberattacks. It is important to remember that SVMs are just one method among many used in IDS, and the choice of algorithm depends on a variety of elements, including the type of data being used, the complexity of the assaults, available computing power, and the particular requirements of the application. The GridSearchCV was run, and the best Hyper-parameters produced were {'C': 10, 'gamma': 'scale', 'kernel': 'rbf'}.

17.3.3 Deep Learning Approach

Deep learning has been a potent strategy for Intrusion Detection Systems (IDS) for fending against the constantly changing terrain of cyberattacks in recent years. Deep learning models can automatically learn complex patterns and representations from enormous volumes of data by utilizing the capabilities of neural networks, which enables them to detect and mitigate various kinds of cyber threats [1]. Deep learning's capacity to learn feature representations directly from raw data, obviating the requirement for manual feature engineering, is one of its significant features. Because it enables the models to efficiently capture both low-level and high-level properties from network traffic, system logs, or other pertinent data sources, this capability is very beneficial for IDS [7]. Deep learning models are excellent at anomaly detection, a crucial component of IDS. These models can learn the predicted patterns and behaviors of a system or network by being trained on a huge volume of typical data [12]. Any departures from these observed trends, such as unexpected behavior or cyberattacks, can therefore be recognized as anomalies. As a result, IDS is given the ability to recognize previously unknown or zero-day attacks, offering proactive defense against new threats. Convolutional neural networks (CNNs) are a popular architecture for IDS that is based on deep learning. CNNs are excellent at finding spatial patterns in data, which makes them suitable for analyzing network traffic or locating image-based cyberattacks. Furthermore, deep learning models can profit from transfer learning strategies, which let them use pre-trained models on comparable tasks or domains to improve performance even with a dearth of labelled data [9]. Deep learning has a lot of potential for IDS and stopping cyberattacks, but it shouldn't be considered a stand-alone remedy. Deep learning-based IDS must be combined with other cybersecurity measures, such as network segmentation, encryption, access control, and routine security updates. Organizations can create a complete defense system that can identify, mitigate, and stop a variety of cyberthreats by integrating these tactics. The results produced by the deep learning models are shown in the Table 17.2. All the models under this approach are trained with 10 epochs.

TabNet. TabNet is a unique neural network architecture that combines features of deep learning and conventional decision trees. It applies feature selection and classification tasks on tabular data using deep neural networks with sparse attention. The TabNet architecture consists of several decision processes, each of which uses a customised Transformer attention mechanism. These decision-making procedures iteratively choose pertinent features and update the input data representations. TabNet

Table 17.2 Comparison of various deep learning based IDS model

Model	Feature selection method	Hyper-parameter	Accuracy (in %)	Time-taken for training (in sec)
TabNet	Correlation matrix	Optuna	90.97	146.07
ResNet			87.36	299.78
VGGNet			99.91	4300.68

can choose focus on particular features, in contrast to conventional deep learning models, which enables it to handle high-dimensional and sparse tabular data successfully. By emphasising crucial details and assigning relative weight to various input variables, TabNet's attention method adds interpretability. It is useful for identifying the underlying causes and symptoms of many different jobs, such as cybersecurity and DDoS attack detection, because to this feature. The best Parameters given by the Optuna are {'n_d': 47, 'n_a': 30, 'n_steps': 6, 'gamma': 1.4979761223647712, 'lambda_sparse': 0.05581785761931777}.

Resnet. Is useful for applications in cyber security because it can recognize complicated patterns and learn deep representations. ResNet models can examine the structural properties of files or executables to distinguish between benign and dangerous software when it comes to malware identification. ResNet can identify subtle and sophisticated malware variants that elude traditional signature-based techniques by utilizing its deep learning capabilities. ResNet's capacity to extract nuanced information from network traffic data makes it useful for intrusion detection systems (IDS). ResNet models may effectively detect aberrant network behavior and different sorts of cyberattacks, such as network intrusion attempts or distributed denial-of-service (DDoS) attacks, by being trained on large-scale datasets of legitimate and malicious traffic. Additionally, ResNet topologies have been used to analyze network traffic, where they discover patterns and irregularities in network communication. ResNet models can identify harmful actions like network scanning and data exfiltration attempts by examining packet-level data. This assists in detecting threats early and helps to guard against security lapses. The best_params given by the Optuna Algorithm were {'learning_rate': 0.0006916757070211176, 'batch_size': 128, 'dropout_rate': 0.23554958940917362}.

VGGNet. The Visual Geometry Group Network, or VGGNet for short, is a deep convolutional neural network design that has found extensive use in computer vision tasks, particularly picture categorization. The ideas of VGGNet can be applied to activities like DDoS attack detection and Intrusion Detection Systems (IDS) even though it was not created expressly for cybersecurity purposes. VGGNet can be used to analyze network traffic data and spot patterns suggestive of attacks or malicious activity in the context of DDoS attacks and IDS. The network traffic data can be sent into the VGGNet architecture after being pre-processed and converted into a suitable format, such as packet-level details or aggregate statistics. Using both low-level and high-level characteristics, VGGNet can learn hierarchical representations of the input data thanks to its deep convolutional layers. This skill is useful for identifying intricate patterns linked to DDoS assaults and separating them from regular traffic. The network can learn the discriminative characteristics by being trained on labelled datasets that include samples of both malicious and benign communication. The transferability of VGGNet is one of its benefits. Pretrained models can be adjusted or used as feature extractors for cybersecurity tasks when they have been trained on large-scale image datasets like ImageNet. The performance on DDoS attack detection and IDS can be improved by utilizing the learnt representations from various visual data sources using this transfer learning approach. The Optuna hyperparameter tuner gave the best params to be {'learning_rate':0.00032413642799984893, 'weight_decay': 1.1050445917650297e-05}.

17.4 Results

The above models were trained and tested on a 72 core CPU with 56 GB Ram and internet speed of 40 Mbps. Tables 17.1 and 17.2 Shows the comparison of various performance metrics of ML and DL based IDS System.

All the models were trained with the best parameters feasible. Starting with the random forest classifier in the machine learning technique, this classifier model can be regarded as scalable because it has the capacity to handle big datasets. Using multi-core computers can substantially shorten the model's training period. However, the number of trees in the model affects how long it takes to train. Regarding this model's adaptability, it may be utilized for a variety of jobs and is renowned for its flexibility. It has the capacity to manage various data kinds, including categorical and numerical data. This model is particularly versatile because it doesn't need a lot of preprocessing. When it comes to the model's real-time processing, this model is not intended to make instantaneous forecasts. The intricacy of the dataset and the number of trees in the forest have a significant impact on the prediction time. Real-time processing is not a good fit for it. False negatives and false positives are another problem that the random forest model faces. However, this problem is entirely dependent on a number of variables, including the dataset's quality, its imbalance, and the categorization threshold.

The Naïve Bayes Classifier model employs a straightforward and computationally efficient approach that is independent of the number of features in the dataset used for training, the naive bayes classifier is typically regarded as being highly scalable. The classifier's flexibility can be constrained in circumstances where there are significant connections between the features since, despite its fame for simplicity, it makes the assumption that features are conditionally independent given the class label. This classifier model's quick training and prediction times make it a good choice for real-time processing. Like any other classifier model, the Bayes model has the problem of false positives and false negatives, but the severity of this problem relies on the calibre of the training data used. Figure 17.3 shows the comparison of various performance metrics.

The SVM struggles with scalability issues for large datasets because the model requires a lengthy training period, especially when the dataset contains a lot of features. This model is quite flexible because it may be applied to many different classification tasks. By utilizing several kernel functions, this model can handle both linearly separable and non-separable data. SVM has a problem with real-time processing because it was not intended to handle it. This is because SVM has a lengthy training period. The goal of SVM is to lessen the problem of false positives and false positives. However, with the proper use of hyperparameters, this problem can be decreased to the point where it is insignificant. SVM seeks to maximize the

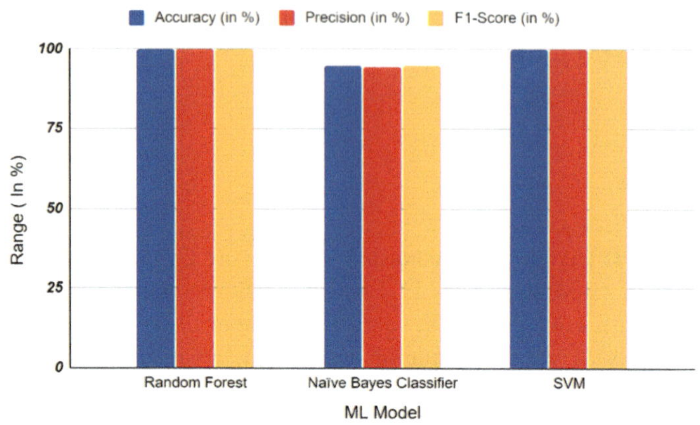

Fig. 17.3 Comparison of machine learning model IDS performance metrics

gap between the classes, however this does not eliminate the problem. Table 17.3 shows the comparison between various metrics.

In the deep learning approach, The TabNet model is built to be highly scalable. Large datasets may be handled effectively by it. It is capable of handling several features that are present in the data. By adopting sparse attention algorithms—which have less computational cost than fully linked models—the scalability can be achieved. The TabNet model can handle a wide range of attributes, including categorical and numerical variables, and is very flexible. Because of its versatile feature transformation and modelling capabilities, TabNet's architecture makes it suited for a variety of datasets. Due to its effective architecture, TabNet also provides real-time processing capabilities. It can forecast outcomes quickly, which qualifies it for real-time processing. By capturing the feature interactions and linkages, the TabNet model seeks to reduce the problem of false negatives and false positives. The degree to which false positives and false negatives are reduced depends on the calibre of the training data.

The ResNet model is scalable and suitable for use with huge datasets. This model was initially created to address issues with picture classification. This flexible paradigm is applied to a variety of computer vision problems. It has demonstrated excellent performance and the capacity to learn from intricate patterns and representations of data. For image classification, object recognition, and picture segmentation,

Table 17.3 Comparison of ML models evaluation metrics

Model	Scalability	Adaptability	Real-time processing	FP & FN
Random forest	Yes	Yes	No	No
Naïve bayes	Yes	Yes	Yes	No
SVM	No	Yes	No	Yes

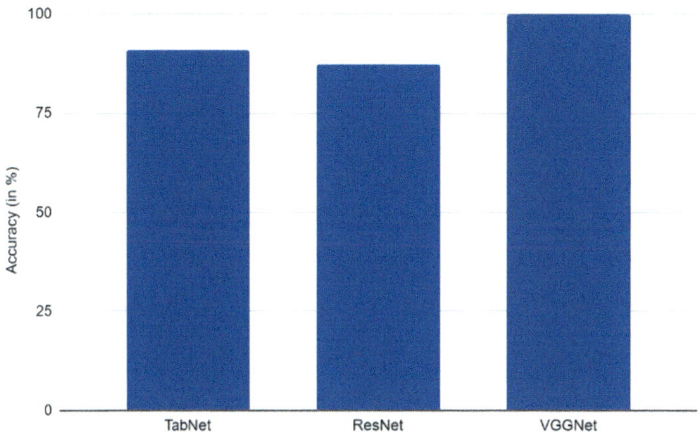

Fig. 17.4 Comparison of accuracy of the DL models

this approach is widely employed. This model can be used to tabular datasets by fine-tuning the architecture and hyperparameters. With this model, real-time processing is challenging because the model takes a long time to train. The depth and scale of the network both enhance the computational complexity of this approach. However, real-time processing for this model can be partially done by altering the amount of the dataset and using network puning. Depending on the detection task and the calibre of the training data, false positives and false negatives are a concern. It can typically produce results with a high degree of precision, which minimizes the problem of false positives and false negatives. But it also depends on how well-balanced and high-quality the dataset is. Figure 17.4 show the comparison of accuracy produced by the DL based IDS system models.

In VGGNet, Due to the deep design and the high computational cost involved in training and deploying this model, this model has scalability issues. This model uses a lot of memory, and as the depth of the model increases, so do the number of parameters. This approach is flexible and can be utilized for many different applications, including object and picture detection. The initial purpose of this approach was for computer vision tasks. Due to the model's computational complexity and quantity of layers, real-time processing can be difficult. The VGGNet model is slower compared to shallower networks because of its deep architecture. Real-time processing can be achieved to some extent through the use of various optimization techniques, network puning, and other methods. Depending on the particular task the model is being used for, there may be a problem with false negatives and false positives. The VGGNet model may generate results with great precision, which mitigates this problem. The dataset's balance and the calibre of the data, nevertheless, can affect this problem. Table 17.4 shows the comparison between the evaluation metrics.

Table 17.4 Comparison of DL models evaluation metrics

Model	Scalability	Adaptability	Real-time processing	False positives and false negatives
TabNet	Yes	Yes	Yes	Yes
ResNet	Yes	Yes	No	No
VGGNet	No	Yes	No	Yes

17.5 Conclusion

Through this proposed work, it has been identified that there are distinct performance traits among several models in the investigation of machine learning and deep learning methodologies. The Support Vector Machine (SVM) classifier has proven to be extremely effective at achieving greater accuracy in the context of machine learning. The Random Forest model, on the other hand, likewise achieves excellent accuracy but necessitates a longer training period. Finally, it can be said that Random Forest generally performs better than SVM in terms of scalability, real-time processing for prediction, handling false negatives and false positives, and managing noisy data. However, SVM stands out in terms of minimizing false negatives and false positives and is well-suited for real time applications because to its adaptability to a wide range of challenges without the need for a lengthy training period. The final decision between these two algorithms is determined by the particular needs and trade-offs of each particular circumstance. In the deep learning approach, it has been identified that VGGNet outperforms other models in terms of accuracy but comes at the expense of more time spent training. In comparison, TabNet shows impressive accuracy with a substantially shorter training period. In conclusion, there are substantial differences between VGGNet and TabNet in terms of scaling, real-time processing, and tabular data adaptation. Compared to VGGNet, which excels at picture classification but may have problems with scalability and real-time processing, TabNet is designed specifically for tabular data and offers improved scalability, faster real-time processing, and increased versatility for tabular datasets. TabNet is an appealing approach for eliminating false negatives and false positives in tabular data contexts due to its ability to precisely capture feature interactions. These findings offer important insights into the choice of machine learning and deep learning models depending on particular application needs, illuminating the trade-offs and benefits connected with each technique.

References

1. Halbouni, A., Gunawan, T.S., Habaebi, M.H., Halbouni, M., Kartiwi, M., Ahmad, R.: Machine learning and deep learning approaches for cybersecuriy: a review. IEEE Access. (2022).
2. Nazario, J.: DDoS attack evolution. Netw. Secur. **2008**(7), 7–10 (2008)

3. Mahjabin, T., Xiao, Y., Sun, G., Jiang, W.: A survey of distributed denial-of-service attack, prevention, and mitigation techniques. Int. J. Distrib. Sens. Netw. **13**(12), 1550147717741463 (2017)
4. Antonakakis, M., April, T., Bailey, M., Bernhard, M., Bursztein, E., Cochran, J., Durumeric, Z., Halderman, J.A., Invernizzi, L., Kallitsis, M., Kumar, D.: Understanding the mirai botnet. In 26th {USENIX} Security Symposium ({USENIX} Security 17) (pp. 1093–1110). (2017).
5. Haripriya, L., Jabbar, M.A.: Role of machine learning in intrusion detection system. In 2018 Second International Conference on Electronics, Communication and Aerospace Technology (ICECA), pp. 925–929. IEEE. (2018).
6. Liu, H., Lang, B.: Machine learning and deep learning methods for intrusion detection systems: a survey. Appl. Sci. **9**(20), 4396 (2019)
7. Ahmad, Z., Shahid Khan, A., Wai Shiang, C., Abdullah, J., Ahmad, F.: Network intrusion detection system: a systematic study of machine learning and deep learning approaches. Trans. Emerg. Telecommun. Technol. **32**(1), e4150 (2021)
8. Das, K., Behera, R.N.: A survey on machine learning: concept, algorithms and applications. Int. J. Innov. Res. Comput. Commun. Eng. **5**(2), 1301–1309 (2017)
9. Wason, R.: Deep learning: evolution and expansion. Cogn. Syst. Res. **52**, 701–708 (2018)
10. Verma, V., Kumar, V.: DoS/DDoS attack detection using machine learning: A review. In Proceedings of the International Conference on Innovative Computing & Communication (ICICC). (2021).
11. Cohen, S.: The evolution of machine learning: past, present, and future. In Artificial intelligence and deep learning in pathology. Elsevier. pp. 1–12 (2021).
12. Mathew, A., Amudha, P., Sivakumari, S.: Deep learning techniques: an overview. Adv. Mach. Learn. Technol. Appl. Proceed. AMLTA **2020**, 599–608 (2021)

Chapter 18
Preserving Our Heritage: Buildings Deep Learning Solutions for Monitoring Cultural Heritage Structures Using Automated Crack Detection

Rajdeep Chatterjee and Biswaroop Nath

Abstract This paper aims to implement state-of-the-art (SOTA) models for real-time crack detection, specifically for use with Unmanned Aerial Vehicles (UAVs). The objective is to facilitate the easy and efficient detection of cracks, thereby enabling the protection of valuable cultural heritage buildings and the prevention of structural damage during adverse weather conditions such as monsoons or floods. In our specific case, YOLOv8 achieved a training time of 0.90 h and yielded impressive mean Average Precision (mAP) scores, including mAP@0.50 (bounding box) of 0.79, mAP@0.50–0.95 (bounding box) of 0.60, mAP@0.50 (mask) of 0.66, and mAP@0.50–0.95 (mask) of 0.26 on the validation dataset. Through our experiments, we determined that YOLOv8 is a superior model, exhibiting comparable speeds to YOLOv5 and accuracy levels on par with YOLOv7. Notably, we selected test images from various online sources and employed our models for crack detection, wherein YOLOv8 outperformed other models. These results firmly establish YOLOv8 as a highly effective and efficient solution for real-time crack detection.

18.1 Introduction

Heritage structures hold immense cultural significance as they embody the history, architecture, and artistic heritage of a particular region. Unfortunately, over time, these buildings can experience cracks due to environmental factors, the passage of time, and structural stress, posing risks to their stability and the safety of occupants. Effective crack segmentation plays a crucial role in evaluating the seriousness of

---

.

R. Chatterjee (✉) · B. Nath
School of Computer Science, KIIT DU, Bhubaneswar, Odisha, India
e-mail: cse.rajdeep@gmail.com

B. Nath
e-mail: biswaroopnath2001@gmail.com

R. Chatterjee
Amygdala AI, Munich, Germany

P. Nanda et al. (eds.), *Intelligent System and Data Analysis*, Smart Innovation, Systems and Technologies 398, https://doi.org/10.1007/978-981-97-5200-3_18

cracks, prioritizing necessary repairs, and ensuring the protection of human lives. Nowadays, modern technologies contribute to the preservation of heritage buildings. It can even be extended to be integrated into an Unmanned Aerial Vehicle (UAVs) to automate the process of detection of cracks [1, 2].

Furthermore, it is crucial to acknowledge that cracks in old buildings can have severe consequences, particularly during heavy rainfall or floods. Such conditions can exacerbate the structural weaknesses caused by cracks, leading to catastrophic collapses and resulting in extensive property damage. Not only does this result in financial losses, but it also poses a significant threat to the safety of individuals residing in the affected building or residing nearby. The collapse of a heritage building can have far-reaching implications, impacting the lives and well-being of those in its vicinity. Hence, it is imperative to address cracks in a timely manner through proper segmentation and subsequent repairs to mitigate the potential risks associated with structural failures and ensure the overall safety of both the buildings themselves and the surrounding community. As of recently in June 26, 2023, a tragic incident occurred in the Ghatkopar area of Mumbai when a ground-plus-three storey building collapsed, resulting in the death of a 94-year-old woman and a 56-year-old man. The collapse took place in Chittaranjan Nagar of Rajawadi Colony. The victims' bodies were discovered in the debris a day later.[1] Similar such incidents have been happening all throughout India and incidents like such highlights the devastating consequences that can occur due to building collapses, emphasizing the importance of ensuring the safety and structural integrity of buildings. This gave us another sense of purpose and motivation to develop and research such technologies that can be used not only to save old heritage buildings and our rich culture but also can save much valuable innocent human lives.

Therefore, the main focus and significant contribution of this work center around offering advanced solutions for the early detection of cracks in old heritage buildings using images of buildings. By utilizing state-of-the-art algorithms, these solutions provide faster, more reliable, and highly accurate detection of cracks. The implementation of these cutting-edge algorithms enables prompt identification of cracks in heritage structures, allowing for timely interventions and necessary repairs. This advancement is crucial in ensuring the preservation and safety of these valuable cultural assets, as it facilitates the proactive management of structural issues before they escalate into major concerns. By employing these latest algorithms, the process of crack detection becomes more efficient and dependable, assisting in the ongoing efforts to safeguard and maintain the structural integrity of old heritage buildings.

The rest of the paper is divided into following sections—Literature Review in Sect. 18.2 , dataset and performance measures in Sect. 18.3, the models are discussed in Sect. 18.4, the results of experiments in Sect. 18.5 and finally conclusion and future work is discussed in Sect. 18.6.

[1] Link: https://www.tribuneindia.com/news/nation/mumbai-building-collapse-2-bodies-recovered-from-debris-520400.

18.2 Literature Review

A plethora of research works are available already for detection of cracks , we cover some of the most note taking and unique and novel approaches for detection of cracks. In [3] the authors proposed automated crack detection techniques utilizing image processing. While accuracy is typically the focus of these approaches, the authors have focused on computation time, which is also crucial for practical applications, given the increasing size of digital images. They focused on a high-speed crack detection method using percolation-based image processing has been introduced, incorporating termination and skip-added procedures to reduce computation time. The percolation process is terminated based on circularity calculations, and subsequent pixels can be skipped based on the circularity of neighboring pixels. The authors shown that experimental results using this approach effectively reduces computation costs. Another interesting similar work for classification of cracks, in [2], the authors have used a machine learning-guided pipeline for accurate classification of road cracks. The proposed approach is compared with state-of-the-art CNN models, demonstrating competitive performance. The use of different feature selection techniques, particularly Orthogonal Matching Pursuit (OMP), yields superior results in terms of feature subset size, classification performance, and training time. The study concludes with a high accuracy rate of around 97% achieved by utilizing OMP feature selection and a Random Forest Classifier tuned with Optuna.

Another interesting work in [4], the authors have proposed a deep-learning model named YOLOv4-FPM based on YOLOv4. They focused on accuracy and robustness of their proposed model. Based on their experimental findings, YOLOv4-FPM demonstrates a mean average precision (mAP) of 0.976, surpassing YOLOv4 by 0.064. Their model achieves a significant reduction in size and parameters, amounting to 18.2% of the original. It processes real-time images at 1000 $\times$ 1000 pixels, achieving a frame rate of 119FPS, which is 20 times faster compared to a recent study. Additionally, YOLOv4-FPM exhibits effective crack detection capabilities across images of various sizes. In another work [5], a supervised deep convolutional neural network is trained to classify image patches in a dataset of 500 images, captured by a low-cost smartphone and sized at 3264 $\times$ 2448 pixels. The authors demonstrated that the proposed deep learning framework, with its learned deep features, outperformed existing hand-crafted feature extraction methods in terms of crack detection performance. A similar interesting paper [6], introduces two pre-processing techniques aimed at eliminating noise for crack detection. Firstly, it employs a subtraction preprocessing method with a smoothed image to eliminate slight variations such as irregular illumination conditions and shading from concrete images. Secondly, a line filter based on the Hessian matrix is employed to enhance line structures associated with cracks. Additionally, thresholding processing is utilized to differentiate cracks from the background. The effectiveness of the proposed method is evaluated through ROC analysis using 50 real images. The experimental findings demonstrate that the proposed approach is successful in detecting cracks in concrete images that contain noise.

All of the aforementioned research studies have employed diverse methods and algorithms. However, in the recent years following these publications, newer and more advanced state-of-the-art (SOTA) models have emerged, offering improved performance and faster processing. Therefore, our contribution primarily revolves around implementing the latest SOTA algorithms, taking advantage of their enhanced capabilities.

18.3 Dataset And Performance Measures

In this section, we will provide a comprehensive discussion on the dataset utilized for training our models as well as the various performance parameters we considered throughout our experiments. During our experiments, we meticulously evaluated the performance of our models using several key performance parameters. These parameters served as indicators of the models' effectiveness and allowed us to gauge their capabilities accurately.

18.3.1 Dataset

The dataset [7] we utilized for training our models comprises annotated images depicting cracks. Specifically, it contains a total of 1551 images. The dataset was sourced from Roboflow and we opted to use version 1 of the dataset. The dataset has already been divided into two portions: 80% for training and 20% for validation. To ensure consistency and fairness in comparing different models, we maintained the same dataset and splitting throughout our experiments.

To provide a visual representation of the training data, Fig. 18.1 showcases the appearance of the images. These images have a resolution of 2560×1440 pixels, presenting a clear view of the crack patterns. These images are already annotated and are ready for training.

By utilizing this annotated dataset, we aimed to train our models effectively in detecting and classifying cracks. The consistent dataset and splitting approach allowed us to perform fair comparisons between different models, enabling us to evaluate their performance accurately and make meaningful observations.

During our experiments, we used Google Colab Notebook environment with Python 3 installed with GPU option enabled for faster training for various models.

18.3.2 Performance Measures

Mean Average Precision (mAP) [8, 9] is a metric used for evaluating object detection models. It utilizes the Intersection over Union (IoU) concept (Fig. 18.2), which mea-

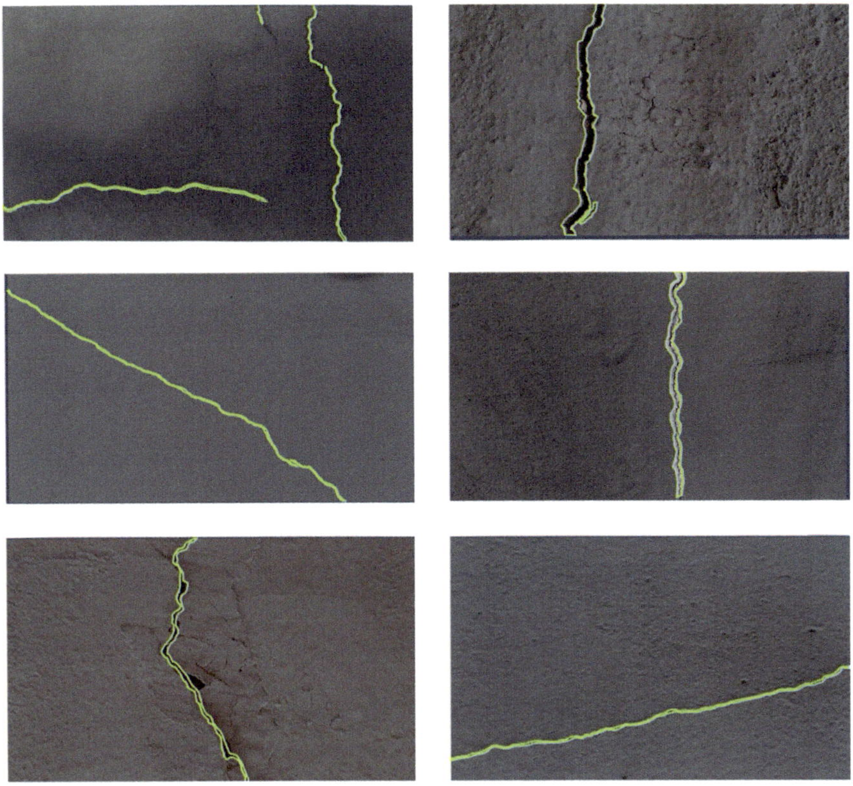

Fig. 18.1 Visualization of the dataset

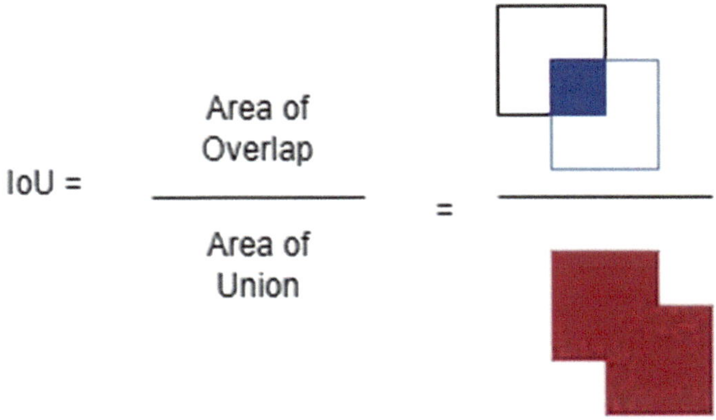

Fig. 18.2 Intersection over union (IoU) concept

sures the overlap between the predicted bounding box and the ground truth bounding box. By comparing the IoU value to a specified threshold, it determines whether the detection is valid (true positive) or not (false positive). A true positive occurs when the IoU is equal to or higher than the threshold, indicating a correct detection. On the other hand, a false positive arises when the IoU is lower than the threshold, representing an incorrect detection. Typically, the threshold is set to values like 50% or 75%, or an average across the range of 50–95%. This allows for a comprehensive evaluation of the model's performance. During our experiments, we find and compare the following parameters for both Box and Mask:

- **mAP@0.50**: mAP@0.50(mean Average Precision) is a metric used in YOLO (You Only Look Once) for evaluating object detection models. It specifically measures the average precision achieved when a predicted bounding box is considered correct if it has an intersection over union (IoU) overlap of at least 50% with the ground truth bounding box. In other words, mAP@0.50 calculates the precision at a fixed IoU threshold of 50% to assess the model's performance in accurately detecting objects.
- **mAP@0.50–0.95**: mAP@0.50–0.95((mean Average Precision)) is a metric used in YOLO (You Only Look Once) for evaluating object detection models. It involves computing precision values at different IoU thresholds ranging from 0.50 to 0.95 and then averaging them. This metric offers a comprehensive evaluation of the model's performance across a wider range of IoU overlaps, capturing its ability to accurately detect objects with diverse levels of spatial overlap with the ground truth boxes.

By thoroughly examining these performance parameters, we were able to gain a comprehensive understanding of our models' capabilities and identify areas for improvement.

18.4 Models

Since, our motive is object detection in real time, YOLO [10–12], which stands for You Only Look Once, is an highly advanced object detection algorithm designed to identify and locate objects within images or video frames. It is renowned for its real-time processing capabilities and high precision. Unlike traditional object detection methods that involve multiple stages, YOLO approaches the task as a single regression problem, resulting in improved speed and efficiency.

The YOLO algorithm (Fig. 18.3) divides the input image into a grid and predicts bounding boxes, object probabilities, and class labels for each grid cell. It can simultaneously predict these attributes for multiple bounding boxes within each cell, enabling it to detect multiple objects in a single pass. YOLO employs a single neural network that simultaneously predicts the classes of objects and their corresponding bounding boxes. In simple terms, the workflow of the algorithm is:

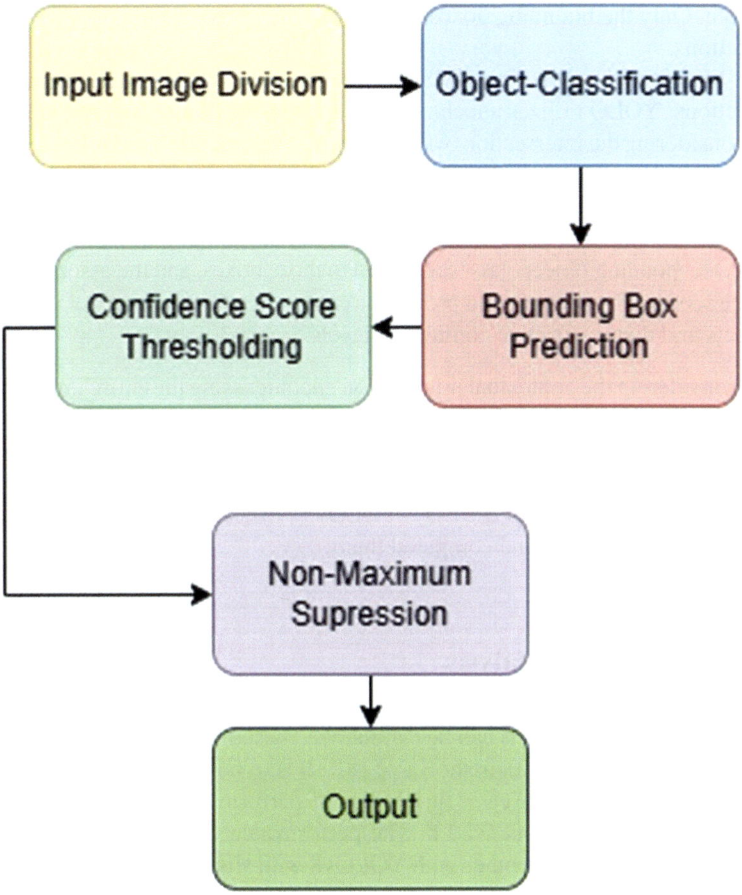

Fig. 18.3 Work-flow diagram of YOLO algorithm

- Input Image Division:The image is partitioned into a grid of cells, with each cell assigned the responsibility of predicting bounding boxes and extracting relevant object information.
- Bounding Box Prediction:In each cell, YOLO makes predictions for multiple bounding boxes, along with their corresponding confidence scores. These predicted bounding boxes are adjusted to accurately enclose the objects present in the image.
- Object Classification: At the same time, YOLO estimates the probabilities of various object classes for each predicted bounding box. This enables the algorithm to determine the types of objects that have been detected.
- Confidence Score Thresholding: YOLO applies a threshold to the confidence scores of the predicted bounding boxes to eliminate predictions with low con-

fidence. Only the bounding boxes that surpass the threshold are regarded as valid detections.

- Non-Maximum Suppression(NMS):In order to eliminate duplicate or overlapping detections, YOLO utilizes a technique called non-maximum suppression (NMS). By considering the intersection over union (IoU) values, NMS identifies the bounding box with the highest confidence among the overlapping ones and selects it as the most accurate detection.
- Output: The output of the YOLO algorithm consists of the chosen bounding boxes, the corresponding object classes assigned to those boxes, and the associated confidence scores. This information provides a comprehensive summary of the detected objects and their respective confidence levels.

By considering the contextual information encompassing the entire image, YOLO effectively captures global information and enhances the accuracy of its predictions. The algorithm finds application in various domains, such as autonomous vehicles, surveillance systems, and real-time object detection on embedded devices. We have focused on latest versions of YOLO—YOLOv5 [13], YOLOv7 [14] and YOLOv8 [15] for our experiments, and compared them.

18.5 Results and Analysis

Our results of various models can be summarized based on the performance metrics (as discussed in Sect. 18.3.2), in the Table 18.1. It is to be noted that these results are on validation-set. It is quite clear that YOLOv5 performs the worst but has a much faster training time of about 0.20 h. The performance of YOLOv8 and YOLOv7 are almost identical to each other with YOLOv8 with slightly greater $mAP_{0.50}$(box): 0.79 $mAP_{0.50-0.95}$(box): 0.60 ,and $mAP_{0.50}$(mask): 0.66 $mAP_{0.50-0.95}$(mask): 0.26. We must keep in mind that YOLOv7 took the most time to train—about 1.47 h while YOLOv8 took 0.90 h which is much faster than YOLOv7 without performance drop. In Table 18.2 we can check the test time on our sample test data which we took from internet resources, consisting of 8 images [16–23] with cracks and different environment situations (Fig. 18.4 visualizes these images). We can easily deduce, that YOLOv7 has a much higher Inference time of 2222.80 ms, while YOLOv8 is not the most fastest, but its various test times are very close to YOLOv5.

Hence, YOLOv8 has much high accuracy as well as speed. Clearly YOLOv8 is the better model which gives high accuracy within minimum possible time. For real-life implementation, YOLOv8, is clearly the superior model, almost having the accuracy of YOLOv7 and near about speeds of YOLOv5.

In Fig. 18.5 depicts the detections of YOLOv5, it is clear that it is unable to detect cracks properly, missing out many areas of visible cracks. The detections of YOLOv7 have been shown in Fig. 18.6, in which we can clearly see, that it performed better than YOLOv5 on most images except for Fig. 18.6e, in which it is unable to detect

Table 18.1 Comparison of model versions

	Model version		
Parameters	YOLOv5	YOLOv7	YOLOv8
mAP@0.50 (box)	0.74	0.75	0.79
mAP@0.50–0.95 (box)	0.49	0.58	0.60
mAP@0.50 (Mask)	0.59	0.65	0.66
mAP@0.50–0.95 (mask)	0.20	0.24	0.26
Training time (h)	0.20	1.47	0.90
No. of epochs	20	20	20

Table 18.2 Comparison of test time of different model versions

	Model version		
Parameters	YOLOv5	YOLOv7	YOLOv8
Preprocess (ms)	0.50	1.30	1.90
Inference (ms)	269.00	2222.80	285.60
NMS per image (ms)	3.40	4.60	14.80

any cracks at all. While, YOLOv8 in Fig. 18.7, we can see it performed exceptionally well in detecting cracks in every images and masking them very accurately. Hence based on our experiments it can easily be concluded YOLOv8 is far superior model not only in terms of speed but also accuracy and masking.

18.6 Conclusion and Future Work

Our aim in this paper is to implement the latest SOTA models for real-time crack detection so that detection of cracks can be done easily without minimal or no human intervention. This technology can be used to automate monitoring and raise red flags for damaged cultural heritage buildings. The same method is also helpful for city authorities to monitor and prevent the crumbling of old buildings during monsoons or floods, thus saving lives and property. From our experiments, it can easily be inferred that YOLOv8 is a much better model, having speeds that are on par with YOLOv5 and accuracy on par with YOLOv7. Even further, we examine the model using real-world random images. Some sample test images are shown in the earlier section. It is empirically and visually established that the YOLOv8 outperforms other models. In our case, YOLOv8 achieves mAP@0.50 (box): 0.79, mAP@0.50: 0.95 (box): 0.60, mAP@0.50 (mask): 0.66, and mAP@0.50–0.95 (mask): 0.26 on the validation dataset in 0.90 h and 20 epochs only. It is clear that YOLOv8 is much better at detecting cracks in real time, as it is both fast and efficient.

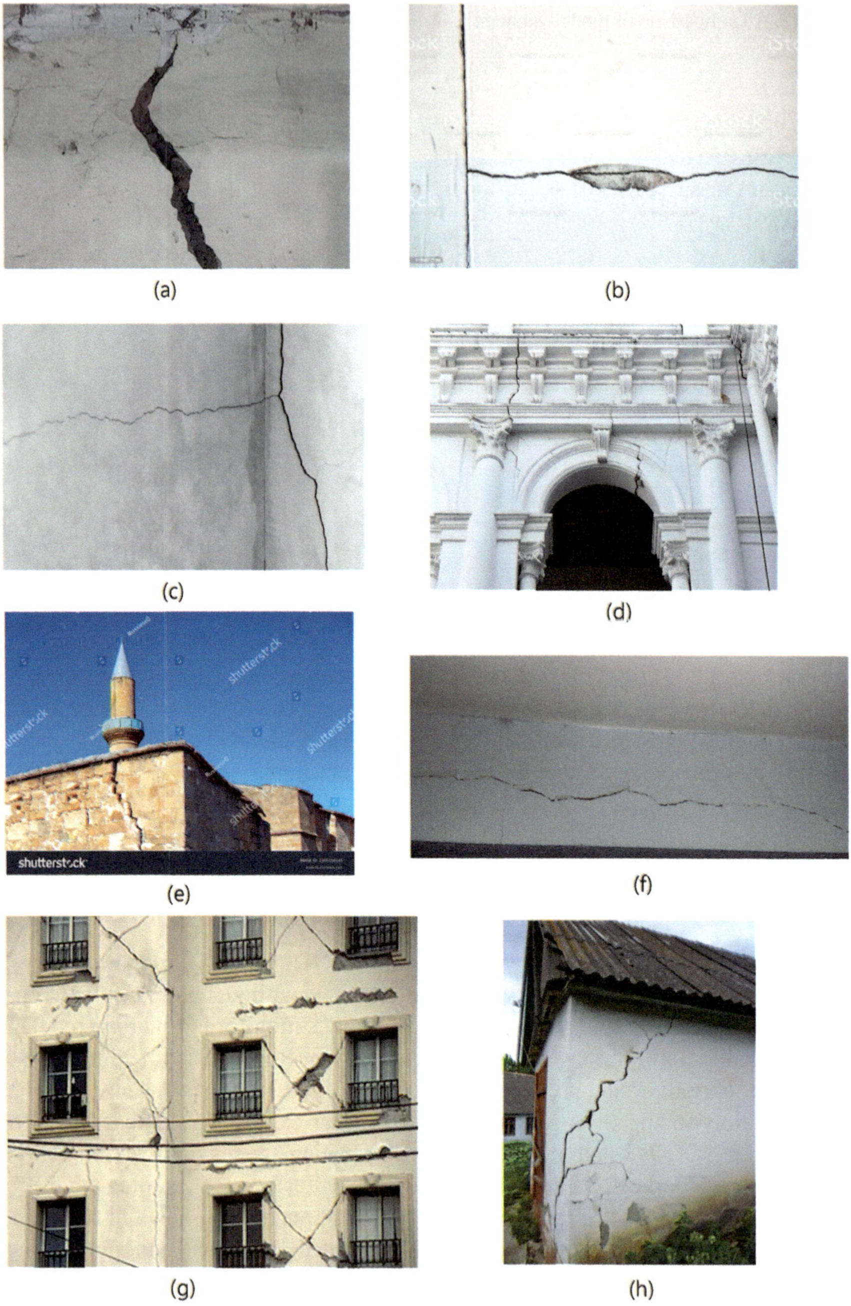

Fig. 18.4 Visualization of sample test images

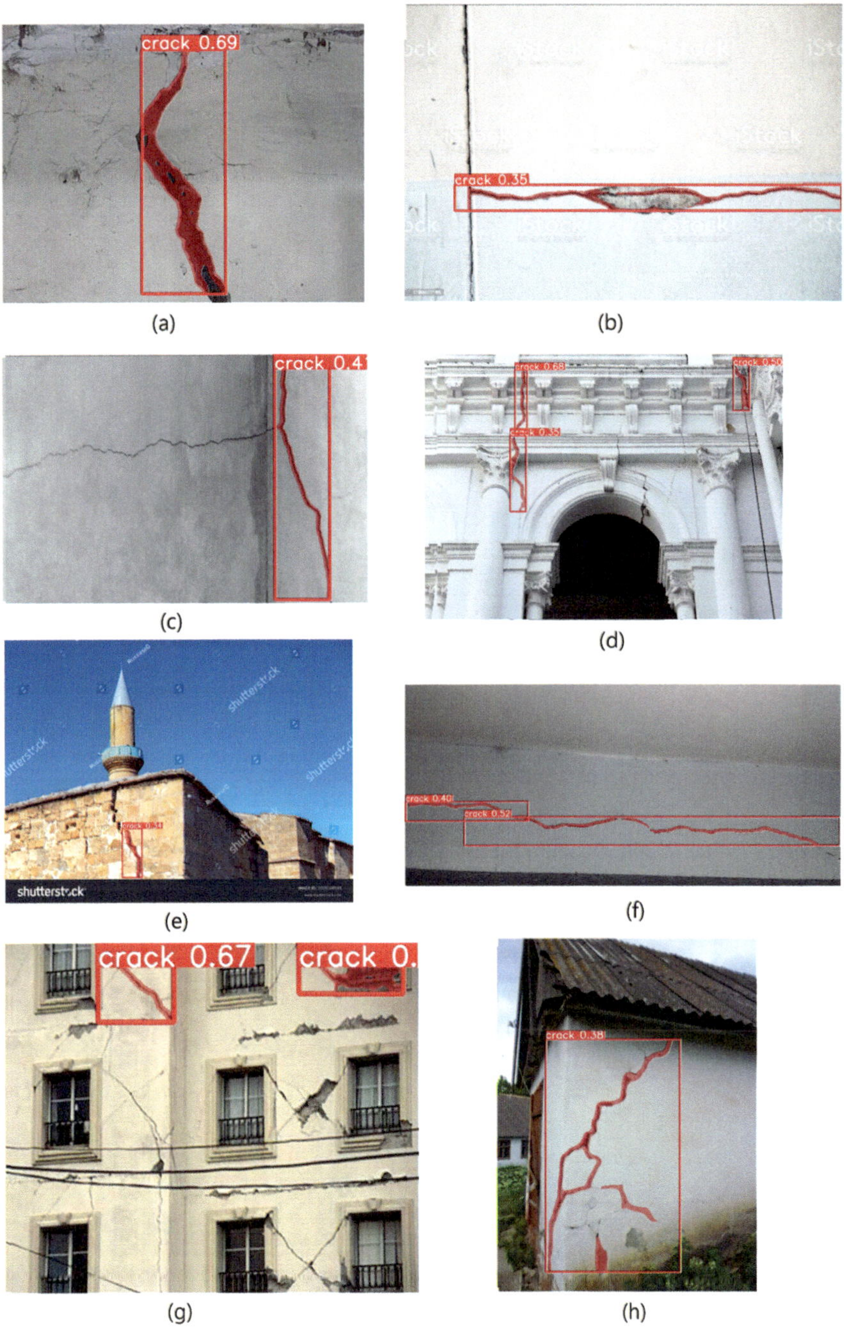

Fig. 18.5 Detections of YOLOv5 on test images

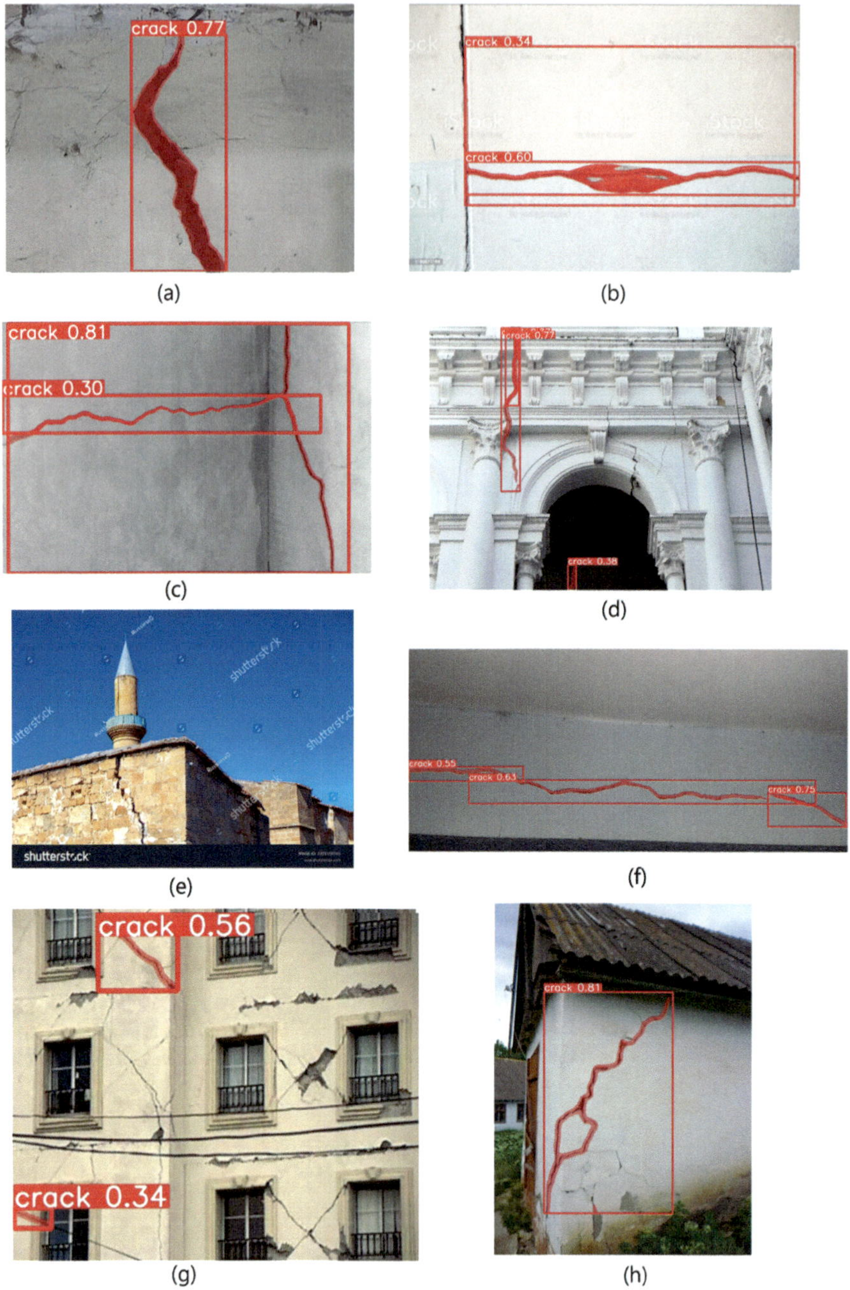

Fig. 18.6 Detections of YOLOv7 on test images

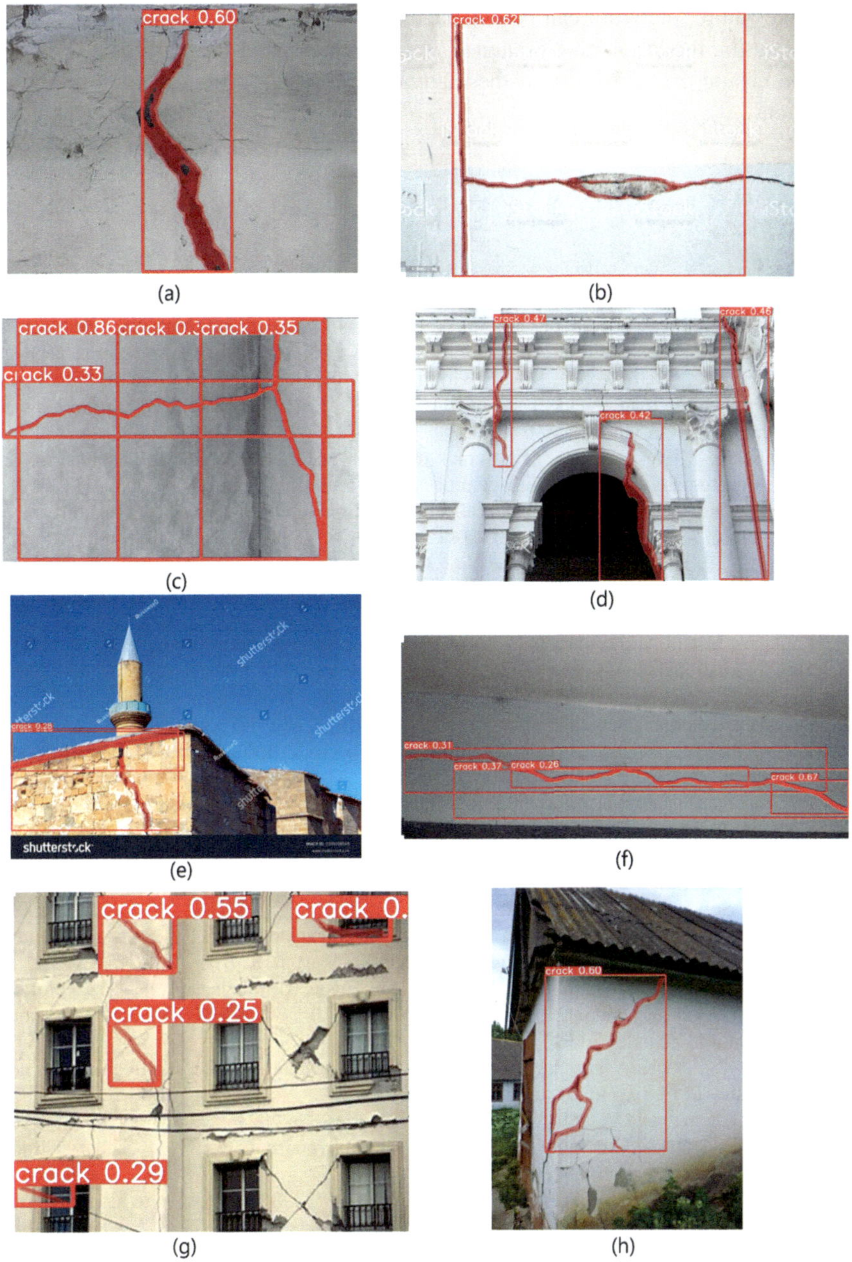

Fig. 18.7 Detections of YOLOv8 on test images

In the future, the work can be extended to real-life practical implementation by deploying the model on an Unmanned Aerial Vehicle, and further research can be done to ensure the scalability of these advanced SOTA models for real-time crack detection.

References

1. Chatterjee, R., Chatterjee, A., Halder, R.: Impact of deep learning on arts and archaeology: an image classification point of view. In: Proceedings of International Conference on Machine Intelligence and Data Science Applications: MIDAS 2020, pp. 801–810. Springer (2021)
2. Chatterjee, R., Chatterjee, A., Roy, S., Gourisaria, M.K.: A fast and effective machine learning approach for road cracks classification. In: 2022 IEEE 19th India Council International Conference (INDICON), pp. 1–6 (2022)
3. Yamaguchi, T., Hashimoto, S.: Fast crack detection method for large-size concrete surface images using percolation-based image processing. Mach. Vis. Appl. **21**(5), 797–809 (2010)
4. Zhenwei, Y., Shen, Y., Shen, C.: A real-time detection approach for bridge cracks based on yolov4-fpm. Autom. Constr. **122**, 103514 (2021)
5. Zhang, L., Yang, F., Zhang, Y.D., and Ying Julie Zhu. Road crack detection using deep convolutional neural network. In: 2016 IEEE International Conference on Image Processing (ICIP), pp. 3708–3712 (2016)
6. Fujita, Y., Mitani, Y., Hamamoto, Y.: A method for crack detection on a concrete structure. In: 18th International Conference on Pattern Recognition (ICPR'06), vol. 3, pp. 901–904 (2006)
7. University. Crack dataset. https://universe.roboflow.com/university-bswxt/crack-bphdr, Dec 2022. Visited on 27 June 2023
8. Beitzel, S.M., Jensen, E.C., Frieder, O.: MAP, pp. 1691–1692. Springer US, Boston, MA (2009)
9. Chatterjee, R., Chatterjee, A., Hafizul Islam, S.K., Khan, M.K.: An object detection-based few-shot learning approach for multimedia quality assessment. Multimedia Syst., 1–14 (2022)
10. Redmon, J., Divvala, S., Girshick, R., Farhadi, A.: Unified, real-time object detection, you only look once (2016)
11. Jiang, P., Ergu, D., Liu, F., Cai, Y., Ma, B.: A review of yolo algorithm developments. Procedia Comput. Sci. **199**, 1066–1073 (2022)
12. Diwan, T., Anirudh, G., Tembhurne, J.V.: Object detection using yolo: challenges, architectural successors, datasets and applications. Multimedia Tools Appl. **82**(6), 9243–9275 (2023)
13. Jocher, G., Chaurasia, A., Stoken, A., Borovec, J., Kwon, Y., Michael, K., Fang, J., Yifu, Z., Wong, C., Montes, D., et al.: Ultralytics/YOLOv5: V7.0-YOLOv5 SOTA Realtime Instance Segmentation. Zenodo (2022)
14. Wang, C.-Y., Bochkovskiy, A., Liao, H.-Y.M.: YOLOv7: trainable bag-of-freebies sets new state-of-the-art for real-time object detectors (2022)
15. Jocher, G., Chaurasia, A., Qiu, J.: YOLO by Ultralytics, Jan 2023
16. Test image-1. https://c.ndtvimg.com/2023-01/60reifp8_house_640x480_11_January_23.jpg, June 2023. Accessed 28 June 2023
17. Test image-2. https://media.istockphoto.com/id/1199572188/photo/concrete-structures-old-buildings-have-cracks-due-to-heavy-loads-for-a-long-time.jpg, June 2023. Accessed 28 June 2023
18. Test image-3. https://images.squarespace-cdn.com/content/v1/5ab7c0657c9327ace0bd3448/1581509446934-7E0MI741KNFBDTFWY5BB/crack.png, June 2023. Accessed 28 June 2023
19. Test image-4. https://static.toiimg.com/photo/msid-66573961/66573961.jpg, June 2023. Accessed 28 June 2023
20. Test image-5. https://c.ndtvimg.com/2023-01/60reifp8_house_640x480_11_January_23.jpg, June 2023. Accessed 28 June 2023

21. Test image-6. https://imgur.com/gallery/ZLUVB6z, June 2023. Accessed 28 June 2023
22. Test image-7. https://dduzkvnw6iy47.cloudfront.net/assets/images/blogs/16191828236082c4e7514dc.jpg, June 2023. Accessed 28 June 2023
23. Test image-8. https://api.gharpedia.com/wp-content/uploads/2018/08/0707020001-08-Cracks-due-to-settlement-of-soil.jpg, June 2023. Accessed 28 June 2023

Chapter 19
Insurance_4_YOU: MCDM Equipped User-Friendly Intelligent System for Personalized Life Insurance Recommendations

Asha Rani, Kavita Taneja, and Harmunish Taneja

Abstract As the middle class in developing nations like ours expands rapidly, the use of recommender systems in the insurance industry is becoming a crucial new field of research. Recommendation systems designed for online platforms like e-commerce, movies, and songs are typically constructed using extensive datasets with the goal of suggesting the most suitable options. However, insurance products possess special attributes that require a different methodology. This paper presents a novel recommendation method for life insurance that aims to provide users with the ability to make informed decisions when choosing personalized and unbiased life insurance products. This paper proposes design and development of 'Insurance_4_YOU', a life insurance recommendation system on the basis of extensive preference specifications by the user. It implements the hybrid FUCOM-TOPSIS method of the MCDM domain to determine personalized insurance policies for the user on the basis of priorities and individual requirements. The prototype is validated by taking live dataset through a designed desktop application that is tailored to provide a user-friendly environment of preference specifications.

19.1 Introduction

As a result of financial liberalization, the insurance market has undergone dramatic change. In recent years, the demand for insurance in the India has skyrocketed, which has led to a proliferation of insurance providers and policies. Instantaneous access to information is the essential component for developing an intelligent environment that adds value for individuals [1]. By offering recommendations tailored to the

A. Rani (✉) · K. Taneja
Panjab University, Chandigarh 160014, India
e-mail: asharana20@gmail.com

H. Taneja
D.A.V. College, Sector 10, Chandigarh 160011, India

P. Nanda et al. (eds.), *Intelligent System and Data Analysis*, Smart Innovation, Systems and Technologies 398, https://doi.org/10.1007/978-981-97-5200-3_19

individual's needs and preferences, recommender systems in the insurance sector protect customers from confusion, false information, commissions paid to brokers, and poor product choices.

Various strategies are used by recommendation systems to propose things or information to users based on their likes, actions, sentiment analysis [2], or similarities to other users. The following are some of the most prevalent approaches used in recommendation systems:

- Collaborative filtering: This strategy recommends things based on comparable users' or objects' interests. It might be item-based or user-based. User-based collaborative filtering finds people who have similar tastes and recommends goods that these users have enjoyed. Item-based collaborative filtering finds and suggests products that are comparable to those indicated interest in by the user. E.g. GroupLens [3], Ringo [4] and Amazon [5].
- Content-based filtering: Suggests items to users based on the qualities or characteristics of the items as well as the user's prior preferences. It entails assessing the content of products and making suggestions based on the user's preferences. E.g. News Dude [6] and LIBRA [7].
- Hybrid recommendation systems: Integrate different recommendation approaches to deliver more accurate and diversified recommendations. A system, for example, may combine collaborative filtering and content-based filtering to capitalize on the merits of both techniques. E.g. CinemaScreen [8] and Fab [9].
- Matrix factorization: Is a technique that breaks down the user-item interaction matrix into lower-dimensional matrices in order to discover latent properties that explain the user-item interactions. It's popular in collaborative filtering [10].
- Demographic filtering: This approach suggests things based on demographic information such as age, gender, geography, or other user characteristics. It's a straightforward method, but it's lacking in customization.
- Information-based filtering: Recommendations are created based on explicit rules or information about the objects and user preferences in this technique. It necessitates domain expertise and may not adequately capture implicit user preferences.
- Association rule mining: This approach detects relationships between things in a dataset and utilizes them to propose goods that are commonly purchased or used together [11].
- Reinforcement learning: This approach models the recommendation process as a sequential decision-making issue and use reinforcement learning algorithms to improve the suggestions over time.
- Context-aware recommenders: To give more relevant and timely suggestions, these systems consider contextual information such as time, location, device, or user scenario.
- Session-based suggestions: These methods provide suggestions based on a user's current session interactions without taking their long-term history into account.

However, the algorithms mentioned above, which are designed for addressing large-scale problems and suggesting the most suitable offers, are not suitable for proposing life insurance policies for the following reasons [12].

- The dimensionality of the data: There are a limited number of insurance products in the market in comparison to the vast array of books, songs, news articles, or films available on internet platforms.
- Complexities and knowledge level: Life insurance policies are more complicated than typical items such as books, movies, services, food, music, and so on. The customer is completely unaware of the complexity of insurance products in various aspects. A detailed feature description is required for effective recommendations. The insurance recommender system will need a more extensive set of attributes representing user requirements in addition to the typical limited collection of demographic information.
- Constraints and frequency of visits: Insurance policies include some restrictions in terms of applicability to the user that must be considered. We cannot propose a retirement plan to a person in need of child plan. Traditional items are not subject to such stringent restrictions. The average person has little engagement with insurance firms. Most individuals find purchasing insurance to be a tedious chore. As a result, most consumers purchase insurance plans through insurance agents.

Recommendations for life insurance consumers must have a high level of credibility. While recommending the wrong movie or song isn't a major concern because the viewer can easily select another alternative from hundreds of movies or songs, promoting the wrong life insurance policy can seriously undermine consumers' faith in their insurance business. The needs of life insurance recommender system mentioned above are well mapped by Multi-Criteria Decision-Making (MCDM) approaches. Additionally, these methods are not constrained by issues like cold start, early rater, scalability, or overspecialization [13].

The current research presents a recommender system that aims to enhance the user experience by suggesting the most suitable life insurance coverage at the right timing. The primary objective of this recommender system is to enhance the efficiency and accuracy of life insurance policy recommendations by offering personalized and unbiased suggestions with minimum complexities.

The utilization of the intelligent recommender system 'Insurance_4_YOU' is facilitated through a desktop application. Our objective is to assist the users who lack expertise and portfolio knowledge of insurance domain but are very well aware of their needs for financial stability and support. Aim of this platform is to free the users from the hidden biases and concerns of agents and companies and present the process of insurance selection in a user-friendly way. This platform delivers the ranked list of life insurance policies based on the preferences specified by the users.

The paper is organized as follows. Section 19.2 presents the related work in life insurance domain recommendations. Section 19.3 demonstrates detailed architecture of 'Insurance_4_YOU' for capturing user demographic information and preference requirements related to different parameters of life insurance products. The methodology of hybrid FUCOM-TOPSIS algorithm is also elaborated in this section. Section 19.4 presents the system experimentation and prototype validation results followed by conclusion in Sect. 19.5

19.2 Related Work

There is a limited body of literature pertaining to recommendation systems in the domain of insurance. In their study, Rokach et al. [12] examine the use of recommender systems in the context of call centers, specifically focusing on the deployment of these systems to facilitate the matching of insurance policy riders with clients by agents. This research focuses around the examination of insurance policies within the realm of healthcare. Abbas et al. [14] introduced a real-time cloud-based knowledge recommendation architecture for the health insurance market. The Multi-Attribute Utility Theory (MAUT) is a decision-making framework employed to select an optimal plan by considering characteristics such as coverage and cost. This selection process involves ranking the available options through the elicitation of preferences. In their study, Xu et al. [11] introduced a recommendation model that focuses on automobile insurance policies and utilizes customer segmentation as a basis. The product recommendation model utilizes the Apriori algorithm in conjunction with Neural Networks. Qazi et al. [15] introduced a recommendation system that utilizes a probabilistic graphical technique to aid agents in suggesting auto, life, property, and umbrella insurance based on customer attributes and portfolio in their work. The study conducted investigations utilizing both Bayesian Networks (BN) and Low Rank Matrix Factorization (LRMF) methodologies. Lesage, Deaconu et al. [16] offer a vehicle insurance recommendation system designed to assist agents in the process of up-selling. The proposed system utilizes a combination of the XGBoost algorithm and the Apriori algorithm in order to determine the appropriate client to whom a recommendation should be provided. Ye Bi et al. [17] proposed a framework Heterogeneous Information Network based Cross Domain Insurance Recommendation System (HCDIR) for insurance domain to deal the issue of cold start and insurance plans' complexities. A real-time web-based prototype proposed by Rahman et al. [18], helps insurance sales professionals provide their clients with the best possible plans. They use a case-based reasoning (CBR) algorithm for recommending the insurance policy based on user features and their matching with users of similar history.

Zibriczky [19] conducted survey of recommendation systems working in insurance and finance domains. The paper establishes the fact that insurance domain requires detailed explicit specification of user requirements and Multi-Criteria Decision-Making (MCDM) methods can be a better option for these type of domains with complex items, indirect interactions and higher constraints.

As today's insurance seeker is quite informed and take a more active part in choosing and analyzing the best policy for them, insurance company websites and web aggregators provide insurance quotations more quickly and easily. They are created to help potential customers in making informed decisions. The main issue is that a prospective insurance seeker seldom understands these internet quotations. Additionally, just a tiny portion of the client's demographic characteristics, such as age, income, gender are taken into considerations. The detailed priorities and requirements are not considered, as a person can choose different policies depending whether the policy required is for retirement, savings or child. Another problem is that, agents' primary goal is often commission rather than their clients' well-being. An agent could take advantage of a customer by offering a lower-quality package for a higher commission. A recommendation model that is comprehensive, impartial, and requires little manual involvement is needed. Although TOPSIS is good at mathematically expressing each option and has high processing effectiveness [20, 21], it lacks a reliable method for eliciting weights. In terms of consistency, Full Consistency Method (FUCOM) gives more reliable results and is remarkably more adaptable and simple for weight elicitation [22]. Hence, the life insurance recommender system using MCDM based hybrid FUCOM-TOPSIS algorithm, is developed in this paper to solve the problems discussed above. With explicit detailed specification of preferences in the prototype, every policy seeker with different preferences is treated as a separate case.

19.3 Insurance_4_YOU: Architecture and Methodology

19.3.1 Prototype Architecture

The proposed application operates as desktop application. The system can be categorized into three modules named as interface, application and policy data model. The graphical user interface is utilized by the user for giving the demographic details and preferences for various policy criteria. The application software facilitates the selection life insurance policies which satisfy the user constraints from policy data model. The hybrid FUCOM-TOPSIS method computes the weights of the criteria and finally computes the closeness coefficient of each eligible life insurance policy and presents a ranked list of recommendation on the interface. The architectural design of the proposed recommender system is depicted in Fig. 19.1.

The prototype interface is presented in Fig. 19.2. Stepwise description of communication of the interface with the policy data model can be summarized as follows:

- User interacts with the system interface and inputs the demographic details such as age, income, and type of policy required. These details serves as constraints for filtering of life insurance policies for which the user is eligible.
- The application module extracts those policies which match with the constraints specified by the user in first step.

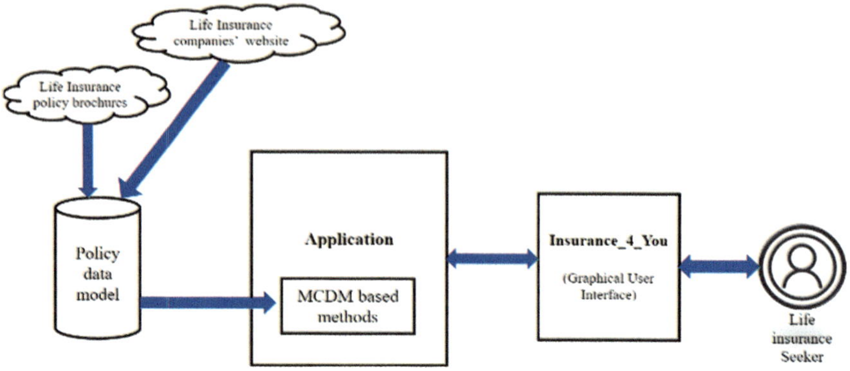

Fig. 19.1 The detailed architecture of Insurance_4_YOU prototype

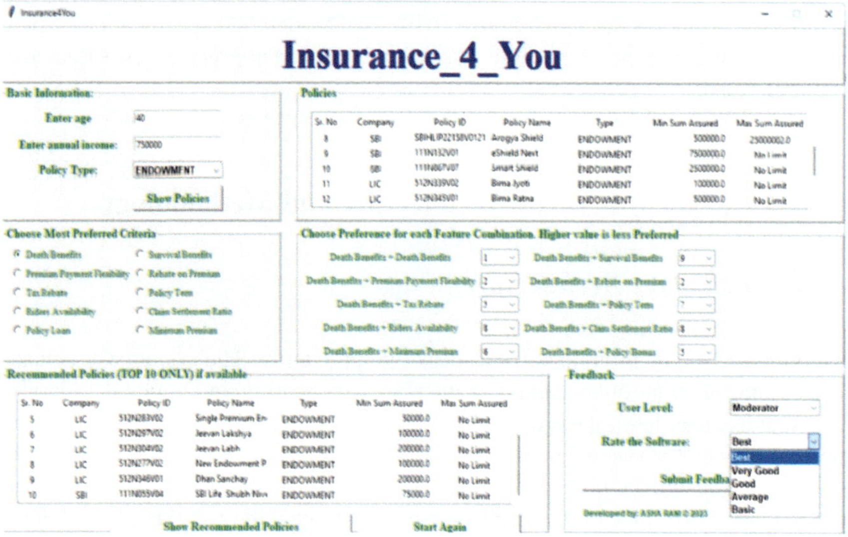

Fig. 19.2 The Graphical User Interface of Insurance_4_YOU prototype

- A list of 10 criteria is displayed on the interface and the user is now required to specify the most important criterion.
- Comparative pairs of the selected most important criterion with all other criteria are displayed on the interface.
- The user is required to give a priority score to each pair. Hence, presents the comparative priority score of each criterion with the most important criterion.
- This priority score of all the criteria is then provided to the application software and by applying hybrid FUCOM-TOPSIS method a ranked list of life insurance policies is displayed on the interface.

- Feedback section is also provided on the interface. It helps in finding the credibility and relevancy of the recommended life insurance policies.

19.3.2 Hybrid FUCOM-TOPSIS Methodology for Preference Elicitation and Ranking of Life Insurance Policies

Insurance_4_YOU methodology can be explained in simplest terms with the help of a pseudo code.

Input:

Demographic features of the user (Age, income, and policy type).

Priority scores of ten life insurance policy criteria (Death_Benefit, Survival_Benefit, Prem_Pay_Flex, Prem_Rebate, Tax_Rebate, Policy_Term, Min_Premium, Avail_Rider, Claim_Setl_Ratio, Policy_Loan).

Performance matrix containing score of each life insurance policy over criteria (dataset of 165 life insurance policies).

Output:

Recommended list of top 10 life insurance policies.

Step 1. Obtain demographic features from the user.

Step 2. Define the list of life insurance policies for which the user is eligible.

Step 3. Obtain objective function from the comparative priority score of life insurance policies.

Step 4. Perform criteria comparisons and obtain constraints.

Step 5. Calculate constraint matrix using constraint 1 and constraint 2

Constraint 1: $^{w_k}/_{w_{k+1}} = \varphi_{k/(k+1)}$

where

$^{w_k}/_{w_{k+1}}$ is ratio of weight coefficients and $\varphi_{k/(k+1)}$ is ratio of priorities of criteria.

Constraint 2: Satisfying the condition of mathematical transitivity

$\varphi_{k/(k+1)} \otimes \varphi_{k+1/(k+2)} = \varphi_{k/(k+2)}$.

where $^{w_k}/_{w_{k+1}}$ is ratio of weight coefficients and $\varphi_{k/(k+1)}$ is ratio of priorities of criteria.

Step 6. Calculate weight matrix satisfying the condition of minimum deviation from full consistency depicted in Eqs. (19.1.1–19.1.4).

Min χ

$$\left| \frac{w_{j(k)}}{w_{j(k+1)}} - \varphi_{k/(k+1)} \right| \leq \chi, \forall j \tag{19.1.1}$$

$$\left| \frac{w_{j(k)}}{w_{j(k+2)}} - \varphi_{k/(k+2)} \otimes \varphi_{k+1/(k+2)} \right| \leq \chi, \ \forall j \tag{19.1.2}$$

$$\sum_{j=1}^{n} w_j = 1 \tag{19.1.3}$$

$$w_j \geq 0, \forall j \tag{19.1.4}$$

Step 7. Calculate normalized decision matrix from the performance matrix of eligible life insurance policies using Eq. (19.2).

$$r_{ij} = \frac{x_{ij}}{\sqrt{\sum_{i=1}^{m} x_{ij}^2}}, \; i = 1\ldots\ldots m, \; j = 1\ldots\ldots.n \tag{19.2}$$

Step 8. Combine weight matrix and normalized decision matrix to obtain weighted normalized decision matrix using Eq. (19.3).

$$v_{ij} = w_j r_{ij}, \; i = 1\ldots\ldots m, \; j = 1\ldots\ldots.n \tag{19.3}$$

Step 9. Determine the ideal and anti-ideal solutions.

The ideal solution (A^*) is calculated using Eq. (19.4)

$$A^* = \{v_{1,}^* v_2^*, \ldots\ldots\ldots, v_n^*\} = \left\{ \left(\max_j v_{ij} \middle| i \in I' \right) . \left(\min_j v_{ij} \middle| i \in I'' \right) \right\}, \quad i = 1\ldots\ldots m, \; j = 1\ldots\ldots.n \tag{19.4}$$

Ideal solution is obtained by taking best performance from the normalized decision matrix.

The anti-ideal solution (A^-) is calculated using Eq. (19.5)

$$A^- = \{v_{1,}^- v_2^-, \ldots\ldots\ldots, v_n^-\} = \left\{ \left(\min_j v_{ij} \middle| i \in I' \right) . \left(\max_j v_{ij} \middle| i \in I'' \right) \right\}, \quad i = 1\ldots\ldots m, \; j = 1\ldots\ldots.n \tag{19.5}$$

Step 9. Calculate relative closeness to the ideal solution using Eq. (19.6).

$$C_i^* = \frac{D_i^-}{D_i^* + D_i^-}, \; i = 1\ldots\ldots m \tag{19.6}$$

Step 10. Perform ranking of the recommended life insurance policies based on value of closeness coefficient.

19.4 System Experimentation and Prototype Validation Results

19.4.1 Policy Data Model

The research considers 165 policies of five companies under six different categories of life insurance policies. The categories taken under consideration are endowment, retirement, unit linked investment plan, child, money back, and term plan. The policy data model created from insurance company websites, insurance experts and policy brochures gives detailed description of policy categorized in 25 features. Among these features four features forms the constraint features which checks the eligibility of a person for a particular policy and helps in generating a list of policies for which the user is eligible. Ten features from the policy data model forms the decision criteria. The user gives the priorities/preferences to these criteria on the basis of personal requirements.

19.4.2 User Data Model

A questionnaire in the Google form was circulated and a dataset of 800 users collected over 38 attributes related to the demographic details, policy requirements and preferences for policy criteria. Some of the characteristics of the dataset can be placed as:

- 43% of the respondents are from the age group of 40–49.
- 26% of the respondents have the monthly income in the range of 45,000–85,000 while for 19% the monthly income is more than 125,000.
- 45% of the respondents are female while the others are males.
- The respondents in higher income group prefer high sum assured as premium payment is not a constraint for them.
- The most important criteria for maximum respondents is benefit on death. The importance of criteria and respondents' preference is shown in Fig. 19.3.
- Also, in terms of ranking the majority of respondents have given first preference to benefit on death and second to benefit on survival. The priority of criteria given by the respondents is shown in Fig. 19.4.
- Both the graphs (Figs. 19.3 and 19.4) are in correspondence with each other, this establish the validity of the dataset.

19.4.3 Prototype Validation

The recommendation is validated using the two test cases. Ten use cases are considered under each test case. The categories of test cases are as follows:

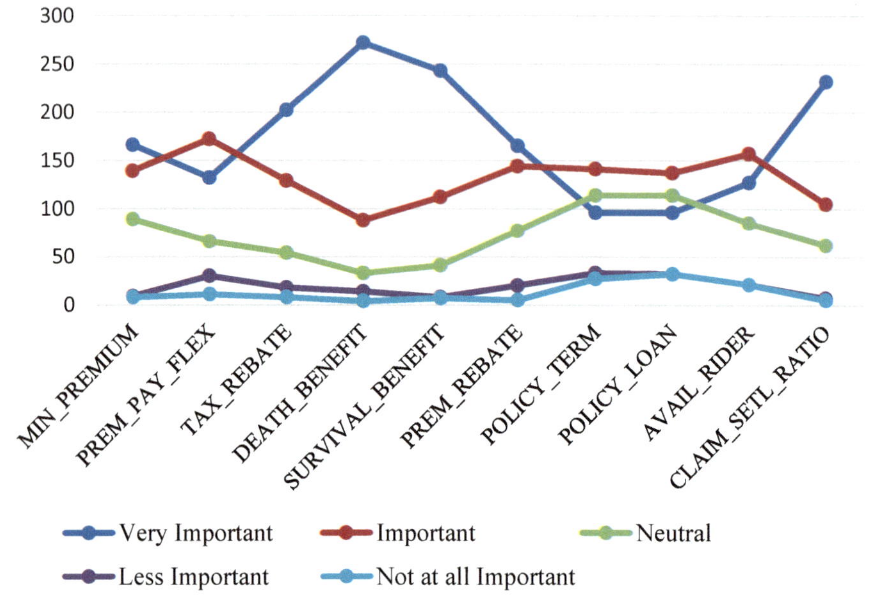

Fig. 19.3 Importance ratings of life insurance policy criteria for various respondents

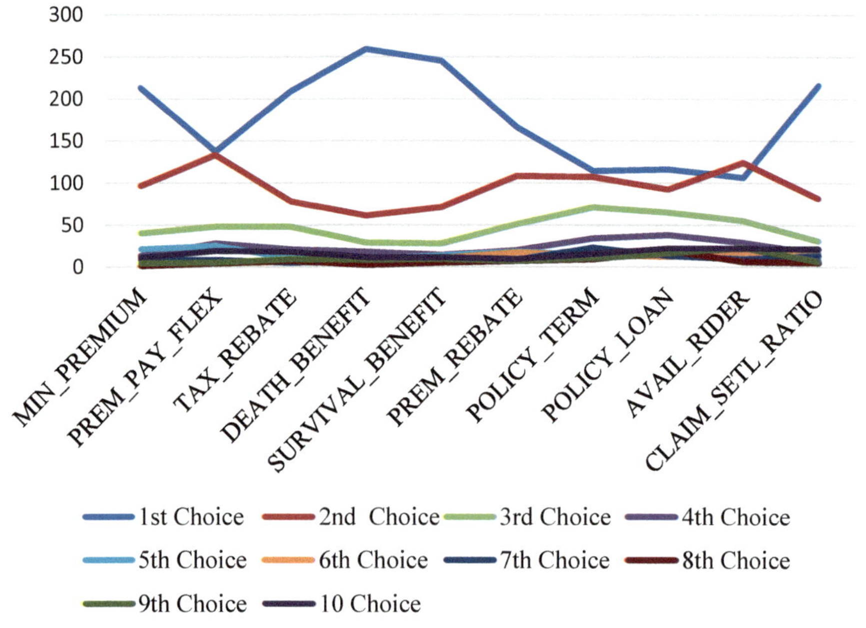

Fig. 19.4 Priority Score of life insurance policy criteria for various respondents

19.4.3.1 Same Demographic Features and Different Criteria Preferences

Consider the demographic feature values as:

Age 40

Annual Income 50000

Policy Type Endowment Plan

Twenty six endowment plans were found to be matched with the above specified demographic features. The result of ten test runs with changed priorities for criteria is conducted and the ranking of the policies obtained is shown in the Fig. 19.5.

The validity and relevancy of our system can be established on the basis of results obtained in Fig. 19.5. It is evident here that when the preference score is changed by the user, the ranking of life insurance policies also change. Our system is quite sensitive toward the preference change and consistent (provide similar results for similar preferences). When the preference score and the priority score is compared it shows the direct effect of priorities on the ranking of life insurance policies.

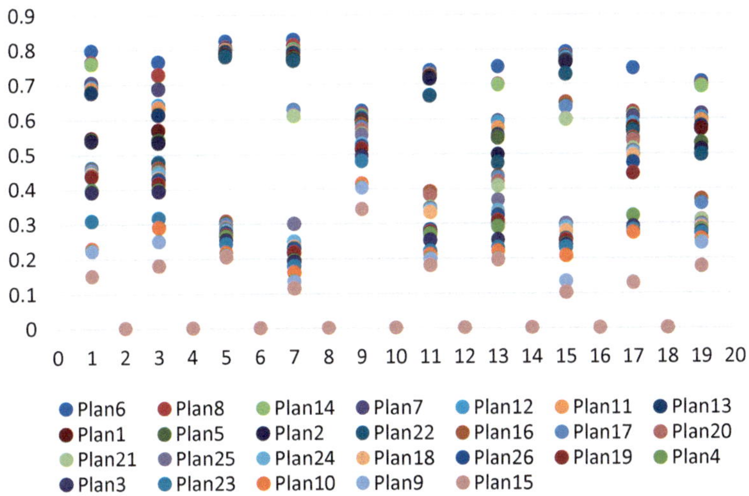

Fig. 19.5 Ranking of plans with respect to changes in the preferences of life insurance policy criteria when demographic features remain unchanged

Table 19.1 Test cases of demographic features and the number of filtered life insurance policies

S. No	Age in years	Income	Policy type	No. of filtered policies
1	5	100,000	Endowment	30
2	18	500,000	Endowment	36
3	25	600,000	Endowment	46
4	45	500,000	ULIP	31
5	56	1,000,000	ULIP	24
6	60	1,200,000	Retirement	26
7	40	2,000,000	Retirement	24
8	20	6,000,000	Money Back	12
9	39	700,000	Term	19
10	18	3,500,000	Term	18

19.4.3.2 Different Demographic Features and Same Criteria Preferences

Ten test cases are created with different demographic specifications under the two types of life insurance policies. The number of filtered life insurance policies extracted from the policy data model, that satisfy the demographic constraints of the user are shown in Table 19.1. These filtered policies are evaluated for performance on policy criteria and the most appropriate is recommended.

19.5 Conclusion

'Insurance_4_YOU' is life insurance recommendation system implementing hybrid FUCOM-TOPSIS method of MCDM in insurance domain. The simplified user preferences for insurance features, which represent the users' current need for insurance purchase taken in a user-friendly interface, serve as the foundation of the suggested system. Preferences for policy criteria are gathered through FUCOM that handles inconsistency and reduce computational complexities. TOPSIS method obtain the criteria weights from FUCOM and interacts with the performance score of policies taken from policy data model to provide the intelligently recommended list of life insurance policies. The Insurance_4_YOU provides a user-friendly and easy way to collect the criteria priorities of different policies by the user through the developed desktop application. The prototype is tested with different demographic features with same criteria preferences and same demographic features with different criteria preferences. When the policy performance score and the criteria priority score are compared, it shows the direct effect of priorities on the ranking of life insurance policies. The feedback is collected through the prototype interface from 400 users, 395 have given 'very good' or above ratings to the life insurance recommender system.

System validation done in terms of relevancy of life insurance policy recommendation, ratings of users and the sensitivity to the changes in policy criteria preference, is giving decent results.

References

1. Jain, P., Taneja, K., Taneja, H.: Convolutional neural network based intelligent advertisement search framework for online english newspapers. Recent Patents Eng. **16**, 4 (2021). https://doi.org/10.2174/1872212115666210715163919
2. Dhuria, S.: Sentiment analysis: an approach in natural language processing for data extraction. Int. J. New Innov. Eng. Technol. **2**(4), 27–31 (2015)
3. Resnick, P., Iacovou, N., Suchak, M., Bergstrom, P., Riedl J.: GroupLens: an open architecture for collaborative filtering of netnews, Proc. 1994 ACM Conf. Comput. Support. Coop. Work. CSCW, pp. 175–186, (1994), https://doi.org/10.1145/192844.192905
4. Shardanand, U., Maes, P.: Social information filtering: algorithms for automating 'word of mouth.' Conf. Human Fact. Comput. Syst. Proceed. **1**, 210–217 (1995)
5. Linden, G., Smith, B., York, J.: Amazon.com recommendations: item-to-item collaborative filtering. IEEE Int. Comput. IEEE Comput. Soc. **3**, 17–80 (2003)
6. Billsus, D., Pazzani, M.J.: User modeling for adaptive news access. User Model. User-Adapted Interact. **10**, 147–180 (2000)
7. Mooney, R.J., Roy, L.: "Content-based book recommending using learning for text categorization," Proceedings of the fifth ACM conference on Digital Libraries, pp. 195–204, (2000), https://doi.org/10.1145/336597.336662
8. Bogers, T., Van Den Bosch, A.: Collaborative and content-based filtering for item recommendation on social bookmarking websites, Vol. 532. (2009).
9. Balabanović, M., Shoham, Y.: Content-based, collaborative recommendation. Commun. ACM **40**(3), 66–72 (1997). https://doi.org/10.1145/245108.245124
10. Zheng, Z., Ma, H., Lyu, M.R., King, I., Member, S.: Collaborative web service QoS prediction via neighborhood integrated matrix factorization. Serv. Comput. IEEE Trans. **6**(3), 289–299 (2013). https://doi.org/10.1109/TSC.2011.59
11. Xu, W., Wang, J., Zhao, Z., Sun, C., Ma, J.: A novel intelligence recommendation model for insurance products with consumer segmentation. J. Syst. Sci. Inf. **2**(1), 16–28 (2017). https://doi.org/10.1515/jssi-2014-0016
12. Rokach, L., Shani, G., Shapira, B., Chapnik, E., Siboni, G.: "Recommending insurance riders," Proceedings of the 28th Annual ACM Symposium on Applied Computing, pp. 253–260, (2013), https://doi.org/10.1145/2480362.2480417.
13. Rani, A., Taneja, K., Taneja, H.: Life insurance-based recommendation system for effective information computing. Int. J. Inf. Retr. Res. **11**(2), 1–14 (2021). https://doi.org/10.4018/ijirr.2021040101
14. Abbas, A., Bilal, K., Zhang, L., Khan, S.U.: A cloud based health insurance plan recommendation system: a user centered approach. Futur. Gener. Comput. Syst. **43–44**, 99–109 (2015). https://doi.org/10.1016/j.future.2014.08.010
15. Qazi, M., Fung, G.M., Meissner, K.J., Fontes, E.R.: "An insurance recommendation system using Bayesian networks," RecSys 2017— Proceedings of the 11th ACM Conference on Recommender Systems, pp. 274–278, (2017), https://doi.org/10.1145/3109859.3109907.
16. Lesage, L., et al.: A recommendation system for car insurance. Eur. Actuar. J. **10**, 377–398 (2020)
17. Bi, Y., Song, L., Yao, M., Wu, Z., Wang, J., Xiao, J.: "A heterogeneous information network based cross domain insurance recommendation system for cold start users," SIGIR 2020— Proceedings of the 43rd International ACM SIGIR Conference on Research and Development in Information Retrieval, pp. 2211–2220, (2020), https://doi.org/10.1145/3397271.3401426.

18. Rahman, S.S.A., Norman, A.A., Soon, K.J.: MyINS: a CBR e-commerce application for insurance policies. Electron. Commer. Res. **5**(1), 373–380 (2006)
19. Zibriczky, D.: Recommender systems meet finance: a literature review. CEUR Workshop Proc. **1606**(5), 3–10 (2016)
20. Ali, Y., Mehmood, B., Huzaifa, M., Yasir, U., Khan, A.U.: Development of a new hybrid multi criteria decision-making method for a car selection scenario. Facta Univ. Ser. Mech. Eng. **18**(3), 357–373 (2020). https://doi.org/10.22190/FUME200305031A
21. Rani, A., Taneja, K., Taneja, H.: Multi criteria decision making (MCDM) based preference elicitation framework for life insurance recommendation system. Turkish J. Comput. Math. Educ. **12**(2), 1848–1858 (2021). https://doi.org/10.17762/turcomat.v12i2.1523
22. Pamučar, D., Stević, Ž, Sremac, S.: A new model for determining weight coefficients of criteria in MCDM models: full consistency method (FUCOM). Symmetry (Basel) **10**(9), 1–22 (2018). https://doi.org/10.3390/sym10090393

Chapter 20
Apriori Analysis of Deep Learning Models on a Multi-class Data Set

Sai Sathwik Kosuru, Y. C. A. Padmanabha Reddy, Perala Venkata Akanksha, Kotagiri Chaitanya, Snehaja Parsi, and Madhu Babu Chunduri

Abstract Deep learning techniques have large scopes in terms of applications and resource requirements, hence calling for a deep analysis of their behavior. Many studies focused on the application part of these models and there is a relatively lesser discussion about the unit-wise behavior. A multi-class dataset was taken and three different models were trained where their performances of training processes are described in terms of runtime characteristics like learning rate, times, etc. An apriori analysis was conducted where the runtime characteristics are investigated for their similarity with pre-defined knowledge of their architectures, as opposed to the posterior analysis which particularly focuses on observations. The work also serves as a framework for defining the logistics of the model training processes and fine-tunes the parameters devised in future for the maximum accuracy of the model. The results were further converted into graphical representations for enhanced delivery of insights. The study is also novel where, as opposed to previous works, in which performances are only tested, the model is evaluated in a design-specific manner dealing with only architectures and the parameters defining them.

S. S. Kosuru · Y. C. A. Padmanabha Reddy · P. V. Akanksha · K. Chaitanya · S. Parsi · M. B. Chunduri
Department of Computer Science and Engineering, B V Raju Institute of Technology, Narsapur, Medak, Telangana 502313, India
e-mail: 21211a05r6@bvrit.ac.in

Y. C. A. Padmanabha Reddy (✉)
Department of Artificial Intelligence & Machine Learning, Chaitanya Bharathi Institute of Technology, Ganidept, 500075 Hyderabad, India
e-mail: padmanabhareddy_cseaiml@cbit.ac.in

P. Nanda et al. (eds.), *Intelligent System and Data Analysis*, Smart Innovation, Systems and Technologies 398, https://doi.org/10.1007/978-981-97-5200-3_20

20.1 Introduction

Deep learning models are the convolutions of Feature analysis and Machine Intelligence Augmentations. While machine-learning models perform similar tasks, that tasks lack Feature analysis feature and the extent of knowledge is less so far. Layer-based approach is implemented in the case of deep learning thereby mimicking the staged learning process of humans. The neural networks are built on the basis of human brain simulations differing from its architecture only in biological aspects. Alternatively, these networks can be termed as virtual brains but yet significantly differ from the biological organ in terms of both structural and dynamic aspects. These models are analyzed just in terms of their performance in all the existing works, thus underscoring the importance of efficiency and precision in their working. Layers classify these models as single layer or multi-layer. Each layer is comprised of filters, collection of neurons, and interconnections. It is to be noted here that latent representations are made through each iteration and these representations are converged/merged to get a perfect output. While these latent representations are created, it is important to observe that the features of the input image is stored and analyzed on an iteration basis. It should even be observed that the creation of a complete picture by the model, even with clarity and perfection, may not satisfy the desired use cases of the model. This is technically termed as **Overfitting**. In other words, particular instances cannot define the performance of the model, thus underscoring the need for generalizing in all aspects.

Conventional Neural Networks are considered as a part of this work which is critical to understanding the fundamentals of the parameter requirements for computer intelligence. Introspection of fundamental aspects may help in design of efficient models subject to effective learning. Current work stresses on runtime conditions of models' functioning and emphasize on parameters affecting models' efficiency. Hard-core solutions can be framed through the proposed findings and diverse range of new architectures can be built on the chassis developed here. The solution also serves as a basis for new models in future in terms of enhanced efficiencies and performances. Existing systems least focused on the general aspects of these models. It should be admitted that the performance is not the same everywhere. Still, it is not risky to have a general intuition about these models' functions and individual effects. Therefore, the current work deals with a particular case of performance in a general manner which is presented in a graphical representation in this paper.

20.2 Literature Survey

20.2.1 Applications

Deep Learning Algorithms are often used in the medical diagnoses [1–4], CCTV footage analysis, Visual forensic sciences, etc. They deal with images and videos(videos may be considered as a collection of a large number of images called frames which are rolled continuously and steadily throughout the duration. Hence, the term videos may not be explicitly considered.). We may also see sparse applications in the scope of machine learning where they deal with mathematical computations and data tables. These models are adapted to multiple scales of images where radiographic images are even analyzed in case of MRI scan analysis in Medical applications. Vehicle counting and tracking is another sound system relevant to Transportation engineering. Agricultural [5] issues like plant disease detection, flower detection, sociological issues like anarchy detection, etc., are already being implemented across the globe with increasing efficiencies along the years. The YOLO models made the field of object detection highly efficient and fast, owing to state of the art frameworks proposed for many years. Surprisingly, fixed architecture have been defied and the novel concepts of dynamic architectures [6, 7], suitable for the current requirements, can be sought out by advanced algorithms like Neural Architecture Search, which is the same that is integrated with YOLO model, resulting in the YOLO NAS model, the latest YOLO model that is able to provide high accuracy as compared to other versions of YOLO. Object detection in turn facilitates the use of mathematical models and structures to segment distinct objects in images, annotating them with bounding boxes. Besides object detection algorithms, relevance of deep learning pre-trained models like GoogleNet, AlexNet, etc. While applications just stress on accuracy [8], there is a need for broad study of internal studies thus underscoring the importance of current work.

20.2.2 Existing Works

Deep learning models are perceived on their performance in many of the existing systems where their performances are seen increased considerably over the previous ones. While various models are used in a single application to assess the suitable one, there is no general intuition as to what the apriori [9] factors state about them. Deep Learning technology is said to constitute the Fourth Industrial Revolution, thus forming an integral part of the **Technical Sustainability**. Deep learning is absolutely higher in efficient on comparison to standard machine learning models. This is true in the cases where we consider advanced machine learning models and even overpowers the best ones. Surveys were also conducted in this context to make models smaller, faster, and efficient [10, 11] thus highlighting the importance of the functioning of these technologies. These surveys are very much relevant to and played a critical role in the current work as it focuses on the same in a different and innovative way. Call for Efficient Deep Learning [12] has been increasing steadily in the recent days with outcome and accuracy-based approaches [13, 14]. Colloquially, it is imminent that accuracy are solely considered while evaluating models in all the existing works. As a result, existing systems work on improving accuracy rather than exploring the fundamental backend structures [14, 15], thus making the process unnecessarily complex and lengthy.

20.3 The Apriori Analysis

20.3.1 Dataset Used for the Study

The apriori analyses require multi-conditional considerations which calls for multiple constraints in the chosen problem. Data and its collection is a critical starting stage for deep learning models and hence, every deep learning model has its roots in the data collected. In this work, we have surveyed on multi-class datasets with minimum allowable number greater than six. There is a high-dimensional contrast among diverse classes in many datasets which makes each class significantly differentiable. Therefore, every model will have a sound performance on such datasets hence, defeating the original purpose of the current research. While there has been a greater number of innovations and researches in the direction of continuous curation of all relevant data in various domains, the problem of consistency in terms of stored features and similarities among various classes is gradually decreasing and the boundaries are rapidly subsidizing. While there has relatively been a lull in simple datasets and fundamental models, such problems seldom find applications, though sparsely. Thus far, datasets are multi-class and multi-labeled, thus widening the scope of the work beyond certain boundaries.

Figure 20.1 illustrates all the dataset-specific details in detail and the description is as follows. Human and plant diseases, food images, Remote sensing image classifications, etc., often contain coarse features, some of which are impossible to discriminate based on visual data. This calls for extremely rugged images for generalizability and increased scalability. Hence, plant disease detection datasets, having ten different class have been considered for the current work and the same were analyzed on a model-comparison basis where around three models are considered for evaluation-based examination. Several components of the collected dataset have some similarities which cannot be differentiated purely based on visual information collected. Similarly, the misclassification in images and semantics cannot be resolved in totality with just a single model. Hence, multiple models are considered for performance-based elucidations and the results are scrutinized for either making an analysis or perform testing of models.

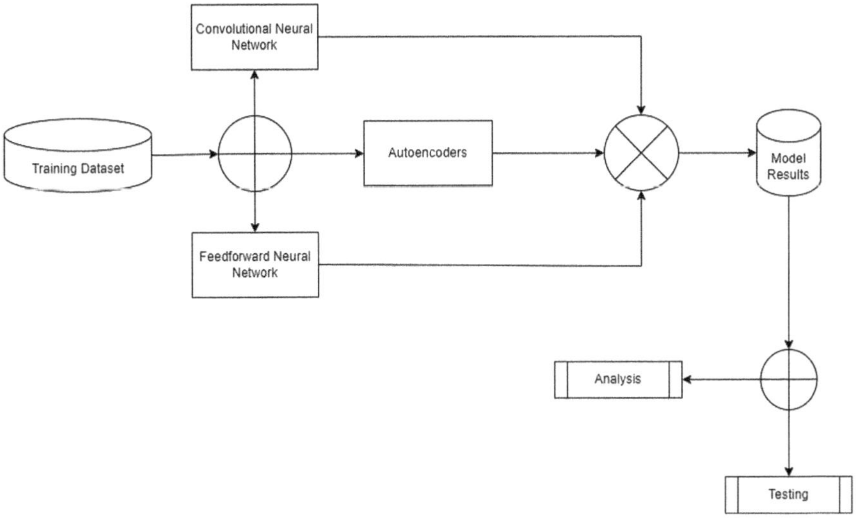

Fig. 20.1 Overview of the contextual dataset-model interactions

20.3.2 Platform Preferences During Analysis

Conventional systems, particularly household PCs, have less computational power where it is not possible to train every model. Moreover, Hardware inconsistencies may dominate the apriori analysis performed here. It is critical to be observed that the apriori analysis involves no hardware considerations and solely rely on functional implementations. Thus, the involvement of a study on platform dependencies may defeat the original purpose of the current research while underscoring the importance of pure-functional analyses in model comparisons. It is worth to be noted that an apriori analysis of deep learning models must include the concept of Hardware considerations since there are non-zero dependencies between function-specific software of models and hardware constraints. Though exploring the web of dependencies involved intricately in models' functions is not a primary focus, it has been necessitated to ensure proper channel of relations between them for efficient analyses, as evident in the current work. In this context, it is necessary to ensure a systematic evaluation of performance metrics across a wide range of platforms thereby calling for the running of same model on different platforms. Since this is an apriori analysis, platform parameters are never considered but model parameters varying as a function of platform-dependent parameters may be tested for a generalized conclusion.

Google Colab Google Colaboratory offers free GPU and deep learning-specific virtual infrastructure with a Jupyter like interface. This service is available as a web-based IDE but with some limitations like not being able to access the physical resources of the computer. Intuitively, some or all of computer vision and related deep learning applications include the use of webcam, speakers, etc., which are generally not supported directly in this IDE. Additionally, it requires no explicit configurations and has many libraries or packages imported and ready to use at hand. Generally, Deep Learning frameworks and heavy algorithms are run on this IDE thereby making it increasingly usable for research purposes. Hence, all the models' trainings' performed as a part of the current work are run in this ide as a primary option or a primary choice.

Jupyter Jupyter is also a web-based ide but it runs on a kernel which is initiated by an Anaconda prompt. It is not directly web-based but utilizes hardware components of the system like Graphical Processing Units and needs installation of certain libraries/packages. Moreover, running deep leaning models on this IDE is often tedious as it levies heavy burden on the systems resources where tight couplings can often cause failures in training processes and jamming of systems. It is even time consuming to run such models because system resources are often disbursed for various other processes and background tasks, hence significantly draining the efficacy of models' training processes when considering the time spent. As mentioned above, to accommodate use of system physical resources flexibly, this has been considered.

Kaggle Kaggle offers ide but to a limited extent. The amount of GPU offered too is limited but as it is data-based application, the ide offers faster training times and low failure rates or ratios. Hence, it has been consulted for few models even though was not initially preferred.

20.3.3 Analysis and Working of Deep Learning Models

The following models are sought out to find probable relations as demonstrated in Fig. 20.2.

Autoencoders Autoencoders are generally used for Generative AI applications and are broadly supervised. They may seem irrelevant in the current context but with appropriate modifications, they may be included. The layered structure supports equal neurons in both input and output layers with hidden layers performing critical functions. Latent representations are initially synthesized reproducing the features thereafter. In this reconstruction phase, errors are examined, accuracy is calculated and finally training is performed. It is worth to be noted that these iterations do not occur in minutes but take several hours for a good accuracy. There are two inbuilt models one of which is used to generate the latent forms using training algorithms

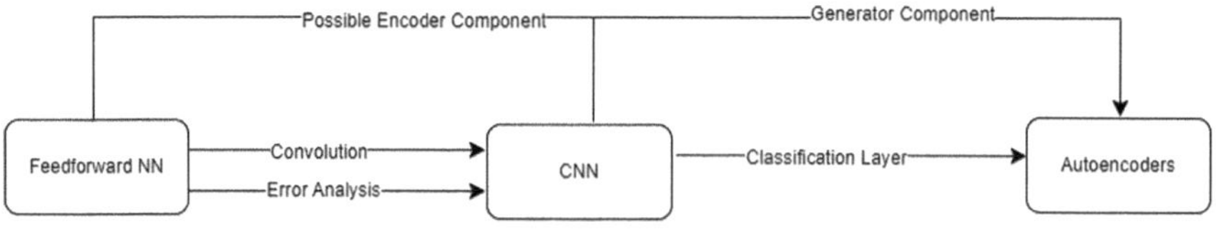

Fig. 20.2 Soughted model relations

and then computing errors to rectify them iteration-wise through the use of discriminator model. While these autoencoders often focus on reconstructions and latent representations, there has been an increasing need for classification-orientation of such models for model comparison purposes. The only difference from the original version would be that a classifying layer is added in the initial parts of the model architecture, thus modifying it into a classification-based model. Alternatively, the model can be used for reconstructing images for enhancing the dataset content in addition to data augmentation and synthetic data techniques.

Feedforward Neural Network This is an elementary model which has three elementary layers: In Data Stream, A single Hidden Layer, and an Out Data Stream layer. There is no role of backtracking or back propagation in this model and the flow of data is simply unidirectional. With appropriate modifications and optional addition of layers, we can continue improving the accuracy on a trial-and-error basis. Upon selecting appropriate hyperparameters, the model can be redefined to produce improved accuracy free of any overfitting. The concept of overfitting is highly dominant here since the number of layers is less and feature extraction may be performed at a higher pace. This leads to increased chances of exploding gradients thereby reducing efficiency gradually.

Convolutional Neural Network This algorithmic implementation is the most common and is treated as the most elementary deep learning neural network. Convolutions are typically found in biological organs like Brain, blood capillaries, and Kidneys where networks of fluid interchanges are mostly found. The same concept is applied on computers where the data/information takes the place of fluids and convolutions provide a more flexible approach for flexible and generalized applicability of the learnt model. Since CNN is found as the most common and preliminary model, the same was investigated for relevance rather than model comparison evaluations in the current work. We used Eq. 20.1 to numerically extract features from different classes on a priority basis. Since the images are similar, we used correlation equations among pixels using three models for efficient performance.

$$G[i,j] = \sum_{n=-1}^{K} \sum_{n=-k}^{K} H[u,v]F[i+u,j+v] \tag{20.1}$$

H and F are filters and pixels, respectively, in 1.

20.4 Results and Discussions

Thus far, there is no fixed benchmark for evaluating the sole responses and individual behaviors of these models. In the current paper, there is a focus on analyzing the models on an ideal basis with just time and performance applications. Model-to-model comparisons are made throughout the work but each model was initialized independent of the another. The results obtained are as shown in Fig. 20.3, and the parameters of training are selected from 1.

Fig. 20.3 Model behavioral plots

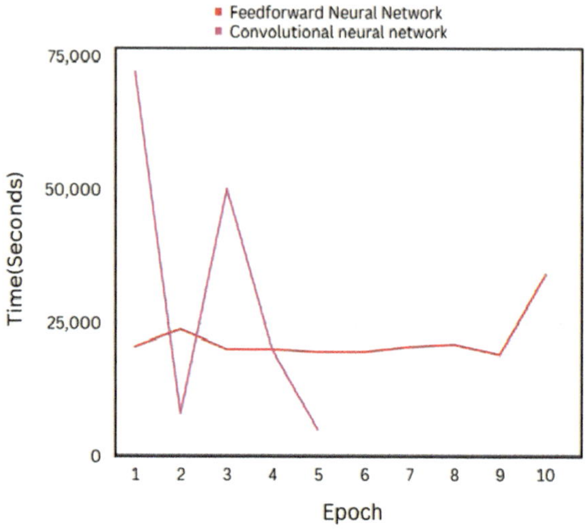

(a) Timing Plots

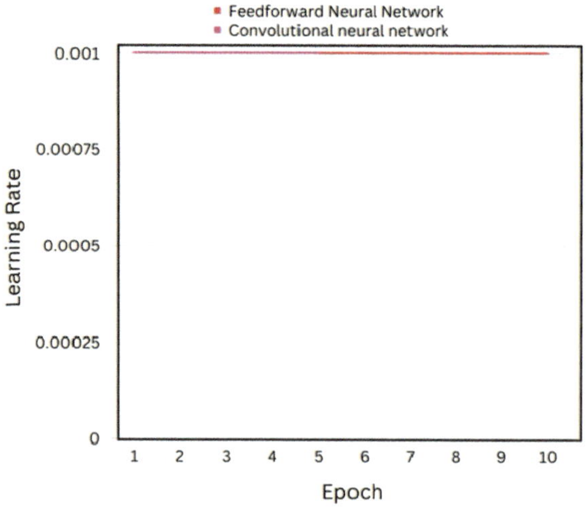

(b) Learning Rate Plots

Fig. 20.3 (continued)

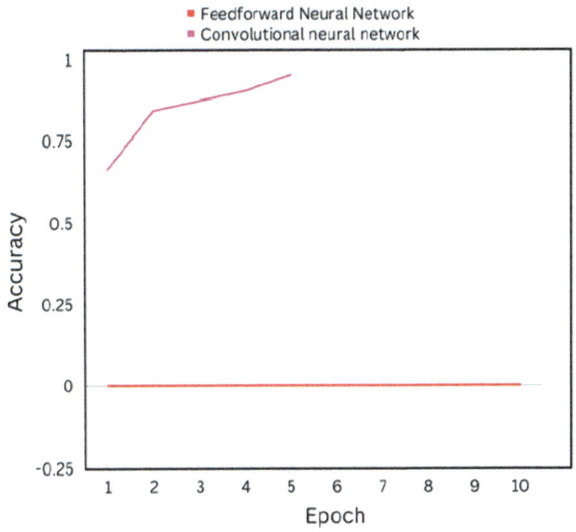

(c) Accuracy Demonstration(Apriori)

20.4.1 *Timestamp Analysis*

Feedforward Neural Network The fastest of the considered models is the Feedforward Neural Network with 48 h for training on a dataset of 1,21,000 images, divided into 10 classes. Hence, multi-class training is not taking a significant amount of time in this model when compared to the others. In this model, there has been a huge emphasis on learning rather than feature extraction as seen in Eq. 20.2.

$$\text{net} = \Sigma_{W_i\,X_j} + b_i \tag{20.2}$$

w_i stands for weights and, b_i stands for bias with x_i as input in Eq. 20.2. Hence, it has taken less time where it just learns a latent version of the original image without delving deep into its features. This is evident when corn leaves are mistaken for pepper leaves owing to poor feature extraction.

Convolutional Neural Network Hyperparameter tuning played an inflecting part in the case of this model's training implementation where 8 sets of parameters are considered for timing analysis. CNN took 2 days for training where only 5 class are considered. For a large number of classes, the model is failing to execute completely wherein it is overfitting easily. The model is not complex in terms of timing considerations but is greatly complex in terms of hyperparameters where hyperparameters are relevant to the concepts of accuracy and had no role to play in time considerations.

Autoencoders Autoencoders are quite complex where conventional IDEs are not supporting longer epochs and larger batches. It is worth noting that this model took multiple attempts for a successful model with repeated failures at approaching results. Autoencoders took around 3 days for initial training where the last epochs had a fatal failure when 10 diverse classes are considered. Hence, 5 classes are only considered where the epochs ran successfully for twelve hours relatively faster than the other models but the space is too high to accommodate the process of model saving. Hence, testing data is not viable even though time is taken at par with other even if partial training is performed.

20.4.2 Accuracy Computation and Learning Rate Determination

Accuracy is variable among the taken models where some models are overfitting too early and very easily. Surprisingly, the learning rate is too fast for dropout methods and regularization techniques to not be effective.

Feedforward Neural Network Epoch wise analysis gives a relatively faster learning rate where the accuracy curve has a very steep slope. Moreover, explicit definition of learning rate seems unnecessary where the accuracy values are sharply increasing with batches. Hyperparameters are not even found to be efficient as the accuracy value begins at 0.9 for almost every set of hyperparameters. First and second epochs are relatively stable in terms of learning rates and the accuracy values are fixed at 0.9 consistently with one-hundredth gradient observed through all the batches, thus indicating a gradual start of accuracy values. While the third epoch starts with overfitting of the model, the model training still progresses even dropout and regularization parameters. Upon testing results, the model is being trained on pixelated values rather than features in this network where the model is not able to differentiate similar images. It is quite intuitive that feedforward neural networks has no backpropagation component and has few layers, hence the neurons are relatively less and they tend to receive all the information at an instance thereby crippling the systematic procedure that may occur in opposite conditions.

Convolutional Neural Network This is an elementary network which supports flexible number of layers, as per user definition. The hyperparameter tuning is critical to improving the accuracy and determining the learning rate of the model. We proceed with defining the numerical amount of layer-wise filters, followed by the consideration of the layers' count. Multiple iterations, aimed at stabilizing accuracy rates from overfitting and improving learning rate were implemented but with marginal difference of results. As opposed to other networks, this network has a moderate learning rate and a gradually increasing accuracy slope. The accuracy, though it starts at 0.823 and progresses slowly in initial epochs, is on a steep rise in the later epochs with accuracy near to 1 in the last epochs. It is interesting to observe that the CNN is not overfitting easily even with increased number of filters. In extreme cases, number of filters used was nearly 8000 but still the model did not overfit. The only problem with this network is that it is overfitting with less data, thus underscoring the potential efficiency of the model.

Autoencoders Autoencoders found approximately stable accuracy (Eq. 20.3) and learning rate variations while consuming a lot of resources for training. Autoencoders training involved a lot of integration issues and configuration problems where layers are not defining the accuracy of the model. Irrespective of the number of layers and defined hyperparameters, Autoencoders are found to be stable and has a very less learning rate. Since autoencoders are just investigated for generating new data sets, in addition to data augmentation and synthetic data techniques, they were not considered for accuracy and learning rate in the current work.

$$L_R = L(x, g(f(x)) + R(h) \tag{20.3}$$

Equation 20.3 is a relation in learning rates where *x*,*g*,*h* are dimensions.

20.5 Discussion of Results

The current apriori analysis results can be summarized as in Table 20.2.

From the results, it is important to observe that timing, accuracy, and learning rate examinations may be considered worthwhile in case of simple models with latest models still requiring a deeper level of understanding. In ideal conditions, these parameters may be taken as sole determiners of model performances, but it is important to note the discrepancies in case of real life conditions where platform inconsistencies and system problems may arise. From the results, it can be stated clearly that model-to-model comparisons can be performed solely based on accuracy computations while their complexities may be examined using learning rates and timing calculations. The final equation of dependency may be framed as Eq. 20.4.

Table 20.1 Considered hyperparameter sets: filter settings

32	32	32	32	32	32
32	64	64	64	64	64
32	64	64	128	128	32
32	64	128	64	64	32
32	64	64	128	64	32

Table 20.2 Apriori analysis results

Parameter	Convolutional NN	Feedforward NN	Autoencoder
Timing	48 Hours	18–44 Hours	72 Hours
Accuracy	0.33–0.98	0.9–1	0.53
Behavior	Steady	Steep	Constant

$$f_{tr}(x) = O(T_{tr}(x)) + h(y_{hw}) * a \tag{20.4}$$

$$T_{tr}(x) \propto h\ (y_{hw}) \tag{20.5}$$

$f_{tr}(x)$ is a function that determines the functional dependencies of all the parameters considered in this analysis where T_{tr} is assumed to be timing calculations, $h(y_{hw})$ is a platform function, and a is the calculated accuracy. This is just a theoretical consideration and mathematical values are never sought in previous works. Key observations include the unexpected non-proportionality between time and learning rate (from 5) and the linear dependency between the functions linking all the sought observations (from 4). Though accuracy is taken as a nonlinear function on a whole, it's behavior is generally observed to be sequential or a step function where it increases gradually with epochs. In time and learning rates' data, they are purely linear functions with significant variations in slopes.

20.6 Conclusion and Future Scope

Preliminary analyses concluded with certain assertions about the computational requirements and output performances of various conventional learning models wherein timestamp, accuracy computation, and Learning rate determination was performed. Adding to this, Platform dependencies and IDE constraints were also considered since deep learning models and computational resources are deeply interlinked during the training period. While there has been a wide difference among the behaviors exhibited by various models, it is important to note the broad results produced by various models on the same dataset. It is even worth noting that irrespective of hyperparameters, the models possess characteristic behaviors, thus highlighting the importance of architectural designs in future models. Hence, the current apriori analysis provides an exemplary overview into the behaviors of the models and model-to-model comparisons on the same dataset on behavioral grounds.

The current research is just an abstracted apriori analysis of models' functioning, which doesn't include the concept of inherent structures and constituent neurons. With latest models and networks being introduced at peak frequencies into the market, it is important to observe the evolutionary behaviors of models and examining their performances. Since, there is no fixed benchmark to determine the performance of these models, an approximate mathematical notation may be devised to employ such measures. Hence, the future work emphasizes on a flexible framework that can adapt to any kind of models in future, thus indicating a need for theoretical findings in the direction of the same.

References

1. Zhang, J., Xie, Y., Wu, Q., Xia, Y.: Medical image classification using synergic deep learning. Med. Image Anal. **54**, 10–19 (2019). https://doi.org/10.1016/j.media.2019.02.010
2. Syed, L., Jabeen, S., Manimala, S., Elsayed, H.A.G.: Data science algorithms and techniques for smart healthcare using IoT and big data analytics. In Studies in Fuzziness and Soft Computing. pp. 211–241. (2019). https://doi.org/10.1007/978-3-030-03131-211
3. Agravat, R., Raval, M.S.: Deep learning for automated brain tumor segmentation in MRI images. In Elsevier eBooks pp. 183–201. (2018). https://doi.org/10.1016/b978-0-12-813087-2.00010-5
4. Kermany, D., Goldbaum, M.H., Cai, W., Valentim, C.C.S., Liang, H., Baxter, S.L., McKeown, A., Yang, G., Wu, X., Yan, F., Dong, J., Prasadha, M.K., Pei, J., Ting, M.Y.L., Zhu, J., Li, C.W., Hewett, S., Dong, J., Ziyar, I., Zhang, K. (2018): Identifying medical diagnoses and treatable diseases by Image-Based Deep Learning, Cell. 172 5 1122-1131.e9. https://doi.org/10.1016/j.cell.2018.02.010
5. Razzak, M.I., Naz, S., Zaib, A.: Deep learning for medical Image Processing: Overview, challenges and the future. In Lecture notes in computational vision and biomechanics. pp. 323–350. (2017). https://doi.org/10.1007/978-3-319-65981-712
6. Gilman, G., Walls, R.J.: Characterizing concurrency mechanisms for NVIDIA GPUs under deep learning workloads. Perform. Eval. **151**, 102234 (2021). https://doi.org/10.1016/j.peva.2021.102234
7. O' Mahony, N., Campbell, S., Carvalho, A., Harapanahalli, S., Hernandez, G.V., Krpalkova, L., Riordan, D., Walsh, J.T.: Deep learning vs. traditional computer vision. In Advances in Intelligent Systems and Computing. pp. 128–144. (2019). https://doi.org/10.1007/978-3-030-17795-910
8. Goswami, S., Singh, S.K.: A simple deep learning based image illumination correction method for paintings. Pattern Recogn. Lett. **138**, 392–396 (2020). https://doi.org/10.1016/j.patrec.2020.08.013
9. Jie, Z., Gang, W.: Intelligence data mining based on improved Apriori algorithm. J. Comput. **14**(1), 52–62 (2019). https://doi.org/10.17706/jcp.14.1.52-62
10. Kiran, T.T.J.: Computer vision accuracy analysis with deep learning model using tensorflow. Int. J. Innov. Res. Comput. Sci. Technol., 8(4). (2020). https://doi.org/10.21276/ijircst.2020.8.4.13
11. Ramchandran, A., Sangaiah, A.K.: Unsupervised anomaly detection for high dimensional data—an exploratory analysis. In Elsevier eBooks. pp. 233–251. (2018). https://doi.org/10.1016/b978-0-12-813314-9.00011-6
12. Borisov, V., Leemann, T., Sessler, K., Haug, J., Pawelczyk, M., Kasneci, G.: Deep neural networks and tabular data: a survey. IEEE Trans. Neural Netw. Learn. Syst. (2022). https://doi.org/10.1109/tnnls.2022.3229161
13. Guex, N., Peitsch, M.C., Schwede, T.: Automated comparative protein structure modeling with SWISS-MODEL and Swiss PdbViewer: a historical perspective. Electrophoresis **30**(S1), S162–S173 (2009). https://doi.org/10.1002/elps.200900140
14. Atanassov, A., Pilev, D.: Pre-trained deep learning models for facial emotions recognition, 2020 International Conference Automatics and Informatics (ICAI), Varna, Bulgaria, pp. 1-6, (2020) https://doi.org/10.1109/ICAI50593.2020.9311334
15. Shrestha, A., Mahmood, A.: Review of deep learning algorithms and architectures. IEEE Access **7**, 53040–53065 (2019). https://doi.org/10.1109/access.2019.2912200

Chapter 21
Comparative Analysis of Emission Prediction Using Machine Learning and Deep Learning

Nitai Shah, Vimal Kumar Pathak, and Ramanpreet Singh

Abstract Monitoring emissions is a critical challenge encountered by numerous industries globally, with the Gas Turbine Industry being particularly affected. The significance of monitoring emissions lies in its ability to assess the performance of these systems and identify their environmental impact. In this paper, we propose multiple machine learning and deep learning models for regression analysis. Our primary objective is to predict Carbon Monoxide and Nitrogen Monoxide emissions from Gas Turbines accurately and compare each model based on their accuracies. By employing machine and deep learning techniques, we aim to develop a robust model that can aid in better understanding emissions patterns and contribute to devising effective strategies for emission reduction. The Random Forest model displayed the lowest mean absolute error at 0.487 and 2.568 for CO and $NO_{x,}$ respectively.

21.1 Introduction

The global demand for electricity continues to grow every year. According to the International Energy Agency (IEA) the global consumption of electricity has nearly doubled from 12,137 TWh to 22,847 TWh between 1999 and 2019 [1]. This upward trend in electricity consumption is predicted to keep rising into the upcoming decades. Despite of our efforts to shift to renewable sources of energy production, we still heavily rely on fossil fuels and combustion to generate roughly 60% of our electricity. The combination of high energy density, abundance and maturity of fossil fuel technology make it easy choice for the production of our electricity. This however comes at a price. The combustion of any hydrocarbon releases several by-products such as

N. Shah (✉) · V. K. Pathak · R. Singh
Department of Mechanical Engineering, Manipal University Jaipur, Jaipur 303007, India
e-mail: nitaishah2001@gmail.com

V. K. Pathak
e-mail: vimalkumar.pathak@jaipur.manipal.edu

R. Singh
e-mail: ramanpreet.singh@jaipur.manipal.edu

P. Nanda et al. (eds.), *Intelligent System and Data Analysis*, Smart Innovation, Systems and Technologies 398, https://doi.org/10.1007/978-981-97-5200-3_21

$CO2$, water vapor, CO, NO_x, $SO2$ and Particulate Matter to name a few. Emissions such as Carbon Monoxide (CO) and various Nitrides of Oxygen (NO_x) are harmful for the atmosphere and are the primary contributors to issues such as air pollution, acid rain and lower Air Quality Indexes (AQI). To combat these issues the Paris Convention was signed by 196 countries to monitor and limit their emissions [2]. According to these accords any power plant having a capacity of 100 MW or above need to continuously monitor their flue gas concentrations of NO_x and CO.

Currently the major approaches to determine emissions according to are based on CEMS or continuous emission monitoring system [3]. In CEMS we are constantly measuring the emissions via CO/ NO_x sensors in real time. PEMS or periodic emission monitoring system is introduced in [3]. In PEMS we are using statistical techniques and previous data to predict the emissions based on certain input parameters that are pertaining to the environment in and around the Gas Turbine. PEMS is advantageous due to its ability to give real-time predictions without using any dedicated sensors for emission monitoring.

The data for emission analysis has been collected from a gas turbine designed to study flue emissions [4]. The goal of this research is to make use of techniques such as linear regression, random forest, gradient boosting and neural networks to perform regression analysis of CO and NO_x emissions. These models are trained on the data and the best model is chosen and analyzed for feature importance. According to the authors of the paper [5] the CEMS and the PEMS model can work in tandem, with the PEMS model raising an alarm if the difference between real world data and predicted data goes above a certain threshold. In the absence of a CEMS a PEMS can work as a backup as well.

Lee et al. [6] conducted a study proposing a gradient boosting model to predict concentrations of PM2.5 in Taiwan. Utilizing a large-scale database from the Environmental Protection Administration and Central Weather Bureau, the database contained data from 77 air monitoring stations and 580 weather stations collected over a 1 year period. The research leveraged past records of PM2.5 along with climatic information from neighboring weather stations, and demonstrated an effective 24-h forecasting results across most air stations. Filho et al. [7] make use of data from diesel/HFO engines used in power plants. Generalization capabilities of linear and non-linear univariate forecasting models on a representative database related to an engine-driven generator in a TPP located in Pernambuco, Brazil are compared. Korpela et al. [8] presented a NO_x emission monitoring technique for natural gas-fired hot water boilers. Their technique uses machine learning model's and process measurements to estimate real-time NO_x emissions. Shakil et al. [9] offered a cost-effective approach which coupled dynamic neural network and sensor data for NO_x and $O2$ measurements. Fichet et al. [10] simulated combustion chemical reactions using a reactor network model and predicted NO_x emissions in gas turbines. Iliyas et al. [11] proposed an inferential sensor using radial basis function (RBF) neural networks for process emission monitoring, effectively estimating emissions based on available process measurements. Lv et al. [12] coupled the least squares support vector machine (LS-SVM) and ensemble models for NO_x emission prediction in coal-fired boilers, enhancing prediction accuracy. Smrekar et al. [13] made use of historical data to develop a multi-step-ahead model that predicts NO_x emissions for coal based boilers.

The organization of this paper is as follows. Section 21.2 introduces the background material's regarding the various model's. Section 21.3 gives a brief description of the dataset along with the pre-processing steps. Section 21.4 outlines the methodology followed and the model parameters and descriptions. The results are displayed in Sect. 21.5 along with the conclusion and future work.

21.2 Background Material

The following section describes the various machine learning and deep learning models employed in this paper. A brief understanding of the model is given along with the mathematical intuition behind each model. The 4 model's chosen for this paper were Linear Regression, Random Forest, Extreme Gradient Boosting and an Artificial Neural Network. Linear Regression was chosen to serve as a baseline metric for the other models. The two ensemble learning models Random Forest and XGBoost were chosen since this class of models often delivers the best results with tabular data. Finally a neural network was chosen as the fourth model since neural networks can best map the non-linear relationships between features and targets.

21.2.1 Linear Regression

The first model utilized is linear regression which is a fundamental and widely used supervised learning algorithm in statistics and machine learning. The goal of this technique is to model the relationship between a dependent variable (target) and one or more independent variables (predictors) by fitting a linear equation to the data. This can be represented by the following equation:

$$y = \beta_0 + \beta_1 x_1 + \beta_2 x_2 + \ldots + \beta_n x_n \tag{21.1}$$

- y represents the dependent variable
- β_0 represents the intercept value
- $\beta_1, \beta_2, \ldots, \beta_n$ represent the coefficients of the independent variables
- $x_1, x_2, \ldots, x_n$ represent the independent variables

Linear regression tries to find the optimal values for the coefficients $\beta_1, \beta_2, \ldots, \beta_n$ that minimize the sum of squared residuals between the predicted values and the actual data points. The Ordinary Least Squares (OLS) method is commonly used to achieve this minimization.

Linear regression is simple yet powerful, making it suitable for various applications, such as predicting house prices, analyzing sales trends, and modeling economic relationships. However, it may not perform well for datasets with non-linear patterns or when there are complex interactions between variables. Linear regression serves as a benchmark to compare our other models to.

21.2.2 Random Forest

The second model utilized is random forest regression which is a versatile and powerful supervised learning algorithm widely used for both classification and regression tasks. This technique combines multiple Decision Trees to make more robust predictions.

The Random Forest algorithm creates each decision tree by randomly selecting a subset of the training data and a random subset of features to use for training. This randomness and the averaging of predictions from multiple trees help reduce overfitting and enhance the algorithm's predictive performance.

For a single decision tree t in the random forest, the prediction can be represented as

$$y_{\text{predicted\_}t} = f_t(x) \tag{21.2}$$

To obtain the final result all predictions from the individual decision trees are averaged as follows:

$$y_{\text{predicted}} = \frac{1}{N} * \sum_{t=1}^{N} y_{\text{predicted\_}t} \tag{21.3}$$

- $y_{\text{predicted\_}t}$ is the predicted output for decision tree t
- $f_t(x)$ represents the output of the decision tree for input feature x
- $y_{\text{predicted}}$ is the overall predicted output
- N is the total number of decision trees in the Random Forest

21.2.3 Extreme Gradient Boosting (XGBOOST)

The third model utilized is XGBoost (Extreme Gradient Boosting). This model is a powerful and popular supervised learning algorithm that falls under the category of gradient boosting. XGBoost is an extension of the traditional gradient boosting algorithm that aims to improve both predictive performance and computational efficiency.

XGBoost has gained widespread adoption due to its accuracy, flexibility, and ability to handle complex datasets. In XGBoost regression, the model can be represented as an ensemble of weak learners, typically decision trees. The final prediction for a given input x is obtained by summing the outputs of all individual decision trees in the ensemble.

$$y_{\text{predicted}} = \sum_{k=1}^{K} f_k(x) \tag{21.4}$$

- K is the total number of decision trees
- $f_k(x)$ is the output of the kth decision tree for the input features

21.2.4 Artificial Neural Networks

The fourth model utilized is an artificial neural network (ANN). This is a powerful predictive model that utilizes interconnected nodes, organized in layers (input, hidden, and output). An activation function is applied to the weighted sum of input and biases for each neuron.

$$z = \sum_{i=1}^{n} w_i x_i + b \tag{21.5}$$

Here, Z is the output of the sum of the weights and biases at a node, where w represents the value of the weight for the ith node and x_i denotes the input. The bias is represented by the term b. Now for each value that is computed we apply the activation function. The output of each layer is passed as the input to the next layer, and so on until we reach the output layer. For the lth layer this is represented as

$$z[l] = W[l] * A[l-1] + b[l] \tag{21.6}$$

$$A[l] = \text{activation\_function}(z[l]) \tag{21.7}$$

Here, $A[l\text{-}1]$ denotes the output of the previous layer. For regression the activation function that is chosen is the ReLU activation function.

$$\text{ReLU}(x) = \max(0, x) \tag{21.8}$$

Here, x is the input to the activation function. For regression the "mean-squared-error" loss function is chosen. The predicted and the true values are compared and the loss is calculated.

$$\text{MSE} = \left(\frac{1}{m}\right) * \sum (\text{true} - \text{predicted})^2 \tag{21.9}$$

To calculate the new weights that minimize the loss function we use the gradient descent algorithm. The formula for gradient descent can be written as

$$W_{x+1} = W_x - n * \left(\frac{\partial \text{Error}}{\partial W_x}\right) \tag{21.10}$$

Here, W_{x+1} denotes the new weight, W_x denotes the current weight, n denotes the learning rate. The term $\frac{\partial \text{Error}}{\partial W_x}$ denotes the partial derivative of the Error with respect to the old weight. Since the neural network is interconnected with many layers and nodes, we need to use the backpropagation algorithm to update the weights and biases.

21.3 Data Description

The following section describes certain statistical measures about the data along with the various features. It also describes the pre-processing pipeline for the data in order for it to be trained by the machine or deep learning models.

21.3.1 Statistical Measures

The data consists of 36,733 rows and 11 columns. Of these 11 columns, 9 are the features and the remaining 2 are the target variables (CO and NO_x).

AT Ambient Temperature
AP Ambient Pressure
AH Ambient Humidity
AFDP Air Filter Discharge Pressure
GTEP Gas Turbine Exit Pressure
TIT Turbine Inlet Temperature
TAT Turbine After Temperature
TEY Total Energy Yield
CDP Compressor Discharge Pressure
CO Carbon Monoxide Concentration
NO_X Nitrogen Oxides Concentration.

The mean obtained is calculated by the following formula

$$\text{mean} = \frac{\sum \text{observations}}{\text{number of observations}} \tag{21.11}$$

21.3.2 Data Pre-processing

Before the data is fed into the models we need to apply scaling to the data. The process of scaling data involves transforming the features of the dataset so that they fall within a specific range or have a specific distribution. This is important so that no inherent biases are introduced in a model because of a features numeric values. For example when we compare feature weights for NO_x predictions we can see that GTEP and TAT have a similar weight. However, from Table 21.1, we can see that GTEP and TAT have very different magnitudes. If scaling hadn't been applied the model might have interpreted TAT to hold a much higher significance due to its numerically larger value.

For scaling we use the Standard Scaler. The Standard scaler applies the following formula to all the features of the dataset

$$x_{\text{scaled}} = (x - \text{mean})/\text{SD} \tag{21.12}$$

Here, X represents an instance of a single feature in the dataset.

Before scaling is applied we need to split the data in to training and testing sets. The training set is used for the model to "train" or learn on. Once the model has trained it has updated its parameters. Now, we can take the test or the "unseen" data

Table 21.1 Some statistical observations of the data

Feature	Minimum	Maximum	Mean	Units
AT	– 6.234800	37.1030	17.712726	°C
AP	985.850000	1036.6000	1013.070165	mbar
AH	24.085000	100.2000	77.867015	%
AFDP	2.087400	7.6106	3.925518	mbar
GTEP	17.698000	40.7160	25.563801	mbar
TIT	1000.800000	1100.9000	1081.428084	°C
TAT	511.040000	550.6100	546.158517	°C
TEY	100.020000	179.5000	133.506404	MWH
CDP	9.851800	15.1590	12.060525	mbar
CO	0.000388	44.1030	2.372468	mg/m3
NO_x	25.905000	119.9100	65.293067	mg/m3

Table 21.2 Random forest model parameters, common for both models

Hyper parameter	Value
Number of estimators	100
Criterion	Gini
Minimum samples split	2
Minimum samples leaf	1

Table 21.3 XGBOOST Model parameters, common for both models

Hyper parameter	Value
Number of estimators	100
Learning rate	0.3
Max depth	6
Sub-sample	1

Table 21.4 ANN parameters, for CO prediction model

Parameter	Value
Layers	5
Neuron architecture	32-16-16-16-1
Activation function	ReLU
Optimizer	Adam

Table 21.5 ANN parameters, for NO_x prediction model

Parameter	Value
Layers	8
Neuron architecture	512-32-32-32-32-32-32-1
Activation function	ReLU
Optimizer	Adam

Table 21.6 CO Emissions result on test set

Model	MAE	R-Squared
Linear regression	0.857	0.567
Random forest	0.487	0.769
XGBoost	0.5377	0.775
ANN	0.644	0.70

and pass it to the model to predict it's accuracy on real world data. It is important for the training data to come from a mixed distribution to avoid as much generalization as possible. If this isn't done accurately then the model may not produce accurate predictions on the test set.

Table 21.7 NO_x emissions result on test set

Model	MAE	R-Squared
Linear regression	5.72	0.53
Random forest	2.568	0.882
XGBoost	2.953	0.8567
ANN	3.92	0.764

21.4 Methodology

This section covers the overall process adapted for this research paper. It describes various objectives in a step by step manner. It also gives a brief description about all the model parameters used in this paper (Fig. 21.1).

21.4.1 Process Description

The data are loaded from a csv file. Then we select all the features (9 columns) with the first target variable. Then, the data are split it to training and testing sets, with a ratio of 80:20. Then scaling is applied to all the features.

It is important to first split the data and then apply scaling. Splitting the data before scaling ensures that the scaling process is performed independently on each subset. If we were to apply scaling first and then split the data, information from the validation or test sets could inadvertently influence the scaling process, leading to data leakage. Data leakage can artificially boost model performance during validation and testing, resulting in an inaccurate assessment of the model's generalization ability. Now the model is called on the data and it is trained.

After training, the model is called on the test data. The predictions are saved and are compared with the actual data. For comparing, the model's two metrics were chosen. These were the MAE (mean absolute error) and the R-Squared score. The formula for MAE is given by

$$\text{MAE} = \frac{1}{n} * \sum_{i=1}^{n} \left| y_i - y_{i\_\text{predicted}} \right| \tag{21.13}$$

- n represents the total number of data points
- y_i represents the actual value from the test set
- $y_{i\_\text{predicted}}$ is the predicted value after applying the model

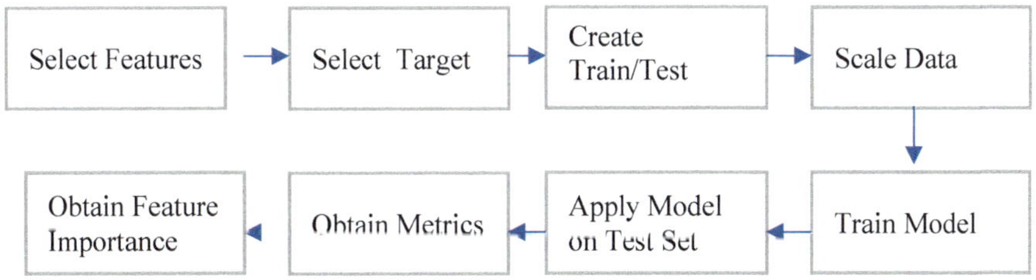

Fig. 21.1 The above flowchart depicts the methodology to train the model, test it on new data and obtain the metrics

Similarly, the formula for R-Squared score can be given as

$$R^2 = 1 - \frac{SS_{res}}{SS_{tot}} \tag{21.14}$$

- SS_{res} is the sum of the squared residuals
- SS_{tot} is the total sum of squares

The R-Squared is a measure of how much better a regression model performs with respect to a simple model that forecasts the mean of the dependent variable for all data points.

21.4.2 Model Parameters

A total of 4 models were chosen to be tested. These were Linear Regression, Random Forest, XGBoost and an Artificial neural network. The model parameters are given as follows.

21.5 Results, Discussion, and Further Work

This section discusses the results obtained by applying the models on the for CO and NO_x results. Scope for further work is discussed in this section.

21.5.1 Results

The models were trained and tested for both Carbon Monoxide Prediction and Nitrogen Oxide Predictions. The results are tabulated as follows.

The Random Forest model produces the least MAE and highest R-squared score for both CO and NO_x predictions. We can visualize the feature importance for both CO and NO_x predictions based on their importance in the Random Forest Model (Figs. 21.2 and 21.3).

21.5.2 Conclusions and Further Work

The CO prediction model using the Random Forest Regressor obtained a 47.63% reduction in MAE compared to the benchmark dataset [5]. A 32.58% increase in

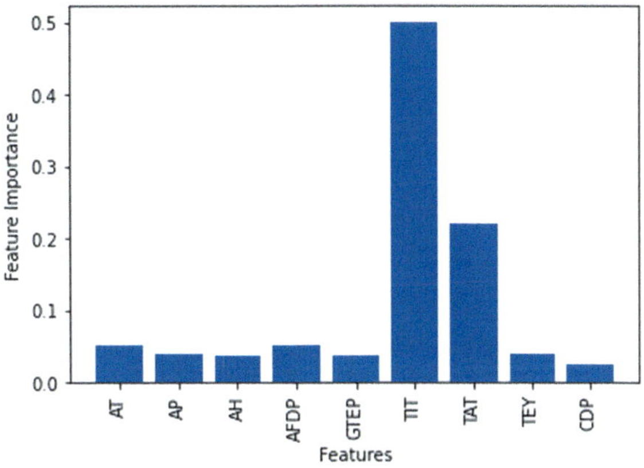

Fig. 21.2 The above figure depicts the feature importance for CO emissions based on the random forest regressor

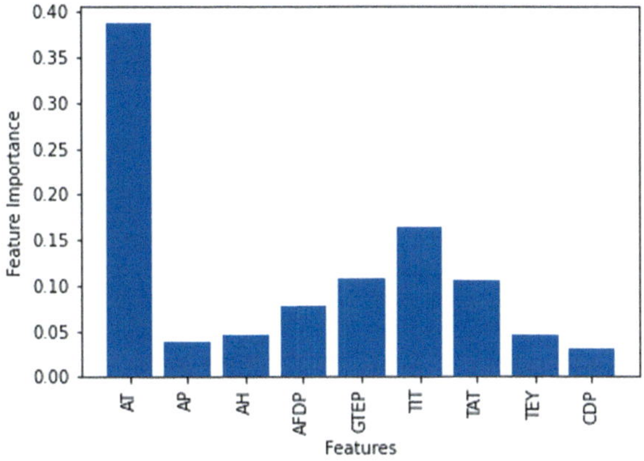

Fig. 21.3 The above figure depicts the feature importance for NO_x emissions model based on the Random Forest Regressor

R-Squared was also observed. Based on the feature weights we can conclude that TIT and TAT hold the most influence on the models predictions.

The NO_x prediction model using the Random Forest regressor obtained a 67.53% reduction in MAE compared to the benchmark dataset [5]. A 37.81% increase in R-Squared was also observed. Based on the feature weights we can conclude that AT, TIT, TAT and GTEP hold the most influence on the models predictions.

Further work can be done in optimizing all the model's to obtain a lower MAE/ R-Square score using hyperparameter tuning. Different feature selection techniques can be employed as well to train the model. Newer datasets can also be formed that take into account CO and NO_x predictions from other gas turbines to improve the generalization of the model in various scenarios. A time series analysis can also be performed to analyze if there are any periodic or cyclic factors that affect gas turbine emissions.

To conclude, this paper applies various machine learning and deep learning models on the benchmark dataset to obtain a reduction in emission error predictions. These models can then be used in power plants as a backup and validation system for CEMS. Moreover, their predictive capabilities enable power plant operators to obtain approximate emissions estimates for planned or under-development facilities, aiding in early emissions management strategies and environmental impact assessments.

References

1. IEA.: Electricity Information: Overview, IEA, Paris https://www.iea.org/reports/electricity-information-overview, License: CC BY 4.0 (2021).
2. Britannica, T.: Editors of Encyclopaedia. Paris Agreement. Encyclopedia Britannica. https://www.britannica.com/topic/Paris-Agreement-2015 (2023, July 13).
3. Swanson, B.G., Lawrence, P.: An alternative approach to continuous compliance monitoring and turbine plant optimization using a PEMS (Predictive Emission Monitoring System). In 18th Industrial Application of Gas Turbines (IAGT) Symposium, Banff, Alberta, Canada. (2009, October).
4. Gas Turbine CO, NO_x Emission Data Set.: UCI Machine Learning Repository. (2019) https://doi.org/10.24432/C5WC95
5. Kaya, H., Tüfekci, P., Uzun, E.: Predicting CO and NO_x emissions from gas turbines: novel data and a benchmark PEMS. Turk. J. Electr. Eng. Comput. Sci.Electr. Eng. Comput. Sci. **27**(6), 4783–4796 (2019)
6. Lee, M., Lin, L., Chen, C.Y., Tsao, Y., Yao, T.H., Fei, M.H., Fang, S.H.: Forecasting air quality in Taiwan by using machine learning. Sci. Rep. **10**(1), 4153 (2020)
7. Siqueira-Filho, E.A., Lira, M.F.A., Converti, A., Siqueira, H.V., Bastos-Filho, C.J.A.: Predicting thermoelectric power plants diesel/heavy fuel oil engine fuel consumption using univariate forecasting and XGBoost machine learning models. Energies **16**(7), 2942 (2023)
8. Korpela, T., Kumpulainen, P., Majanne, Y., Häyrinen, A.: Model based NO_x emission monitoring in natural gas fired hot water boilers. IFAC-PapersOnLine **48**(30), 385–390 (2015)
9. Shakil, M., Elshafei, M., Habib, M.A., Maleki, F.A.: Soft sensor for NO_x and O2 using dynamic neural networks. Comput. Electr. Eng.. Electr. Eng. **35**(4), 578–586 (2009)
10. Fichet, V., Kanniche, M., Plion, P., Gicquel, O.: A reactor network model for predicting NO_x emissions in gas turbines. Fuel **89**(9), 2202–2210 (2010)
11. Iliyas, S.A., Elshafei, M., Habib, M.A., Adeniran, A.A.: RBF neural network inferential sensor for process emission monitoring. Control. Eng. Pract.Pract. **21**(7), 962–970 (2013)
12. Lv, Y., Liu, J., Yang, T., Zeng, D.: A novel least squares support vector machine ensemble model for NO_x emission prediction of a coal-fired boiler. Energy **55**, 319–329 (2013)
13. Smrekar, J., Potočnik, P., Senegačnik, A.: Multi-step-ahead prediction of NO_x emissions for a coal-based boiler. Appl. Energy **106**, 89–99 (2013)

Chapter 22
Disease Detection in Beans Leaves Using Transfer Learning

Riya Rawal, Anuradha Chug, and Amit P. Singh

Abstract A bean plant is rich in protein, carbs, zinc, and iron. Numerous bean varieties, including kidney beans, are high in antinutrients, which prevent the body's enzymes from performing some functions. Beans that are high in phytates and phytic acid prevent the metabolism of vitamin D and hinder bone formation. But some maladies such as angular leaf spot, bean rust, halo blight, white mold, bacterial wilt, etc., along with some pests, namely aphids, armyworms, etc., have a repercussion on its production. The intent of this work is to compare the results of customized CNN on augmented and original data with transfer learning methods, such as VGG-16, ResNet50, and EfficientNet, that can identify plant diseases at an early stage. The beans dataset available on Tensor Flow is used for the experiment. In comparison, the highest accuracy of 98.75% was given by ResNet50 model.

R. Rawal (✉) · A. Chug · A. P. Singh
USICT, GGSIPU, Dwarka, New Delhi 110078, India
e-mail: riya.1.phd.usict.2022@ipu.ac.in

A. Chug
e-mail: anuradha@ipu.ac.in

A. P. Singh
e-mail: amit@ipu.ac.in

P. Nanda et al. (eds.), *Intelligent System and Data Analysis*, Smart Innovation, Systems and Technologies 398, https://doi.org/10.1007/978-981-97-5200-3_22

22.1 Introduction

India's economy has been heavily influenced by agriculture. Modern farming methods should be applied to improve agriculture. One such technology is the use of image processing with transfer learning for the early discovery of leaf maladies in bean plants. Bean seed belongs to several Leguminosae plants, which are grown for their green pods and used as vegetables for human and animal consumption and soil development. For its green pod, it is a significant vegetable crop grown throughout India. Compared to other pod vegetables sold in the Indian market, they are more nutrient-dense. Since the bean is a day-neutral plant, it often requires a chilly temperature and is grown during the Rabi season. It can be grown in northern India from December to February and August to September. The beans are planted in rows by hand or using a seed drill. Rabi seeding may not provide sufficient sprouting if the temperature is low or cold. Beans should be picked when they are still tender to avoid developing strands. Dry beans should be given time to reach full maturity. Numerous types of soils are suitable for growing beans but Indian beans grow well on loamy soils, whether they are clay or silty loam. An ideal soil pH range for growing beans is between 5.5 and 6.0. Soil should be rich in biodegradable matter for optimal production. The most effective way for growing beans is drip irrigation. Leafhoppers, aphids, and glasshouse whiteflies are the major pests and angular leaf spots and bean rust are the major diseases. Brown and yellow spots, early and late blistering, and other common bacterial, fungal, and viral diseases are all present in plants. Image processing commonly is utilized to survey the disease's beset region and distinguish any contrasts in the color of the region. The farmer has difficulty manually identifying the symptoms and types of diseases using his vision when using conventional technology to detect disease. It is difficult to spot diseases early on. The proposed model makes use of transfer learning and image processing to classify and identify diseases.

22.2 Literature Review

Deep learning-based models have been used in a number of studies [1–3] for the identification and classification of leaf disease in plants of various crop species. In order to obtain quicker training times and greater accuracy, Elfatimi et al. [4] have recommended a profound learning methodology to recognize and analyze bean leaf illness utilizing an open dataset of plant leaves and the MobileNet model. Following the application of each design autonomously, the outcomes were compared and it was found that the MobileNet model had the highest classification performance. The proposed model's normal classification precision for two unhealthy and one healthy categories is more than 97% on the training dataset and more than 92% on the test dataset. Singh et al. [5] on ten species of plants have described a method for the automatic recognition and categorization of plant leaf diseases using a genetic algorithm. The proposed system's average classification accuracy is 97.6%. Devaraj et al. [6] have proposed an automatic computer vision-based system for classifying and detecting red spider mite disease in bean crops using image processing and sending information to farmers using mobile computing. The proposed system's average accuracy is 84%. Gangwar et al. [7] have used Inceptionv3 and diverse classifiers, including logistic regression, SVM, and neural networks, to achieve the most cutting-edge solution with the highest precision of 99.4% with the goal to detect and classify grape diseases. Mallick et al. [8] have created a profound learning-based method to distinguish the mung bean bug and illness by employing transfer learning that has effectively distinguished six diverse sorts of maladies and four sorts of bugs out of sound and influenced leaves accumulated in different seasons with a normal precision of 93.6%. Tiwari et al. [9] have trained a multilayer convolutional neural network architecture on a collection of images of plant leaves from six different crops in 27 different categories where five-fold cross-validation and testing are done to thoroughly assess the generated model in varying parameters and the empirical outcomes appeared on concealed pictures with perplexed background conditions, a normal cross-validation precision of 99.58% and a normal test precision of 99.199% are achieved. Selvaraj et al. [10] have employed three different architectures: ResNet50, InceptionV2, and MobileNetV1. Because of its efficiency, the Single Shot Multibox (SSD) model with the MobileNetV1 and faster RCNN with ResNet50 and InceptionV2 was decided for the object detector model architecture. At that point, the models were retrained employing a transfer learning approach to create an AI-based banana illness and pest discovery system to help banana peasants. Six diverse models were made from 18 different categories, with a precision of more than 90% in most of the models tried. In [11] Garg et al. have developed a real-time plant disease detection and prevention mobile application "CROPCARE". The composition consists of three imperative parts: IoT sensor-based monitoring (ISM), cloud-based decision model generation (CDMG), and user interaction system—mobile vision. The inadequate features that are present in the images are found using the Bicubic interpolation approach in conjunction with data augmentation. The SRCNN model receives the augmented dataset in order to transform a low-resolution image into a high-resolution one. The decision model is created using the super-resolution convolution network (SRCNN) and the pre-trained model MobileNetV2. InceptionV3, MobileNetV2, and ResNet34 with and without data augmentation were used in the experiment, and MobileNetV2 with data augmentation provided the best accuracy of 96.12%. The suggested approach also offers suggestions that inform farmers on the current state of the soil, the weather, disease prevention techniques, etc. The farmers would find it beneficial since it supports both Hindi and English glossaries. Table 22.1 shows an assessment of the discussed methods.

Table 22.1 Assessment of the available methods

Author	Plants	Classifiers	Accuracy (%)
Elfatimi et al.	Beans	MobileNet	More than 90
Singh et al.	Multiple plants	MDC, K means, SVM	95.71
Devaraj et al.	Beans	Multiclass classification	84
Gangwar et al.	Grapes	Transfer learning	More than 90
Mallick et al.	Indian mung bean	Vanilla model and MobileNetV2	93.65
Tiwari et al.	6 crops	MobileNetV2 and DenseNet121	99.58
Selvaraj et al.	Banana	ResNet50, InceptionV2, MobileNetV1	More than 90
Garg et al.	Multiple plants	SRCNN and MobileNetV2	96.72

22.3 Materials and Methods

The techniques, datasets, and the method proposed are described in detail in this section.

22.3.1 *Dataset*

TensorFlow datasets is an accumulation of accessible datasets with TensorFlow or alternate Python ML systems. All datasets are disclosed as tf.data. Datasets, empowering efficient and high-performance input pipelines. Tensorflow Datasets (TFDS) is available in two packages: the stable version, which is issued every few months, and others, which are released every day, and contain the most recent versions of the datasets. Beans dataset includes 1296 images of beans captured in the field with mobiles. It is divided into two disease classes (angular leaf spot and bean rust) and one healthy class. Researchers from Uganda's National Crops Resources Research Institute (NaCRRI) annotated the data secured by the Makerere AI research lab. The dispersal of the count of images in each class is illustrated in Table 22.2. Figure 22.1 depicts a handful of the photos of leaves from the dataset.

Table 22.2 Total images for each category

Category	Number of images
Angular leaf spot	432
Bean rust	436
Healthy	428
Total	1296

Fig. 22.1 Sample images of bean leaves of various classes along with their labels

22.3.2 Data Augmentation

One of the most challenging tasks in image classification and segmentation is boosting these models' generalization ability. The difference between a model's performance when evaluated against training data and never-before-seen data is called generalizability. Models with weak generalizability have overfitted the training data. Data augmentation employs current information to make reformatted copies of datasets utilized to deliberately extend the training set. It includes making little alterations to the data or generating advanced data spots by employing deep learning. Authentic information is the origin of augmented data, with a couple of small refinements. To extend the degree and assortment of the training set within the sample of image augmentation, spatial and color space changes (inverting, scaling, editing, radiance, and sharpness) are enforced. Image augmentation employs various methods on images like reverting, pruning, rotating, expanding, and zooming at random. But the performance of the model can be obstructed when numerous transformations are enforced to the same picture. Kernel filters: sporadically alter the image's clarity or blurriness. Blending and combining various pictures is known as an "image mixture."

Fig. 22.2 Augmented images of bean leaves

22.3.3 Convolutional Neural Network (CNN)

CNN is a deep learning network architecture that promptly learns from the input. CNNs are especially useful for recognizing patterns and elements, classes, and groups in images. CNN is capable of handling millions of layers, which together can be used to recognize different angles or highlights [12] of a picture. Every training picture is subdued to channels at multiple resolutions, and the result of each convolved picture is fed to the taking layer. Starting with principal characteristics like luminosity and outlines, channels get more perplexed until they find attributes to precisely recognize the item. An input layer, a yield layer, and various covered-up layers in between them altogether constitute a CNN. These layers perform activities on the information to procure characteristics that are interesting to the information. Convolution, activation, and pooling are the three most prevalent layers. Convolution runs an arrangement of convolutional channels through the crude pictures, actuating diverse perspectives of the pictures with each channel. Rectified linear unit (ReLU) keeps positive values while exchanging negative values to zero. It empowers speedier and more proficient training. Because the activated characteristics are passed over to the following layer, this is often cited as activation (Fig. 22.2).

22.3.4 VGG16

A convNet is a form of an artificial neural network, also known as a convolutional neural network. VGG16 [13] is a type of CNN broadly respected as one of the finest computer vision models accessible nowadays. Despite having an enormous number of hyperparameters VGG works with a convolution layer of a 3×3 channel with stride 1 and repeatedly utilizes the same buffer and maxpool layer of a 2×2 channel with stride 2. It has 2 fully connected (FC) layers and a Softmax at the end for yield. The 16 in VGG16 implies the truth that it has 16 stacked levels. This prototype has overall 21 layers disintegrated into 13 convolutional rows, 5 max pooling rows, and 3 dense rows, but exclusively sixteen learnable parameter layers. The input vector measurement for VGG16 is 224, 244 with 3 RGB channels. This network is quite hefty, with approximately 138 million characteristics. The advantage of using numerous smaller layers rather than a single big layer is that the convolution layers are accompanied by more nonlinear activation layers, which improves the decision functions and allows the network to converge rapidly. VGG16 is an object identification and classification design that can recognize 1000 pictures from a particular batch with a precision of 92.7%. It could be a prevalent image classification strategy that functions well with transfer learning.

22.3.5 *EfficientNet*

EfficientNet [14] is a tuning technique that uses an effective compound coefficient to evenly adjust all depth (amount of network levels), width (number of network channels), and resolution variables. ConvNets are frequently scaled up to improve precision. Although it is important to balance all of the network dimensions, this can be accomplished by simply scaling each of the three dimensions with a uniform ratio. To evenly adjust network breadth, depth, and resolution a compound coefficient is employed in the EfficientNet. In general, when the input picture is bigger, then the network requires more layers as well as channels to grow the responsive field and acquire finer-grained designs, and this same approach is backed by the compound method. The foundational Efficient NetB0 network is built on the capsized bottleneck surplus pieces of MobileNetV2and squeeze-and-excite sections. The leftover blocks are used to create information patterns in deep neural networks. Residual blocks use skip connections to join the beginning and the ending of a convolutional block. The channels are wide at the start of the convolutional block, then shrink as the depth of the block increases until they become wide again due to the additional information. Inverted residual blocks, on the other hand, follow the inverse structure of regular residual blocks. Squeeze-and-excite block (SE) is a building component for CNNs that perform dynamic feature channel-wise recalibration to optimize the interdependencies between channels. This entails that the network dynamically assigns higher weights to the most important channels instead of weighing all channels equally. The network dimensions are specified as follows:

$$\text{depth}: \ d = \alpha^{\Phi} \tag{22.1}$$

$$\text{width}: \ w = \beta^{\Phi} \tag{22.2}$$

$$\text{resolution}: r = \gamma^{\Phi} \tag{22.3}$$

where Φ is the compound coefficient and with the help of tiny grid searches the constants α, β, and γ can be discovered. The user-specified coefficient Φ controls the number of accessible resources.

22.3.6 ResNet 50

Residual network (ResNet) [15] is a type of convolutional neural network (CNN) having 50 layers created by Kaiming et al., their 2015 publication where for image perception a deep residual learning network was used. Residual neural systems are artificial neural systems that frame networks by superimposing residual pieces. The bottleneck design is used for the building block in the 50-layer ResNet. A residual bottleneck block employs 1×1 convolutions to limit the number of variables and matrix multiplications. This allows for significantly quicker layer training. A pre-trained neural network can classify photos into 1,000 different article categories. As a result, the neural network learned detailed feature representations of many images. The input size of the image for this model is 224×224. The ResNet-50 model is a five-staged model with each stage containing a convolution and an identity block. Each convolutional block contains three convolutional layers, similar to the identity block. In this model, trainable parameters available are over 23 million. ResNet employs skip connections which are created by stacking convolution layers and incorporating the original input into the convolution block's output used for feeding input from the previous to the next layer and also aids in mitigating the vanishing gradient issue.

22.4 Research Methodology

In this work, as depicted in Fig. 22.3 a customized CNN model with the following characteristics is used: batch size 32, image size 256, three RGB channels, and twenty epochs. The dataset is divided, with 80% being utilized for training and the rest for testing the prototype. The model is compiled by applying the Adam optimizer. Images are rotated randomly by 20% and flipped randomly in the horizontal and vertical orientations for data augmentation. As a result of the testing discoveries, the efficacy of the models was assessed by applying the commonly used execution measures classification precision.

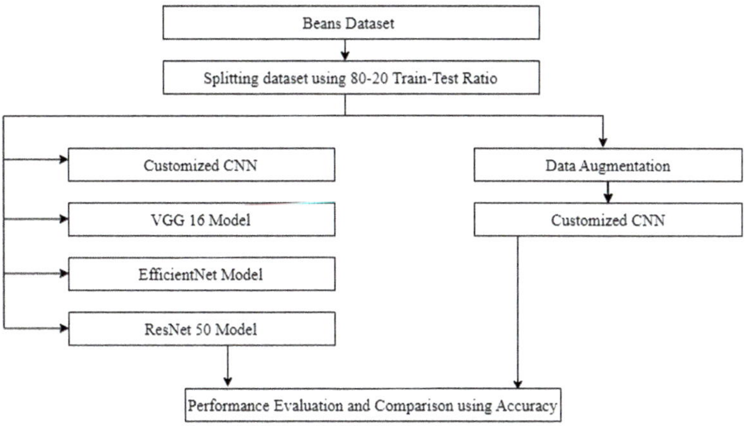

Fig. 22.3 Research methodology

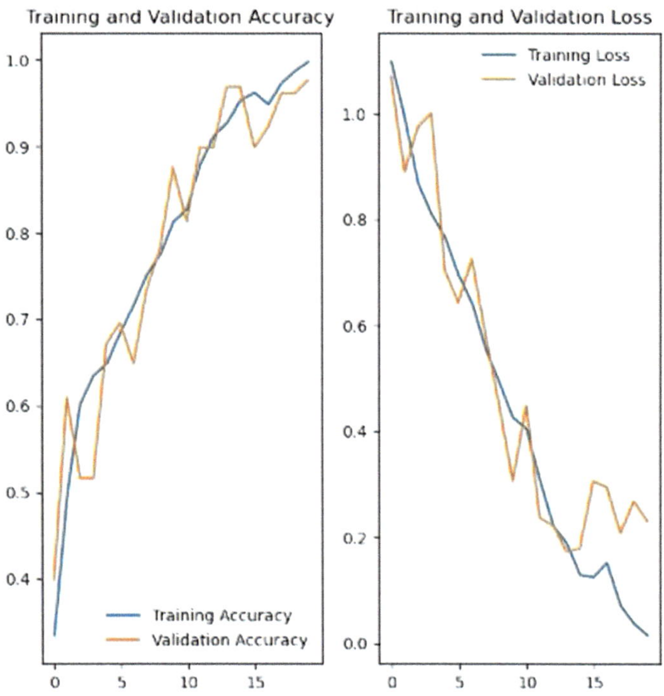

Fig. 22.4 Accuracy and loss plots of training and validation of CNN

Table 22.3 Comparison of classification accuracy

S. No.	Model used	Accuracy
1	CNN	97.66
2	CNN (with augmentation)	86.62
3	VGG16	95.63
4	EfficientNet	98.12
5	ResNet50	98.75

22.5 Results and Discussion

This section focuses on the outcomes of the suggested approach when applied to the beans dataset. The accuracy of the simple and augmented data was computed, and the results were compared. Accuracy and loss graphs were created for training and validation. In this experiment, the confidence of the images evaluated was also determined. Table 22.3 compares the models based on the computed accuracy of the dataset using different transfer learning models. The accuracy and loss plots for training and validation of the models employed in this study are shown in Figs. 22.4, 22.5, 22.6, 22.7, and 22.8. Figures 22.9 and 22.10 show the prediction and confidence of the testing images using CNN with and without augmentation. When the experiment is run on the original dataset, the ResNet50 transfer learning model has the best accuracy (98.75%) while the VGG16 transfer learning model has the lowest (95.63%). When the data was enhanced and new images were added to the existing dataset, the accuracy fell to 86.62%, the lowest of all. The use of visualization helps the user comprehend how the model works. It simplifies the evaluation of underfitting and overfitting and the modification of hyperparameters. The *x*-axis in the accuracy and loss graphs represents the number of epochs (20 in this illustration) and the *y*-axis represents the accuracy or loss %. The metric used to depict whether the machine learning model's output is valid or not that gives a number between 0 and 1 and is used by every prediction is the confidence score. The better the score, the more confident the model is in its prediction. Figure 22.9 shows that the photos are accurately predicted and have a very high confidence score. In contrast, while the pictures in Fig. 22.10 are accurately predicted, their confidence level is less. This indicates that when compared to the augmented dataset, the photos without augmentation are appropriately predicted.

22.6 Conclusion and Future Work

The current study examines the data augmentation strategy as well as transfer learning models. In comparison with the other models, the accuracy computed for the enhanced dataset is the lowest. Even while the model without augmentation had a greater training accuracy, it was near to overfitting. As a result, with real data, such a model is more likely to fail. Although the training accuracy was reduced by

Fig. 22.5 Accuracy and loss plots of training and validation of CNN with augmentation

augmentation, the model got more accurate. Thus, while working with fewer data, data augmentation approaches can be more successful. When data augmentation is conducted on a small dataset with fewer training samples, there is a considerable difference in validation accuracy. This study can be improved in the future by using more transfer learning models and fine-tuning the hyperparameters. Feature selection approaches can also be implemented, and the same work can be done on different crop datasets.

Fig. 22.6 Accuracy and loss plots of training and validation of EfficientNet model

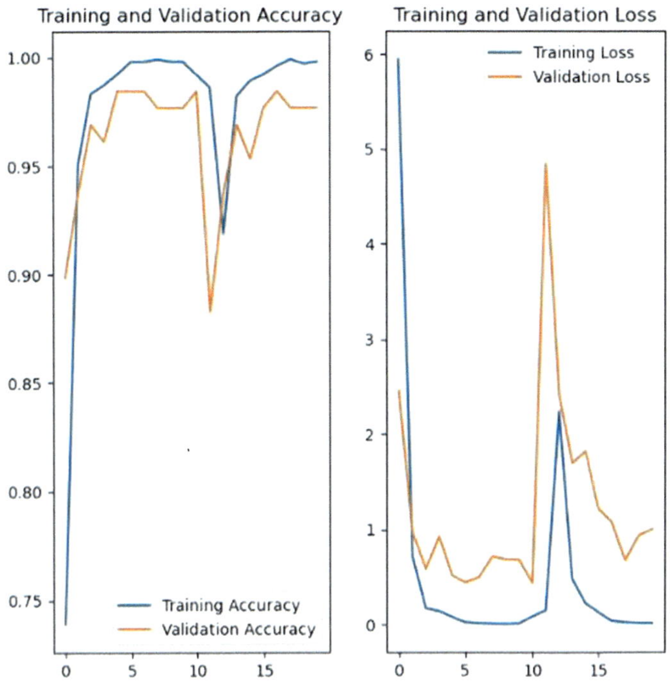

Fig. 22.7 Accuracy and loss plots of training and validation of VGG16 model

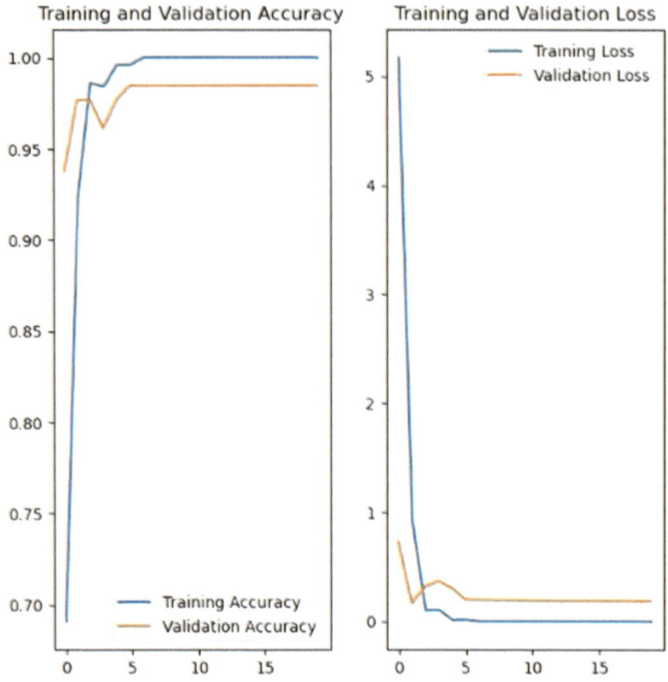

Fig. 22.8 Accuracy and loss plots of training and validation of ResNet50 model

Fig. 22.9 Prediction and confidence of images of CNN model with no augmentation

Fig. 22.10 Prediction and confidence of images of CNN model with augmentation

Acknowledgements We are thankful to the Division of Science and Technology (DST) for their monetary cooperation while carrying out this analysis beneath the project named "Application of Internet of Things (IoT) in Agriculture Sector", DST/Reference No. T-319/2018-19.

References

1. Li, K., Lin, J., Liu, J., Zhao, Y.: Using deep learning for Image Based different degrees of ginkgo leaf disease classification. Information **11**(2), 95 (2020)
2. Hu, W.J., Fan, J., Du, Y.X., Li, B.S., Xiong, N., Bekkering, E.: MDFC–ResNet: an agricultural IoT system to accurately recognize crop diseases. IEEE Access **8**, 115287–115298 (2020)
3. Singh, K., Kumar, S., Kaur, P.: Detection of powdery mildew disease of beans in India: a review. Orient J Comput Sci Technol **9**(3), 226–234 (2016)
4. Elfatimi, E., Eryigit, R., Elfatimi, L.: Beans leaf diseases classification using mobilenet models. IEEE Access **10**, 9471–9482 (2022)
5. Singh, V., Misra, A.K.: Detection of plant leaf diseases using image segmentation and soft computing techniques. Inform process Agric **4**(1), 41–49 (2017)
6. Devaraj, P., Arakeri, M.P., Vijaya Kumar, B.P.: Early detection of leaf diseases in beans crop using image processing and mobile computing techniques. Adv Comput Sci Technol **10**(10), 2927–2945 (2017)
7. Gangwar, N., Tiwari, D., Sharma, A., Ashish, M., Mittal, A.: Grape leaf disease classification using transfer learning. Int Res J Eng Technol **7**(3), 1 (2020)
8. Mallick, M.T., Biswas, S., Das, A.K., Saha, H.N., Chakrabarti, A., Deb, N.: Deep learning based automated disease detection and pest classification in Indian mung bean. Multimed Tools Appl **82**(8), 12017–12041 (2023)
9. Tiwari, V., Joshi, R.C., Dutta, M.K.: Dense convolutional neural networks based multiclass plant disease detection and classification using leaf images. Eco. Inform. **63**, 101289 (2021)

10. Selvaraj, M.G., Vergara, A., Ruiz, H., Safari, N., Elayabalan, S., Ocimati, W., Blomme, G.: AI-powered banana diseases and pest detection. Plant Methods **15**, 1–11 (2019)
11. Garg, G., Gupta, S., Mishra, P., Vidyarthi, A., Singh, A., Ali, A.: CROPCARE: an intelligent real-time sustainable IoT system for crop disease detection using mobile vision. IEEE Int Things J **10**(4), 2840–2851 (2021)
12. Abade, A., Ferreira, P.A., de Barros, V.F.: Plant diseases recognition on images using convolutional neural networks: a systematic review. Comput. Electron. Agric. **185**, 106125 (2021)
13. Simonyan K, Zisserman A (2014) Very deep convolutional networks for large-scale image recognition. arXiv preprint arXiv:1409.1556
14. Tan M, Le Q (2019) Efficientnet: rethinking model scaling for convolutional neural networks. Int Conf Mach Learn. PMLR
15. He K, Zhang X, Ren S, Sun J (2016) Deep residual learning for image recognition. In: Proceedings of the IEEE conference on computer vision and pattern recognition. pp 770–778)

Chapter 23
A Novel Architecture for a Medical Image Recognition System Using Deep Learning-Based Multiple Regression Evaluation

Jyoti Parashar, Kamal Upreti, Iram Fatima, Divya Gangwar, Ashok Kumar Saini, and Prashant Vats

Abstract Models based on machine learning are optimization models that collect data, assess it, and deliver the reports required by specialists and management to make the best decisions. The application of contemporary machine learning allows the organization to quickly analyze photographs, differentiate voices assist in providing customer service, assess the information that is at hand, and uncover connections to aid in decision-making processes. The results of this investigation use quantitative methodologies to collect data and analyze it using mathematical procedures such as regression modeling as well as analysis of variance. Deep learning techniques applied to digital imaging, particularly in medical treatment, can increase picture quality, aid in modeling, aid in making the best possible diagnosis, and successfully address demands from patients. To analyze the hypothesis, investigators intend to utilize statistical approaches such as descriptive data analysis, regression evaluation, and analysis of variance (ANOVA). The authors employ the purposive sample approach

J. Parashar
Department of Computer Science & Engineering, Dr. Akhilesh Das Gupta Institute of Technology & Management, New Delhi, India
e-mail: jyoti.parashar123@gmail.com

K. Upreti (✉)
Department of Computer Science, Christ University, Delhi NCR Campus, Ghaziabad, India
e-mail: kamalupreti1989@gmail.com

I. Fatima
Department of AI & ML, Dr. Akhilesh Das Gupta Institute of Technology & Management, New Delhi, India

D. Gangwar
Department of Management, Babu Banarasi Das Institute of Technology & Management, New Delhi, India
e-mail: divyagangwar2013@gmail.com

A. K. Saini · P. Vats
Department of Computer Science & Engineering, School of Computer Science & Engineering, Manipal University Jaipur, Jaipur, Rajasthan, India

P. Nanda et al. (eds.), *Intelligent System and Data Analysis*, Smart Innovation, Systems and Technologies 398, https://doi.org/10.1007/978-981-97-5200-3_23

to choose respondents from the healthcare industry. Purpose sampling is a non-probability sampling approach. Researchers collected data from 193 respondents working at hospitals that are privately owned in Southern Asia. As stated by the study, all factors, including efficiently meeting patient needs, have a probability value of under 0.05, indicating that they are statistically noteworthy. Following the study, the coefficient of variance (R squared) is 0.744, or 74.4%. According to the study, there is a high association between better image quality and ML-based digital picture identification systems. The recognition of patterns and the application of artificial intelligence to computerized recognition of pictures also have a close link.

23.1 Introduction

The rise of new technologies such as machine learning, artificial intelligence (AI), as well as deep learning is revolutionizing several areas of economics and associated operations. Machine learning technologies have been tremendously beneficial in the area of image processing, particularly in the realm of huge picture analysis and delivering understandable output. Machine learning assists users in separating images to recognize an image in a more apparent form. It should be understood that the application of automated learning-based techniques for processing has increased in contemporary society as users require improved classification, the process of segmentation, and overall image recognition [1]. In the healthcare business, e.g., healthcare professionals and nursing are trying to dissect photos to discover diseases and other abnormalities so that patients may receive proper medical treatment. On the opposite end of the spectrum, as picture distribution becomes more complicated, many programs are being used to provide users with essential clarity about the image and allow them to make the proper decision. As a result, users perpetually concentrate on improving categorization to improve the accuracy and impact of the photos so that useful details can be determined, and multiple components may be analyzed [2]. The involvement with machine learning and the incorporation of various upgraded techniques in methods such as machine learning aid in the creation of improved image processing, which aids in generating clear conclusions by analyzing the difficulties. Machine learning algorithms assist in signal processing for engineering and focus on improving video imagery processing so that it may be applied in a variety of sectors such as technology, health care, and vehicular accident preventative measures, among others. Interior structures hidden behind the skin and bones are made apparent through imaging for medical purposes, providing clinicians with more information regarding the way to manage and evaluate patients. The common practice is to develop an understanding of normal physiological and anatomical characteristics to potentially discriminate between different anomalies in illness (Fig. 23.1).

Image capture is the first phase that frequently involves preprocessing, such as scaling. Following that is picture enhancement, which is one of the simplest and most aesthetically appealing components of the digital processing of images. The

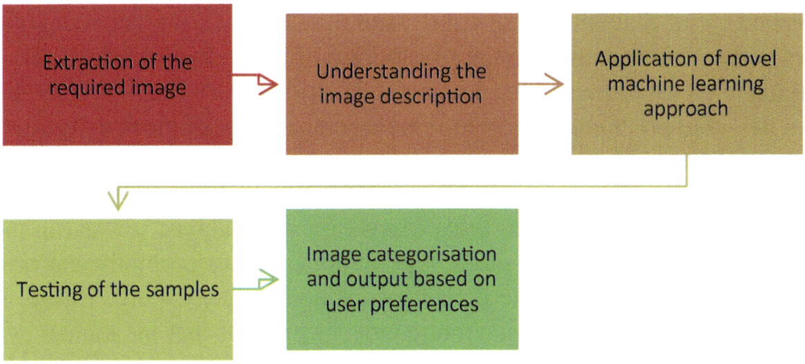

Fig. 23.1 Block diagram representing the digital image processing

goal of enhancement methods is to reveal hidden features or merely to highlight particular desirable parts of an image. Improving the look of a picture by adding contrast is a frequent example of boosting. It is critical to note that the discipline of processing images known as enhancements is very subjective. The next stage, picture restoration, is also concerned with improving the appearance of an image. Picture repair, as opposed to augmentation, is quantitative since restoration procedures usually focus on quantitative or statistical representations of photographic degeneration [3]. Enhancement, on the contrary hand, is reliant on discretionary individual preferences to determine what constitutes an effective augmentation outcome. Furthermore, color image processing has significantly enhanced the use of digital pictures on the Internet.

Multi-resolution processing represents images at multiple state resolution settings. Compression refers to techniques for reducing the size of a picture while retaining as much of its quality as feasible. Picture components that aid in the representation and description of form are extracted during morphological processing. The first stage in segmentation is to divide the image into multiple sections [4]. An image is represented and described using a variety of approaches, including boundary representation, which emphasizes outer form components such as vertices and inflection, and regional representation, which emphasizes inner qualities such as texturing and skeletal components. The technique of recognition entails assigning labels to an item based on its properties. The knowledge base manages the link between modules as well. A knowledge base is used by a graphic processing appliance to hold knowledge regarding a problem.

23.2 Related Work

Recognition of images evolved through phases such as recognition of text, computerized editing, image recognition, and recognition of targets. In general, if a particular sector has a need that existing technologies cannot supply, emerging technology is produced to fill that need. The same thing may be stated for the recognition of pictures software. The advancement of this kind of technology is intended to allow algorithms to create more reliable data on an individual basis and to tackle the challenge of unidentified or insufficient information. The image of the body is a subset of computerized images. The origin of the information is recognition of image models can aid in the diagnosis of numerous diseases [5]. The models may be trained to scan MRI or X-ray images, as well as other kinds of visual outputs, to pick up medical problems. These facts can help medical practitioners make promptly and precisely diagnoses that can help with the patient's treatment strategy. Furthermore, it may boost their ability to detect minute malignant constituents that the human eye may overlook. This early and accurate diagnosis would improve the efficacy of healthcare facilities and service providers, reduce the time required for imaging and rescreening, and improve doctors' ability to give a timely and precise therapeutic approach.

Any picture analysis must include structural analysis. The research and understanding of structure is the foundation of the human intellect. However, the degradation of human cognition has resulted in an increasing demand for technology to replace or augment the normal operation of the human brain, alongside the introduction of personal computers and the emergence of artificially intelligent systems. This will enable you to recognize the photographs on your computer. Only data classification is used to do sample analysis [6]. It is a field of study that is closely related to mathematics. Many concepts are found on statistics and probability. Structural analysis is classified into three types: mathematical framework analysis, combinatorial physics, and interpretation. They are correct because digital image processing is identical to biological imagery. Image analysis algorithms include data collecting, preliminary processing, distribution and selecting procedures, categorization structure, and classification argumentation. Data extraction employs pictures to communicate information such as sound and light as photonic data. It entails altering important details about an issue that a machine may detect in some way [7]. The beginning of processing frequently refers to the stages of the drying process, getting clean, and rotating the image processing, which enhances the fundamental features of the picture. Whenever the structure itself becomes apparent, its components must be delivered and chosen. The pictures we view are distinct. If we are to adequately examine them, we must first understand what these things are [8]. Buying one of these properties is a catastrophic idea. The image's general qualities tend to export it. The overall qualities tend to output widely recognized images; moreover, relevant functions are extracted based on the selection of essential functions. Furthermore, the extraction and selection of various functions are vital for picture recognition so that they can supply consumers with the services they require.

The construction of the light field encourages the collection of the whole radiation in various directions, thus resulting in containing the general updated conventional screen and these will assist in recording the two-dimensional nature light projection on the backdrop, which includes the edge area, through the use of increased constraints. This allows you to do things like refocus and change viewpoints after taking photographs. However, it also entails several technological hurdles in image processing system collection and standardization [9]. The investigation of an arrangement between essential studies on a basic issue aids in concentrating on static light fields. The use of many dense networks makes management easier. Finally, the development of effective strategies for the compression and dispersion of light fields is a significant catalyst in many applications.

Employing a variety of strategies for coping, the visual system of humans can sense an extensive spectrum of light intensity levels, including exceptionally bright to extremely dark environments that we encounter in our real-life surroundings regularly. Current imaging approaches, however, are still limited in their ability to meet or implement a wide variety of situations. Wide Color Gamma (WCG) is frequently used to boost the resolution of HDR images. HDR photography has reached maturity [10]. Extending HDR to video, on the other hand, faces the scientific issue of delivering low-cost, high-quality solutions that influence the whole visual processing process, encompassing content capture, reproducing tones, color supervision, computer programming, and projections.

The concepts of quality of experience (QoE) are connected to the user's general contentment with the application they are using for digital photography. Quality management is inextricably tied to both subjective and objective quality evaluation methodologies [11]. Measures of optically perceived quality that utilize human visual perception models are models that were created as a consequence of years of study. Deep learning technologies have recently been developed that allow for important applications to the issue. Although these perceptual quality measurements are effective, several critical issues must be addressed [7]. To begin, when utilized for video recordings, the suitable detection measurement only applies to certain cameras, ignoring time samples. Second, while color is a significant component, no universally approved label quality standard addresses color directly. The whole supervision of learning applications is regarded as the most important part of the process of machine learning. This necessitates the use of critically tagged data and information; this allows for the use of completely labeled technology, which aids in time-saving computing mechanisms [12]. In certain cases, the material may be difficult to comprehend; thus, the use of deep learning technologies and other related techniques aids in efficiently meeting the demands [13]. These characteristics often include presenting the general data so that the mathematical framework can predict the output efficiently, assisting in patterning interpretation, as well as offering greater clarity and understanding (Table 23.1).

23.3 Proposed Model

The first surviving image of real-world characteristics was produced in 1827; nevertheless, the Lumiere brothers invented the cinematographic film throughout the late 1900s and made it available for public viewing. However, with more than 20 years of general development, although the most recent upgrade has affected the latter half of the twentieth century, this has permitted the creation of new technologies

Table 23.1 Literature review related to machine learning in digital imaging

Elaziz et al. [1]	Hybrid feature extraction	Various data are extracted from the information available which includes alphabets, symbols, characteristics, etc. Neural networks are applied to classify the information and appropriate patterns are extracted. The individuals and users are trained to apply the appropriate tool so that clear images can be obtained, and zoning of the images can be ascertained for better clarity and analysis
Jing and Tian [12]	Neural network	This method is highly useful in analyzing the mathematical model and supports in analyzing the image more clearly for effective diagnosis. This model applied two different layers which support parameter sharing, enhanced interaction, and better representation
Nguyen et al. [14]	KNN	It is based on the nearest distance of the classes, and KNN enables computing the best weight for analyzing the correct point so that the image quality can be enhanced
Liu et al. [15]	Support vector mechanism	SVM is applied to classify the data through regression problems, and SVM also enables in recognizing of the pattern effectively, recognizing the data, and enhancing feature extraction for better quality
Hosny et al. [16]	Light fields and volumetric imaging	In this method, the 7D method is applied which supports light field magnification as it enables collating the radiance from different directions. Hence, it possesses better picture clarity than the traditional model of 2D imaging

ineffectively, paving the path for the digital revolution to take place [17]. The development of critical imagery permitted the use of film-based cameras to be phased out, and the digital transformation in the image business began in 1990. Because of the introduction of associated innovations, there has been a proliferation in digital process-enabled image processing and different complicated processing has been used to obtain better- and higher-quality images.

JPEG is a type of picture file format that is saved via lossy compression. As a result of compression, the process of the resulting image is a trade-off between storage space and the quality of the image. Users can adjust the compression level setting to minimize the amount of storage while keeping the appropriate degree of quality. Image quality is minimally affected by 10:1 compression. The more aggressive the degree of compression setting, the faster the photograph's quality drops. The JPEG format (Joint Photographic Experts Group) was developed in early 1992 as a standard for storing pictures, while Moving Picture Experts Group (MPEG) was developed for coding moistures and accompanying sounds. By ensuring interpretability, these technologies have proven extremely important and have characterized the success of company executives, users, and other stakeholders [18]. It is essential to remember that JPEG is the most used picture format tile because it provides higher image quality. A JPEG image is seen as a series of segments, each with a marker at the beginning.

Each of the markers commences with a 0xFF byte including a marker indicator that determines its kind. Different payloads are monitored per what kind of indicator. With the introduction of smartphones in early 2010, there was an evolutionary shift in how images were processed that enabled the integration of cameras with exceptional resolution in smartphones, causing an upheaval in the handheld device category. The introduction of cameras with high resolutions in smartphones was viewed as a threat to digital cameras in 2010, therefore higher-powered cameras and other features were developed to improve image quality. A significant solution was also implemented, which began the comprehensive deployment of vital networks. Furthermore, the proliferation of 3G and 4G networks has effectively aided in the transmission of different enormous volumes of medical files. This has revolutionized the methods for communicating and the sharing of data and photographs at any juncture in time, allowing consumers to critically catch the various images and interchange them with their connections. The technologies known as 3G and 4G are associated with wireless communications standards. Mobile communications is one of the industries that is always changing to give faster and more reliable Internet access on the go capabilities. Each new technology significantly improves effectiveness as well as versatility over its predecessor. 3G and 4G are commonly separated from their counterparts in terms of technological compatibility with an information transmission rate, capacity, data processing design, connectivity variety, and so on. Third Generation (or 3G) refers to the development of specialized mobile devices for providing high-quality information and internet services. The fourth Generation, or 4G, provides a lot of possibilities for faster and more improved Internet connection experiences, as well as multiple connections to it.

Since the early 2010s, deep learning technologies have been widely used to leverage comprehensive processing for varied data to address essential features effectively. Computer image recognition is defined as the total process of stimulating photographs to recognize patterns and give high-quality, error-free images to businesses, and other stakeholders. With the emergence of machine learning approaches, the complicated veneer to human beings recognizing patterns has been substituted with technology that processes images rapidly and cheaply [14]. Machine learning is defined as the total process of gathering essential information from various sources, enabling coordination with multiple systems, and is used in a variety of disciplines including the field of engineering, health care, banking, the retail industry, and other sectors. The diversity and accessibility of data in the era of the digital uprising are tremendous, and users must use it efficiently to develop significant interpretations, foresee patterns, and make swift judgments [13]. The application of machine learning is commonly employed in the discipline of health care to capture information about patients, assess their medical condition pattern, illness kind, and medicine delivered to patients. Furthermore, analyzing various reports such as X-rays, CT scans, and other relevant procedures assist medical practitioners in making accurate diagnoses.

The study's main goal is to examine the major variables affecting medical practitioners and others in employing KNN for improved image processing, namely for X-rays, in the field of healthcare business (Fig. 23.2).

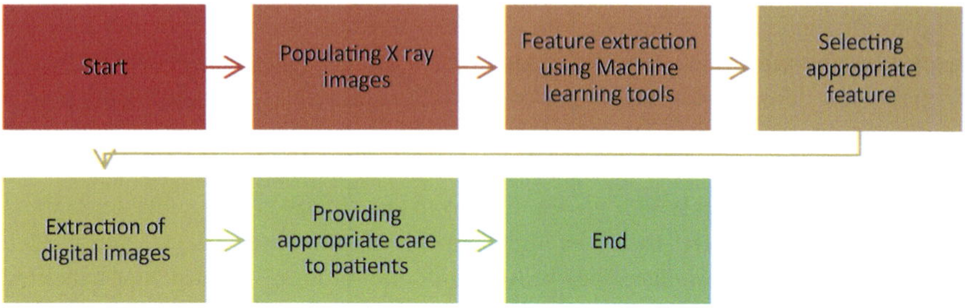

Fig. 23.2 Proposed model

With the advancement of modern technologies, the level of complexity of the processing of images, recognition of patterns, and comprehension has increased, and the ability of humans to recognize them has also grown in expense, time, and effort. As a result, the use of machine learning allows for the analysis of images, providing users with a clear picture for performing analyses, estimating patterns, and assisting in successful decision-making [10, 19, 20]. As a result, the present research is more focused on analyzing the consequences of machine learning algorithms to interpret images in healthcare facilities, primarily X-rays, so that healthcare professionals can efficiently analyze the information and perform accurate decisions so that appropriate drugs can be stipulated to the patients as well as assistance in enhancing their mental and physical well-being [21].

The suggested approach employs unique methods for machine learning such as hybrid extracting features, convolutional neural network training, K Nearest Neighbors, as well as support vector mechanisms. In the current context, the use of machine learning techniques for interpreting X-rays in the medical services business is rising as medical professionals must concentrate on successfully analyzing digital pictures, estimating patterns, and providing better medical assistance to patients [22].

23.4 Methodology

The purpose of the study is to focus on analyzing the major factors influencing medical practitioners and others in using the KNN for better image processing mainly for X-rays in the healthcare industry. The implementation of machine learning in the diagnostics process like X-rays and other aspects has been transforming the medical industry as practitioners can more easily map the image, assist in performing quick analyses on reports, and provide better counsel to patients for a faster recovery [23].

The writers often explore the issue using a descriptive approach to study since it allows for an efficient evaluation of any number of variables. The investigators focus on correctly and methodically evaluating the function of machine learning techniques for the successful processing of images in the medical care business; for this objective, a complete questionnaire is produced and distributed to participants

[24]. The authors of the paper use the technique of purposive sampling to collect data obtained from those who responded in the medical profession. The technique of purposeful sampling is a type of non-probability sampling technique that allows study participants to use their discretion in selecting respondents to the sample for the survey because the purpose of this research is to analyze the most important factors of machine learning's efficacy for improving the use of digital imaging in the healthcare industry, particularly for X-ray detection [25]. The study participants have gathered data from almost 193 responses, which are going to be analyzed. The Likert rating system is designed to represent the important closed-ended questionnaire [1—Does not support; 2—Less support; 3—Neutral; 4—Support; 5—Strongly support].

As a result, this study employs a quantitative research method. As a result, this study employs a quantitative method of inquiry to gather data as well as analyze it using statistical approaches such as regression modeling and analysis of variance [19]. The use of machine learning in computerized imaging, particularly in the healthcare sector, helps to improve picture quality, create patterns, make suitable diagnoses, and efficiently meet the requirements of patients [26, 27]. The study authors propose to evaluate the assumption using statistical approaches such as qualitative analyses, regression modeling, and ANOVA (also known as analysis of variances).

23.5 Results and Discussions

23.5.1 Descriptive Analysis

This part is concerned with giving thorough data analysis based on information gathered from respondents. The descriptive analysis, the regression evaluation, and the test of chi-square are all part of the investigation (Table 23.2).

The demographic information variables of the sample's population fall within the better spectrum, and the average of the standard deviation of the demographic information is less, suggesting that the information is close to the mean value; additionally, the degree of skewness is almost 0 for the majority of the variables related to demographics, indicating that the information is distributed normally. Descriptive

Table 23.2 Descriptive analysis

Descriptive stats	Gender	Age	Qualification	Current designation	Experience
Mean	1.46	2.24	1.90	1.93	2.67
Std. deviation	0.50	0.90	0.67	0.75	1.47
Skewness	0.16	0.41	0.11	0.12	0.40
Kurtosis	− 2.00	− 0.53	− 0.74	− 1.19	− 1.23

analysis, which is also often known as sophisticated statistics or descriptive mathematics, is the practice of using statistical approaches to put into words or characterize a set of data. One of the most popular methods of data assessment is descriptive analysis, which is known for producing understandable conclusions from uninterpreted facts. The primary purpose of descriptive analysis is to attempt to characterize or summarize the data. Even though it does not offer forecasts, it may be quite valuable in professional situations. This is mostly because descriptive statistics make data easier to digest, which may make it easier for researchers to act on it (Table 23.3).

In addition, 54.9% of responders have an experienced professional medicine degree, 27.5% have an advanced degree in nursing, and 17.6% have radiology accreditation and associated courses. 44% of these individuals are now working as medical professionals and Doctor of Medicine, 31.6% as nurses, and 24.4% as radiologists in the medical business. 28% of those who responded have fewer than 5 years of expertise, 25.9% have 5–10 years of expertise, 18.7% have more than 20 years of expertise, 15.5% have 10–15 years of expertise, and the rest of the respondents have

Table 23.3 Percentage analysis

	Frequency	Percent
Gender		
Male	104	53.9
Female	89	46.1
Age		
21–30 years	39	20.2
30–40 years	89	46.1
40–50 years	44	22.8
Above 50 years	21	10.9
Qualification		
Post-graduation	53	27.5
MBBS	106	54.9
Others	34	17.6
Designation		
Nurses	61	31.6
Medical practitioner	85	44
Radiologist	47	24.4
Experience		
Less than 5 years	54	28
5–10 years	50	25.9
10–15 years	30	15.5
15–20 years	23	11.9
More than 20 years	36	18.7

Table 23.4 Role of machine learning in X-ray imaging

ML in imaging	Values in %
Respondents disagreeing most	7.3
Respondents disagreeing	9.3
Respondents neutral	16.6
Respondents agreeing	37.3
Respondents agreeing most	29.5

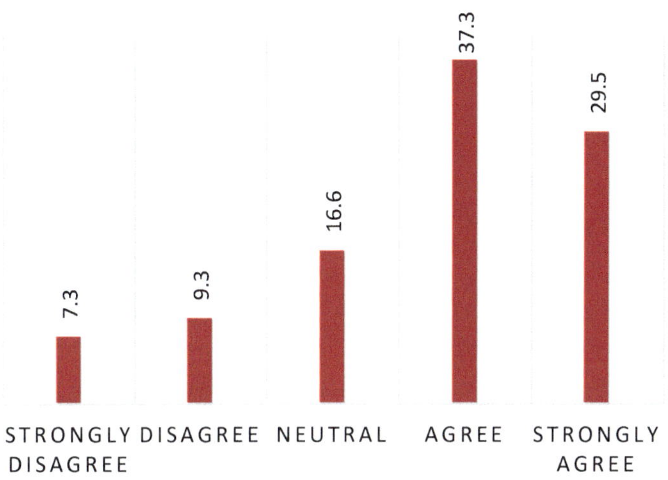

Fig. 23.3 Chart representing the role of machine learning in X-ray imaging

15–20 years of expertise. Table 23.4 gives the findings of our analysis of the function of machine learning in the field of X-ray imaging (Fig. 23.3).

According to the Table 23.4 analysis, 29.5% of the participants are strongly in agreement with the assertion that artificial intelligence plays a significant role in interpreting X-ray images in the field of healthcare industry, and 37.3% agree with the statement. As a result, roughly 66.3% of those surveyed are extremely optimistic that the use of machine learning plays a critical role in X-ray scanning; nevertheless, 16.6% of participants were neutral to the declaration, 9.3% disagreed with the assertion, and 7.3% strongly disagreed with it.

23.5.2 *Regression Analysis*

Regression analysis quantifies the correlation among a pair of variables, namely independent variable X along with dependent variable Y. The regression analysis examines how closely the two variables are connected to help generate superior business strategies, judgments, and forecasts. Regression analysis involves a set of

Table 23.5 Regression weights

Regression	*B*	Std. error	*t*	Sig
(Constant)	0.078	0.17	0.47	0.64
Better image quality	0.319	0.10	3.18	0.001
Identification of pattern	0.329	0.09	3.53	0.001
Addressing patient requirements effectively	0.016	0.08	0.19	0.85
Supportive in making a proper diagnosis	0.247	0.09	2.73	0.01
R squared	0.744			
F data	140.480			
Sig	0.000			

methods from statistics that are used to evaluate the relationships between a dependent variable and any number of independent variables. It may be used to mimic the long-term connection among variables and assess the degree to which each of the connections connecting them has become became acquainted (Table 23.5).

The generic regression model is:

$Y =$ Constant $+ a \times$ 1st variable $+ b \times$ 2nd variable $+ c \times$ 3rd variable $+ d \times$ 4th variable

Based on analyses from Table 23.5, the regression weights are being stated as:

Better image quality is 0.319; identification of pattern is 0.329; addressing patient requirements effectively is 0.016 and supportive in making a proper diagnosis is 0.247. Hence, the regression equation is stated as follows:

Y (ML in Digital Image recognition) $= 0.078 + 0.319 \times$ better image quality $+ 0.329 \times$ Identification of pattern $+ 0.016 \times$ Addressing patient requirements effectively $+ 0.247 \times$ Supportive in making proper diagnosis.

The analysis further states the R squared value is 0.744 or 74.4%; hence, the significance value of all the variables except addressing patient requirements effectively is less than 0.05, hence they are statistically significant.

23.5.3 *Test of Hypothesis Using Analysis of Variance Test*

1st test

Null: There is no significant relationship between improved image quality and the use of machine learning techniques in digital image identification. ANOVA is an analysis of statistics approach that separates methodical and random elements to account for apparent aggregated variation within the data set in question. The offered collection of information has been statistically impacted by methodical but not unpredictable causes. Analysts utilize the ANOVA test to assess the effect of independent variables on the variable that is the dependent variable in a model-regression study. The ANOVA test allows you to evaluate more than one group at once to discover whether there may be a relationship connecting them. The *F* statistic, also known as the *F*-ratio, is the result of thc ANOVΛ formula that allows for the evaluation of many sets of data to determine variability within and between samples (Table 23.6).

According to the findings, there is a substantial relationship between improved image quality and methods that use machine learning in the digital recognition of pictures.

2nd test

Null: There is no significant relationship between the recognition of patterns and the use of machine learning techniques in digital image classification (Table 23.7).

According to the findings, there is a strong link between the recognition of patterns and the use of machine learning techniques in computerized image recognition.

3rd test

Null: There is no significant relationship between aiding appropriate diagnosis and the use of machine learning techniques in digital image identification (Table 23.8).

According to the findings, there is a strong link between assisting with appropriate diagnosis and the use of machine learning techniques in digital picture identification.

4th test

Null: There is no significant link between assisting in effectively addressing patient needs and the use of learning algorithms in digital picture identification (Table 23.9).

According to the findings, there is a substantial relationship between assisting in effectively responding to patient needs with the use of machine learning techniques in digital picture identification.

Table 23.6 ANOVA between better image quality and application of machine learning in digital image recognition

ANOVA	Data value	Overall sig
Among the groups	288.01	0.001
F data	151.84	

Table 23.7 ANOVA between the identification of patterns and application of machine learning in digital image recognition

ANOVA	Data value	Overall sig
Among the groups	288.01	0.001
F data	150.45	

Table 23.8 ANOVA between supporting proper diagnosis and application of machine learning in digital image recognition

ANOVA	Data value	Overall sig
Among the groups	205.98	0.001
F data	118.02	

Table 23.9 ANOVA between supporting in Addressing patient requirements effectively and application of machine learning in digital image recognition

ANOVA	Data value	Overall sig
Among the groups	219.56	0.001
F data	150.7	

Table 23.10 Potential of ML in creating better image quality for diagnosis

Quality aspects	Values in %
Respondents disagreeing most	8.3
Respondents disagreeing	8.8
Respondents neutral	16.1
Respondents agreeing	35.2
Respondents agreeing most	31.6

The results of the data evaluation were thoroughly collated in this part of the paper, and the regression model indicates that major variables such as improved image excellence, pattern recognition, and assistance with establishing an accurate diagnosis are extremely important in the utilization of machine learning methods for improving the processing of digital images in the healthcare industry [28, 29]. Medical practitioners have stated that machine learning (ML) has been incredibly influential in determining the contours of the X-ray, which has helped provide an accurate evaluation (Table 23.10).

According to the investigators, most medical professionals and others have said that the use of machine learning aids in the creation of higher picture quality, as well as in the processing and detection of patterns [30–33]. Furthermore, the crucial variables are highly significant in statistical terms, suggesting that the application of machine learning techniques is beneficial in improving the quality of images in the field healthcare industry [34–37].

23.6 Conclusion

In health care, physicians and nurses communicate photographs to diagnose illnesses as well as additional problems so that patients are given appropriate medical care. On the other hand, as picture distribution becomes more complicated, numerous apps are being developed to assist users in obtaining the clarity of the required image and making the correct selection. As a result, users are always working to enhance the ranking to increase the accuracy and impact of the photographs, allowing significant information to be identified and many elements to be analyzed. The importance of machine learning and the use of numerous sophisticated algorithms in machine learning methods allow for improved image processing, which aids in making informed judgments and clear issue assessments. Machine learning algorithms aid in signal processing in the field of technology and are focused on improving video image processing for use in a wide range of fields, including technology, medicine, traffic accident prevention, and many others. As a result, the machine learning application may analyze the image, show people how the analysis should be done, evaluate the model, and aid in successful decision-making. As a result, this study focuses on analyzing the impact of machine learning algorithms on the processing of images in healthcare facilities related to X-rays, so that a medical professional may efficiently analyze information and perform a proper analysis, prescribe appropriate medications, and support patients while improving their general well-being and health. The application of machine learning has a huge influence on the treatment of patients in medical delivery infrastructures.

References

1. Elaziz, M.A., et al.: New machine learning method for image-based diagnosis of COVID-19. PloS ONE **15**(6), e0235187 (2020)
2. Chachan, F.A.: Study the effect of using digital processing on the original image statistically. Int. J. Eng. Technol. **7**(3.36), 229–233 (2018)
3. Vats, P., et al.: A novel approach for detection of intracranial tumor using image segmentation based on cellular automata. In: Intelligent Sustainable Systems: Selected Papers of WorldS4 2021, vol. 2. Springer Singapore (2022)
4. Kaushik, S., et al.: A comprehensive analysis of mixed reality visual displays in context of its applicability in IoT. In: 2022 International Mobile and Embedded Technology Conference (MECON). IEEE (2022)
5. Zhu, B., et al.: Achieving the carbon intensity target of China: a least squares support vector machine with mixture kernel function approach. Appl. Energy **233**, 196–207 (2019)
6. Chaudhury, S., et al.: A novel approach to classifying breast cancer histopathology biopsy images using bilateral knowledge distillation and label smoothing regularization. Comput. Math. Methods Med. **2021**, 1–11 (2021)
7. Vats, P., Biswas, S.S.:. Big data analytics in real time for enterprise applications to produce useful intelligence. In: Data Wrangling: Concepts, Applications and Tools, vol 187 (2023)
8. Yan, W., et al.: Research on nonlinear distorted image recognition based on artificial neural network algorithm. J. Interconnect. Netw. **22**(Supp06), 2148002 (2022)

9. Sharma, N., et al.: A robust framework for governing blockchain-based distributed ledgers during COVID-19 for academic establishments. In: ICT with Intelligent Applications: Proceedings of ICTIS 2022, vol. 1, pp. 35–41. Springer Nature Singapore, Singapore (2022)
10. Yardley, L., Bishop, F.L.: Using mixed methods in health research: benefits and challenges. Br. J. Health Psychol. **20**(1), 1–4 (2015)
11. Upreti, K., Arora, S., Sharma, A. K. Pandey, A. K., Sharma, K. K., Dayal, M.: Wave height forecasting over ocean of things based on machine learning techniques: an application for ocean renewable energy generation. IEEE J. Ocean. Eng. https://doi.org/10.1109/JOE.2023.3314090 (2022)
12. Jing, L., Tian, Y.: Self-supervised visual feature learning with deep neural networks: a survey. IEEE Trans. Pattern Anal. Mach. Intell. **43**(11), 4037–4058 (2020)
13. Eltoukhy, M.M., et al.: Computer aided detection of mammographic mass using exact Gaussian–Hermite moments. J. Amb. Intell. Hum. Comput. 1–9 (2018)
14. Nguyen, Long D., et al. "Deep CNNs for microscopic image classification by exploiting transfer learning and feature concatenation. In: 2018 IEEE International Symposium on Circuits and Systems (ISCAS). IEEE (2018)
15. Liu, L., et al.: Image recognition technology based on machine learning. IEEE Access (2020)
16. Hosny, K.M., Darwish, M.M.: A kernel-based method for fast and accurate computation of PHT in polar coordinates. J. Real-Time Image Proc. **16**, 1235–1247 (2019)
17. Paygude, P.: et al.: Image processing using machine learning. IJSDR2009078 Int. J. Sci. Dev. Res. (2020)
18. Dufaux, F.: Grand challenges in image processing. Front. Sig. Process. **1**, 675547 (2021)
19. Kim, H., Sefcik, J.S., Bradway, C.: Characteristics of qualitative descriptive studies: a systematic review. Res. Nurs. Health **40**(1), 23–42 (2017)
20. Chen, J., Chen, L., Shabaz, M.: Image fusion algorithm at pixel level based on edge detection. J. Healthcare Eng. **2021** (2021)
21. Arora, A., et al.: A comprehensive study on social network analysis for digital platforms to examine and solve the behavioral patterns of everyday routines. In: ICT Systems and Sustainability: Proceedings of ICT4SD 2022, pp. 13–21. Springer Nature Singapore, Singapore (2022)
22. Kushwah, J.S., et al.: A Comprehensive System for Detecting Profound Tiredness for Automobile Drivers Using a CNN. In: Advanced Computing and Intelligent Technologies: Proceedings of ICACIT 2022, pp. 407–415. Springer Nature Singapore, Singapore (2022)
23. Upreti, K., Sheng-Lung, P., Ramdas, K.P., Prasun, C., Al-Alshaikh Halah, A., Sharma, A. K., Poonia R.C.: A multi-model unified disease diagnosis framework for cyber healthcare using IoMT- cloud computing networks. J. Disc. Math. Sci. Crypt. **26**(6), 1819–1834 (2023). https://doi.org/10.47974/JDMSC-1831
24. Lambert, V.A., Lambert, C.E.: Qualitative descriptive research: an acceptable design. Pacific RIM Int. J. Nurs. Res. **16**(4), 255–256 (2012)
25. Vats, P.: A novel study of fuzzy clustering algorithms for their applications in various domains. In: The 4th Joint International Conference on Information and Communication Technology, Electronic and Electrical Engineering (JICTEE), pp. 1–6. IEEE (2014)
26. Bhagat, A.D., et al.: A survey of cloud architectures: confidentiality, contemporary state, and future challenges. In: 2022 3rd International Conference on Issues and Challenges in Intelligent Computing Techniques (ICICT). IEEE (2022)
27. Yang, L., et al.: Suggestive annotation: a deep active learning framework for biomedical image segmentation. In: Medical Image Computing and Computer Assisted Intervention—MICCAI 2017: 20th International Conference, Quebec City, QC, Canada, September 11–13, 2017, Proceedings, Part III 20. Springer International Publishing (2017)
28. Upreti, K., et al.: Enhanced algorithmic modelling and architecture in deep reinforcement learning based on wireless communication Fintech technology. Optik **272**, 170309 (2023)
29. Haque, M., et al.: A systematic meta-analysis of blockchain technology for educational sector and its advancements towards education 4.0. Educ. Inf. Technol. 1–27 (2023)

30. Kumar, N., Upreti, K., Upreti, S.,Shabbir Alam, M., & Agrawal, M.: Blockchain integrated flexible vaccine supply chain architecture: Excavate the determinants of adoption. Hum. Behav. Emerg. Technol. 1–12 (2021). https://doi.org/10.1002/hbe2.302
31. Efimenko, M., Ignatev, A., Koshechkin, K.: Review of medical image recognition technologies to detect melanomas using neural networks. BMC Bioinf. **21**, 1–7 (2020)
32. Bhatnagar, S., Dayal, M., Singh, D., Upreti, S., Upreti, K., Kumar, J.: Block-hash signature (BHS) for transaction validation in smart contracts for security and privacy using blockchain. JMM **19**(04), 935–962 (2023)
33. Singh, P., et al.: Cloud-based patient health information exchange system using blockchain technology. In: Information and Communication Technology for Competitive Strategies (ICTCS 2021) Intelligent Strategies for ICT, pp. 569–577. Springer Nature Singapore, Singapore (2022)
34. Upreti, K., et al.: IoT-based control system to measure, analyze, and track basic vital indicators in patient healthcare monitoring system. In: International Conference on Advanced Communication and Intelligent Systems. Springer Nature Switzerland, Cham (2022)
35. Kapula, P.R., et al.: The block chain technology to protect data access using intelligent contracts mechanism security framework for 5g networks. In: 2022 2nd International Conference on Advance Computing and Innovative Technologies in Engineering (ICACITE). IEEE (2022)
36. Vats, P., et al.: A hybrid approach for retrieving geographic information in wireless environment using indexing technique. In: ICT Analysis and Applications, pp. 145–155. Springer Singapore (2022)
37. Upreti, K., et al.: Advancing pancreatic cancer diagnosis with artificial neural networks: current research and future prospects. In: 2023 7th International Conference on Intelligent Computing and Control Systems (ICICCS). IEEE (2023)
38. Guo, Y., et al.: Deep learning for 3d point clouds: a survey. IEEE Trans. Pattern Anal. Mach. Intel. **43**(12), 4338–4364 (2020)
39. Rana, A., Valenzise, G., Dufaux, F.: Learning-based tone mapping operator for efficient image matching. IEEE Trans. Multim. **21**(1), 256–268 (2018)
40. Galdi, C., et al.: Light fields for face analysis. Sensors **19**(12), 2687 (2019)
41. Duan, L., et al.: Video coding for machines: a paradigm of collaborative compression and intelligent analytics. IEEE Trans. Image Process. **29**, 8680–8695 (2020)
42. Ibrahim, R.A., Elaziz, M.A., Lu, S.: Chaotic opposition-based grey-wolf optimization algorithm based on differential evolution and disruption operator for global optimization. Exp. Syst. Appl. **108**, 1–27 (2018)
43. Zhao, W., Zhang, Z., Wang, L.: Manta ray foraging optimization: an effective bio-inspired optimizer for engineering applications. Eng. Appl. Artif. Intell. **87**, 103300 (2020)
44. Wang, B., Chen, L.L., Wang, M.: Novel image segmentation method based on PCNN. Optik **187**, 193–197 (2019)
45. Dartmann, G., Song, H., Schmeink, A., eds.: Big Data Analytics for Cyber-Physical Systems: Machine Learning for the Internet of Things. Elsevier (2019)
46. O'Mahony, N., et al.: Deep learning vs. traditional computer vision. In: Advances in Computer Vision: Proceedings of the 2019 Computer Vision Conference (CVC), Volume 1 1. Springer International Publishing (2020)

Chapter 24
Mobile Controlled Wheeled Robotic Arm—A Future of Smart Education

X. Anitha Mary, Gurram Aakash, Vijila Gnanaraj, Bhupathiraju S. S. M. Karthikeya Varma, and Kocherlakota Snigdha

Abstract In this paper, we discuss the design and implementation of a cutting-edge mobile-controlled wheeled robotic arm specifically tailored for educational purposes. The prime objective is to develop an innovative robotic arm that incorporates a wide array of sensors, actuators, and a highly efficient microcontroller. The wireless control mechanism facilitated by a mobile device further enhances the functionality of the proposed robotic arm. Notably, the incorporation of a wheeled base imparts exceptional mobility to the system, while the arm itself exhibits remarkable capabilities in executing diverse tasks, rendering it an exceptionally versatile educational platform. A significant aspect highlighted in this article is the emphasis on affordability, which plays a pivotal role in advancing research pertaining to robotic control. Additionally, the utilization of state-of-the-art 3D printing technology is explored for the construction of the robotic arm, enabling rapid prototyping and cost-effectiveness. The paper offers an in-depth elucidation of the hardware components and interface, elucidating their individual functionalities. The robotic arm software control is meticulously designed over Android-based application, where the arm functionalities are presented through code-based designs.

24.1 Introduction

The recent years have witnessed an impressive upswing in the prevalence of robotics within the educational sphere, with mobile-controlled robotic arms taking center stage. This research manuscript introduces a specially crafted mobile-controlled wheeled robotic arm aimed at enhancing the educational experience. This innovative mechanical appendage, equipped with an array of sensors, actuators, and a microcontroller, provides a multi-dimensional platform for comprehensive learning in the domains of robotics, electronics, and programming. This conceptualized robotic arm is embedded with wireless control abilities, enabling effortless management through

X. A. Mary (✉) · G. Aakash · V. Gnanaraj · B. S. S. M. K. Varma · K. Snigdha
Department of Robotics Engineering, Karunya Institute of Technology and Sciences, Coimbatore, Tamilnadu, India
e-mail: anithamary@karunya.edu

P. Nanda et al. (eds.), *Intelligent System and Data Analysis*, Smart Innovation, Systems and Technologies 398, https://doi.org/10.1007/978-981-97-5200-3_24

a mobile device. Its user-centric and instinctive interface facilitates seamless interaction with the system. The arm's unique feature is its wheeled base, promoting ease of movement and relocation in diverse learning spaces like classrooms or laboratories. The arm is also capable of executing varied tasks ranging from object handling to intricate maneuvers. We delve into the detailed design and execution of the mobile-controlled wheeled robotic arm, laying bare the complex software and hardware elements. The research also comprises a scrupulous analysis of results obtained from carefully orchestrated experiments to gauge the performance of the robotic arm and affirm its pedagogical benefits. In essence, the proposed wheeled robotic arm presents an engaging and distinctive pathway for immersive learning in robotics and the practical application of robotic design and control principles. It is indeed a hands-on experiential tool, assisting in unraveling the complexities of the world of robotics.

Additionally, the results derived from meticulously conducted experiments are expounded upon to evaluate the robotic arm's performance and substantiate its educational efficacy. Overall, the proposed robotic arm provides a distinctive and captivating opportunity for acquiring knowledge in robotics and fostering practical expertise in the realms of robot design and control. It offers a hands-on experiential avenue to explore the intricacies of robotics.

Numerous robotic control systems incur substantial costs owing to the utilization of high-precision actuators and the custom machining of intricate components. To expedite the progress of robotic control research, it is advised that significant strides can be made by substantially reducing the price of robotic arms while retaining their valuable performance characteristics [1].

Enhanced cost-effectiveness can contribute to broader adoption, consequently expediting advancements in the field. Robotic arms can be evaluated along various dimensions, encompassing backlash, payload capacity, speed, repeatability, compliance, human safety, and cost. In the realm of robotics research, certain dimensions hold greater significance than others. Notably, for tasks involving object grasping and manipulation, high repeatability and minimal backlash assume pivotal importance [2].

The developed robotic arm was constructed utilizing 3D printing technology and fabricated with a polyacidic filament, specifically polylactic acid (PLA). This material was chosen for its advantageous attributes, including lightweight nature, high durability, cost-effectiveness, and ease of machining. The robotic arm model employed in this study remains faithful to the exact specifications outlined in [reference]. Noteworthy variants of the robotic arm include keypad controlled, voice control, gesture control, and other innovative control mechanisms Nonetheless, conventional industrial robots predominantly employ the conventional teaching process for programming, which remains a laborious and time-intensive task demanding technical proficiency [3]. Many prototype robotic arms are typically tethered or wired-controlled. In this study, we have implemented an integrated voice and remote control system within an application developed using MIT app inventors. The primary objective of this project is to enable seamless robot movement triggered by voice commands received from the operator. In situations where voice commands are not audible, an alternative method involving remote control access to the robot is implemented. The primary focus of this paper is to devise methodologies that facilitate users in controlling and programming robots.

Over the years, a multitude of advanced techniques have been devised and implemented as algorithms to optimize the functioning of robotic arms, enabling them to execute pick-and-place operations with enhanced efficiency and precise adherence to specific requirements or application constraints. Several examples comprise of joystick controlled industrial robotic system [4], Six degrees of freedom (6DOF) voice-controlled grasping robotic arm utilizing ultrasonic sensor technology [5], robotic arm controlled based on gyro-accelerometer utilizing an AVR microcontroller [6], efficient and viable gesture-controlled robotic arm [7], development of teaching pendant optimized for robotic arm application in diverse verticals [8].

The subsequent sections of this article are organized as follows: Sect. 2 presents an in-depth analysis of the hardware components and interface, encompassing a comprehensive elucidation of each individual element and their seamless integration. Section 3 delves into intricate aspects of system design, encompassing the modeling of the robotic arm and the control application. Section 4 provides an extensive discussion on system testing and the ensuing results. Subsequently, concluding remarks are presented, followed by an exploration of the future scope of this research endeavor.

24.2 Related Work

Bhargava [9] developed a robotic arm that possesses 5 degrees of freedom (5DOF), managed via an Arduino Uno microcontroller. The user interaction takes place through a group of potentiometers, which provide the input signals for the arm operations. The arm structure includes four rotational joints and an end effector, all driven by servomotors for facilitating the rotary movement. The Arduino software provides fine-tuned control of the servomotors, resulting in rotations that mirror the rotational displacement of the potentiometer shafts. Notably, the definition of a robot greatly depends on the characteristics and dynamics of the interconnected links that define its structure and movements. Gautam [10] penned a comprehensive review article that sheds light on the evolution and development of industrial robotic arms.

One of the primary objectives of this research is to address the challenges associated with handling objects situated at a considerable distance from human workers. To overcome this issue, an advanced and selective operational control methodology for the robotic arm has been devised. This approach enables precise manipulation and efficient placement or retrieval of objects, thereby enhancing the arm's capabilities in remote object-handling scenarios. The successful development of the robotic arm has resulted in precise movement control. Nevertheless, the expense associated with cable replacement necessitates a strategic approach to minimize friction on the table. Therefore, optimizing the arm's design to reduce friction is of paramount importance, as it directly contributes to prolonging the maintenance cycle.

Elfasahany [11] presents a robot arm for light material tasks. It uses acrylic for durability and servo motors with encoders for precise movement. The design focuses on four degrees of freedom and utilizes a commercially available gripper.

Fu and Bhavsar [12] a manipulator project featured a flexible and mobile automatic machine comprising a main control system, MQTT server, and web application. It utilized Arduino and ESP8266 for servo control and data transmission. MQTT, with an EMQTT broker, enabled remote control via cloud-based MQTT servers. The web app, developed in HTML5 with the Canvas tag, allowed user interaction and calculated joint angles. Challenges included protocol selection and addressing baud rate compatibility between Arduino and ESP8266.

Yusoff et al. [13] presented the development of a wireless mobile robot arm for pick-and-place tasks, utilizing an Arduino Mega platform and wireless PS2 controller. Performance assessment encompassed speed, distance, and load capacity. Its purpose was to safely and efficiently handle distant and hazardous objects. The project overview included system architecture, mechanical and electrical design, and software development. Key components featured were the Arduino Mega 2560 microcontroller and Sony PS2 wireless controller, with results and analysis covering velocity, load capacity, wireless controller application, servo motor, and circuit analysis.

24.3 Hardware and Interface

24.3.1 Microcontroller: Arduino Uno

Arduino microcontroller (Fig. 24.1) is utilized to regulate the robot's actions. This Arduino receives command signals transmitted from a mobile application via a Bluetooth connection. Subsequently, the Arduino processes these signals and sends corresponding commands to the motor driver and YwRobot power supply, enabling precise control over the robotic arm's movements.

24.3.2 Servo Motor

Specialized motors known as servomotors drive the movements of robotic joints and limbs. These motors are designed to rotate to a specific angle and maintain that position, making them an ideal choice for applications requiring precision. Due to their compact and lightweight nature, servomotors are a perfect fit for smaller robotic implementations. Their operation can be governed using pulse width modulation (PWM) signals (Fig. 24.2).

Fig. 24.1 Microcontroller: Arduino Uno. *Source* [14]

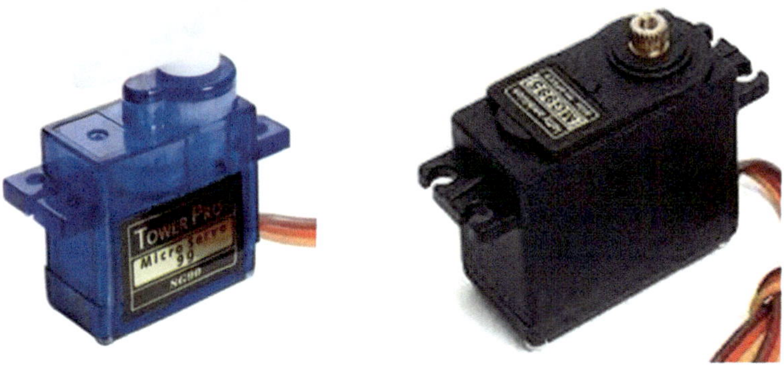

Fig. 24.2 Servo motor. *Source* [15]

24.3.3 DC Motor

Direct current (DC) motors, due to their capacity to convert electrical energy into mechanical movements, are commonly employed in robotics and automation. They come in an assortment of sizes and power ratings, offering various control methods including PWM signals and H-bridge circuits (Fig. 24.3).

24.3.4 HC-05 Bluetooth Module

The HC-05 Bluetooth module (Fig. 24.4) is a wireless communication device that facilitates Bluetooth protocol-based exchanges between microcontrollers and other

Fig. 24.3 DC motor. *Source* [16]

Fig. 24.4 HC-05 Bluetooth module. *Source* [17]

devices. Its compact size, cost-effectiveness, and ease of use make it a popular choice for a myriad of applications. It supports data transmission, reception, and wireless control of devices.

24.3.5 L298N Motor Driver

The L298N motor driver is a widely used H-bridge module for managing DC and stepper motors. It accepts digital signals from a microcontroller for easy control and caters to high current and voltage requirements of motors. The L298N's reliability and user-friendly nature make it a preferred choice for automation and robotics projects (Fig. 24.5).

Fig. 24.5 L298N motor driver. *Source* [18]

24.3.6 YwRobot Breadboard Power Supply

The YwRobot breadboard power supply is a compact module that provides regulated DC power to a breadboard. It is designed to work with a 9 V battery or DC power supply and can deliver up to 700 mA of current at 3.3 V or 5 V. As a dependable source of power for microcontrollers and other breadboard devices, the YwRobot power supply is ideal for prototyping and experimental stages (Fig. 24.6).

Fig. 24.6 YwRobot breadboard power supply *Source* [19]

24.4 System Design

24.4.1 Mobile Robot

The performance of the robot is measured over smoothness and its flexibility, and thus, the motor robot chassis is designed of lightweight yet strong materials which offers high resilience. The wheels are built to withstand the mass of the robot and also to offer a solid grip on the ground. To give stability and keep the robot from toppling over, the four wheels are arranged in a rectangle. There are four 12 V DC motors, L298N motor driver, which is in charge of regulating each motor's speed and direction, is attached to these motors. Dual H-bridge motor drivers like the L298N allow for simultaneous control of two motors. The same 12 V DC source that powers the motors also powers the motor driver.

The mobile robot's brain is governed by an Arduino Uno board. It gets instructions from the smartphone app and converts them into signals that drive the motors. The L298N motor driver receives commands from the controller that are processed by the control algorithm to regulate the speed and direction of the motors using input signals from the app. An external battery pack that can be recharged powers the mobile robot. The Arduino Uno and L298N motor driver are linked to the battery pack, which supplies the energy required to run the motors and move the mobile robot.

Control signals to the Arduino UNO board can be transmitted through voice commands or keypad signals. Voice commands are processed by speech recognition algorithms, which transcribe the user's spoken instructions into digital signals decipherable by the Arduino board. Alternatively, keypad signals can be dispatched using the smartphone app's graphical user interface (GUI), granting the user the ability to select specific movements and functions for the mobile robot. By leveraging the app, users can swiftly modify the mobile robot's speed and direction, simplifying navigation across varied terrains.

24.4.2 Robotic Arm

The mobile-controlled wheeled robotic arm, a complex and intricate system, is designed for precise and controlled arm movements. This is accomplished through the utilization of high-precision servo motors, advanced control algorithms, and real-time control signals. This section outlines the robotic arm's system design and its interaction with the mobile app to facilitate these functions. The robotic arm structure is composed of five segments. Alongside these, it includes a mechanical gripper for object manipulation. The design incorporates six actuators: three MG995 servo motors and three SG90 servo motors. The MG995 motors are responsible for arm movements, while the SG90 motors govern the wrist motion of the robotic arm real-time control signals. In this section, we'll go through the robotic arm's system design and how the mobile app interacts with it to enable these actions (Fig. 24.7).

Each joint of the arm is connected to a single actuator, which is a servo motor.

The system makes use of two YwRobot power supply modules to power the servo motors (Fig. 24.8). Three MG995 servo motors are powered by one module, and three SG90 servo motors are powered by the other module. An Arduino UNO microcontroller board, which gets command signals from a mobile app, manages these power supply modules. The servo motors' efficient operation depends on the power supply modules. They supply the servo motors with the necessary voltage and current. The built-in safety measures in the YwRobot power supply modules guard against overvoltage and overcurrent to safeguard both the user and the robotic arm.

The user can utilize keypad signals to direct the robotic arm's movements using the mobile app. These control signals are then processed by the Arduino UNO microcontroller board and sent to the YwRobot power supply modules, which in turn drive the servo motors. The robotic arm's many components are moved by the servo motors, which enable it to carry out a variety of diverse movements and manipulations.

The gripper force analysis involves calculating the gripper force using the specified formula.

$$\mu \times Fg \times nf = W \tag{24.1}$$

where μ the is co-efficient of the friction of the part surface against the finger contact surface, Fg is the force exerted by gripper, nf is the number of fingers for contacting, and W is the size of object to be grasped (Table 24.1).

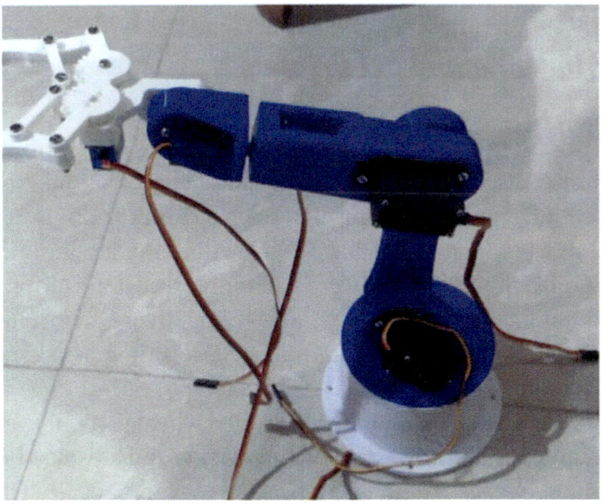

Fig. 24.7 3D-printed robotic arm

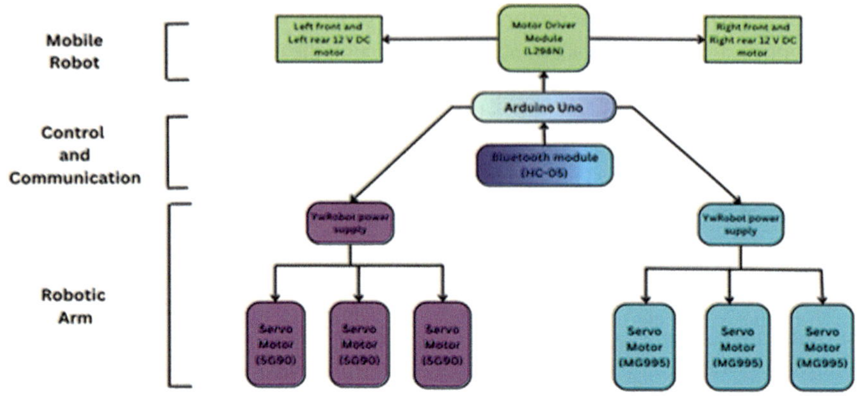

Fig. 24.8 Block diagram of the mobile-controlled wheeled robotic arm

Table 24.1 Parameters and considered values

Parameters	Values
nf	2
μ	0.85
W	50

Based on these values, $\mu \times nf$ is 0.85 * 2 = 1.70, and thus, 1.7 $Fg = 50$, which yields Fg as 29.4 N. Therefore, the force exerted by gripper while grasping an object is approximately 29.4 N.

24.4.3 3D Modeling and 3D Printing

The advent of 3D printing technology has seen rapid integration across diverse fields, fueled by its cost-effectiveness and the simplicity of automating its processes. Compared to traditional manufacturing methods, 3D printing offers economical alternatives, streamlines production by dispensing with complex tooling, and supports varying production volumes. Moreover, it promotes customization by facilitating simple software modifications. These benefits collectively contribute to its ubiquitous application across industries, positioning 3D printing as a crucial tool in the manufacturing domain [20].

The fabrication process of the mobile-controlled wheeled robotic arm was initiated with the 3D design of its various components. Using the advanced Creality CR-10 3D printer, all parts of the Arduino robotic arm were produced. This printer, with its extensive build volume, is ideal for creating large components like those required for the robotic arm. In addition, its high resolution ensures the printed parts are of excellent quality, including the precise spur gears for opening and closing the gripper.

As the servo motor rotates, the gears respond in kind, enabling the gripper to close or open based on the motor's rotation direction. This setup offers a straightforward and effective method for controlling the gripper's movements. The gripper's design enables a firm grip to handle objects of varying shapes and sizes.

The robotic arm's comprehensive design imparts necessary robustness and strength. The rigidity of the design ensures that the arm can bear the weight of the objects it manipulates. Leveraging 3D design and printing techniques ensures the arm's high quality and precision. Furthermore, the use of high-grade filament ensures the arm's durability, allowing it to withstand the inevitable wear and tear from repeated usage.

24.4.4 Mobile App Development

We have used the MIT app inventor which is a platform to develop the customized applications according to the application of the user [21]. This platform was developed by the Google and is now maintained by the Massachusetts Institute of Technology (MIT). For the easy programming and development, the building blocks are used in spite of conventional complex coding as shown in Fig. 24.9.

After coding the app is converted into zip file which can be shared among the users easily. This app is compatible with any Android mobile. The communication between the Android application and the robot is done by using Bluetooth module (HC-05). The range of the Bluetooth is 10 m in closed areas and beyond 10 m in the open areas.

The app interface is shown in Fig. 24.10 where at the top we have the connect and disconnect options and also to indicate where the app is connected to Bluetooth or not. Then followed by mobile robot control and robotic arm control, at the end the speed control.

24.4.4.1 Keypad Control

The keypad control is used to control the robotic arm and mobile robot. There are four keys available to operate the mobile robot which makes the mobile robot to move forward, backward, left and right directions, which helps to carry the robotic arm.

24.4.4.2 Voice Control

The voice control is used to move the mobile robot using voice commands. It uses the Google Speech Recognition technology to get the audio input from microphone of the mobile and then convert the speech into desired text it is done by Google speech-to-text converter API, this text is used as control or command statement

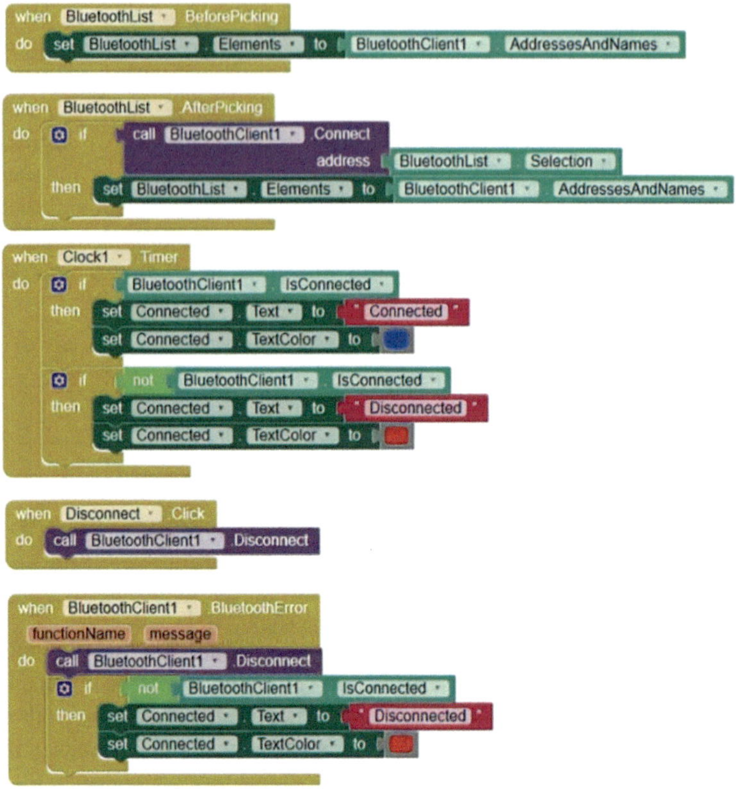

Fig. 24.9 Code blocks of app development

by the Arduino to control the mobile robot. The spoken language interface must be English language. The commands we used are left, right, front, back, and stop (Fig. 24.11).

24.5 System Test and Result

The educational mobile-controlled robotic arm underwent thorough design and rigorous testing to evaluate its performance, mobility speed, force handling, and balance during movements. The mobile app interface was extensively tested and found to operate flawlessly without errors or malfunctions.

Upon activation, the robotic arm's motors immediately move to the preset home position, controlled by signals from the mobile app via Bluetooth. The arm consistently executes commands with precision during detailed movement tests, evaluating

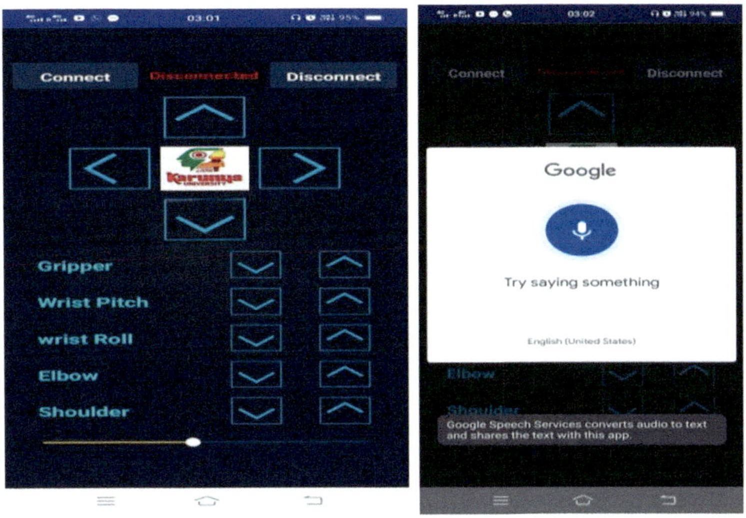

Fig. 24.10 App interface

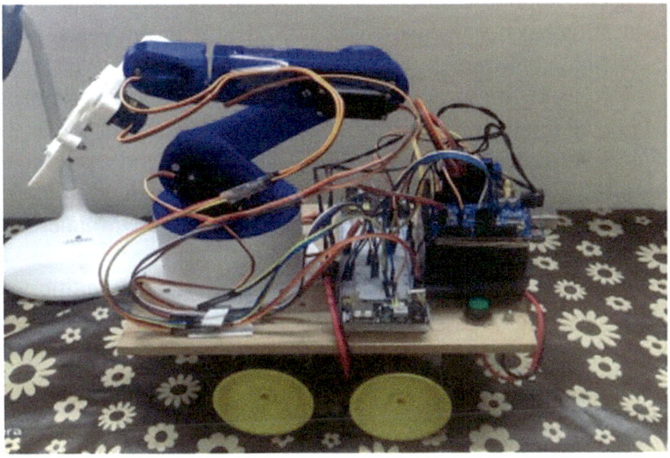

Fig. 24.11 Integrated wheeled mobile robot

each motor's effectiveness, and accuracy. Experiments on the waist motor confirmed smooth base swiveling, while the shoulder motor seamlessly lifted and lowered the arm.

The elbow motor flexed and extended the arm without issues. Waist-roll and waist-pitch motors efficiently rotated the arm around its axis. The gripper motor flawlessly opened and closed the gripper. The wheeled robot's mobility was tested for high precision and speed while managing its weight, including the integrated robotic arm.

The mobile robot met all parameters, exhibiting good speed and precise navigation for efficient task completion. Its ability to navigate challenging terrains makes it versatile for outdoor and industrial use. The robot maintained impressive balance, crucial for stability, even with the robotic arm. These results confirm its exceptional performance, especially in terms of speed and size.

The future prospects of this project are to implement the forward kinematics and inverse kinematics algorithm [22] to the robot developed and to use the robot for the basic pick-and-place operations in the industry which can be a low-cost automation for the modern technologies like Industry 4.0 applications.

Designing a more sophisticated control system that enables instantaneous decision-making based on environmental information such as localization, obstacle avoidance, and object recognition which makes the mobile robot act autonomously [23].

24.6 Conclusion

This research has demonstrated the feasibility and effectiveness of using a mobile-controlled wheeled robotic arm for educational purposes. The importance of technology in education has become increasingly evident in recent years, and this study provides a valuable contribution to this field. By utilizing additive technology and modern software tools, we were able to significantly reduce development periods, thereby making the production process more efficient and cost-effective. The robot's links and joints were designed using 3D CAD software and manufactured using printing, resulting in a stiff enough robot for pick-and-place industrial applications. PLA filament material was an excellent choice because it is environmentally friendly and has great strength and durability.

The robotic arm can be used for educational purposes in addition to pick-and-place tasks. Due to its mobile-controlled feature, learning is more interactive, which boosts student engagement and helps them remember what is being taught. People of all ages and ability levels can utilize the user interface designed for the operator to program robot motions since it is simple to use and friendly.

During the design and development of the robotic arm, a range of challenges emerged. Achieving meticulous control over the robot's movements and ensuring the structural integrity of the arm were the major ones. These issues were effectively addressed by choosing appropriate sensors and motors. Additionally, the arm's construction involved the use of lightweight, 3D-printed materials, which further contributed to resolving these challenges.

In the near future, the robotic arm is set to be used in educational settings. The plan is to gather real-time feedback and human responses during its use. This valuable information will be used to enhance and refine the robotic arm in the ongoing development process.

References

1. Quigley, M., Asbeck, A., Ng, A.: A low-cost compliant 7-DOF robotic manipulator. In: IEEE International Conference on Robotics and Automation, Shanghai, China, pp. 6051–6058, (2011). https://doi.org/10.1109/ICRA.2011.5980332
2. Pawar, V., Bire, S., More, S., More, K., Mule, R.: Review on design and development of robotic arm generation-1. Int. J. Innov. Sci. Res. Technol. **3**(3), 527–529 (2018). www.ijisrt.com. ISSN 2456-2165
3. Gayatri Devi, V., Vijetha, T.: Mems controlled robotic arm with gestures. IJSDR **1**(6) (2016). ISSN: 2455-2631
4. Rahman, R., Rahman, M.S., Bhuiyan, J.R.: Joystick controlled industrial robotic system with robotic arm. In: IEEE International Conference on Robotics, Automation, Artificial-intelligence and Internet-of-Things (RAAICON), Dhaka, Bangladesh, pp. 31–34 (2019). https://doi.org/10.1109/RAAICON48939.2019.18
5. Wang, Z., Chen, D., Xiao, P.: Design of a voice control 6dof grasping robotic arm based on ultrasonic sensor, computer vision and alexa voice assistance. In: 10th International Conference on Information Technology in Medicine and Education (ITME), Qingdao, China, pp. 649–654 (2019). https://doi.org/10.1109/ITME.2019.00150
6. Bhuyan, A.I., Mallick, T.C.: Gyro-accelerometer based control of a robotic Arm using AVR microcontroller. In: 2014 9th International Forum on Strategic Technology (IFOST), Cox's Bazar, Bangladesh, pp. 409–413 (2014). https://doi.org/10.1109/IFOST.2014.6991151
7. Atre, P., Bhagat, S., Pooniwala, N., Shah, P.: Efficient and feasible gesture controlled robotic arm. In: 2018 Second International Conference on Intelligent Computing and Control Systems (ICICCS), Madurai, India, pp. 1–6 (2018). https://doi.org/10.1109/ICCONS.2018.8662943
8. Fukui, H., Yonejima, S., Yamano, M., Dohi, M., Yamada, M., Nishiki, T.: Development of Teaching Pendant Optimized for Robot Application, pp. 72–77 (2009). https://doi.org/10.1109/ARSO.2009.5587070
9. Bhargava, A., Kumar, A.: Arduino controlled robotic arm 2017. In: International conference of Electronics, Communication and Aerospace Technology (ICECA), Coimbatore, India, pp. 376–380 (2017). https://doi.org/10.1109/ICECA.2017.8212837
10. Gautam, R., Gedam, A.: Review on development of ındustrial robotic arm. Indus. Res. Dev. **04** (2017)
11. Elfasakhany, A., Yanez, E., Baylon, K., Salgado, R.: Design and development of a competitive low-cost robot arm with four degrees of freedom. Modern Mech. Eng. **01** (2011). https://doi.org/10.4236/mme.2011.12007
12. Fu, S., Bhavsar, P.C.: Robotic arm control based on ınternet of things. In: 2019 IEEE Long Island Systems, Applications and Technology Conference (LISAT), Farmingdale, NY, USA, pp. 1–6 (2019). https://doi.org/10.1109/LISAT.2019.8817333
13. Yusoff, M.A.K., Samin, R.E., Ibrahim, B.S.K.: Wireless mobile robotic arm. Proc. Eng. **41**, 1072–1078 (2012). ISSN 1877-7058
14. https://www.myinnovation.in/product-page/arduino-uno-r3-development-board-atmega328p-board-compatible-model
15. https://www.electronicwings.com/arm7/servo-motor-interfacing-with-lpc2148, https://www.amazon.in/MG995-Metal-Servo-Pro-360-Degree/dp/B0983359LD
16. https://www.amazon.in/Themisto-Motor-Multipurpose-Brushed-Applications/dp/B0728K974M
17. https://robocomp.in/product/hc-05-bluetooth-module-with-ttl-output/
18. https://www.amazon.in/Techleads-Driver-Module-Stepper-Arduino/dp/B07N4M2K98
19. https://m.indiamart.com/proddetail/breadboard-power-supply-module-3-3v-5v-solderless-bread-board-mb102-erase-20878067348.html
20. Tamimi, H., Abuamara, F., Al-Baik, O.: Utilization of 3D printing technology: a case study of UAE. In: 2021 8th International Conference on Computing for Sustainable Global Development (INDIACom), New Delhi, India, pp. 426–429 (2021)

21. N.P., M.V., M.S.K., R.L., V.V.: Voice ordered home mechanization system with deployment of Arduino Uno and MIT App ınventor. In: 2021 Second International Conference on Electronics and Sustainable Communication Systems (ICESC), Coimbatore, India, pp. 218–222 (2021). https://doi.org/10.1109/ICESC51422.2021.9532719
22. Dahari, M., Tan, J.-D.: Forward and inverse kinematics model for robotic welding process using KR-16KS KUKA robot. In: 2011 Fourth International Conference on Modeling, Simulation and Applied Optimization, Kuala Lumpur, Malaysia, pp. 1–6 (2011). https://doi.org/10.1109/ICMSAO.2011.5775598
23. Jafar, F.A., Yokota, K., Suzuki, Y., Matsuoka, T.: Autonomous mobile robot self-localization based on environmental visual features. In: 4th International Conference on Autonomous Robots and Agents, Wellington, 2009, pp. 245–250 (2009). https://doi.org/10.1109/ICARA.2000.4803

Chapter 25
FDA-TO-FCA: Flow Direction Algorithm-Based Task Offloading in Fog-Cloud Architecture

Prashant Shukla, Sudhakar Pandey, Shivani Gupta, and Deepika Agrawal

Abstract With the development of a lot of delay-sensitive applications, it is very difficult to perform all the tasks locally because our devices have less computation power. Traditional Cloud Server (CS) has large computing power and storage but it is very far from IoT devices that's why it consumes a lot of energy and take a huge time to execute a task. For these types of applications, local devices and CSs are not good choices. So, to overcome this delay-sensitive problem, the Fog computing (FC) concept was introduced. In FC, we can offload our task to Fog Nodes (FN) that are close to the user device and take less execution time and energy than a CS. To minimize the execution time for delay-sensitive applications, we have modeled our problem and converted our solutions from a bounded real space to a finite discrete space that uses the Flow Direction Algorithm (FDA). Our algorithm performance is 12 % better than cloud-only, 6 % better than Fog-only, and 6 % better than IoT-only on average. In this, the task can execute locally or on a FN or CS. The proposed approach performed better to reduce the execution time.

25.1 Introduction

In recent years, there is a lot of growth in delay-sensitive applications like online gaming, augmented reality, image, and signal processing, etc. Our Internet-of-Things (IoT) devices are limited in terms of battery, computation, and storage capacity. So, a lot of research is going on to speed up the task processing time, and task offloading is one of its upcoming solutions [1, 2]. In task offloading, end-user devices (EDs) can save time by offloading the task in Fog Node (FN) or Cloud Server (CS) then FN or CS process that task and provide the result to the MDs [3]. The distance between the

P. Shukla (✉) · S. Pandey · D. Agrawal
Department of IT, NIT, Raipur, Chhattisgarh 492010, India
e-mail: pshukla.phd2019.it@nitrr.ac.in

S. Gupta
Department of CSE, IIT, Patna, Bihar 801106, India

P. Nanda et al. (eds.), *Intelligent System and Data Analysis*, Smart Innovation, Systems and Technologies 398, https://doi.org/10.1007/978-981-97-5200-3_25

CS and the MD is large so it increases energy consumption and cost that influences real-time applications. So, in recent years, storage capability and computation are partially migrated from the CS to the FN. Fog computing (FC) moves storage and computing resources close to EDs [4]. FC brings closer the EDs where the data is generated and the computing capacity and data storage instead of relying on a central location that can be at a great distance from the ED. The main reason behind this is that real-time data does not suffer latency issues that can affect an application's performance [4, 5]. The main advantages of Fog-Cloud Architecture (FCA) are the following: (1) In comparison with local computing, FCA can overcome the problem of limited computing capacity in user equipment. (2) If we directly offload our tasks to the CS, it will create large latency because of the long distance between the ED and the CS [6, 7].

In this paper, we have taken 3 types of network topology and N different tasks and we have 3 options to perform a task like we can perform a task locally on FN or CS. If we can perform the task locally, then we have to calculate only the execution time, but user EDs are limited in terms of battery and computing capacity. If we offload tasks in the FN or CS, we have to calculate the transmission and execution times. The main contribution of this paper is to minimize the cost of task offloading in FC using the Flow Direction Algorithm (FDA). The rest of the paper is arranged as given: Sect. 2 defines the system model for our problem. Section 3 is about the problem formulation. Section 4 defines the notations. Section 5 defines the FDA algorithm. Section 6 defines the methodology to solve the task offloading problem. Section 7 reports the simulation results. Section 8 concludes the work.

25.2 Related Work

Shuyue et al. [8] have taken multiple dependent tasks and tried to minimize the execution cost and task completion time in multi-access edge computing using the queue-based multi-objective particle swarm optimization algorithm. When the task is created first, it calculates the priority of the task according to the execution time for every task and then four types of the condition have checked if the ED has enough resources to execute that task, then it is executed in the local ED if not then it offloads the task in nearby FN. If that FN is also overloaded or unable to execute that task is offloaded to another FN. If the above three conditions are failed, then it is finally offloaded to the CS for execution. According to the decision, the execution cost and task completion time have been calculated. Wang et al. [9] has taken EDs that are surrounded by two or more FNs. ED forms a utility function and shares it with the nearest FN and according to the receive function, it applied the PROMOT algorithm to obtain the best offloading proportion and transmission power and sent this result back to the EDs then according to that result ED offload the tasks. Ahmed et al. [10] divided delay-sensitive and computation-intensive tasks into sub-tasks and it is offloaded into multiple Mobile Edge Computing (MEC) servers using a sequential offloading scheme which is used to minimize the latency and failure probability of offloading the tasks using the Genetic Algorithm.

Joshi et al. [11] solved the latency and energy consumption problem by proposing a new approach where according to the delay, energy, and communication constraints tasks get partitioned. When the user enters a new service area, then the VM migration algorithm is applied to optimize the service delay by copying the task and VM requirements into the migration queue. Zhu et al. [12] proposes the multi-objective optimization problem and solved the task offloading problem in joint cloudlet deployment to reduce the use of energy, delay in response of tasks, and the number of deployed cloudlets. Chen et al. [6] solved the task offloading problem in an ultra-dense network by dividing it into two sub-problems of task placement (where the task should be processed) and resource allocation (how many resources should be allocated to each task) according to the solution of these two sub-problems offloading is performed to reduce the delay. Also, in the recent literature [13, 14], we can see a variety of approaches have been used in various task offloading algorithms. But for solving offloading problem, author had rarely used the meta-heuristic approach. In this paper, we have proposed task offloading using the FDA to minimize the offloading cost.

25.3 System Model

In this part, we have established the offloading cost of a workflow, which takes into account task execution time, network latency when transferring data, and energy use when processing tasks and data by servers and IoT devices. The overall utility for offloading a particular workflow, or the amount of time it takes to complete it from the moment of the request until the final output data is returned to the asking IoT device, is specifically the sum of all the jobs' energy costs. The utility relies on the underpinning FCA network is shown in Fig. 25.1, which includes several CS and FN. We have represented the underlying network using the following parameters:

- $D_R(l_i, l_j)$ denotes the Data Rate, i.e., time taken to transmit a unit of data from location l_i to location l_j across the network.
- $D_E(l_i)$ denotes the Energy Consumption for Data, i.e., energy consumed for handling a unit of data at the location l_i.
- $V_R(l_i)$ denotes the Task Rate, i.e., time taken run a unit task at the location l_i.
- $V_E(l_i)$ denotes the Energy Consumption for Task, i.e., energy consumed for running a unit task at the location l_i .

We have defined a workflow (w) containing N tasks as:

$$W = \{V, D\}, \tag{25.1}$$

where, V denotes the task sizes and D denotes the data dependency between consecutive tasks defined as:

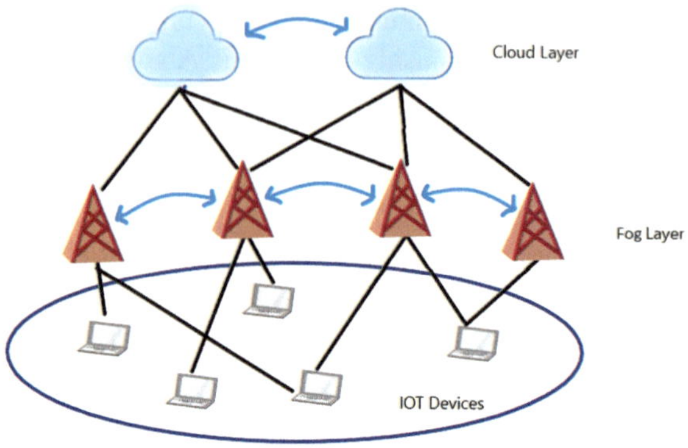

Fig. 25.1 Network topology

$$V = \{(v_i) : 1 \leq i \leq N\} \tag{25.2}$$

$$D = \{(d_i) : 0 \leq i \leq N\} \tag{25.3}$$

We have represented an offloading solution (L) for a workflow containing N tasks as:

$$L = \{(l_i) : 1 \leq i \leq N\} \tag{25.4}$$

Since the workflows are generated by IoT devices only, we assume the initial location to be at the IoT device. Hence, $l_0 = 0$, we have defined the offloading cost $C(W, L)$ as the summation of overall delay and energy consumption for offloading a workflow W using offloading solution L.

$$\text{Cost}(W, L) = \sum_{i=1}^{N}(C_i^I + C_i^T + C_i^O) + D_R(l_N, l_0), \tag{25.5}$$

where C_i^I is cost contributed by the input data of the ith task, C_i^O is cost contributed by the output data of the ith task, and C_i^T is cost contributed by the ith task itself.

$$C_i^I = d_{i-1} \times (D_R(l_{i-1}, l_i) + D_E(l_i)) \tag{25.6}$$

$$C_i^T = v_i \times (V_R(l_i) + V_E(l_i)) \tag{25.7}$$

$$C_i^O = d_i \times D_E(l_i) \tag{25.8}$$

25.4 Problem Formulation

In this paper, in terms of restricted battery capacity and computing power of EDs, we consider the following problem: (1) Task offloading decides the processing of tasks locally or in the FN or in the CS, and in which FN or CS, the task should be placed. (2) We first formulate the task offloading problem of the FCA. Table 25.1 depicts the notations used in this chapter. Without loss of generality, we can take n number of tasks, L is the number of locations of infrastructure, W is the set of workflows, π is the solution space, and C is the cost. So, using this information, we have to calculate the minimum cost for task offloading. Given a workflow W, find the optimal offloading solution L that minimizes the offloading cost C. Objective: Our aim is to minimize the task offloading cost. Hence,

$$\min_{L}\{C(W, L)\} \tag{25.9}$$

25.5 Methodology

Applications running on IoT devices generate workflows that are a set of programs or tasks that require computation resources. Through the process of offloading, IoT devices can run their workflows remotely by offloading tasks into the FN or CS. In this paper, we have performed the task offloading with 10–100 tasks. For these tasks, we have estimated the optimal offloading cost and compared it with other costs. We have taken population size equal to the number of tasks.

Table 25.1 Notations used

Notations	Description
F	Solution as flow
β	Number of neighbors
W	Weight ranges between 0 and infinity
N	Position of neighbor
δ	Neighborhood radius
V	Flow velocity vector
R	Random vector between 0 and 1 with uniform distribution
S	Random scalar between 0 and 1 with normal distribution
Slope	Slope vector between neighbor and current position of the flow
F_{μ}	Best flow
F_{rand}	Random flow
l_b	Lower limit of the decision variables
u_b	Upper limit of the decision variables
d	Dimensions of problem

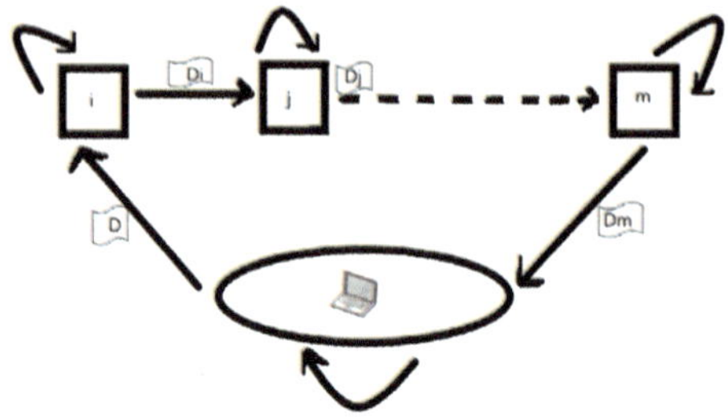

Fig. 25.2 Task offloading methodology

It is assumed that workflow is generated in the IoT device which is considered as 0 location. Here, we can perform total $(1 + E + C)$ actions means we can have a total of $(1 + E + C)$ locations where we can put our tasks for execution, where E is the number of FNs and C is the number of CSs. In Fig. 25.2, we have taken dependent tasks when 1st task is completed its execution at the given location (i) it will generate output with new data size (D_i) then it will be sent to the next task location (j) and then its execution will start, in this way we perform the execution. At the end when all the tasks will be executed, then the final output data will be sent back to the IoT device. Using this scenario, we calculated offloading cost using the FDA algorithm and the best-estimated cost and best solution are returned.

Input parameters		Output parameters
Number of tasks	[10,100]	Estimated best solution
Number of neighbors	8	Estimated best cost

Figure 25.3 shows different types of network topology. In this paper, we have taken three types of network topology. The first network topology is a very simple one with one ED, one FN, and one CS, the second one with one ED, three FNs, and two CSs, and in the third network topology, we have taken one ED device, five FNs, and two CSs. In this, IoT device is always taken as location 0 where the data will be generated and the FN will be taken as the location from 1 to the number of FNs available. Let's say there are k FNs available so we have taken 1 to k FNs. Each FN has to consider as a unique location. For the CS, we took location from $(k + 1)$ to the number of CSs available.

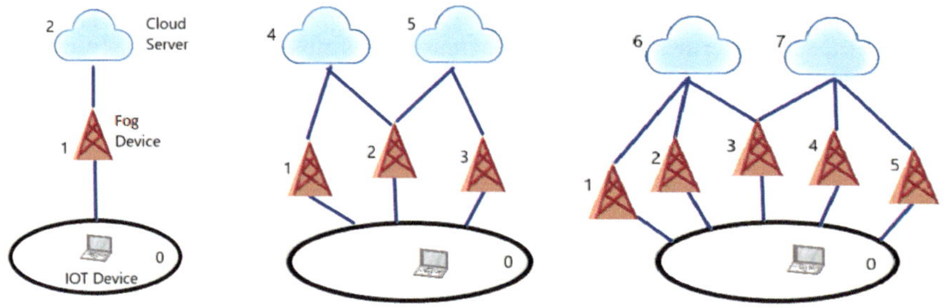

(a) Topology-1 (b) Topology-2 (c) Topology-3

Fig. 25.3 Different network topologies

25.5.1 Flow Direction Algorithm (FDA)

FDA is a meta-heuristic approach that is based on the natural process of rainfall. If we divide the drainage basin into nine small cells, then D8 method can be used to determine the direction of runoff. In this method for each cell, we consider 8 neighboring cells in each direction and each cell has some height and distance from the adjacent cell according to these parameters, the slope is calculated and the runoff moves to the direction of the highest slope. After calculating the direction of the flow, a value is assigned to each cell that represents the number of cells that flow into that cell. Using this approach, the basin outlet point will be assigned the largest number. If the height of a cell is less than the neighboring cell, then it is observed as a hole and according to that, it should be filled [15]. Using the following equation, we generate our random solution:

$$F_i = l_b + R * (u_b - l_b) \tag{25.10}$$

The position of the neighboring solution is calculated using the given relation:

$$N = F_i + S * \delta \tag{25.11}$$

If the value of δ is small, then it can trap into the local optimum, and if the value of δ is large, then there is a higher probability that it finds near to the optimal solution so the balance between these two is very important. In this algorithm first, we search globally when we are near to the optimal solution then we search locally.

$$\delta = R * (F_{\text{rand}} - F_i) * (F_{\mu} - F_i) * W \tag{25.12}$$

In the above equation, the first term shows that flow F moves to random position (R). For the second term as the iteration increases, F is closed to F_μ and the Euclidean distance between these two is reduced to zeros. Therefore, the local search is down. The W in the third term is calculated as follows:

$$W = ((1 - (ii_{\max}))^{(2*S)}) * (R * (ii_{\max})) * R \tag{25.13}$$

There is a high variation in the values W that helps in escaping from local optimum. Flow velocity vector is calculated using the given relation:

$$V = S * \text{Slope} \tag{25.14}$$

The vector of the slope of ith flow relative to the jth neighbor is calculated using the following equation:

$$\text{Slope} = (\text{Cost}(F) - \text{Cost}(N_\mu))/(F_i - N_\mu) \tag{25.15}$$

The given equation is used to calculate the new position of the flow:

$$F_{\text{new}} = F_i + V * (F_i - N_\mu)/(F_i - N) \tag{25.16}$$

If the cost function of the new flow is less than the current flow, then it will move in the same direction otherwise it will move in the dominant slope direction that can be shown using the given equation:

$$\text{If Cost}(F_r) < \text{Cost}(F) : F_{\text{new}} = F_i + S * (F_r - F_i) \tag{25.17}$$

$$else F_{\text{new}} = F_i + 2S * (F_\mu - F_i) \tag{25.18}$$

Algorithm 1 Task Offloading using FDA

Require: Hyper-Parameters $(d, l_b, u_b, \alpha, \beta, i_{max})$
Randomly initialize a population (F) containing α number of offloading solutions Select the minimum cost solution F_μ from the population,
$F_\mu = \underset{i\in(1,\alpha)}{argmin}[Cost(F_i)]$
For $i = 1$ to i_{max}
Sample an n-dimensional random vector $(\mathcal{R})$ from a known uniform distribution and a random scalar $(\mathcal{S})$ from a known normal distribution as:
$\mathcal{R} \sim \mathcal{U}_{(0,1)}$
$\mathcal{S} \sim \mathcal{N}_{(0,1)}$
Set $i_r = i/i_{max}$ and calculate W as: $W = (1 - i_r)^{2\mathcal{S}} \cdot ((i_r\mathcal{R}) \times \mathcal{R})$
For each solution $F_i \in F$ **do**
Choose $n \sim \mathcal{U}_{(0,1)}$ and generate a random solution F_{rand}.
Calculate δ as: $\delta = n(F_{rand} - F_i) \times (F_\mu - F_i) \times W$
Create a set (N) of β neighbors as: $N = [F_i + n\delta : n \sim \mathcal{U}_{(0,1)}]$
Find the minimum cost neighboring solution (N_μ) as: $N_\mu = \underset{i\in(1,\beta)}{argmin}[Cost(N_i)]$
If $Cost(N_\mu) < Cost(F_i)$
Update velocity vector (V) as: $V = n \times \frac{Cost(F_i)-Cost(N_\mu)}{F_i-N_\mu}$
Update solution vector (F_i) as: $F_i^* \leftarrow F_i + V \times \frac{F_i-N_\mu}{F_i-N_i}$
else
Choose a random solution $(F_r \in F)$
If $Cost(F_r) < Cost(F_i)$ **then**
Update solution vector (F_i) as: $F_i^* \leftarrow F_i + r \times (F_r - F_i) : r \sim \mathcal{N}_{(0,1)}$
Else
Update solution vector (F_i) as: $F_i^* \leftarrow F_i + 2n \times (F_\mu - F_i) : n \sim \mathcal{U}_{(0,1)}$
End if
End if
End for
Update the population (F) using updated solutions (F^*) as:
$F_i \leftarrow \underset{i\in(1,\beta)}{argmin}[Cost(F_i), Cost(F_i^*)]$
End For

25.6 Result and Discussion

This experiment is performed using 3 types of network topology in which the first network topology has one ED, one FN, and one CS. The second one with one ED, three FNs, and two CSs. The third network topology we have taken one ED, five FNs, and two CSs. Using these 3 network topologies, we applied the FDA algorithm to perform task offloading in the FCA. In this, we have taken 10 to 100 tasks and performed 100 iterations. For different numbers of tasks, different costs like estimated cost, device cost, FN cost, and CS cost are shown in the graph. Device cost means the total cost when all the tasks run on the ED, FN cost means the total cost when all the tasks run on the FN, and CS cost means the total cost when all the tasks run on the CS. Here V is the task size and D is the data size.

	V	D
Range	[1,100]	[1,60]

Figure 25.4a demonstrates the decrease in the overall cost of offloading the task during iteration, and Fig. 25.4b depicts the variations in weight values during iterations. Figures 25.5a, 25.6a, and 25.7a demonstrate the increase in the overall cost of offloading with the increase in the number of tasks in network topology 1, 2, and 3, respectively, and Figs. 25.5b, 25.6b, and 25.7b depict the variations costs with estimated best cost in network topology 1, 2, and 3, respectively. The above figures show an illustration of the difference between the estimated best cost and other costs, where Est_{cost} shows the estimated best cost, D_{cost} is the device cost, E_{cost} is the minimum of all the FN costs, and C_{cost} is the minimum of all the CS costs. In all three experiments, our estimated cost and efficiency are better than all the other solutions.

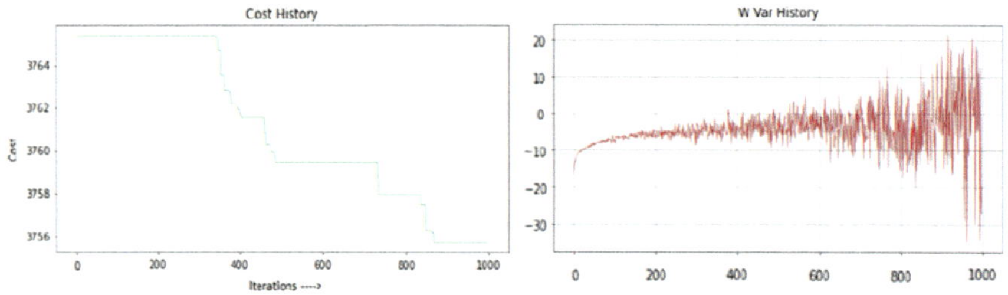

(a) Decrement in cost (b) Variation in weight

Fig. 25.4 Cost and weight graph

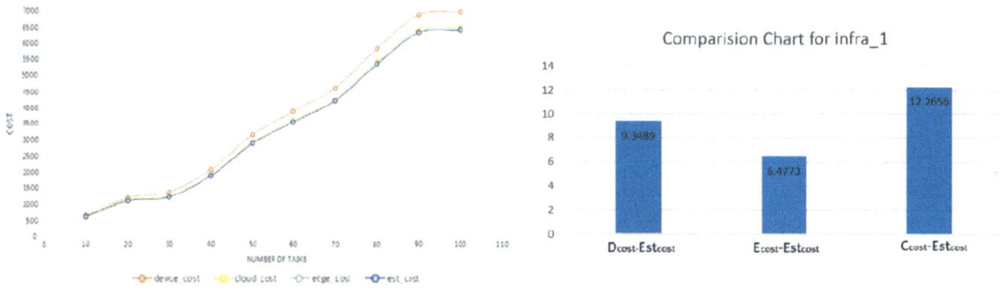

(a) Cost Graph for Network Topology 1

(b) Difference in costs and estimated best cost in network topology 1

Fig. 25.5 Cost graph and difference in cost and estimated best cost for topology-1

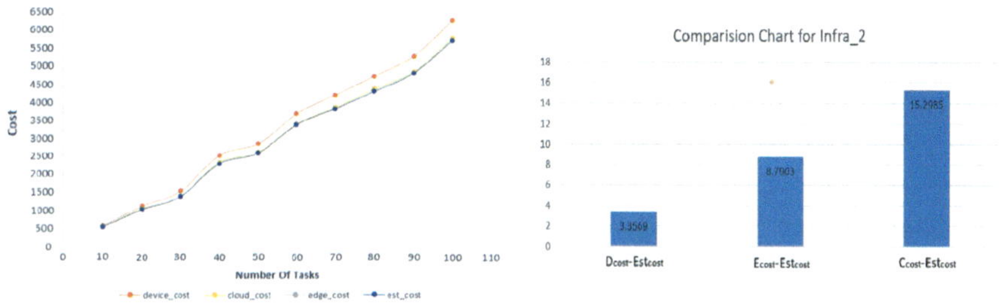

(a) Cost Graph for Network Topology 2

(b) Difference in cost and estimated best cost in network topology 2

Fig. 25.6 Cost graph and difference in cost and estimated best cost in topology 2

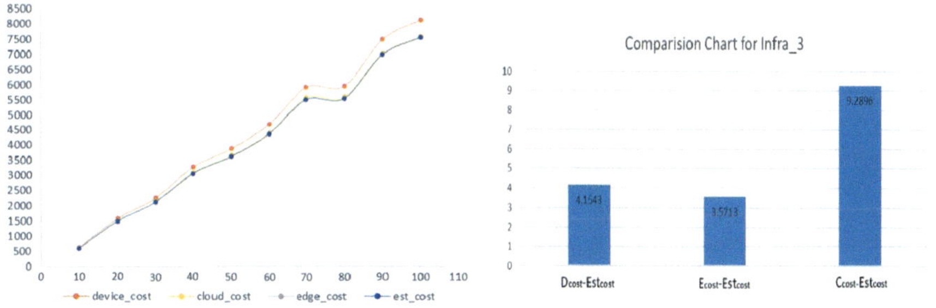

(a) Cost Graph for Network Topology 3

(b) Difference in cost and estimated best cost in network topology 3

Fig. 25.7 Cost graph and difference in cost and estimated best cost in topology 3

25.7 Conclusion

In this paper, we investigate the task offloading problem in the FCA. There are a lot of delay-sensitive applications and we have limited resources and battery life for the user equipment. So, task offloading in the FCA is one of the important solutions to this problem. In this paper, we applied task offloading in 3 types of workflows that contain one IoT device that generates different tasks and a different number of FNs and CSs in which we have to offload the task so that we minimize our cost. For this, we applied the FDA algorithm to minimize the computation time and we are getting better results. Our algorithm performance on average is 12% better than cloud-only, 6% better than Fog-only, and 10% better than IoT-only. In future work, we will consider task offloading problems in a more complex environment with the mobility of users.

25.8 Statements and Declarations

The authors declare that they have no known competing financial interests or personal relationships that could have separated to influence the work reported in this paper.

References

1. Shukla, P., Pandey, S.: MAA: multi-objective artificial algae algorithm for workflow scheduling in heterogeneous fog-cloud environment. J. Supercomput. 1–43 (2023)
2. Sun, J., Gu, Q., Zheng, T., Dong, P., Valera, A., Qin, Y.: Joint optimization of computation offloading and task scheduling in vehicular edge computing networks. IEEE Access **8**, 10466–10477 (2020)
3. Li, Z., Zhu, Q.: Genetic algorithm-based optimization of offloading and resource allocation in mobile-edge computing. Information **11**(2), 83 (2020)
4. Huynh, L.N., Pham, Q.V., Pham, X.Q., Nguyen, T.D., Hossain, M.D., Huh, E.N.: Efficient computation offloading in multi-tier multi-access edge computing systems: a particle swarm optimization approach. Appl. Sci. **10**(1), 203 (2019)
5. Liu, C.F., Bennis, M., Poor, H.V.: Latency and reliability-aware task offloading and resource allocation for mobile edge computing. In: 2017 IEEE Globecom Workshops (GC Wkshps), pp. 1–7. IEEE (2017)
6. Chen, M., Hao, Y.: Task offloading for mobile edge computing in software defined ultra-dense network. IEEE J. Sel. Areas Commun. **36**(3), 587–597 (2018)
7. Yang, L., Zhang, H., Li, M., Guo, J., Ji, H.: Mobile edge computing empowered energy efficient task offloading in 5G. IEEE Trans. Veh. Technol. **67**(7), 6398–6409 (2018)
8. Ma, S., Song, S., Yang, L., Zhao, J., Yang, F., Zhai, L.: Dependent tasks offloading based on particle swarm optimization algorithm in multi-access edge computing. Appl. Soft Comput. **112**, 107790 (2021)
9. Wang, J., Wu, W., Liao, Z., Sherratt, R.S., Kim, G.J., Alfarraj, O., et al.: A probability preferred priori offloading mechanism in mobile edge computing. IEEE Access **8**, 39758–39767 (2020)

10. Al-Habob, A.A., Dobre, O.A., Armada, A.G., Muhaidat, S.: Task scheduling for mobile edge computing using genetic algorithm and conflict graphs. IEEE Trans. Veh. Technol. **69**(8), 8805–8819 (2020)
11. Joshi, V., Patil, K.: Delay-energy aware task offloading and VM migration policy for mobile edge computing. Wirel. Pers. Commun. **123**(4), 3309–3326 (2022)
12. Zhu, X., Zhou, M.: Multiobjective optimized cloudlet deployment and task offloading for mobile-edge computing. IEEE Internet Things J. **8**(20), 15582–15595 (2021)
13. Shukla, P., Pandey, S., Hatwar, P., Pant, A.: FAT-ETO: fuzzy-AHP-TOPSIS-Based efficient task offloading algorithm for scientific workflows in heterogeneous fog–cloud environment. In: Proceedings of the National Academy of Sciences, India Section A: Physical Sciences, pp. 1–15 (2023)
14. Shukla, P., Pandey, S., Agarwal, D.: An efficient offloading technique using DQN for MEC-IoT networks. In: 2023 6th International Conference on Information Systems and Computer Networks (ISCON), pp. 1–7. IEEE (2023)
15. Karami, H., Anaraki, M.V., Farzin, S., Mirjalili, S.: Flow direction algorithm (FDA): a novel optimization approach for solving optimization problems. Comput. Ind. Eng. **156**, 107224 (2021)

Chapter 26
Comparison of Segmentation Algorithms for Extraction of Stone from Kidney X-ray Image

N. Nisha, Vijaypal Singh Dhaka, and Deepak Sinwar

Abstract Healthcare systems are gaining popularity because of the advancement of new technology aiding healthcare professional for diagnosis of disease from medical modalities. The detection and classification of disease using AI and its technologies are active research areas in the last decade. Radiologist usually detects stone from the kidney using different imaging like CT scan, MRI scan, and X-ray. For the detection of stone in kidney X-ray images, various segmentation and classification methods are used. These different classification and segmentation methods are used to reduce errors and time of healthcare professional in diagnosis of stone in X-ray image. In this paper, comparison of different algorithms like thresholding algorithm, contour-based algorithm, deep learning algorithms like U-Net, U-Net++, V-Net, V-Net++, and Transformer Net are used for segmentation of stone from kidney X-ray images. The thorough comparison of different deep learning architectures in diagnosing stone from kidney X-ray images is also made in this paper. Most of the papers work on classification of stone from different medical modalities before segmentation which cause data leakage problem. In this research paper, comparisons of different segmentation algorithms help in achieving higher accuracy and reduce the data leakage problem.

N. Nisha (✉)
Department of Computer and Communication Engineering, Manipal University Jaipur, Jaipur, India
e-mail: vasudeva.nisha1@gmail.com

V. S. Dhaka
Director of School of Computing and Intelligent Systems, Department of Computer and Communication Engineering, Manipal University Jaipur, Jaipur, India

D. Sinwar
Department of IoT and Intelligent Systems, Manipal University Jaipur, Jaipur, India

P. Nanda et al. (eds.), *Intelligent System and Data Analysis*, Smart Innovation, Systems and Technologies 398, https://doi.org/10.1007/978-981-97-5200-3_26

26.1 Introduction

The kidney stone disease is one of the most frequent urological diseases that affect on the 1–13% of population of the world. This disease affects on one people among ten people in the world. Stone in kidney is formed by formation of crystal in urine due to chemical imbalance or family hereditary. The stones are made up from calcium oxalate, struvite uric acid, and cystine stones. There are different types of stone for kidney, ureter, and bladder area, i.e., nephrolithiasis stone form in kidney, cystolithiasis stone forms in bladder, and ureterolithiasis stone forms in ureter. These stones are formed in human body due to lifestyle, diet, and abnormalities of metabolism. Due to stone formation, patients are suffering from other diseases also, like discomfort, back pain, fever, and nausea [1]. It is important to provide appropriate medical dialysis of kidney stone at early stage. Biomedical imaging techniques are very important for diagnosis and research process. Medical modalities show all the interior organs in the body for diagnosis purpose. The most reliable imaging techniques for diagnosing kidney stone are computed tomography (CT), ultrasonography, and X-ray. Most of the hospitals use CT scan images of kidney stone due to the three-dimensional structure of these images, but ultrasonography and X-ray imaging are also used for detection of stone because of low cost and availability of machine in all region, i.e., urban as well as rural as compared to CT scan. For segmentation and classification of stone from all these imaging modalities, different artificial intelligence and computer vision algorithms have been applied which play an important role [2]. Among all these algorithms, segmentation of medical images is very important and difficult task for detection of stone. A lot of research has been done for segmentation of stone in CT, ultrasound, and MRI images. But the segmentation of stone in KUB X-ray images is under-explored. The segmentation of stone from the X-ray image is difficult for medical experts due to 2D structure of X-ray images. Sometimes small size stones are not visible in X-ray images, and some stones are not visible in X-ray images. The motivation of this paper is to find out best algorithm for segmenting the stone from X-ray images that take less computational time and gives improved accuracy. The main contribution of paper is highlighted as follows:

1. Comprehensive discussion of traditional and deep learning models.
2. Prepare the dataset comprising labeled kidney X-ray images.
3. Comparative analysis of the different techniques for stone segmentation and finding out the best algorithm.
4. Implementation and result analysis of all techniques based on evaluation matrix.

The rest of the paper is organized as follows. Section 26.2 discusses the existing work for segmentation of stone from medical modalities. The architecture of traditional methods and deep learning methods for the segmentation of stone from the kidney X-ray images are implemented in the Sect. 26.3. The comparison analysis from the performance matrix is explained in Sect. 26.4. The whole paper is discussed in Sect. 26.5. Finally, Sect. 26.6 concludes this paper.

26.2 Related Work

There are different segmentation techniques related with kidney like kidney segmentation, cyst area segmentation, kidney tumor segmentation, and kidney cortex segmentation. As the identification and detection of stone in particular region of kidney is very time-consuming task by radiologist. For detecting the region of interest advanced automatic segmentation tools would be very supportive for radiologist in less time and accuracy. Artificial intelligence-based different segmentation methods like thresholding-based, contour-based, and other deep learning algorithms achieve good results and accuracy with detection of stone from X-ray image. Researcher used different medical image processing techniques for study stone detection from kidney X-ray images. Author proposed fully connected convolutional neural network for classification of stone in medical images [3]. Another author extends this fully connected architecture with other deep learning algorithms. These deep learning algorithms can use for classifying the data based on healthy or stone category. Author [4] developed computer-aided diagnosis algorithm with deep learning architecture for detection of stone from kidney X-ray images automatically and evaluation model performance metric. For detection of stone using CAD algorithms, all dataset collected from different sources is divided into training and testing datasets. Author [5] proposed a new method, i.e., combination of thresholding and deep learning algorithms for detection of object in X-ray images. Author et al. [6] developed vino methods for classification of composition of kidney stones by dual energy imaging methods. In [7], author proposed a method in which U-Net algorithm is used for segment the organ in KUB X-ray images.

26.3 Methodology

In this paper, segmentation methods like thresholding, contour-based, U-Net, U-Net++, V-Net, V-Net++, and Transformer Net are used for segment the stone from kidney X-ray images. These methods trained on X-ray images dataset and calculated its performance matrix like accuracy, loss value, and recall value. The performance matrix result assists in finding of best method by comparing accuracies of all the methods.

26.3.1 Dataset

The dataset for this kidney X-ray images collected from different hospitals of India for study purpose. This dataset contains 1324 kidney X-ray images out of which 662 are in healthy category and remaining 662 are in stone category. All the images are in JPEG, PNG, or JPG format. After collecting all images, created mask image manually using label-studio tool and saved all image as PNG image. The mask images are now added into dataset of kidney X-ray images, i.e., 662 images and 662 masks in healthy category, 662 images, and 662 masks in stone category.

26.3.2 Preprocessing

All the images of the dataset are of different format and different size, so this will show error when it reads through different methods as an input at the time of training. For solving this problem, all images in dataset are converted into same size of 256 × 256. For preprocessing, in the first step converted all the images into same image format, here all mask images are in PNG format and original images are in different image format. In the second step, all the images, original as well as mask images converted into grayscale image format for saving time. Now all images are of (256 × 256 × 3) converted into (256 × 256 × 1).

Figure 26.1 shows the original image of size (256 × 256 × 3) of healthy and stone images, grayscale images of the sizes (256 × 256 × 1), and mask images of healthy and stone.

26.3.3 Segmentation Techniques for Stone Detection

There are number of methods for segmentation of stone from the medical modalities. But in this paper, segmentation techniques are applied on X-ray images for segmentation of stones like thresholding technique, contour-based method, U-Net, U-Net++, V-Net, V-Net++, and Transformer Net deep learning algorithms.

26.3.3.1 Thresholding Methods

Thresholding methods are used for segmentation of stone in kidney X-ray images. There are number of thresholding methods for segmenting stone from X-ray image [8].

(1) Global thresholding

Global thresholding is simple thresholding segmentation method which is used for segmentation of stone from X-ray image. The goal of this method is to separate stone from the surrounding background image based on thresholding value as shown in Fig. 26.2.

(2) Adaptive Thresholding

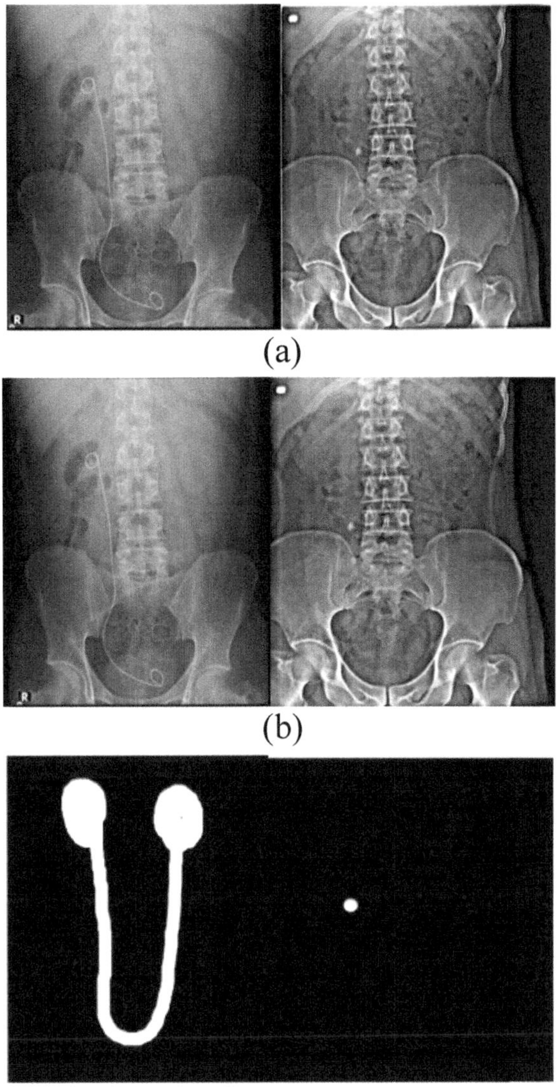

(c)

Fig. 26.1 **a** Original image of healthy and stone, **b** grayscale images of healthy and stone, and **c** mask images of healthy and stone

Adaptive thresholding method is also used for segmentation of stone from the X-ray image. In this method, first divides the image into smaller regions and then thresholding value applies on each region individually for better results as shown in Fig. 26.3.

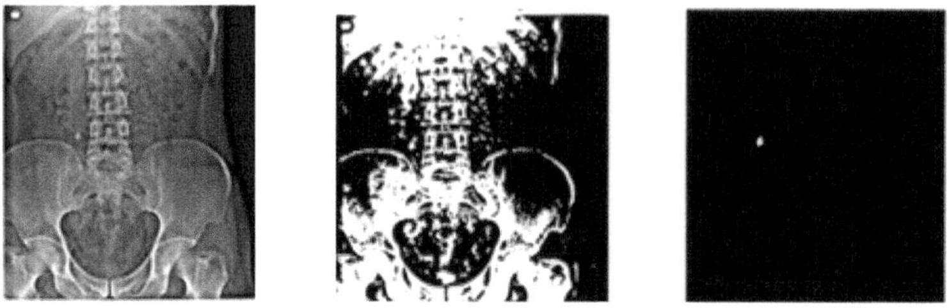

Fig. 26.2 Segmentation of stone using global thresholding method

Fig. 26.3 Segmentation of stone using adaptive thresholding

(3) Edge-based Thresholding

Edge-based thresholding is a segmentation technique that segment the stone from the background kidney X-ray image by detecting of edge. The edges represent the boundaries of object in the image. After detection of edge in the image, thresholding value is applied on detected edges for segment stone from kidney X-ray image as shown in Fig. 26.4.

(4) Region-based thresholding

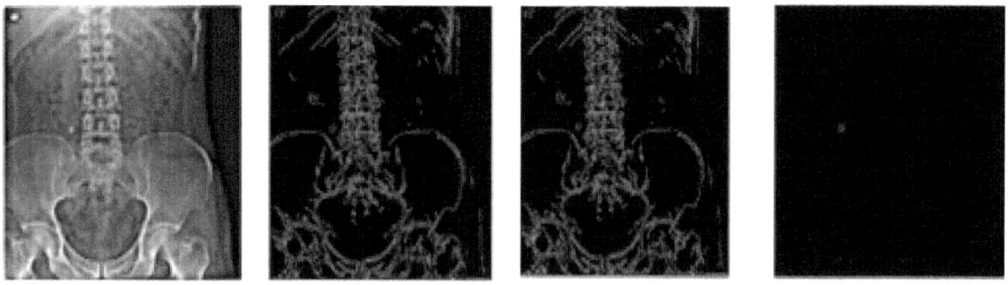

Fig. 26.4 Segmentation of stone using edge-based thresholding

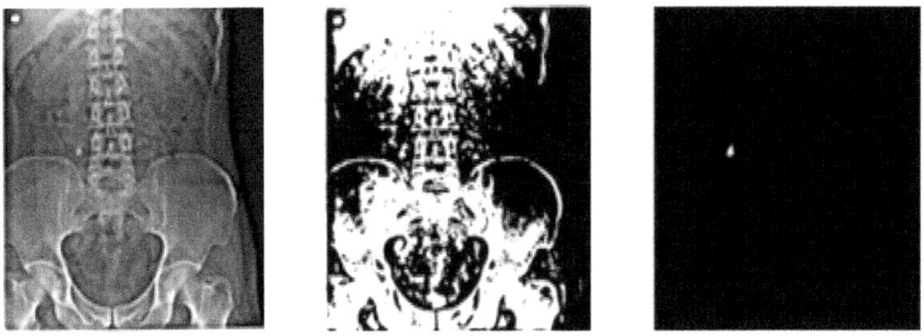

Original X-ray image   Binary Mask   Segmented Image

Fig. 26.5 Segmentation of stone using region-based thresholding

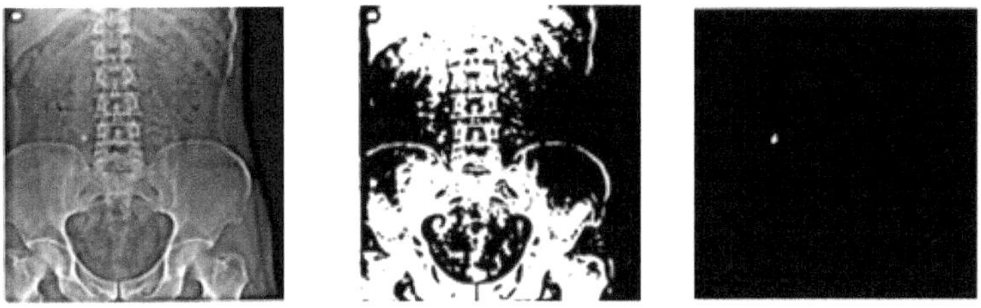

Original X-ray image   Thresholded Image   Segmented Image

Fig. 26.6 Segmentation of stone using morphological thresholding

Region-based thresholding method is another segmentation technique that identifies pixel of same properties and grouped them for considering as a kidney stone region or background region. The technique basically works on similarity of pixel in particular region, pixel intensities then thresholding value is applied on region for extracting the stone from the X-ray image as shown in Fig. 26.5.

(5) Morphological thresholding

Morphological thresholding technique is also the segmentation technique that uses morphological operation for segment stone from kidney X-ray image. It identifies the area of interest, size, and similarity of pixel for segment stone from the background image. After calculating the pixel, thresholding value is applied for extracting stone from the image with accurate results as shown in Fig. 26.6.

26.3.3.2 Contour-Based Segmentation

Contour-based technique is another segmentation technique that uses contour analysis and contour detection for segmenting stone from the kidney X-ray image [9].

This method segments the stone from background image by detecting the boundaries of stones in KUB area. The boundaries of the stone are calculated by edge detector algorithms after that analysis of image is done for getting accurate results.

26.3.3.3 Deep Learning-Based Segmentation

There are number of algorithms available for stone segmentation in KUB X-ray images like U-Net, U-Net++, Transformer Net, SegNet, DeepLab, V-Net, FCN, ResNet. But in this paper, segmentation techniques are used for extraction of stone from X-ray images. In all these algorithms, different layers consist of convolution block, encoder, and decoder block. These encoder and decoder block consist of number of filters, pooling layer and skip connection for segmenting stone accurately. After segmentation, performance matrix is calculated for finding best algorithm for stone segmentation.

26.4 U-Net Algorithm

The U-Net algorithm is a deep learning segmentation algorithm and its architecture is fully connected convolutional network architecture that is used in various medical imaging applications [10]. This is a U-shaped architecture which consists of encoding and its corresponding decoding path. There are three steps in U-Net (1) encoder (2) bottleneck (3) decoder. At the encoder step, the input image is passed through a series of convolutional layer followed by ReLU activation function and max pooling for reduce the dimension of image and for increasing the features from the input image. The bottleneck step is at the bottom of U-Net which contains higher number of features and information about the input image. Whereas the decoder step concatenates the features from the encoder block using skip connection for increasing number of features at every stage for accurate image segmentation [11, 12].

Figure 26.7 shows the architecture of U-Net, in this architecture input image pass into two-dimensional convolutional layer with 32 convolution filters, 3×3 kernel size followed by size of 2×2 max pooling layer. After that input image pass into the process of encoder and decoder for finding output. The output layer uses the kernel of size 1×1, number of classes and sigmoid function as activation function. After defining the model, training of model is done for finding out training parameters, loss, accuracy, and validation results.

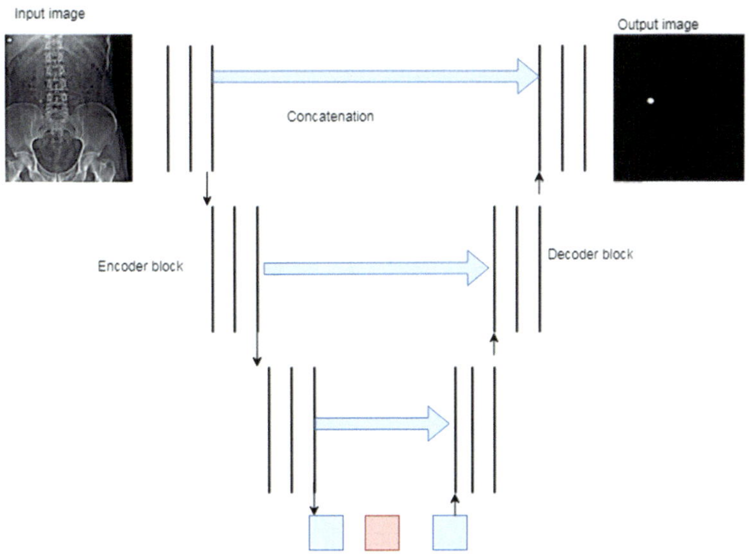

Fig. 26.7 U-Net architecture

26.5 U-Net++

U-Net++ architecture is extension of U-Net architecture as shown in Fig. 26.8. This nested architecture is created by recursively connecting the skip connection of U-Net. This process allows the model to extract more features from the image of different resolution and scale, which helps in accurately segmenting the object from the image [13]. This has same encoder and decoder block as in U-Net but there is nested connection between the encoder and decoder block at different resolutions. This nested feature in encoder and decoder block gives more accurate features and accuracy.

26.6 V-Net

This architecture is also the extension of U-Net architecture. This is segmentation algorithm mainly used in medical field based on the concept of fully connected architecture. The main aim of this algorithm is to label and identify the region of interest in data automatically. This architecture also consists of encoder and decoder block like in U-Net, encoder part consists of multiple convolutional layers for extraction of features from the image whereas the decoder part consists of transpose of convolution for finding the output accurately. In between the encoder and decoder block, skip connections are used for allowing the network to use features from the different levels [14]. After defining the model architecture, training of model is required for

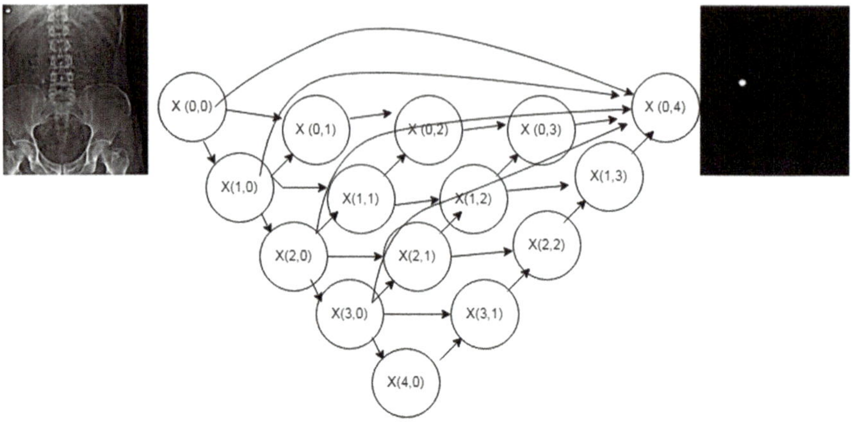

Fig. 26.8 Architecture of U-Net++

finding out loss and accuracy of defined model. The V-Net architecture is shown in Fig. 26.9.

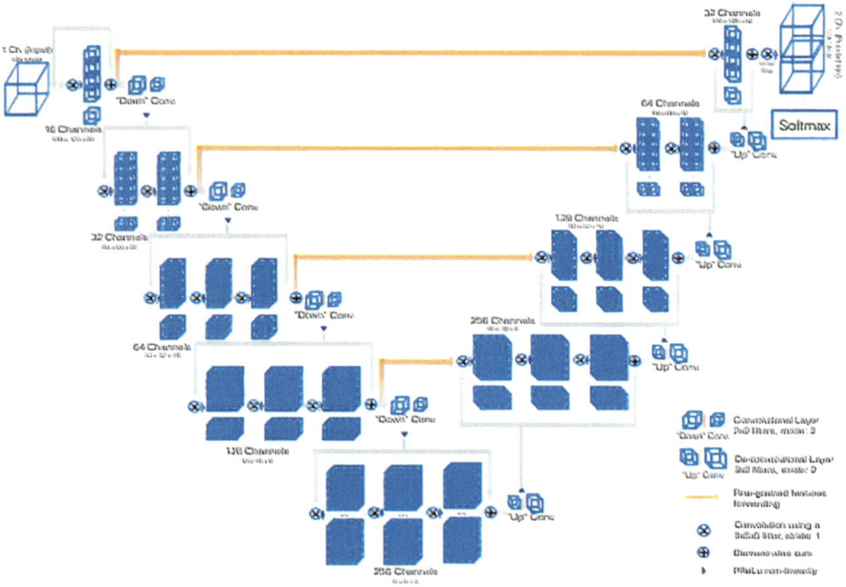

Fig. 26.9 Architecture of V-Net [14]

26.7 V-Net++

This architecture is the new version of V-Net architecture which has the features of V-Net architecture with some additional features to improve the accuracy and efficiency of the model [3]. In this model, multiple resolution approach is used in the skip connections at each level for extracting more features of different resolution in the image. After feature extraction, path aggregation and attention gate are used for generating the compressive representation and suppressive noise in the image for achieving better results.

26.8 Transformer Net

The Transformer Net is another segmentation algorithm consisting of encoder and decoder blocks same as previous architecture. The important part of this model is image patching, positional encoding, and its self-attention mechanism. In image patching, input image is divided into small number of patches for collecting more information of feature in the image. After image patching, position encoding is done for encoding the patch for capturing feature within the image. Then self-attention process is required for extracting relationship between the different patch in the image and encoding the information of the image exactly. After completion of this process, fully connected feed-forward network is used in transforming and representing the features [3].

26.9 Result and Analysis

Comparison of different methods and algorithms for semantic segmentation as thresholding-based method, contour-based method, and number of deep learning algorithms like U-Net, U-Net++, V-Net, V-Net++, and Transformer Net are discussed in this paper. This section compares the performance of model explored in this paper. On training of these entire models on the kidney dataset and obtained the following results as loss, recall, accuracy, and precision values. Tables 26.1, 26.2, and 26.3 show the performance of different segmentation algorithms. Figures 26.10, 26.11, 26.12, 26.13, 26.14, 26.15, and 26.16 show the performance graph of different segmentation algorithms.

Table 26.1 Performance comparison of various thresholding techniques

Method	Accuracy	Precision	Recall	$F1$-score
Global thresholding	0.9987	1.0	0.3531	0.5219
Adaptive thresholding	0.9989	1.0	0.4124	0.584
Edge-based thresholding	0.9984	1.0	0.1327	0.2344
Region-based thresholding	0.9988	1.0	0.3813	0.5521
Morphological thresholding	0.9987	1.0	0.3672	0.5371

Table 26.2 Performance of contour-based segmentation

Method	Precision	Recall	$F1$-score	Accuracy
Contour-based	0.8829	0.9318	0.9058	0.93185

Table 26.3 Comparison of accuracy analysis

Model	Accuracy	loss	Validation accuracy	Validation loss
U-Net	0.6614	0.1930	0.1008	0.5102
U-Net++	0.9583	nan	1.0000	nan
V-Net	0.6614	0.1938	0.1008	0.5108
V-net++	0.9034	0.2590	0.7708	0.6080
Transformer Net	0.9588	0.6102	0.9768	0.3012

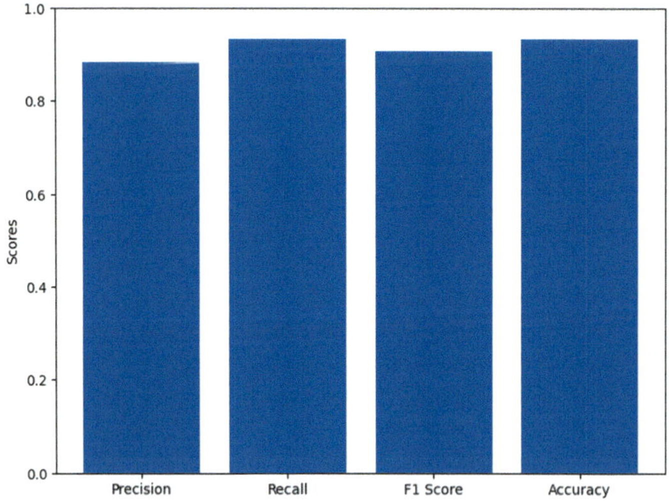

Fig. 26.10 Performance graph of contour-based segmentation

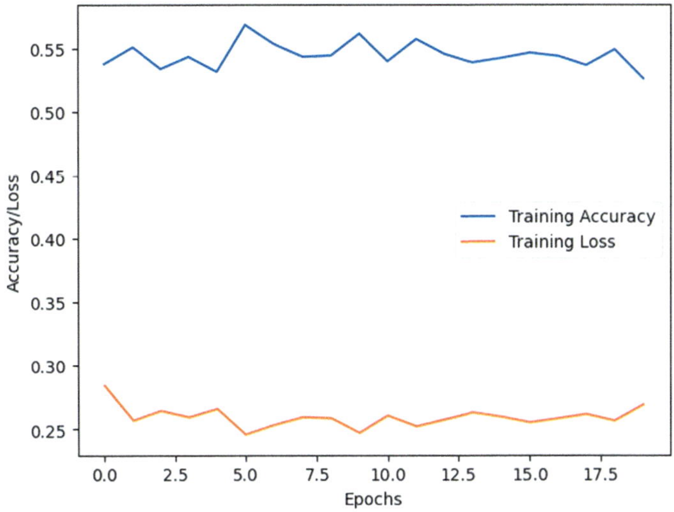

Fig. 26.11 Graphical representation of training accuracy/loss versus epochs of U-Net

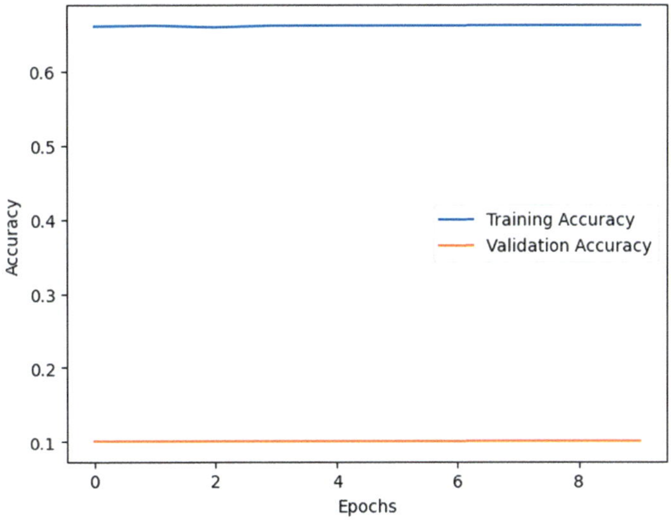

Fig. 26.12 Graphical representation of accuracy of U-Net

26.10 Discussion

In this paper, different segmentation algorithms are applied on the same dataset for finding out the best algorithm. For finding out performance, different functions are used like precision, recall, dice coefficient, and accuracy. In all models, dataset

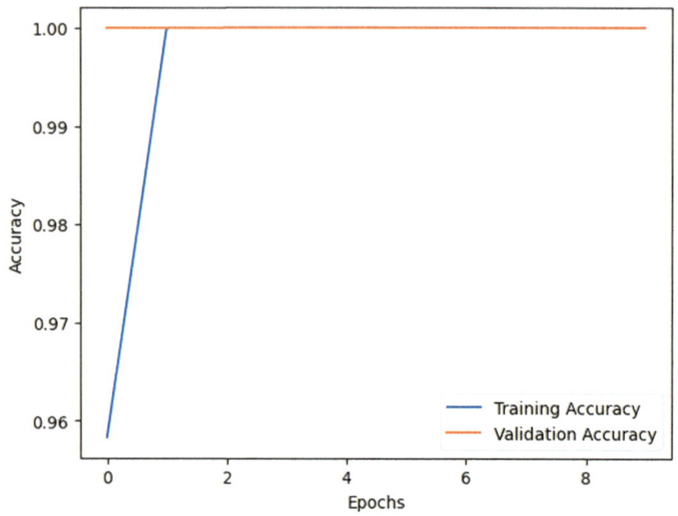

Fig. 26.13 Graphical representation of accuracy of U-Net++

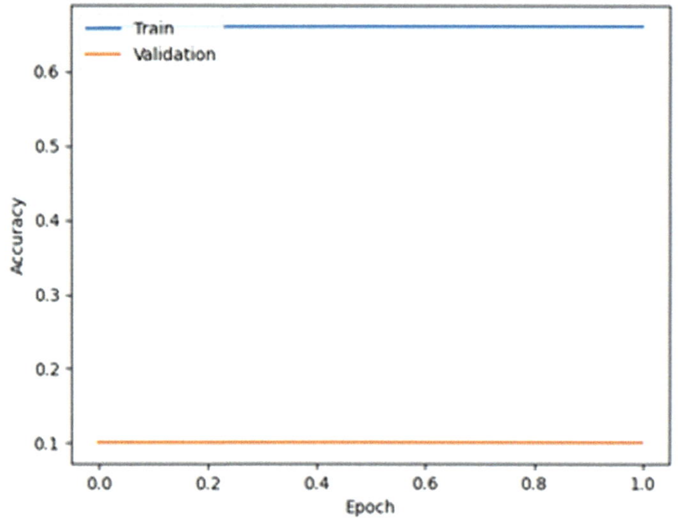

Fig. 26.14 Graphical representation of accuracy of V-Net

is divided into training and test dataset using train- test split method in the ratio of 80:20 after dividing the dataset the model is trained on training dataset and test dataset is used for evaluating the performance of the model. The training of the model is depending upon the number of epochs, activation function, batch size, optimizer, and loss function. The algorithms performance analyses are shown into Tables 26.1, 26.2, 26.3 and Figs. 26.10, 26.11, 26.12, 26.13, 26.14, 26.15, and 26.16. The performance

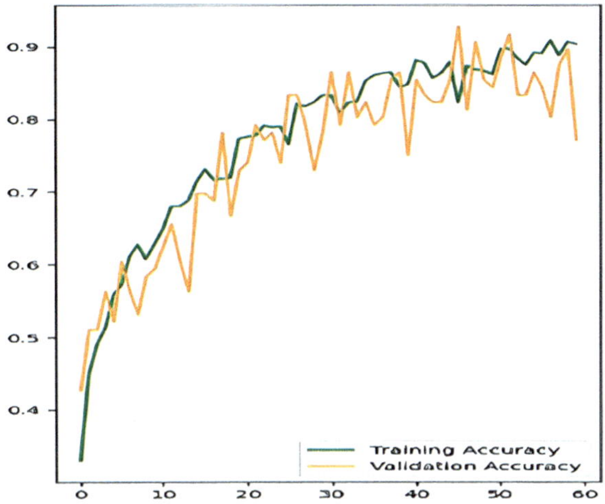

Fig. 26.15 Graphical representation of accuracy of V-Net++

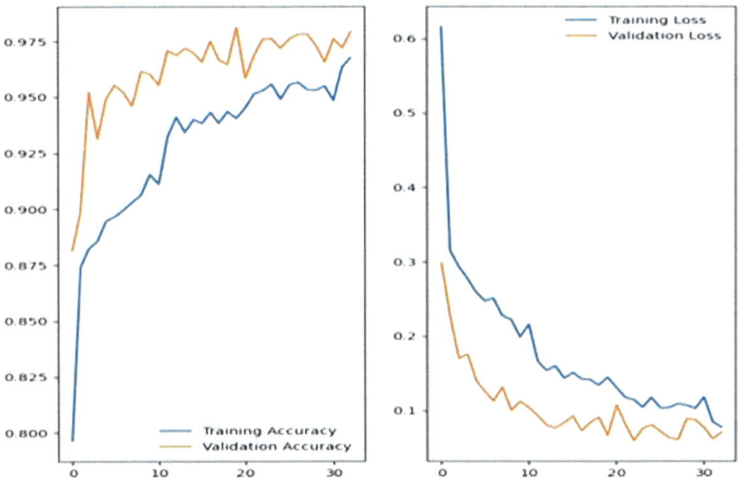

Fig. 26.16 Graphical representation of Transformer Net

table shows the results of different algorithm, by which we can find out the best algorithm for the segmentation of stone in kidney X-ray image.

26.11 Conclusion

The segmentation algorithms are used to segment stone from the kidney X-ray image by dividing the whole image into small pixels and extracting all the features from the stone images. These algorithms help professionals in finding out the stone in particular region in less time with more accuracy. Segmentation plays important role in classification of stone also as after performing segmentation the accuracy of classification results higher. The thresholding method and contour-based methods are traditional segmentation techniques whereas the other models are the deep learning models which help in segmenting stone automatically in less time. By observing the results of these algorithms on the same dataset, the U-Net++ deep learning model gives higher accuracy as compared to others for segmentation of stone from the kidney X-ray images.

References

1. Li, D., Xiao, C., Liu, Y., Chen, Z., Hassan, H., Su, L., Huang, B.: Deep segmentation networks for segmenting kidneys and detecting kidney stones in unenhanced abdominal CT images. Diagnostics **12**(8), 1788 (2022)
2. Chavan, M., Varadarajan, V., Gite, S., Kotecha, K.: Deep neural network for lung image segmentation on chest X-ray. Technologies **10**(5), 105 (2022)
3. Salpea, N., Tzouveli, P., Kollias, D. Medical image segmentation: a review of modern architectures. In: European Conference on Computer Vision, pp. 691–708. Springer Nature Switzerland, Cham
4. Kobayashi, M., Ishioka, J., Matsuoka, Y., Fukuda, Y., Kohno, Y., Kawano, K., et al.: Computer-aided diagnosis with a convolutional neural network algorithm for automated detection of urinary tract stones on plain X-ray. BMC Urol. **21**(1), 1–10 (2021)
5. Ge, Y., Zhang, Q., Sun, Y., Shen, Y., Wang, X.: Grayscale medical image segmentation method based on 2d&3d object detection with deep learning. BMC Med. Imaging **22**(1), 33 (2022)
6. Kahani, M., Tabrizi, S.H., Kamali-Asl, A., Hashemi, S.: A novel approach to classify urinary stones using dual-energy kidney, ureter and bladder (DEKUB) X-ray imaging. Appl. Radiat. Isot. **164**, 109267 (2020)
7. Rani, G., Thakkar, P., Verma, A., Mehta, V., Chavan, R., Dhaka, V.S., et al.: KUB-UNet: segmentation of organs of urinary system from a KUB X-ray image. Comput. Methods Programs Biomed. **224**, 107031 (2022)
8. Bhargavi, K., Jyothi, S.: A survey on threshold based segmentation technique in image processing. Int. J. Innov. Res. Dev. **3**(12), 234–239 (2014)
9. Hemalatha, R., Thamizhvani, T., Dhivya, A.J.A., Joseph, J.E., Babu, B., Chandrasekaran, R.: Active contour based segmentation techniques for medical image analysis. Med. Biol. Image Anal. **4**(17), 2 (2018)
10. Ronneberger, O., Fischer, P., Brox, T.: U-net: Convolutional networks for biomedical image segmentation. In: Medical Image Computing and Computer-Assisted Intervention–MICCAI 2015: 18th International Conference, Munich, Germany, October 5–9, 2015, Proceedings, Part III 18, pp. 234–241. Springer International Publishing (2015)
11. Preedanan, W., Kumazawa, I., Kondo, T., Junichiro, I.: Urinary stones segmentation in abdominal X-ray images based on U-Net deep learning model and data augmentation techniques. In: 2020 IEEE 5th International Conference on Signal and Image Processing (ICSIP), pp. 118–123. IEEE (2020)

12. Preedanan, W., Suzuki, K., Kondo, T., Kobayashi, M., Tanaka, H., Ishioka, J., et al.: Urinary stones segmentation in abdominal X-Ray images using cascaded U-Net pipeline with stone-embedding augmentation and lesion-size reweighting approach. IEEE Access **11**, 25702–25712 (2023)
13. Zhou, Z., Rahman Siddiquee, M.M., Tajbakhsh, N., Liang, J.: Unet++: a nested u-net architecture for medical image segmentation. In: Deep Learning in Medical Image Analysis and Multimodal Learning for Clinical Decision Support: 4th International Workshop, DLMIA 2018, and 8th International Workshop, ML-CDS 2018, Held in Conjunction with MICCAI 2018, Granada, Spain, September 20, 2018, Proceedings, vol. 4, pp. 3–11. Springer International Publishing (2018)
14. Milletari, F., Navab, N., Ahmadi, S.A.: V-net: fully convolutional neural networks for volumetric medical image segmentation. In: 2016 Fourth International Conference on 3D Vision (3DV), pp. 565–571. IEEE (2016)

Chapter 27
Sentiment Classification in Code-Mixed Indo-Aryan Languages: A Transformer-Based Survey

Saikat Roy and Jatinderkumar R. Saini

Abstract Sentiment classification (SC) of user-generated comments (UGC) on social media and e-commerce websites written in code-mixed languages is a challenging task in natural language processing (NLP). In current days, Transformer-based models with attention mechanism are achieving state-of-the-art (SoTA) results for NLP tasks. When many authors have explored NLP tasks for code-mixed Dravidian languages using Transformer models, the languages belonging to the Indo-Aryan language family still have not been addressed to a large extent. The present work focuses on SC task on eight code-mixed Indo-Aryan languages such as Assamese–English, Bengali–English, Hindi–English, Marathi–English, Nepali–English, Odia–English, Punjabi–English and Urdu–English. In this present survey, we have studied in detail the existing code-mixed datasets of Indo-Aryan languages which have addressed SC tasks using Transformer-based models, the evolution of Transformer-based models for Indian Languages, research progress, analysis, challenges and research scopes during the evaluation of Indo-Aryan code-mixed dataset using Transformer-based models.

27.1 Introduction

In recent years, due to a surge in social media platforms such as Twitter, Facebook and other microblogging websites, people often like to communicate their feelings, emotions, hates, sarcasm about product, movies and foods on social media platforms in Unicode (monolingual) text [1, 2]. Advanced research on NLP tasks in English has already been explored to large extent. People living in bilingual and multilingual societies like India, it is common phenomenon to mix (or switch) linguistic

S. Roy
Department of Computer Science, National Defence Academy, Pune, India

J. R. Saini (✉)
Symbiosis Institute of Computer Studies and Research, Symbiosis International (Deemed University), Pune, India
e-mail: saini_expert@yahoo.com

P. Nanda et al. (eds.), *Intelligent System and Data Analysis*, Smart Innovation, Systems and Technologies 398, https://doi.org/10.1007/978-981-97-5200-3_27

units of different languages (codes) at different levels such as intra-sentence level, inter-sentence level or word level/morpheme level [3] code-mixing/code-switching. Though 'code-mixing' refers to the mixing of codes in the same sentence written in a single script (intra-sentential), 'code-switching' refers to the mixing of codes in two different sentences written in a single script. Some papers have used code-mixing and code-switching interchangeably. An example of code-mixing and code-switching is given below written in Hindi–English Romanized script.

Code-Mixing: 'Is Mobile ka battery backup bekar he' (Written in Romanized script, containing Hindi and English language). Translation in English: 'Battery of this mobile is worthless'.

Code-switching: 'This mobile screen is superb. Magar is ka dam bahut kam he' (Written in Romanized script, one sentence in English language and another sentence Hindi language). Translation in English: 'This mobile screen is superb. But the price is very less'.

There is possibility that mixing of several types of scripts (Mixed Script) where native script (Devanagari script, Eastern Nagari script, etc.) in Unicode can be mixed with Latin (English) script, Romanized script. This mixing of scripts is also termed as code-mixing by some authors. So code-mixing is the umbrella term under which code-switching, mixed scripts might be considered. Many authors have studied the reasons behind the appearance of code-mixing such as lexical needs, linguistic motivation and emotional needs [4–6]. The frequency of usage code-mixing types in social media has been studied where it has been found inter-sentential switching occurs more than intra-sentential switching [7, 8]. Due to informal nature of writing system of code-mixed languages, it incurs following challenges to NLP tasks such as language identification, sentiment classification and hate speech detection:

- Since there is no standard writing system, spelling variations in Romanized text are very common challenge, e.g., 'ye mobile bahut beautiful he' is written in code-mixed Romanized script in Hindi-English language. Translation in English, 'This mobile is very beautiful'. 'bahut' word is in Hindi language which means 'very' in English. As there is no standard spelling rule 'bahut' may be found as 'bahot', 'bahott', etc., in different comments/posts.
- Understanding repetition of words or letters for expressing exaggeration of emotion and normalization of words such as 'Is mobile ka picture quality is superbbbbbbb!!!' is written in Hindi–English code-mixed language. Translation in English, 'Picture quality of this mobile is superb'.
- Understanding the context of unconventional acronyms such as 'gd ni8', 'Tx' which means 'good night', 'Thanks' respectively in English and SC task of short length, incorrect grammatical posts.
- Extracting sentiments embedded in emojis and emoticons given in posts/comments.

Indo-Aryan languages, which are a branch in Indo-Iranian languages in Indo-European language family, consist of languages such as Assamese, Bengali, Gujrati, Hindi, Kashmiri, Marathi, Maithili, Nepali, Odia, Punjabi, Sindhi, Urdu, etc. In

Table 27.1 Comparison of existing code-mixed surveys with present work

	[11]	[14]	[13]	[12]	Present work
Number of code-mixed Indian languages	03	01	04	03	08
Models considered	Lexicon-based, ML-based and deep learning-based	ML-based, DNN-based	Lexicon-based, ML-based	ML-based, DNN-based	Transformer-based Models
NLP Tasks	Sentiment classification	Hate speech Detection	Sentiment classification	Sentiment classification	Sentiment classification

India, Indo-Aryan languages are spoken by around 900 million people.[1] A large number of research works have already been carried out with respect to sentiment classification task in Indo-Aryan languages using Transformer-based models [9, 10] compared to code-mixed Indo-Aryan languages. So less explored Indo-Aryan code-mixed languages have motivated us to make a Transformer-based survey. Code-mixed Dravidian languages (Kannada, Malayalam, Tamil, Telugu) datasets using Transformer-based model have been explored largely through NLP contest by Forum of Information Retrieval Evaluation (FIRE)-2020 and FIRE-2021. So, the purpose of this survey is significant for future research community in manifolds. We have chosen Transformer-based models because this is the state-of-the-art language model to solve NLP tasks such as sentiment classification and hate speech detection.

There are existing surveys on code-mixed Indian languages such as [11] explored on the progress of research on sentiment classification task of code-mixed Indian Languages such as Hindi–English, Tamil–English and Bengali–English languages. The survey of [12] was extension to the survey conducted by [11] where different machine learning (ML) techniques and deep neural network (DNN) models were discussed and results on sentiment classification in code-mixed ILs were summarized. Mahadzir et al. [13] reviewed sentiment classification task on code-mixed ILs and non-ILs as well as they highlighted challenges and future scope of research related to sentiment classification task. Rani et al. [14] focused on details of dataset used for hate speech detection of code-mixed Hindi–English text, techniques used, performances of models and corresponding challenges. The present work focuses survey on Transformer-based model for evaluating sentiment classification on eight code-mixed Indo-Aryan languages such as Assamese–English, Bengali–English, Hindi–English, Marathi–English, Nepali–English, Odia–English, Punjabi–English and Urdu–English. Table 27.1 describes how our work is different from existing survey works in sentiment classification task in ILs.

The rest of the paper is organized as follows. In Sect. 27.2, we have discussed details of presently available code-mixed datasets in Indo-Arya languages for SC

[1] https://en.wikipedia.org/wiki/Indo-Aryan_languages.

task. In Sect. 27.3, we have discussed in brief about different state-of-the-art Transformer models. In Sect. 27.4, we have discussed sentiment classification addressed by different authors on Indo-Aryan languages using Transformer-based models and in Sect. 27.5, we have concluded our survey.

27.2 Datasets for Code-Mixed Indo-Aryan Languages

In this section, we have described details of code-mixed datasets for sentiment classification in Indo-Aryan languages. To the best of our knowledge, we have found the following code-mixed datasets which have been evaluated by different authors using Transformer-based models. Except the following datasets, there are other code-mixed datasets in Indo-Aryan languages, but we have not discussed them in this present study as they have not been evaluated using Transformer-based models.

27.2.1 Bengali–English, Hindi–English and Urdu–English

SAIL-2017 dataset [15]**.** This dataset has been obtained from shared task of Sentiment Analysis of Indian Languages (SAIL) at International Conference on Natural Language Processing (ICON)-2017 for code-mixed Bengali–English and Hindi–English language pair. Authors had collected tweets from twitter using Twitter4j API[2] and later manually removed spam, incomplete, unknown language tweets. Comments/posts are in Romanized Script. SAIL-2017 dataset for both Bengali–English and Hindi–English is sentiment class-wise moderately distributed highly code-mixed corpus. Complexity (Code-Mixing Index (CMI)) of code-mixing of the entire Bengali–English corpus is 135.34 and code-mixing of entire Hindi–English corpus is 113.13 [16]. The higher the CMI value, the higher is code-mixing complexity. There are 24.72% English tokens, 52.56% of Romanized Bengali tokens, nearly 21% Universal token and 1% of emoticons in Bengali–English corpora and there are 39.03% English tokens, 41.22% of Romanized Hindi tokens, nearly 20% Universal token and negligible amount emoticons in Hindi–English corpora [17]. Dataset statistics has been described in Table 27.2.

Hindi–English Code-Mixed (HECM) dataset [18]**.** Author collected sentences (3879 sentences) from popular Facebook pages of Indian action, and politician and then manually annotated the dataset with three levels of polarity, viz. positive, negative, and neutral. Authors manually pre-processed the collected comments, so that single sentence Roman script comments having less than 50 words are kept in the dataset. Due to sarcastic nature of some sentences, inter-annotator agreement is moderate (Cohen Kappa Coefficient: 64.00% [19]. This dataset is imbalanced as

[2] http://twitter4j.org/en/.

Table 27.2 Dataset details (%)

Dataset name	Social media	Size	Polarity distribution
SAIL-2017 Bengali–English code-mixed dataset [15]	Twitter	5.5 K tweets	Positive: 40.44 Neutral: 27.21 Negative: 32.34
SAIL-2017 Hindi–English code-mixed dataset [15]	Twitter	18.5 K tweets	Positive: 31.43 Neutral: 45.70 Negative: 22.85
Hinglish SentiMix dataset [20]	Twitter	20 K tweets	Positive: 33.08 Neutral: 37.46 Negative: 29.46
Hindi–English code-mixed (HECM) dataset [18]	Facebook	3879 sentences	Positive: 35 Neutral: 50 Negative: 15
Urdu–English code-mixed MultiSent Dataset [21]	Twitter	20.5 K tweets	Positive: 35.10 Neutral: 16.63 Negative: 48.27

negative polarity samples are less compared positive polarity and neutral polarity samples. Details of dataset have been provided in Table 27.2.

Hinglish SentiMix dataset [20]**.** Authors of this annotated dataset released two datasets—Hinglish (Hindi–English code-mixed) corpora of size 20 K and Spanglish (Spanish–English) corpus of size 19 K. Authors had collected tweets from Twitter using Twitter API.[3] This dataset was obtained from Sentiment Analysis on Code-mixed Tweets-shared Task 9 of International Workshop on Semantic Evaluation 2020 (SemEval-2020). In this survey, we have focused in Hinglish corpus. Sentiment class-wise distribution of Hinglish corpus is moderately balanced. Authors had provided word-level and sentence-level annotation. Inter-annotator agreement was 55%. The average CMI is nearly 25%. Dataset statistics is available at Table 27.2.

Urdu–English Code-Mixed MultiSent[4] dataset [21]**.** Authors collected tweets on General Election of 2018 of Pakistan from twitter and after manually removing short length posts of one word, dataset has been reduced to 20,735 tweets. However, along with code-mixed Urdu-English Romanized language comments, there are monolingual English, and Romanized Urdu language comments in this dataset. Authors have not discussed overall complexity code-mixing in the dataset and inter-annotator agreement during labelling tweets into three sentiment polarity, viz. positive, negative, and neutral. Dataset statistics has been given in Table 27.2.

[3] https://developer.twitter.com/.

[4] https://github.com/haroonshakeel/multisenti.

27.3 Transformer-Based Models

To resolve several NLP tasks, manual feature extraction in lexicon-based approach suffers from contextual understanding of ambiguous words in text whereas deep learning-based architectures are capable of extracting subtle features of text automatically through its hidden layers. But convolution-based neural network does not have any memory capacity, so sequential neural network such as recurrent neural network model (RNN) has been introduced but RNN suffers from vanishing gradient problem due to long-term dependency in text. Long short-term memory (LSTM) developed for handling vanishing gradient problem but later attention mechanism was introduced where word embeddings of the most relevant part of the input sequence text were fed into the encoder part of Transformer [22], a self-supervised deep learning-based encode-decoder model which can tackle long-term dependency in text. Transformer is built by stacking encode-decoder modules where each module is developed by stacking feed-forward and attention layers. Transformer has gained popularity for its superiority in automatic extracting contextual features even in code-mixed environment.

27.3.1 Bidirectional Encoder Representation from Transformers (BERT)

BERT [23] is a self-supervised, multilayer, bidirectional, Transformer-based, multi-headed attention model that encodes contextual feature from text sequence with word-level and subword-level embeddings. Model learns the context from both sides of the sentence and generates better representation for polysemantic words in text which LSTM lacks. Pretrained language models are available in 'Hugging Face' model Repository.[5] BERT architecture has 12 layers, 768 hidden states, 12 attention heads, and 110 M parameters. During tokenization, each sentence and tokens are numbered, so that at the time input representation, it is built by related tokens, sentence, position embedding can be considered. In BERT pretraining, masked language modelling (MLM) and next sentence prediction (NSP) are used. In MLM, random tokens are masked and later, masked tokens are predicted. In NSP, next sentence is predicted using existing sentence connections.

27.3.2 MBERT (BERT Base-Multilingual-Case)

Multilingual BERT [23], a variant of BERT, is self-supervised pretrained language model using large monolingual Wikipedia corpora on 104 languages with shared

[5] https://github.com/huggingface/transformers.

word vocabulary. It does not take manually labelled input text for modelling and model automatically generates labelled text. mBERT is a 12-layer Transformer model. During model training, there is no way to mark which language is fed, it only marked by input IDs. This transfer learning [24] strategy where model is pretrained with large unlabelled text, later labelled text is fed into model to fine tune the model with downstream NLP task.

27.3.3 DistilBERT

DistilBERT [25] is smaller version of BERT as DistilBERT has 40% less (distilled) parameter than BERT but keeps 97% of its language understanding, so it extracts word embedding quicker than BERT. DistilBERT and BERT have the same architecture. This model has 6 layers, 12 attention head, and 134 million parameters.

27.3.4 Robustly Optimized BERT Pretraining Approach (RoBERTa)

Basic architecture of BERT and RoBERTa [26] is same but RoBERTa has large number of parameters compared to BERT to enhance the performance of RoBERTa. RoBERTa is trained on larger amount of data compared to BERT. Pretraining of RoBERTa uses only MLM technique. Unlike static masking (same mask in every epoch of training) in BERT model, dynamic masking in RoBERTa emerges as robust model. The Byte-Pair Encoding (BPE) [27] has enhanced encoding of large corpus efficiently.

27.3.5 XLM-RoBERTa (Cross-Lingual Model RoBERTa)

Since XLM_RoBERTa [28] is a cross-lingual pretrained language model which has been trained using 100 different languages with shared word vocabulary of different languages. It is suitable for multilingual downstream NLP tasks such as question answering, sentiment classification and sequence labelling. XLM_RoBERTa is conjugation of XLM [29] and RoBERTa. It has been trained on large (2.5 TB) Common Crawl Data [30]. It has performed well for all multiple cross-lingual benchmarks. This model consists of 12 layers, 8 attention heads, and 270 M parameter size. This model is trained with dynamic MLM technique where some random tokens of the corpus are masked, so that model can predict the masked tokens. With this MLM technique same token is not mask in every epoch so that model evolves with improved hyperparameter. This model uses Byte-Pair Encoding for encoding (tokenizing) corpora. BPE is hybrid encoding scheme which encode at subword level.

27.3.6 ALBERT

ALBERT (A Lite BERT) [31] is a smaller version of BERT with same training environment as BERT. ALBERT is pretrained with 12-layers, 768 hidden layers, 12 attention heads, and 12 M parameters.

27.3.7 IndicBERT

IndicBERT [32] model is pretrained using large corpus of 12 Indian languages (Assamese, Bengali, English, Gujarati, Hindi, Kannada, Malayalam, Marathi, Oriya, Punjabi, Tamil, and Telugu) and English language using MLM technique. IndicBERT is a variant of ALBERT model. This model is suitable to create word embedding for Indian languages because this model is pretrained with news, article, blogposts written in Indian languages. IndicBERT has very small number of trainable parameters compared BERT because IndicBERT follows parameter sharing across different hidden layers during training.

27.3.8 Multilingual Representation for Indian Languages (MuRIL)

MuRIL [33] is a BERT model pretrained with monolingual corpus of 17 Indian languages including English including English, (Assamese, Bengali, Gujarati, Hindi, Kannada, Kashmiri, Malayalam, Marathi, Nepali, Oriya, Punjabi, Sanskrit, Sindhi, Tamil, Telugu and Urdu) translated and transliterated Roman script using Wikipedia, PMINDIA, Dakshina corpora. MuRIL has been trained with MLM and Translated Language Modelling (TLM) technique.

27.3.9 XLNet

BERT model pretrained with MLM technique but it has the disadvantage that token prediction in BERT is irrespective of token dependency. So, XLNet [34] is introduced to overcome this challenge, but XLNet suffers from positional discrepancy in pretraining and fine-tuning. XLNet avoids the limitation of BERT of neglecting predicted token dependency by conjugating the ideas of Transformer-XL [35] and autoregressive model which estimate the probability distribution of text corpus. Instead of predicting token in sequential order of MLM, XLNet uses permuted language modelling (PLM) where tokens are predicted in random order in a sentence.

27.3.10 MPNet

MPNet [36] removes the limitations of BERT (no dependency among predicted tokens) and XLNet (inconsistent regarding positional information of a sentence) and introduces pretraining using MLM and permutated language modelling (PLM). MPNet uses auxiliary position as input into PLM to get final sentence position as well as predicted token dependency.

27.3.11 CharacterBERT

CharacterBERT [37] is a variant of BERT, and CharacterBERT has introduced Character-CNN module which focuses on character representation in text instead of word or subword representation. After the successful representation of large corpus in BERT, several other variants of BERT have been developed as per the need of the NLP tasks or to remove the drawbacks of the previous Transformer-based model. Instead of adopting complex WordPiece tokenization of BERT model which is more suitable for complex NLP task, Character-CNN represents word by consulting characters of the word. CharacterBERT is a smaller version of BERT because CNN module for embedding is much smaller than WordPiece tokenizer of BERT. CharacterBERT is more suitable for medical domain task.

27.4 SC in Code-Mixed Indo-Aryan Language Pair

Sentiment classification in NLP is the task of identifying the orientation of feeling/ emotion ('positive', 'negative', 'neutral', 'mixed feeling'), i.e., polarity embedded in textual information [38, 39]. Many times, we use the term 'sentiment analyses', and 'opinion mining' interchangeably with 'sentiment classification'. Automatic sentiment classification becomes more challenging task for code-mixed ILs compared to monolingual text due to informality in code-mixed text environment. In this survey, we use the term sentiment classification (SC). In this survey, we have found three code-mixed Indo-Aryan language pairs which have evaluated the code-mixed dataset using Transformer-based models. In the following subsections, we will discuss on research progress, research gaps and challenges in sentiment classification in code-mixed IL pair with respect of Transformer-based model.

27.4.1 Bengali (Bn)–English (En)

For SC task, [17] has evaluated small size code-mixed Bengali–English corpus—SAIL-2017 using BERT along with other deep learning, and ML models. Since evaluation of the corpus using BERT base, multilingual, uncased model (pretrained with multilingual Unicode language along with formal English text) has not responded well for unknown tokens of code-mixed corpus/out of vocabulary words (OOV), authors have developed BERT base, domain, uncased for domain-specific learning by introducing unknown tokens of code-mixed corpus in BERT base, uncase. Quality of embeddings generated from BERT base, domain, uncased were not suitable due to meagre amount of dataset instances. Multilingual BERT (mBERT) model which is pretrained with multilingual Unicode languages, has given poor result on code-mixed corpus as accuracy reached to less than 50%. BERT model has slightly outperformed deep learning models, and result has been established using tenfold cross-validation of test dataset by authors.

Challenges. $\text{BERT}_{\text{base, domain, uncased}}$ has incorporated domain-specific custom embeddings of Bengali–English code-mixed corpus, so it is anticipated that $\text{BERT}_{\text{base, domain, uncased}}$ outperforms $\text{BERT}_{\text{base, uncased}}$ model. Improvement on performance of $\text{BERT}_{\text{base, domain, uncased}}$ model is a challenge. Development of large size (more than 20 K sentences) Bengali English code-mixed corpus [40] for SC task is required to establish the superiority of Transformer-based models.

Research Gaps. [17] have explored using $\text{BERT}_{\text{base, uncased}}$ and $\text{BERT}_{\text{base, domain, uncased}}$, but reported result (weighted average F1-score) is not significantly high as it is anticipated from the high-end BERT model. It has slightly outperformed deep learning-based models. So, there is sufficient scope to study BERT variants such as XLM-RoBERTa, MuRIL with large code-mixed balanced Bengali–English corpus which has not yet been touched by any author. Insufficient amount of corpus discourages future researchers to explore Transformer-based models on code-mixed Bengali language, otherwise fine-grained sentiment classification of user posts/comments in social media of code-mixed Bengali language is the upcoming research area.

27.4.2 Hindi (Hi)–English (En)

On code-mixed Hindi–English SAIL-2017 dataset, BERT [17, 41] model has performed very poor because of presence of unknown tokens of code-mixed languages. Since automatic context understanding requires large-scale corpus, [17, 41] have reported low F1-score on HECM dataset using BERT. Since large Transformer model like XLM-RoBERTa demands large-scale corpus for context embedding, [41] has revealed the result by ensemble of XLM-RoBERTa and context-aware sentence embeddings on SAIL-2017 dataset, but they have not achieved any significant improvement over BERT by adding sentence embedding because XLM-RoBERTa already generates context-dependent tokens. Evaluating small dataset such as HECM using Transformer-based model is not compatible. On the other hand, on Hinglish SentMix data (larger than HECM, SAIL-2017 dataset), [42, 43] have explored BERT, mBERT, RoBERTa, DistilBERT and it has been revealed that BERT, mBERT, DistilBERT have given better F1-score compared to RoBERTa because RoBERTa model is larger than BERT, mBERT, DistilBERT. So large-scale corpus is suitable for evaluation by RoBERTa, XLM-RoBERTa demands. Advantage of large Transformer-based model is in its efficient context embedding algorithm.

Challenges. On SC task, development large-scale, balanced, standard code-mixed corpus for Hindi language is current need for evaluating the dataset using large Transformer-based model. Improvement on performance of Transformer-based model is still challenge for code-mixed Hindi language. Existing datasets are cleaned and filtered, so SC task on real-time posts/comments using Transformer-based models are challenging task because real-time posts are noisy (spelling variations, repetitive words, incomplete words, emojis, emoticons) in most of the times. Since natural data is mostly imbalanced, it is challenging task to develop Transformer-based model which can handle imbalances in the dataset.

Research Gaps. Since XLM-RoBERTa, XLNet, IndicBERT, MuRIL model have not been explored in details for code-mixed Hindi language, researchers can demonstrate multi-task learning (MTL) [44] approach by fine-tuning a single Transformer model to hard parameter sharing. Fine-grained analysis such aspect-based sentiment analysis using Transformer-based model is an interesting research work which has not yet been touched by any author.

27.4.3 Urdu (Ud)–English (En)

On MultiSent dataset on SC task, authors [21] have shown that even mBERT can give better result, if model is trained with large code-mixed corpus whereas large size model would fail when trained with small size corpus. Authors [21] have explored mBERT and XLM-RoBERTa (XLM-R), and XLM-RoBERTa has outperformed mBERT as usual (see Table 27.3).

Challenges. Though XLM-RoBERTa has achieved slightly higher performance than mBERT, still it is a challenge to improve performance of XLM-RoBERTa. Development of balanced dataset for sentiment classification is challenging task but handling of imbalanced dataset with spelling variations, emojis, and emoticons is even more challenging task. It is a challenge to pre train Transformer-based models with large corpus so that model can capture all the nuances of the corpus for downstream NLP tasks.

Research Gaps. Performance of XLM-RoBERTa may be explored with respect to the classification of sentiment of customers post/comments on different features

Table 27.3 Evaluation of CM Indo-Aryan languages corpus using BERT variants model

Ref. (code-mixed language pair)	Transformer model(s)	Dataset	$F1$-score (%)
[17] (Bn–En)	$\text{BERT}_{\text{base, uncased}}$	SAIL-2017 Bengali–English code-mixed dataset	65.60
[17] (Hi–En)	$\text{BERT}_{\text{base, uncased}}$	SAIL-2017 Hindi–English code-mixed dataset	62.10
[17] (Hi–En)	$\text{BERT}_{\text{base, uncased}}$	Hindi–English code-mixed (HECM) dataset	67.00
[41] (Hi–En)	BERT	Hindi–English code-mixed (HECM) dataset	59.00
[41] (Hi–En)	BERT	Hindi–English code-mixed SAIL-2017 dataset	48.33
[42] (Hi–En)	mBERT RoBERTa DistilBERT	Hinglish SentiMix dataset	67.70 63.50 68.70
[43] (Hi–En)	BERT	Hinglish SentiMix dataset	68.00
[21] (Ud–En)	mBERT XLM-RoBERTa	MultiSent dataset	69.00 71.00

objects on code-mixed Urdu language. BERT variants can be experimented with extended version of MultiSent dataset. Instead of finding the output from outer layer of Transformer, feeding the concatenated output of the hidden layer into another deep neural network, and getting the prediction from outer layer of deep neural network using SoftMax classifier might be an interesting research work [45, 46].

As our objective of this survey is to follow the research progress and evolution of Transformer-based model in Indo-Aryan language family, we have displayed F1-score (%) of evaluation of code-mixed dataset using BERT variants model only in Table 27.3.

27.5 Conclusion

In this survey, we have found code-mixed corpus (HECM, SAIL-2017, MultiSent) in Bengali–English, Hindi–English, and Urdu–English languages which have been evaluated using Transformer-based models by authors. There is scope to explore BERT variants like IndicBERT, MuRIL, and XLNet with current dataset. Since high-end model like Transformer already has been shown supremacy in NLP tasks using efficient contextual embedding which is not possible by traditional lexicon-based or machine learning-based algorithms, it is required to develop large corpus (more than 20 K sentences) on existing Indo-Aryan code-mixed Indian Languages, so that BERT variants remain compatible with the input corpus size. To the best of our knowledge, we have found Indo-Aryan code-mixed languages like Assamese–English, Odia–English, Marathi–English, Nepali–English, and Punjabi–English which have not yet been touched by SoTA Transformer-based model on SC task because absence of any code-mixed dataset for SC tasks, so there is huge research scope to the research community to develop code-mixed balanced benchmark corpus for SC task, evaluate the corpus using XLM-RoBERTa, IndicBERT, MuRIL, etc. For code-mixed Bengali–English, Hindi–English, and Urdu–English languages, the corpus is required to augment to realized SoTA result from Transformer-based models.

References

1. Raj, N.V., Saini, J.R.: Loyalty score generation for customers using sentimental analysis of reviews in e-commerce. In: Dutta, P., Chakrabarti, S., Bhattacharya, A., Dutta, S., Shahnaz, C. (eds.) Emerging Technologies in Data Mining and Information Security. Lecture Notes in Networks and Systems (LNNS), vol. 490, pp. 461–473. Springer, Singapore (2023)
2. Barve, Y., Saini, J.R., Pal, K., Kotecha, K.: A Novel evolving sentimental bag-of-words approach for feature extraction to detect misinformation. Int. J. Adv. Comput. Sci. Appl. **13**(4) (2022)
3. Chakravarthi, B.R., Priyadharshini, R., Muralidaran, V., Jose, N., Suryawanshi, S., Sherly, E., McCrae, J.P.: DravidianCodeMix: sentiment analysis and offensive language identification dataset for Dravidian languages in code-mixed text. Language Resour. Eval. **56**, 765–806 (2022)
4. Poplack, S., Walker, J.: Pieter Muysken, bilingual speech: a typology of code-mixing. Cambridge University Press, Cambridge (2000) pp. xvi+306. J. Linguist. **39**(3), 678–683 (2003)
5. Myers-Scotton, C.: Bilingual speech, a typology of code-mixing. Language **78**(2), 330–333 (2002)
6. Dewaele, J.-M.: Emotions in multiple languages, 1st edn. Palgrave Macmillan UK, London (2010)
7. Jakob, J.C.: The analysis of code switching and code mixing phenomenon in English Facebook group of youth community. J. English Teach. Appl. Linguist. **2**, 1–7 (2020)
8. Hidayat, T.: An analysis of code switching used by facebookers. Sekolah Tinggi Keguruan dan Ilmu Pendidikan (STKIP) Siliwangi Bandung (2008)
9. Kaur, J., Saini, J.R.: A study and analysis of opinion mining research in Indo-Aryan, Dravidian and Tibeto-Burman language families. Int. J. Data Mining Emerg. Technol. **4**(2), 53–60 (2014)
10. Rani, S., Kumar, P.: A journey of Indian languages over sentiment analysis: a systematic review. Artif. Intell. Rev. **52**, 1415–1462 (2019)
11. Ahmad, G.I., Singla, J., Nikita, N.: Review on sentiment analysis of Indian languages with a special focus on code mixed Indian languages. In: 2019 International Conference on Automation, Computational and Technology Management (ICACTM), pp. 352–356. IEEE, London, UK (2019)
12. Ahmad, G.I., Singla, J., Ali, A., Reshi, A.A., Salameh, A.A.: Machine learning techniques for sentiment analysis of code-mixed and switched Indian social media text corpus—a comprehensive review. Int. J. Adv. Comput. Sci. Appl. **13**(2) (2022)
13. Mahadzir, N.H., Omar, M.F., Nawi, M.N.M., Salameh, A.A., Hussin, K.C.: Sentiment analysis of code-mixed text: a review. Turk. J. Comput. Math. Educ. (TURCOMAT) **12**(3), 2469–2478 (2021)

14. Rani, P., Suryawanshi, S., Goswami, K., Chakravarthi, B.R., Fransen, T., McCrae, J.P.: A comparative study of different state-of-the-art hate speech detection methods in Hindi-English code-mixed data. In: Proceedings of the Second Workshop on Trolling, Aggression and Cyberbullying, pp. 42–48. European Language Resources Association (ELRA), Marseille, France (2020)
15. Patra, B.G., Das, D., Das, A.: Sentiment analysis of code-mixed Indian languages: an overview of SAIL_Code-mixed shared task @ICON-2017. ArXiv. abs/1803.06745 (2018)
16. Das, A., Gambäck, B.: Identifying languages at the word level in code-mixed Indian social media text. In: Proceedings of the 11th International Conference on Natural Language Processing, pp. 378–387. NLP Association of India, Goa, India (2014)
17. Jamatia, A., Swamy, S.D., Gambäck, B., Das, A., Debbarma, S.: Deep learning based sentiment analysis in a code-mixed English-Hindi and English-Bengali social media corpus. Int. J. Artif. Intell. Tools **29**(5), 2050014 (2020)
18. Joshi, A., Prabhu, A., Shrivastava, M., Varma, V.: Towards sub-word level compositions for sentiment analysis of Hindi-English code mixed text. In: Proceedings of COLING 2016, the 26th International Conference on Computational Linguistics: Technical Papers, pp. 2482–2491. The COLING 2016 Organizing Committee, Osaka, Japan (2016)
19. Cohen, J.: A coefficient of agreement for nominal scales. Educ. Psychol. Meas. **20**(1), 37–46 (1960)
20. Patwa, P., Aguilar, G., Kar, S., Pandey, S., Pykl, S., Gambäck, B., Chakraborty, T., Solorio, T., Das, A.: SemEval-2020 Task 9: overview of sentiment analysis of code-mixed tweets. In: Proceedings of the Fourteenth Workshop on Semantic Evaluation, pp. 774–790. International Committee for Computational Linguistics, Barcelona (online) (2020)
21. Younas, A., Nasim, R., Ali, S., Wang, G., Qi, F.: Sentiment analysis of code mixed Roman Urdu-English social media text using deep learning approaches. In: 2020 IEEE 23rd International Conference on Computational Science and Engineering (CSE), pp. 66–71. IEEE, Guangzhou, China (2020)
22. Vaswani, A., Shazeer, N., Parmar, N., Uszkoreit, J., Jones, L., Gomez, A.N., Kaiser, Ł. ukasz, Polosukhin, I.: Attention is all you need. In: Guyon, I., Luxburg, U. Von, Bengio, S., Wallach, H., Fergus, R., Vishwanathan, S., Garnett, R. (eds.) Advances in Neural Information Processing Systems, vol. 30. Curran Associates, Inc. (2017)
23. Devlin, J., Chang, M.-W., Lee, K., Toutanova, K.: BERT: pre-training of deep bidirectional transformers for language understanding. In: Burstein, J., Doran, C., Solorio, T., (eds.) Proceedings of the 2019 Conference of the North American Chapter of the Association for Computational Linguistics: Human Language Technologies (NAACL-HLT), 2019, Minneapolis, MN, USA, June 2–7, 2019, vol. 1: Long and Short Papers, pp. 4171–4186. Association for Computational Linguistics, Stroudsburg, PA, USA (2019)
24. Bengio, Y.: Deep learning of representations for unsupervised and transfer learning. In: Proceedings of the 2011 International Conference on Unsupervised and Transfer Learning Workshop, vol. 27, pp. 17–37. JMLR.org, Washington, USA (2011)
25. Sanh, V., Debut, L., Chaumond, J., Wolf, T.: DistilBERT, a distilled version of BERT: smaller, faster, cheaper and lighter. arXiv preprint arXiv:1910.01108 (2019)
26. Liu, Y., Ott, M., Goyal, N., Du, J., Joshi, M., Chen, D., Levy, O., Lewis, M., Zettlemoyer, L., Stoyanov, V.: Roberta: a robustly optimized bert pretraining approach. arXiv preprint arXiv: 1907.11692 (2019)
27. Sennrich, R., Haddow, B., Birch, A.: Neural machine translation of rare words with subword units. In: Proceedings of the 54th Annual Meeting of the Association for Computational Linguistics, vol. 1: Long Papers, pp. 1715–1725. Association for Computational Linguistics (ACL), Berlin, Germany (2016)
28. Conneau, A., Khandelwal, K., Goyal, N., Chaudhary, V., Wenzek, G., Guzmán, F., Grave, E., Ott, M., Zettlemoyer, L., Stoyanov, V.: Unsupervised cross-lingual representation learning at scale. In: Proceedings of the 58th Annual Meeting of the Association for Computational Linguistics. pp. 8440–8451. Association for Computational Linguistics (ACL), Online (2020)

29. Lample, G., Conneau, A.: Cross-lingual language model pretraining. In: Proceedings of the 33rd International Conference on Neural Information Processing Systems (NIPS), pp. 7059–7069. Curran Associates Inc., Red Hook, NY, USA (2019)
30. Smith, J.R., Saint-Amand, H., Plamada, M., Koehn, P., Callison-Burch, C., Lopez, A.: Dirt cheap web-scale parallel text from the common crawl. In: Proceedings of the 51st Annual Meeting of the Association for Computational Linguistics, vol. 1: Long Papers, pp. 1374–1383. Association for Computational Linguistics (ACL), Sofia, Bulgaria (2013)
31. Lan, Z., Chen, M., Goodman, S., Gimpel, K., Sharma, P., Soricut, R.: ALBERT: a lite BERT for self-supervised learning of language representations. In: 8th International Conference on Learning Representations (ICLR) 2020. OpenReview.net, Addis Ababa, Ethiopia, April 26–30, (2020)
32. Kakwani, D., Kunchukuttan, A., Golla, S., Gokul, N.C., Bhattacharyya, A., Khapra, M.M., Kumar, P.: IndicNLPSuite: monolingual corpora, evaluation benchmarks and pre-trained multilingual language models for Indian languages. In: Findings of the Association for Computational Linguistics: EMNLP 2020, pp. 4948–4961. Association for Computational Linguistics (ACL), Online (2020)
33. Khanuja, S., Bansal, D., Mehtani, S., Khosla, S., Dey, A., Gopalan, B., Margam, D.K., Aggarwal, P., Nagipogu, R.T., Dave, S., Gupta, S., Gali, S.C.B., Subramanian, V., Talukdar, P.P.: MuRIL: Multilingual representations for Indian languages. arXiv preprint arXiv:2103.10730 (2021)
34. Yang, Z., Dai, Z., Yang, Y., Carbonell, J., Salakhutdinov, R., Le, Q. V: XLNet: generalized autoregressive pretraining for language understanding. In: Proceedings of the 33rd International Conference on Neural Information Processing Systems (NIPS), pp. 5753–5763. Curran Associates Inc., Red Hook, NY, USA (2019)
35. Dai, Z., Yang, Z., Yang, Y., Carbonell, J.G., Le, Q.V., Salakhutdinov, R.: Transformer-XL: attentive language models beyond a fixed-length context. In: Korhonen, A., Traum, D.R., Màrquez, L. (eds.) Proceedings of the 57th Conference of the Association for Computational Linguistics, ACL 2019, vol. 1: Long Papers, pp. 2978–2988. Association for Computational Linguistics (ACL), Florence, Italy (2019)
36. Song, K., Tan, X., Qin, T., Lu, J., Liu, T.-Y.: MPNet: masked and permuted pre-training for language understanding. Adv. Neural. Inf. Process. Syst. **33**, 16857–16867 (2020)
37. El Boukkouri, H., Ferret, O., Lavergne, T., Noji, H., Zweigenbaum, P., Tsujii, J.: CharacterBERT: reconciling ELMo and BERT for word-level open-vocabulary representations from characters. In: Proceedings of the 28th International Conference on Computational Linguistics, pp. 6903–6915. International Committee on Computational Linguistics (ICCL), Barcelona, Spain (Online) (2020)
38. Cambria, E., Schuller, B., Xia, Y., Havasi, C.: New avenues in opinion mining and sentiment analysis. IEEE Intell. Syst. **28**(2), 15–21 (2013)
39. Liu, B.: Sentiment analysis and opinion mining. Springer Nature (2022)
40. Tareq, M., Islam, M.F., Deb, S., Rahman, S., Mahmud, A.A.: Data-augmentation for Bangla-English code-mixed sentiment analysis: enhancing cross linguistic contextual understanding. IEEE Access. **11**, 51657–51671 (2023)
41. Pradhan, R., Sharma, D.K.: An ensemble deep learning classifier for sentiment analysis on code-mix Hindi-English data. Soft. Comput. **27**, 1–18 (2022)
42. Bhange, M., Kasliwal, N.: Hinglish NLP at SemEval-2020 Task 9: fine-tuned language models for Hinglish sentiment detection. In: Proceedings of the Fourteenth Workshop on Semantic Evaluation, pp. 934–939. International Committee for Computational Linguistics (ICCL), Stroudsburg, PA, USA (2020)
43. Parikh, A., Bisht, A.S., Majumder, P.: IRLab_DAIICT at SemEval-2020 Task 9: machine learning and deep learning methods for sentiment analysis of code-mixed tweets. In: Proceedings of the Fourteenth Workshop on Semantic Evaluation, pp. 1265–1269. International Committee for Computational Linguistics (ICCL), Stroudsburg, PA, USA (2020)
44. Hande, A., Hegde, S.U., Priyadharshini, R., Ponnusamy, R., Kumaresan, P.K., Thavareesan, S., Chakravarthi, B.R.: Benchmarking multi-task learning for sentiment analysis and offensive

language identification in under-resourced Dravidian languages. arXiv preprint arXiv:2108.03867 (2021)
45. Ou, X., Li, H.: YNU@ Dravidian-CodeMix-FIRE2020: XLM-RoBERTa for multi-language sentiment analysis. In: Forum for Information Retrieval—FIRE (Working Notes), pp. 560–565 (2020)
46. Zhu, Y., Dong, K.: YUN111@ Dravidian-CodeMix-FIRE2020: sentiment analysis of Dravidian code mixed text. In: Forum for Information Retrieval—FIRE (Working Notes), pp. 628–634 (2020)

Chapter 28
Secure File Sharing System Using Hybrid Authentication and Secured Key Management Algorithm

Aryan Dokania, Anshul Bamb, and J. Dheeba

Abstract When web file sharing is used, users can safely receive and share any files or documents over the web using any of their preferred web browsers. Though the existing cryptographic algorithms perform well in end-to-end encryption, they are not resistant to attacks that extract information from the computer system, such as passwords and encryption/decryption keys. In this paper, a hybrid method is proposed for secure and efficient end-to-end encryption. This paper focuses on secure key exchange, efficient data encryption, and strengthening of passwords, which is achieved by a combination of four cryptographic algorithms. The proposed hybrid cryptographic method is secure against several attacks, including man-in-the-middle attacks and side-channel attacks. The proposed method offers many advantages over traditional end-to-end encryption methods, including security, efficiency, and resistance to side-channel attacks. Extensive analytical and experimental results show that the proposed method is efficient enough to be used in a file sharing system.

28.1 Introduction

Due to the extensive use of digital platforms by professionals, there is a need to transmit information safely and securely. Trade, e-commerce, hospitals, and banks need to exchange files privately with their clients regularly. Files shared on the internet are illegally being copied, stolen, or modified on the network, requiring more data protection. The basic requirement of any cryptographic system is confidentiality, authentication, and integrity, which is achieved using the basic cryptographic algorithm like RSA (Rivest, Shamir, Adleman) [1] and DES (Data Encryption Standard) [2]. SMTP is most commonly used by users for file sharing, where the communication goes to different untrusted routers. Because standard SMTP email does not offer a method for encrypting data from the sender's end to the recipient's end, users have looked for options in addition to SMTP email to transfer secret and safe information.

A. Dokania · A. Bamb · J. Dheeba (✉)
School of Computer Science and Engineering, Vellore Institute of Technology, Vellore, India
e-mail: dheeba.j@vit.ac.in

P. Nanda et al. (eds.), *Intelligent System and Data Analysis*, Smart Innovation, Systems and Technologies 398, https://doi.org/10.1007/978-981-97-5200-3_28

As the internet has developed, cloud-based file sharing services [3, 4] have emerged, promising to enable users to safely share data. These services enable file sharing between users using secure SMTP. Through an SSL/TLS connection, the sender can upload a file to the service's server using their web browser. While being transferred from the sender's web browser to the service's server, the file is encrypted. The file is immediately encrypted by the service using a symmetric encryption method once it reaches the service's server. The service also stores the encryption key. The service utilizes the encryption key to decode the file when the receiver wants it. The file can then be downloaded from the server of the service by the recipient using their web browser over an SSL/TLS connection.

Numerous scholars have presented their work on file sharing in a secured format. The cipher text was created by Agrawal and Pal [5] using the 8-bit ASCII code value and some straightforward calculations like logical NOT and basic double division. The textbook data was encrypted and decrypted by the authors using a reliable, efficient, and symmetric-key-grounded technique. The use of cover techniques, deceptive techniques, or precise processes can all be used to impose encryption. Data security or concealment from unauthorized reading or modification is the goal of encryption. Abutaha et al. [6] dealt with transmission of images securely over the internet using an end interface crypto system. The suggested strategy outperforms other models in terms of attacks by 1.5 times. The strategy employed a chaotic system with characteristics resembling randomness and no periodicity. The image is divided into a sequence of 4 bytes by the parallel chaos crypto system, and the resulting sequence is XORed with the randomized and non-periodic bytes generated by two non-recursive filters to create the encrypted image. The method has demonstrated resistance to statistical and brute force attacks with a significant level of unpredictability and uncertainty.

Ma and Zhang [7] designed an encryption system for encrypting private information in an IoT environment, which is stored in the relay nodes. According to experimental findings, the suggested system encrypts and decrypts more quickly, has better cipher text conversion efficiency, and sustains fewer network attacks while encrypting for the same amount of time. Rahman et al. [8] used a keyword monoalphabetic substitution algorithm for encrypting data with a key which was from a list of keys. The encrypted message and the recipient's mailbox are both kept in the database together with the key's serial number. The recipient chooses the key value after logging into his or her own account and uses the key to decode the encrypted communication to reveal the original message. Every node has a list of their friends, and transactions happen through friends, according to Jain and Kumar [9]. They counteract attacks like Sybil assaults, collusion, and self-centered nodes by utilizing a weighted graph system and reputation system. Most of the works presented has a strong encryption and decryption algorithm but lacks in key management and side-channel attack. In this paper, we are focusing on,

- Building a more secured system utilizing a hybrid model for encryption and key management.
- Providing solution for side-channel attack

The rest of the paper is organized as follows, Sect. 28.2 describes the framework for a secure file sharing system, Sect. 28.3 gives the result and discussion of the proposed system and Sect. 28.4 gives the conclusion.

28.2 Framework for a Secured File Sharing System

In this section, we describe our hybrid model for a secure file sharing system. We consider a system where there are users who transfers files. There is a service server belonging to applications to which the users need to communicate in a secure manner. Our model utilizes a hybrid algorithm by combining the following cryptographic techniques.

(i) ECDHE (Elliptic Curve Diffie–Hellman Key-Exchange Protocol) is used to create a secure shared secret between the sender and receiver. ECDHE is a key-exchange protocol that uses elliptic curve cryptography to establish a shared secret between two parties. ECDHE is more efficient than traditional key-exchange protocols, such as RSA, and it is also more resistant to side-channel attacks.
(ii) 256-Bit AES–GCM symmetric-key encryption and decryption are used to encrypt and decrypt the data. AES is a symmetric encryption algorithm that uses a 256-bit key. GCM is a mode of operation for AES that provides both confidentiality and integrity protection.
(iii) PBKDF2 (Password-Based Key Derivation Function 2) is used to strengthen the password. PBKDF2 is a password hashing function that is used to strengthen passwords. PBKDF2 works by repeatedly hashing the password with a salt and a work factor. This makes it more difficult for attackers to crack passwords.
(iv) ECIES (Elliptic Curve Integrated Encryption Scheme) and ECC (Elliptic Curve Cryptography) are used to encrypt the symmetric key. ECIES is an encryption scheme that uses elliptic curve cryptography to encrypt data. Elliptic curves are employed in public-key cryptography known as ECC. ECC is more efficient than traditional public-key cryptography, such as RSA, and it is also more resistant to side-channel attacks.

Figure 28.1 shows the overall framework for the proposed secured file sharing system. The diagram also shows the following components:

- Sender: The sender is the user who wants to send the encrypted data.
- Receiver: The receiver is the user who wants to receive the encrypted data.
- Elliptic curve public and private keys: The sender and receiver generate their own elliptic curve public and private keys. The public key is used to encrypt data, and the private key is used to decrypt data.
- ECIES: ECIES is an encryption scheme that uses elliptic curve cryptography to encrypt data.
- ECC: ECC is a type of public-key cryptography that uses elliptic curves.

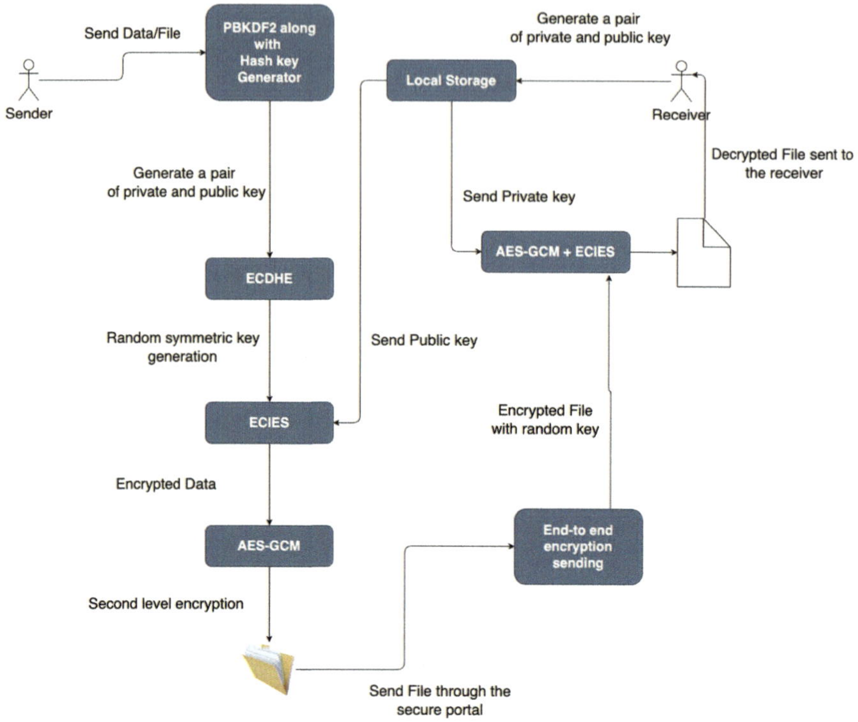

Fig. 28.1 Framework of the proposed system

- Random symmetric key: The sender generates a random symmetric key. This key is used to encrypt and decrypt the data.
- Encrypted symmetric key: The sender encrypts the random symmetric key using the receiver's public key. This encrypted symmetric key is sent to the receiver.
- Decrypted symmetric key: The receiver uses their private key to decrypt the encrypted symmetric key. This decrypted symmetric key is used to encrypt and decrypt the data.
- AES-GCM: AES–GCM is a mode of operation for AES that provides both confidentiality and integrity protection.
- Data: The data that the sender wants to send to the receiver.
- PBKDF2: PBKDF2 is a password hashing function that is used to strengthen the password that is used to encrypt the symmetric key.

Both the sender and the recipient create their own unique public and private elliptic curve keys. A random symmetric key is encrypted using ECIES by the sender using the recipient's public key. The recipient receives the encrypted symmetric key from the sender after that. The symmetric key is decrypted by the recipient using their private key. AES–GCM is used by the sender and receiver to encrypt and decrypt the data. PBKDF2 is used to hash the password, together with a salt and work factor.

The symmetric key is encrypted using the hash value that is produced. The suggested architecture is resistant to a wide range of threats. First, the method is immune to man-in-the-middle attacks because to the usage of ECDHE and ECC. Second, it becomes more challenging for attackers to crack passwords when PBKDF2 is used. Third, the usage of AES–GCM safeguards the data's integrity and secrecy. The suggested architecture is effective enough to be implemented in a system for file sharing. The size of the data being encrypted, the type of hardware being used, and the cryptographic methods being employed are some of the variables that affect how well the algorithm performs. We do think that our technique is effective enough to be implemented in a file sharing system, though. Below pseudo-code in Figs. 28.2 and 28.3 depicts the approach.

28.3 Security Analysis and Countermeasures

The experiments were done by conducting a series of security attacks and testing the efficiency of the hybrid algorithm. We tried to break the algorithm by using a variety of methods, including man-in-the-middle and side-channel attack. This section concisely studies the countermeasures against different types of attacks.

28.3.1 Man-in-the-Middle Attack

A cyberattack known as a "man-in-the-middle" occurs when an attacker intercepts and reroutes communications between two parties [10]. This can allow the attacker to eavesdrop on the communications, modify the communications, or even impersonate one of the parties.

Countermeasures: The proposed hybrid algorithm is a key-exchange protocol that combines symmetric and asymmetric encryption. Symmetric encryption is used to encrypt the data, while asymmetric encryption is used to exchange the keys. As a result, even if an attacker has access to one of the keys, it will be more challenging for them to defeat the encryption. Also, second-level encryption of the random key by AES–GCM removes the possibility of guessing the key as it has been locked with private key security. Another layer of added security is the use of the PBKDF2 hashing mechanism which is used to hash the keys, thus making the system more secure. The hybrid algorithm can be used to mitigate man-in-the-middle attacks in a number of ways. First, the use of asymmetric encryption makes it more difficult for an attacker to impersonate one of the parties. This is because the attacker would need the private key for the party they are impersonating, which is typically very difficult to obtain. Second, the use of symmetric encryption makes it more difficult for an attacker to eavesdrop on the communications. This is because the attacker would need to have the key to decrypt the data, which is typically only known to the two parties involved

```
  function EncryptFile(filename: string; password: string):
string;
  var
      sharedSecret: string;
      symmetricKey: string;
      ciphertext: string;
      publicKey: string;
      encryptedSymmetricKey: string;
  begin

      // Generate a random salt.
      salt := GenerateRandomSalt();

      // Hash the password with the salt.
      hashedPassword := HashPassword(password, salt);

      // Generate a shared secret using ECDHE.
      sharedSecret := GenerateSharedSecret(hashedPassword);

      // Generate a symmetric key using PBKDF2.
      symmetricKey := GenerateSymmetricKey(sharedSecret);

      // Open the file for reading.
      file := OpenFile(filename, "rb");

      // Read the file into a buffer.
      buffer := ReadFile(file);

      // Encrypt the buffer using AES-GCM.
      ciphertext := EncryptData(buffer, symmetricKey);

      // Close the file.
      CloseFile(file);

      // Get the public key of the receiver.
      publicKey := GetPublicKey(receiver);

      // Encrypt the symmetric key using the public key.
      encryptedSymmetricKey:= EncryptData(symmetricKey, pub-
licKey);

      // Return the encrypted symmetric key and ciphertext.
      result := encryptedSymmetricKey, ciphertext;
  end;
```

Fig. 28.2 Pseudo-code for encryption

in the communication. We evaluated the effectiveness of the hybrid algorithm in mitigating man-in-the middle attacks by conducting a series of experiments. In our experiment, we attempted to intercept and modify the communications between the two parties using a variety of attack methods. Our experiments showed that the hybrid algorithm was effective in mitigating man-in-the-middle attacks. It is more difficult

```
  function DecryptFile(encryptedSymmetricKey: string; cipher-
text: string; password: string): string;
  var
      sharedSecret: string;
      symmetricKey: string;
      plaintext: string;
  begin
      // Decrypt the symmetric key using ECIES.
      symmetricKey   :=   DecryptSymmetricKey(encryptedSymmet-
ricKey, password);
      // Decrypt the data using AES-GCM.
      plaintext := DecryptData(ciphertext, symmetricKey);
      // Return the plaintext.
      result := plaintext;
  end;
```

Fig. 28.3 Pseudo-code for decryption

for an attacker to pretend to be one of the parties when asymmetric encryption is used, and it is more difficult for an attacker to eavesdrop on communications when symmetric encryption is used.

28.3.2 Side-Channel Attack

A sort of attack known as a side-channel attack takes advantage of data that is revealed by a cryptographic device such as timing information, to gain an advantage, while it is functioning normally [11]. This information can be used to gain access to the device's secrets, such as encryption keys.

Countermeasures: The proposed algorithm is a combination of different algorithms that are used to solve a single problem as it will make it really hard for the attacker to gain information even related to keys as there is a random key generator which is also hashed by ECIES and AES–GCM making it impossible. In the context of side-channel attacks, the proposed algorithm can be used to improve the security of a cryptographic device by making it more difficult for an attacker to exploit side-channel information as the file is locked by the private key of the receiver and needs to again go through AES–GCM and ECIES decrypter. The system is made more secure by using the PBKDF2 hashing algorithm to hash the keys, which adds another degree of protection. A type of attack called a side-channel attack makes use of information that a cryptographic device reveals while it is operating properly. The attacker can then use this information to infer the secret information that is being processed by the device.

In addition to conducting attacks, we also conducted a number of security tests. These tests included:

- Resistance to known attacks: We tested the algorithm against a variety of known attacks, including brute force attacks, side-channel attacks, and differential cryptanalysis and found that the algorithm was resistant to all of these attacks.
- Resistance to unknown attacks: We also tested the algorithm against unknown attacks. We did this by generating a large number of random keys and encrypting a large amount of data. We then tried to break the algorithm by using a variety of methods. We could not break the algorithm, even with the most powerful attacks.

These tests show that our algorithm is very secure against both known and unknown attacks.

As shown in Table 28.1, the hybrid algorithm combines security, performance, and usability in the best possible way.

- Security: To achieve extremely high security, our approach combines symmetric-key and public-key encryption. Symmetric-key cryptography is used to encrypt the data, while public-key cryptography is used to exchange keys. This group of algorithms offers strong defense against numerous threats.
- Speed: Our algorithm is very fast, even for large amounts of data. This is because it uses a combination of efficient algorithms, such as elliptic curve cryptography and AES.
- Ease of Use: Our algorithm is easy to use. It is implemented in a variety of programming languages, and there are many tutorials available online.

Overall, the proposed algorithm is the best choice for applications that require high security, speed, and easc of use which is demonstrated in the case study. We compared the hybrid algorithm and Diffie–Hellman [12, 13] in terms of security, performance, and usability. The research paper compared the hybrid algorithm and Diffie–Hellman in terms of security, performance, and usability. The results proves that the hybrid algorithm was more secure than Diffie–Hellman, but it was also slower and also found that the hybrid algorithm was more user-friendly than Diffie–Hellman [14]. Symmetric encryption was used to encrypt the data, while asymmetric encryption was used to exchange the keys. This made it more difficult for an attacker to break the encryption, even if they knew one of the keys.

Table 28.1 Comparison of the hybrid algorithm with other existing algorithms

Algorithm	Security (%)	Speed (%)	Ease of use (%)
Proposed hybrid algorithm	99	98	95
RSA	95	90	90
Diffie–Hellman	98	99	90
Elliptic curve cryptography (ECC)	97	99	95
Triple DES	90	90	95
AES	95	99	95

Diffie–Hellman is a key-exchange protocol that only uses asymmetric encryption. This made it more vulnerable to attack [15]. In particular, Diffie–Hellman was susceptible to a man-in-the-middle attack, where an attacker could intercept the keys and decrypt the data [16]. The hybrid algorithm was more secure and user-friendly than Diffie–Hellman. However, it was also slower. The choice of which algorithm to use depended on the specific needs of the application. The hybrid algorithm should be used if security was the most vital factor. If performance was the most crucial factor, then Diffie–Hellman should be used. The hybrid algorithm should be used if usability was the most important factor.

28.4 Conclusion

The suggested architecture is resistant to a wide range of threats. First, the method is immune to man-in-the-middle attacks because of the usage of ECDHE and ECC. Second, it becomes more challenging for attackers to crack passwords when PBKDF2 is used. Third, the usage of AES–GCM safeguards the data's integrity and secrecy. The suggested architecture is effective enough to be implemented in a system for file sharing. The size of the data being encrypted, the type of hardware being used, and the cryptographic methods being employed are some of the variables that affect how well the algorithm performs. The algorithm can be recommended be used for applications where security was the most crucial factor. In conclusion, the proposed hybrid algorithm that combines four different cryptographic algorithms: ECDHE, 256-Bit AES–GCM symmetric-key encryption and decryption, PBKDF2, and ECIES and ECC offers a number of advantages over traditional end-to-end encryption methods.

References

1. Rivest, R.L., Shamir, A., Adleman, L.: A method for obtaining digital signatures and public-key cryptosystems. Commun. ACM **21**(2), 120–126 (1978)
2. Standard, D.E.: Data encryption standard. In: Federal Information Processing Standards Publication, p. 112 (1999)
3. Indu, I., Anand, P.R., Shaji, S.P.: Secure file sharing mechanism and key management for mobile cloud computing environment. Indian J. Sci. Technol. **9**(48), 1–8 (2016)
4. Thabit, F., Alhomdy, S., Al-Ahdal, A.H., Jagtap, S.: A new lightweight cryptographic algorithm for enhancing data security in cloud computing. Global Trans. Proc. **2**(1), 91–99 (2021)
5. Agrawal, E., Pal, P.R.: A secure and fast approach for encryption and decryption of message communication. Int. J. Eng. Sci. 11481 (2017)
6. Abutaha, M., Amar, I., AlQahtani, S.: Parallel and practical approach of efficient image chaotic encryption based on message passing interface (MPI). Entropy **24**(4), 566 (2022)
7. Ma, H., Zhang, Z.: A new private information encryption method in internet of things under cloud computing environment. Wirel. Commun. Mob. Comput. **2020**, 1–9 (2020)
8. Rahman, M.M., Akter, T., Rahman, A.: Development of cryptography-based secure messaging system. J. Telecommun. Syst. Manage. (2016)

9. Jain, A., Kumar, S.: FriendShare: a secure and reliable framework for file sharing on network. J. Netw. Comput. Appl. **120**, 1–16 (2018)
10. Kumar, A., Hameed, S.: Diffie Hellman stand the test of time (protocol's limitations, applications and functional divergence). Int. J. Comput. Appl. **176**, 1–5 (2020). https://doi.org/10.5120/ijca2020920380
11. Mihailescu, M.I., Nita, S.L.: Elliptic-curve cryptography. In: Pro Cryptography and Cryptanalysis with C++23. Apress, Berkeley, CA (2023). https://doi.org/10.1007/978-1-4842-9450-5_9
12. Rescorla, E.: Diffie-hellman Key Agreement Method (No. rfc2631) (1999)
13. Cheon, J.H.: Security analysis of the strong Diffie-Hellman problem. In: Advances in Cryptology-EUROCRYPT 2006: 24th Annual International Conference on the Theory (2006)
14. Visconti, A., Mosnáček, O., Brož, M., Matyáš, V.: Examining PBKDF2 security margin—case study of LUKS. J. Inf. Secur. Appl. **46**, 296–306 (2019). ISSN 2214-2126. https://doi.org/10.1016/j.jisa.2019.03.016
15. Kim, K., Choi, S., Kwon, H., Kim, H., Liu, Z., Seo, H.: PAGE—practical AES-GCM encryption for low-end microcontrollers. Appl. Sci. **10**(9), 3131 (2020). https://doi.org/10.3390/app10093131
16. Ullah, S., Zheng, J., Din, N., Hussain, M.T., Ullah, F., Yousaf, M.: Elliptic curve cryptography; applications, challenges, recent advances, and future trends: a comprehensive survey. Comput. Sci. Rev. **47**, 100530 (2023). ISSN 1574-0137. https://doi.org/10.1016/j.cosrev.2022.100530
17. Ahirwal, R.R., Ahke, M.: Elliptic curve diffie-hellman key exchange algorithm for securing hypertext information on wide area network. Int. J. Comput. Sci. Inf. Technol. **4**(2), 363–368 (2013)
18. Sung, B.Y., Kim, K.B., Shin, K.W.: An AES-GCM authenticated encryption crypto-core for IoT security. In: 2018 International Conference on Electronics, Information, and Communication (ICEIC), pp. 1–3. IEEE (2018)
19. Josefsson, S.: PKCS# 5: password-based key derivation function 2 (PBKDF2) test vectors (No. rfc6070) (2011)
20. Visconti, A., Mosnáček, O., Brož, M., Matyáš, V.: Examining PBKDF2 security margin—case study of LUKS. J. Inf. Sec. Appl. **46**, 296–306 (2019)
21. Gayoso Martínez, V., Hernández Encinas, L., Queiruga Dios, A.: Security and practical considerations when implementing the elliptic curve integrated encryption scheme. Cryptologia **39**(3), 244–269 (2015)
22. Kapoor, V., Abraham, V.S., Singh, R.: Elliptic curve cryptography. Ubiquity **2008**(May), 1–8 (2008)
23. Mahto, D., Khan, D.A., Yadav, D.K.: Security analysis of elliptic curve cryptography and RSA. In: Proceedings of the World Congress on Engineering, vol. 1, pp. 419–422 (2016)
24. Mahto, D., Yadav, D.K.: Performance analysis of RSA and elliptic curve cryptography. Int. J. Netw. Secur. **20**(4), 625–635 (2018)
25. Mahto, D., Yadav, D.K.: RSA and ECC: a comparative analysis. Int. J. Appl. Eng. Res. **12**(19), 9053–9061 (2017)

Chapter 29
IPDD: Intelligent Parkinson Disease Detection System

Manish Singhal, Vinita Tiwari, and Ashish Sharma

Abstract Worldwide, millions of peoples are affected by complex neuro degenerative disorder, that is known as PD (Parkinson Disease). In recent years, it is very critical for accurate diagnosis of PD at early stage for implementing timely interventions and improving patient life. There have been growing interest in developing intelligent systems for Parkinson's disease detection, harnessing the potential of cutting-edge technologies such as artificial intelligence, machine l earning, and sensor-based data analysis. This research paper presents an in-depth exploration of an Intelligent Parkinson Disease Detection System (IPDDS) designed for early diagnosis and quality level monitoring of PD using two most accurate machine learning techniques—SVM and Random Forest. The presented work relates the use of various performance metrices like Accuracy, Confusion Matrix, Precision, etc. Standard Dataset having voice samples of human's having Parkinson Disease and not having PD is used to validate the various performance parameters and predict the PD or non-PD.

29.1 Introduction

29.1.1 Objective

In this research we discuss the recent advances in understanding Parkinson's disease. Here we also wish to cover the sources of data, areas of agreement and controversy, and areas timely for developing research. The study also discusses the management

Vinita Tiwari and Ashish Sharma—These authors contributed equally to this work.

M. Singhal (✉) · V. Tiwari
Electronic and Communication, Indian Institute of Information Technology, Kota 325003, Rajasthan, India
e-mail: write2singhal@gmail.com

V. Tiwari
e-mail: vinitatiwari@iiitkota.ac.in

A. Sharma
Computer Science and Engineering, Manipal University, Jaipur 303007, Rajasthan, India

P. Nanda et al. (eds.), *Intelligent System and Data Analysis*, Smart Innovation, Systems and Technologies 398, https://doi.org/10.1007/978-981-97-5200-3_29

options available for movement complications and nonmovement symptoms at both stages, i.e., early and later stage of diagnosis [1]. This work presents the diagnosis of Parkinson's disease considering the use of machine learning. The work suggests that machine learning models have shown better diagnostic performance with processing of data acquisition. This work presents an innovative approach for developing an intelligent Parkinson's disease detection system that combines voice analysis. The proposed system has the potential to revolutionize the field of PD diagnosis and monitoring, improving the quality of life for those affected by this challenging neurodegenerative disorder. The paper concludes by suggesting that future work could focus on developing and validating machine learning models in clinical settings using locally recruited human participants, as well as exploring new data modalities of Parkinson's disease [2].

29.1.2 Parkinson Disease and Its Types

PD (Parkinson's disease) can be categorized into different types based on their clinical features and causes. The main categories of PD can be considered as follows:- (1) Idiopathic PD: It is categorized as most common PD accounting 85–90% of all cases and Characterized with gradual onset of symptoms, including rigidity,tremors, bradykinesia (slowness of movement), and postural instability. (2) Atypical PD: It has some additional symptoms with rapid progression in PD. It is having group of neurodegenerative disorders. (3) Drug-induced PD: This type of Parkinsonism is caused by certain medications that block dopamine receptors in the brain. It is usually reversible once the medication is stopped. (4) Vascular PD: This type of Parkinsonism is caused by a series of small strokes that damage the brain's blood vessels, leading to symptoms similar to Parkinson's disease. (5) PD due to other causes: This includes Parkinsonism caused by head trauma, infections, toxins, or other underlying medical conditions [3]. Figure 29.1 represent the most identical diseases.

29.2 Prakinson's State-of-the-Art Related Work

The study [4] presents a study on Parkinson's disease in a country region population. The study provides insights into the prevalence, incidence and mortality of PD and examines association between PD and various demographic and clinical factors. Entire Norwegian population database has been used in this study. The study also found that PD is associated with increased mortality, particularly in older age groups. Parkinson's disease affects over 1% of individuals who are 65 years old or older is characterized by a progressive depletion of dopamine in the brain for further research on this topic.

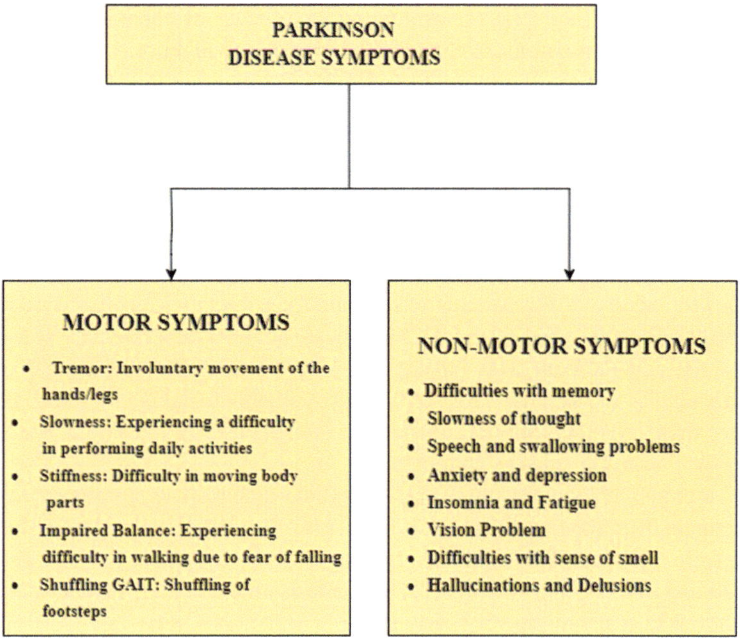

Fig. 29.1 Classification of Parkinson's disease using various symptoms

In [5], the levodopa challenges are administered in which participants having their history of usual morning levodopa. In this study, the levodopa equivalent doses were calculated through standard conversions. In standard conversions, firstly the patients are assessed with "MED OFF" condition for 60–90 min then levodopa intake in "MED ON" condition. In both of the conditions, the following parameters were assessed: motor performance, severity of dyskinesias, voice quality, respiratory support for speech, voice variability, voice stability and the rate of speech.

In this study [6], the speech patterns of the PD patients and healthy controls were measured using a head-mounted condenser microphone placed approximately 5 cm from the subject's mouth. In this study, sampling of speech signals has been taken at 48 kHz with resolution of 16-bit. Samples were taken with reference to single session with speech specialist of individual patients.

It is described in [7], that in vocal impairment, 78% of early untreated PD occurs. Deficits in all speech characteristics were found in only 13.04 % of PD subjects, while 26.09% showed deficits only in prosody, 17.39% only in articulation, and 21.74%.

Classical signs of Parkinson disease have been introduced in [8] which includes hypokinesia, bradykinesia, resting tremor and muscle rigidity. It is also discovered that 90% patients having PD will also develop speech disorders and swallowing.

This work [9] is a comprehensive review of speech and voice disorders commonly experienced by individuals with Parkinson's disease (PD) and the various treatment options available. The review discusses the limitations of evidence-based medicine and processes (EBM and EBP) in evaluating the efficacy of behavioural treatments for PD. The authors recommend that clinical practice should not be based solely on the amount of randomized controlled trials (RCTs) or other evidence. The review concludes that speech treatment can improve overall quality.

In [10] abnormal and asymmetric functions of PD has been considered. Additionally, the study highlights the importance of assessing and treating respiratory and control of phonatory deficits in individuals with PD, as these deficits are associated with abnormal laryngeal somatosensory function.

This work [11] is a study on volume perception in Parkinsonism speech. The study examines the ability of PD patients to regulate their speech volume and perceive it accurately. The study used a David Clarke headset and a Marantz tape recorder to collect data, and the data were subsequently edited and analysed using the KAY Elemetrics CSL4300 system. This study offers valuable insights into the challenges faced by those with hypophonic dysarthria and may be used to improve speech therapy for Parkinson's disease patients.

The findings [12] suggest that prior to clinical diagnosis, a significant decrease in fundamental frequency of free speech occurs. Two individuals with Parkinson's disease were included. The potential significance of these findings for early detection and treatment of PD is to provide useful biomarkers for evaluating the pharmacologic interventions efficacy in the very early stage.

The article [13] discusses the early signs of impairment in PD and how they manifest in vocalization. It covers the characteristic changes in voice on perceptual and acoustic analyses, as well as the physiological and anatomical correlates of these changes. The article also explores the potential of early detection of voice dysfunction in PD to lead to better treatment outcomes.

The study [14] mentions that the speech signal is characterized using signal processing algorithms to extract clinically useful features of PD progression. The study also mentions that an algorithm of robust feature selection is used to select the most parsimonious model, and that linear and nonlinear regression techniques.

29.3 Proposed Framework

In this section we are proposing a novel framework for processing of speech signal for detection of PD for the affected persons only and try to analyze the proposed system performance using some of the important performance matrices like accuracy, $F1$ score, Precision, etc. Below figure represents the framework. The objective of this proposed work is to develop a robust machine learning framework for distinguishing between healthy individuals and patients with PD using Voice data. By leveraging state-of-the-art machine learning techniques, such as deep learning and feature engineering, we aim to identify relevant biomarkers that can accurately discriminate between the two groups. Additionally, we will explore the potential of using different machine learning models and feature selection methods to enhance the performance and interpretability of our classification system (Fig. 29.2).

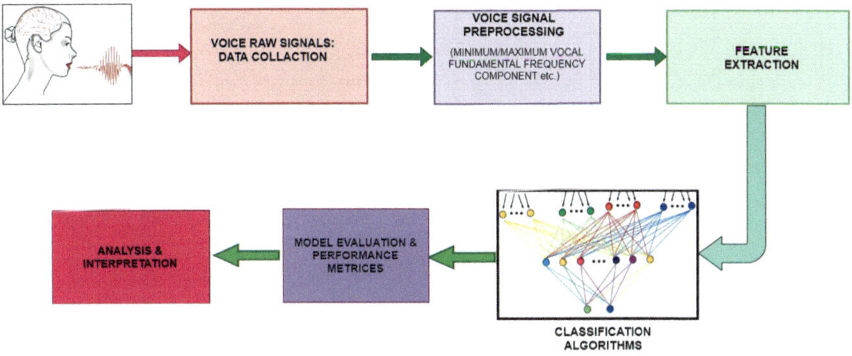

Fig. 29.2 Framework: voice-based PD detection framework

29.3.1 Data Collection and Preprocessing

UCI Machine Learning Repository PD Detection Dataset.

- Data set consists of Biomedical Voice Measurements from a total of 31 patients among which 23 have PD.
- Data is in the form of ASCII CSV format.
- There are around 6 recording per patient.

29.3.2 Feature Extraction

Data was present int the form of ASCII CSV format. Particular voice measure is represented by each column and 195 voice recordings are represented by row in the table. Attributes like Signal fractal scaling exponent, Minimum vocal fundamental frequency, Maximum vocal fundamental frequency were extracted and present for each patient in the CSV format. These extracted features will act as inputs to our predictive models based on machine and deep learning.

29.3.3 Machine Learning Algorithms

We propose a comprehensive investigation of two widely used machine learning algorithms for distinguishing between healthy individuals and PD patients using UCI data set. Our primary objective is to develop a robust classification framework capable of accurately identifying PD based on voice features. To achieve this, we explore the performance of two popular ML algorithms. We seek to determine the effectiveness of these ML algorithms in accurately classifying PD based on our data.

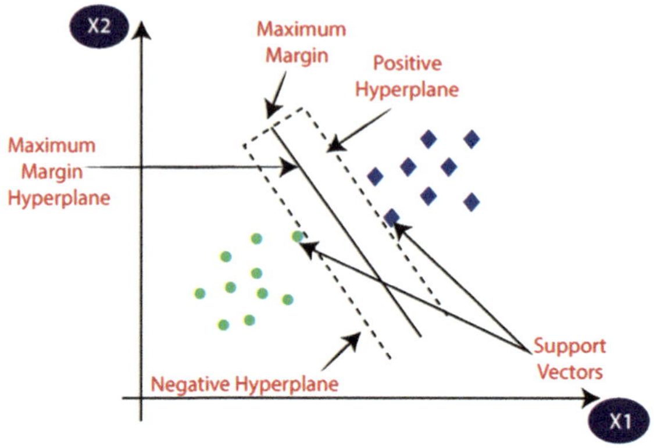

Fig. 29.3 Framework: support vector machine

Support Vector Machine [15]

- It is a binary classification algorithm which is used to find best hyperplane to separate different class data points with maximum margin (Fig. 29.3).

Random Forest [16]

- Random forests are an ensemble learning method that combines multiple decision trees to improve the accuracy and robustness of predictions.
- It randomly selects a subset of features to build each tree, which helps in improving accuracy.
- It uses multiple decision trees to make predictions (Fig. 29.4).

29.3.4 Model Evaluation and Performance Metrics

To measuring the effectiveness of proposed framework, we are evaluating the performance of the trained models using appropriate most important and widely used performance metrics. In our work we have used 8 metrics for classifications which include accuracy, precision, recall, $F1$ score, MAPE, MSE, RMSE, MAE. Using these performance metrics, we can identify the most suitable algorithm, feature extraction methods for our data.

29.3.5 Analysis and Interpretation

Upon evaluating the performance of our proposed framework for distinguishing between PD patients and healthy individuals using voice data, we obtained insightful

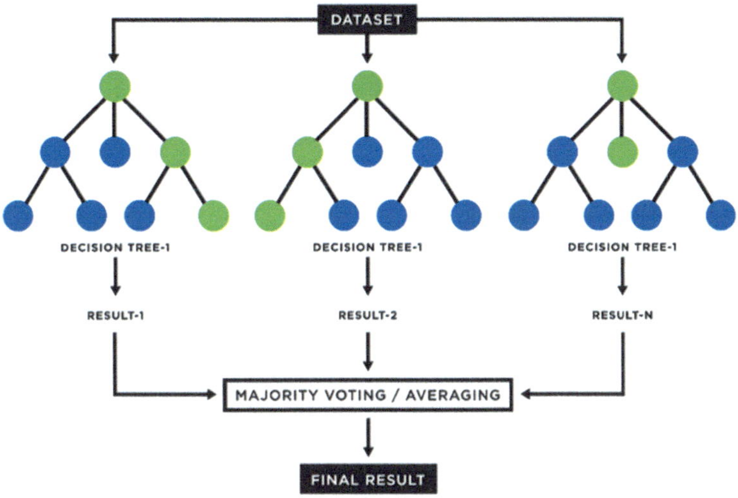

Fig. 29.4 Framework: random forest structure

results. The analysis of the experimental outcomes provides valuable insights into the efficacy and potential of ML algorithms and DL architectures in PD classification.

29.4 Result Analysis

This section aims to provide an in-depth examination and interpretation of the findings obtained through data analysis and experiments. This summary serves as an overview of the major discoveries and highlights the main outcomes of the study. We deduce conclusions using performance metrics like Mean Absolute Error, Accuracy, $F1$ Score, Precision, Recall. We can effectively communicate the significance and implications of their findings, encouraging further exploration and understanding within the academic community.

29.4.1 Performance Metrics

The performance metrics were calculated using a rigorous evaluation process. Tables 29.1 and 29.2 represent the performance metric obtained by implementing ML and DL algorithms on the UCI Machine Learning Repository Parkinson's Disease Detection Dataset.

Table 29.1 Model evaluation metrics

Model	Accuracy	$F1$ score	Precision	Recall
Random forest	0.99856	0.50000	0.66666	0.40000
SVM classifier	0.98695	0.33333	1.00000	0.20000

Table 29.2 Model evaluation metrics

Model	MSE	RMSE	MAE	MAPE
Random forest	0.57142	0.75592	0.57142	Inf.
SVM classifier	0.71428	0.84515	0.71428	10.987697

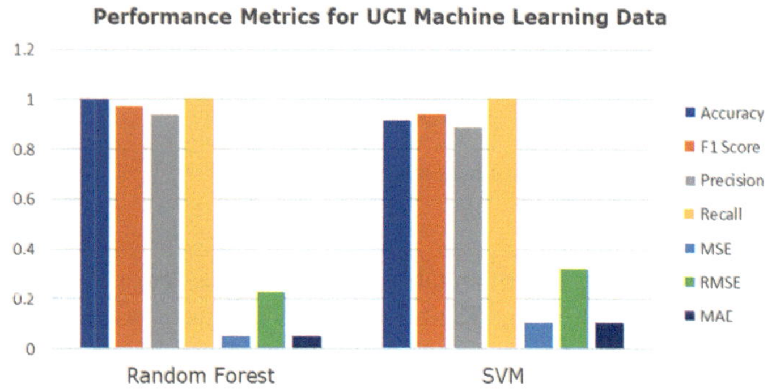

Fig. 29.5 Performance metrics for UCI machine learning data

Performance parameters calculated for both machine learning algorithms are shown Fig. 29.5 in form of bar graph to compare the matrices used for random forest and SVM.

29.4.2 *Confusion Matrices*

Figures 29.6 and 29.7 illustrate the confusion matrices for both machine learning algorithm on the all Features Labelled Voice Dataset. The confusion matrix provides insights into the performance of the algorithms by showing the FN (false negative), FP (false positive), TN (true negative), (TP) true positive classifications.

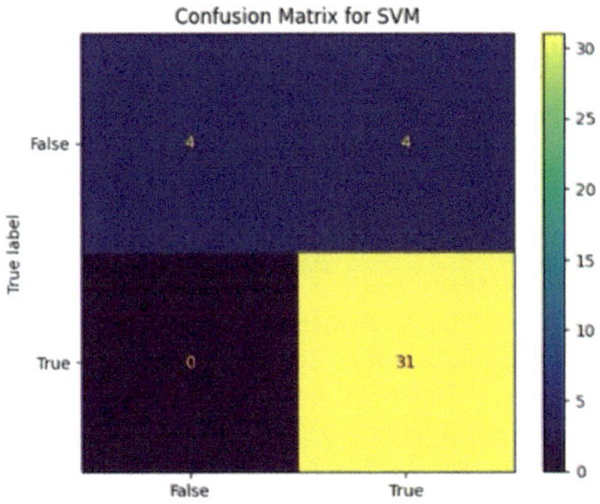

Fig. 29.6 Confusion matrix for SVM using UCI dataset

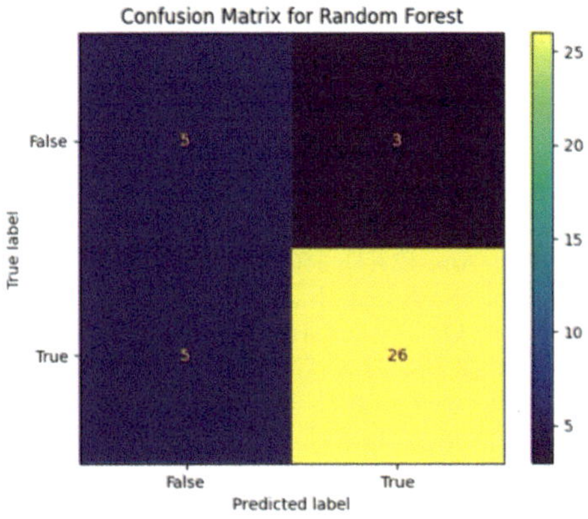

Fig. 29.7 Confusion matrix for random forest using UCI dataset

29.5 Conclusion

Intelligent Parkinson Disease Detection System has been explored in this work with potential of machine learning and artificial intelligence in the early diagnosis and monitoring of Parkinson's disease. A comprehensive investigation of two widely used machine learning algorithms for distinguishing between healthy individuals and PD patients using UCI dataset has been carried out. The analysis of the experimental outcomes provides valuable insights into the efficacy and potential of machine learning algorithms and deep learning architectures in PD classification. The proposed analysis concludes that machine learning algorithms and deep learning architectures have the potential to aid in the early diagnosis and monitoring of Parkinson's disease, and encourages further exploration and understanding within the academic community.

References

1. Davie, C.A.: A review of Parkinson's disease. Br. Med. Bull. **86**(1), 109–127 (2008). https://doi.org/10.1093/bmb/ldn013, https://academic.oup.com/bmb/article-pdf/86/1/109/1078190/ldn013.pdf
2. Mei, J., Desrosiers, C., Frasnelli, J.: Machine learning for the diagnosis of Parkinson's disease: a review of literature. Front. Aging Neurosci. **13** (2021). https://doi.org/10.3389/fnagi.2021.633752
3. Ouhmida, A., Raihani, A., Cherradi, B., Lamalem, Y.: Parkinson's disease classification using machine learning algorithms: performance analysis and comparison. In: 2022 2nd International Conference on Innovative Research in Applied Science, Engineering and Technology (IRASET), pp. 1–6 (2022). https://doi.org/10.1109/IRASET52964.2022.9738264
4. Brakedal, B., Toker, L., Haugarvoll, K., Tzoulis, C.: A nationwide study of the incidence, prevalence and mortality of Parkinson's disease in the Norwegian population. NPJ Parkinson's Dis. **8**(1), 19 (2022)
5. Fabbri, M., Guimarães, I., Cardoso, R., Coelho, M., Guedes, L.C., Rosa, M.M., Godinho, C., Abreu, D., Gonçalves, N., Antonini, A.: Speech and voice response to a levodopa challenge in late-stage Parkinson's disease. Front. Neurol. **8**, 432 (2017)
6. Rusz, J., Tykalová, T., Novotnỳ, M., Ržička, E., Dušek, P.: Distinct patterns of speech disorder in early-onset and late-onset de-novo Parkinson's disease. NPJ Parkinson's Dis. **7**(1), 98 (2021)
7. Rusz, J., Cmejla, R., Ruzickova, H., Ruzicka, E.: Quantitative acoustic measurements for characterization of speech and voice disorders in early untreated Parkinson's disease. J. Acoust. Soc. Am. **129**(1), 350–367 (2011)
8. Sapir, S., Ramig, L., Fox, C.: Speech and swallowing disorders in Parkinson disease. Curr. Opin. Otolaryngol. Head Neck Surg. **16**(3), 205–210 (2008)
9. Ramig, L.O., Fox, C., Sapir, S.: Speech treatment for Parkinson's disease. Expert Rev. Neurotherapeutics **8**(2), 297–309 (2008)
10. Hammer, M.J., Barlow, S.M.: Laryngeal somatosensory deficits in Parkinson's disease: implications for speech respiratory and phonatory control. Exper. Brain Res. **201**, 401–409 (2010)
11. Ho, A.K., Bradshaw, J.L., Iansek, R.: Volume perception in Parkinsonian speech. Movement Disorders Official J. Movement Disorder Soc. **15**(6), 1125–1131 (2000)
12. Harel, B.T., Cannizzaro, M.S., Cohen, H., Reilly, N., Snyder, P.J.: Acoustic characteristics of Parkinsonian speech: a potential biomarker of early disease progression and treatment. J. Neurolinguistics **17**(6), 439–453 (2004)
13. Ma, A., Lau, K.K., Thyagarajan, D.: Voice changes in Parkinson's disease: what are they telling us? J. Clin. Neurosci. **72**, 1–7 (2020)
14. Tsanas, A., Little, M., McSharry, P., Ramig, L.: Accurate telemonitoring of Parkinson's disease progression by non-invasive speech tests. Nat. Proc., 1 (2009)
15. Sood, T., Khandnor, P.: Classification of Parkinson's disease using various machine learning techniques. In: Advances in Computing and Data Sciences: Third International Conference, ICACDS 2019, Ghaziabad, India, 12–13 Apr 2019, Revised Selected Papers, Part I 3, pp. 296–311. Springer (2019)
16. Awad, M., Khanna, R., Awad, M., Khanna, R.: Support vector machines for classification. In: Efficient Learning Machines: Theories, Concepts, and Applications for Engineers and System Designers, 39–66 (2015)

Chapter 30
Performance Analysis of the Quantum Schemes in the Resource-Constrained IoT Network

Diksha Chawla and Pawan Singh Mehra

Abstract Recently, there has been significant expansion and the discovery of vast opportunities in communication enabled by the Internet of Things (IoT). When implementing IoT-enabled communication, it is crucial to give priority to security as a key area of focus. In network communication that links diverse sources, the generation and exchange of keys will become notably complex because of their constantly changing characteristics. As per analysis, lightweight cryptography is suitable for low-resource smart devices. Nevertheless, depending on the security of traditional cryptographic techniques is vulnerable to potential breaches by emerging quantum computing algorithms. The motivation for our analysis of quantum-based schemes for IoT communication frameworks stems from recent quantum threats to IoT security. Therefore in this paper, we analyzed the IoT security concerns and quantum-based solutions. The main objective of our work is to analyze the quantum-enabled schemes for securing IoT communication framework, the technological shift from classical to quantum cryptography and performance analysis of quantum-enabled schemes. This paper highlights the technical obstacles in schemes based on quantum technology and also performs security analysis by comparing classical and quantum-enabled authentication protocols for secure IoT communication.

30.1 Introduction

The connectivity of IoT applications through the internet has provided a futuristic vision of connected living, as represented in Fig. 30.1. Such widespread usage of IoT applications results in enormous amounts of information that must be processed and handled effectively. IoT is a growing technology that is continuously gaining attention year after year. The idea is that all devices are connected to provide a diverse range

D. Chawla (✉) · P. S. Mehra
Delhi Technological University, Rohini, Delhi 110042, India
e-mail: diksha03.chawla@gmail.com

P. S. Mehra
e-mail: pawansingh@dtu.ac.in

P. Nanda et al. (eds.), *Intelligent System and Data Analysis*, Smart Innovation, Systems and Technologies 398, https://doi.org/10.1007/978-981-97-5200-3_30

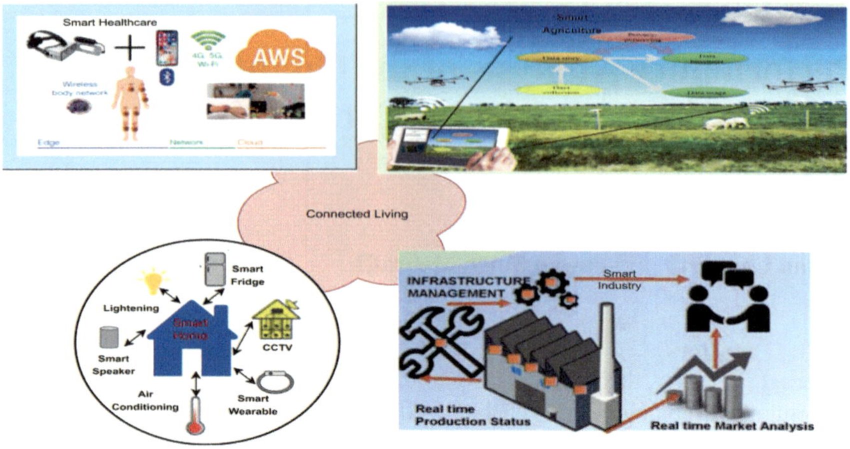

Fig. 30.1 Vision of connected living

of automated applications. These applications include smart grids, smart health care systems, and smart cyber-physical systems. The data generated from these devices is encrypted using classical cryptographic methods and considered unbroken [1]. The cryptography mechanism aims to help ensure the confidentiality, privacy and authentication of the users and data in the IoT ecosystem.

Quantum computing breaks the security of classical authentic and secure communication. Quantum computers are based on the laws of quantum mechanics. It can accomplish tasks such as reverse computing and evaluating the secret key in no time. The ability of quantum computers to solve complex mathematical problems in milliseconds applies to various classical encryption algorithms. It is due to quantum superposition that can hold both 0 and 1 states simultaneously and hence can perform parallel computation. The most secure RSA-encryption we use for financial transactions is RSA-2048 encryption. The quantum computer running Shor's algorithm with stable 4099 qubits can break RSA-2048 encryption in 10 s. Most of the internet security today relies on public key cryptography. The public key encryption algorithms used today are RSA, DSA, DH, ECDH, and ECDSA. The computational complexity of these algorithms is based on integer factorization and discrete logarithmic problems. Organizations such as the military, banks, and security organizations want to keep their data secure and confidential for the long term, say, 25 years. These organizations require adopting quantum cryptography techniques. QC uses quantum superposition and entanglement, which provide quantum parallelism. The QM principle, such as the Heisenberg uncertainty principle, is utilized to develop an indestructible cryptosystem. In the realm of quantum computing, the uncertainty principle asserts that making measurements disrupts the state of the quantum system. Hence, if an adversary tries to access information, the system collapses. In addition, photon polarization is used to make communication more secure. This nonstop run of photons, along with the QM principles, makes QC an appealing option for corroborating data security and conquering eavesdropping.

Based on the preceding discussion within this paper, we examine different quantum-enabled approaches for resource-limited IoT. To the best of the authors, the comparison of recent quantum-enabled schemes for resource-constrained IoT frameworks is not yet presented. Our work includes the analysis of security issues concerning IoT applications, classical cryptographic solutions and its comparison with quantum-enabled solutions for resource-constrained IoT.

30.1.1 Contributions and Outline

Our work's primary contributions can be summarized as follows:

1. Through an examination of IoT communication security concerns, we delve into the specific areas where these concerns are applicable.
2. A review of major classical authentication techniques for IoT security has been provided to identify and discuss the role of quantum in secure IoT-enabled smart applications.
3. Detailed discussions of the analysis of quantum-enabled schemes are investigated.
4. In this paper, we conducted a comparative analysis of quantum authentication and key distribution schemes with the specific goal of evaluating their effectiveness in a quantum context.
5. We are comparing the overall security achieved by quantum-enabled protocols with that of classical cryptographic schemes used in other authentication protocols.

The following sections of the paper are organized in the following manner: Sect. 30.2 provides an overview of current quantum authentication schemes and initiatives. Section 30.3 delves into the fundamentals of quantum cryptography. Section 30.4 explores conventional authentication methods. A detailed examination of quantum schemes is presented in Sect. 30.5, while Sect. 30.6 analyzes their performance. Section 30.7 addresses the technical challenges associated with implementing quantum-enabled schemes. Section 30.8 focuses on the security aspects achieved by the proposed protocol. Additionally, Table 30.1 elucidates commonly used abbreviations and symbols.

30.2 Literature Survey

The IoT has enabled novel applications, such as intelligent cities and remote healthcare. Considering E-health care applications, huge data, such as COVID-19 variants patient data, are generated daily. Smart healthcare applications are more vulnerable

Table 30.1 Abbreviations/symbols

Abbreviations/ Symbols	Description	Abbreviations/ Symbols	Description	Abbreviations/ Symbols	Description
QC	Quantum Computing	$\|V\rangle$	BRA-KET notation	ECC	Elliptic curve cryptography
PQC	Post-quantum Cryptography	$\|\Psi\rangle$	Wave-function	AES	Advanced Encryption Standard
RSA, DSA, DH	Rivest-Shamir-Adleman, Digital Signature, Diffie-Hellman	$\otimes$, $\|$	Tensor-Product Concatenation	BB84	Bennett and Gilles Brassard in 1984

to attacks due to the massive sensitive data collected daily. Cheng et al. [2] discussed how quantum computing affects various cryptographic protocols. Latif et al. [3] discussed mechanism based on controlled alternate quantum walks (CAQWs). They worked on video encryption using QW analogous to a simple random walk. They also proposed a scenario for implementing secure S-boxes [4] for 5G-IoT based on one-walker.

Furthermore, sensitive data is generated from applications such as smart E-healthcare information and financial transactions. These applications are not secure while transferring on 5G-cloud to store and process data remotely [14]. The proposed model provides mutual authentication and privacy of data by using QC for user registration and mutual authentication based on ECC for resource-constrained IoT environments. The UAV communications for monitoring and surveillance across borders where the data captured by them is of utmost importance. Hence, Sharma et al. [1] discussed the requirement for a secure communication network between UAVs. Mehra et al. [15] discussed the authentication protocol for securing IoT communication from attacks such as forward secrecy and reply attacks. Nagdewani et al. [16] discussed the home-based security and proposed home security system with an accuracy of 97.66%. Chawla et al. [17] discussed the need for quantum-based authentication protocols and proposed a quantum layer for secured IoT architecture. Table 30.2 shows the comparative analysis of quantum schemes for IoT communication. Considering the latest advancements in classical cryptographic methods and quantum schemes, we evaluate the performance of recent quantum schemes in comparison with state-of-the-art techniques.

30.3 Quantum Background

30.3.1 Quantum Key Distribution

In contrast to public key exchange schemes based on RSA and ECC, QKD schemes are based on the law of physics. The protocol achieves security by encoding information in quantum states of light. The distinctive attributes of quantum key distribution (QKD), rooted in the principles of quantum mechanics, encompass the Heisenberg uncertainty principle, the impossibility of cloning, and the phenomenon of quantum entanglement. These fundamental quantum mechanics laws make it impossible for an adversary to eavesdrop on a QKD system without perturbing the quantum state. QKD protocols leverage the Heisenberg uncertainty principle and quantum entanglement as their foundation.

Table 30.2 Comparative analysis of secure IoT communication techniques

Author	Objective	Quantum cryptography?	Algorithm	Security analysis	Results	Demerits
Ralegankar et al. [5]	Quantum cryptography	Yes	Classical attacks	Yes	Quantum UAV benefits	Implementation challenges are not discussed
Malina et al. [6]	Privacy enhancing techniques	No	PQC	Yes	Classical and PQ attacks on IoT communication-layered architecture are analyzed	PQC schemes are not covered
Broadment and Schaffner [7]	Quantum money	Yes	QKD	No	Quantum cryptography and its enabling technologies	Application security is not considered
Gupta and Dutta [8]	QKD protocol and ECC	Yes	QKD	Yes	Solutions of COW and ECC are analyzed	MIMA, DoS
Arul et al. [9]	QK-GRID Generation	Yes	QK-GRID	Yes	Quantum key GRID	IoT-specific applications not discussed
Mitchell [10]	5G-QC impact	No	PQC	No	5G AKA protocol is proposed	Challenges and open issues are not considered
Routray et al. [11]	QC robust security for IoT	Yes	QKD	Yes	Discussed quantum-safe protocols for IoT communication	Major threats are not analyzed
Bhatt and Sharma [12]	Integration of QC with IoT	Yes	BB84	No	QC and quantum computing are suitable from a commercial perspective	Integration challenges are not analyzed
EL-Latif et al. [13]	Quantum protocol for data protection	No	QWHF-1 and QWHF-2	Yes	Quantum walks (QWs) for generating cryptographic keys	High processing and cost

30.3.2 Superposition

A quantum computer with one qubit provides us with two-dimensional working space. For example, for 10 qubits, we get 1024 dimensions. Qubits represent a different set of possible values for example, three qubits represent eight possible values: $|\mathbf{000}\rangle$, $|\mathbf{001}\rangle$, $|\mathbf{010}\rangle$, $|\mathbf{011}\rangle$, $|\mathbf{100}\rangle$, $|\mathbf{101}\rangle$, $|\mathbf{110}\rangle$, $|\mathbf{111}\rangle$. The probability of collapsing qubit is $|\boldsymbol{\alpha}_{00}|^2$, $|\boldsymbol{\alpha}_{01}|^2$, $|\boldsymbol{\alpha}_{10}|^2$, $|\boldsymbol{\alpha}_{11}|^2$.

30.3.3 Quantum Entanglement

Qubits can be entangled directly, physically, or indirectly using quantum gates. The two qubits Bell-states are represented as in Eq. (30.1):

$$(|00\rangle_{AB} + |11\rangle_{AB})\backslash\sqrt{2} \tag{30.1}$$

When Alice conducts a measurement on her qubit within system A, it causes her qubit to undergo a collapse to a specific state $|\boldsymbol{a}\rangle_A$ as she observed. Simultaneously, Bob's qubit within system B also experiences a collapse, resulting in the state $|\boldsymbol{a}\rangle_{\boldsymbol{B}}$. Hence, both simultaneous measurements yield the same outcome. Additionally, entanglement is a property of the multiqubit system that can describe their collective state using the kronecker product (Tensor) as in Eq. (30.2):

$$\begin{aligned}|00\rangle &= \begin{bmatrix}1\\0\\0\\0\end{bmatrix} = \begin{bmatrix}1\\0\end{bmatrix} \otimes \begin{bmatrix}1\\0\end{bmatrix};\ |01\rangle = \begin{bmatrix}0\\1\\0\\0\end{bmatrix} = \begin{bmatrix}1\\0\end{bmatrix} \otimes \begin{bmatrix}0\\1\end{bmatrix}\\ |10\rangle &= \begin{bmatrix}0\\0\\1\\0\end{bmatrix} = \begin{bmatrix}0\\1\end{bmatrix} \otimes \begin{bmatrix}1\\0\end{bmatrix};\ |11\rangle = \begin{bmatrix}0\\0\\0\\1\end{bmatrix} = \begin{bmatrix}0\\1\end{bmatrix} \otimes \begin{bmatrix}0\\1\end{bmatrix}\end{aligned} \tag{30.2}$$

30.4 IoT Application Security Concerns with Quantum Solutions

The IoT applications such as smart homes, smart agriculture, and smart industrial IoT also suffer from impersonation, forgery, eavesdropping, and secret key attacks. Therefore, in this section, we discussed IoT-enabled specific application concerns:

30.4.1 Smart Homes

The IoT-based smart home systems provide many benefits to users by utilizing WSNs [18]. The currently used classical cryptographic techniques based on public key cryptography will not sustain the future. The more smart homes make user life convenient, the more people depend on smart remote devices. Simple encryption methods are insufficient to secure smart home applications from preventing quantum attacks. Due to the accessibility of multiple users, mutual authentication is required before login into the device.

30.4.2 Industrial IoT

The recent advancement in IIoT is based upon sensor network communication connected through the internet and creates security concerns. The technology controls many types of machinery and supervises critical applications of power plants, smart cities and health care systems. Post-quantum security for intelligent infrastructure has been analyzed. The authors argued that the use of post-quantum secure encryption algorithms may become necessary for ensuring the long-term security of data in compliance with legal mandates and national or international regulations.

30.4.3 E-Healthcare

In the twenty-first century, the e-healthcare application came forward to address the shortcomings of classical healthcare systems [19]. The growing utility of quantum computers imposes additional demands on healthcare institutions to employ encryption methods resilient against quantum threats. While existing encryption protocols would require traditional computers numerous millennia to decipher, quantum computers can achieve this feat in a significantly shorter timeframe. Chawla et al. [20] proposed a quantum-enabled solution to secure patient health data from classical and futuristic quantum attacks.

30.5 Overview of Classical Authentication Schemes

30.5.1 Public Key Encryption

The encryption is based on the mathematical framework, where public and private keys are generated randomly. Anyone with access to public keys can encrypt data, but only authorized users with private keys can decrypt the information. The researchers

[21] also focus on ECC due to its advantages in lightweight cryptography [22]. In IoT, alternative approaches such as hyper-elliptic curve cryptography, NTRU and Blue Ray were proposed, which are based on the shorter key size.

30.5.2 Private Key Encryption

The same key is used at both ends to send and receive messages [15]. The major challenge is to share secret keys with legitimate users. It is difficult to identify eavesdroppers in the communication channel; therefore, ensuring that keys are shared with only legitimate users is a difficult task. Based on the AES cryptographic primitive, the author [23] proposed a lightweight security enhancement scheme. Their work is based on a dynamic-key generation process based on time to provide authentication. RFID security is also analyzed to protect user location information.

30.6 Analysis of Quantum Cryptography Authentication Schemes

This section considers the research findings and discussion of quantum schemes in the context of a resource-constrained IoT environment. This section shows the trends of previous studies considering quantum cryptography solutions for securing IoT communication frameworks.

30.6.1 Quantum Systematic Analysis

The authors [12] proposed the weaknesses of classical encryption techniques discussed the drawbacks of classical encryption techniques and discussed quantum computing that employs a flow of photons for data transmission, and these photons possess a characteristic called "spin." There exist four fundamental spin orientations: horizontal, vertical, 45° diagonal and negative 45° diagonal. In the realm of physics, the Heisenberg uncertainty principle is a captivating concept, asserting that it is impossible to precisely measure all the attributes of a particle without causing a disturbance to its existing state. The recent impact of quantum on IoT security followed by considering how cryptography is used in 5G and by exploiting backwards-compatibility features of the 5G security system has also been discussed [10]. However, the author is unable to provide light on new cryptographic protocols which can resist quantum attacks.

Furthermore, the pursuit of data security dates back to the Spartan era, and today, unmanned aerial vehicles (UAVs) are gaining widespread popularity across various fields like agriculture, military and versatile mobility. Through an examination of [5], the paper sheds light on current security challenges in real-world scenarios and underscores the significance of quantum cryptography in safeguarding information compared to conventional solutions. The authors discussed the role of quantum coin flipping, quantum commitment, and position-based quantum cryptography for secure UAV authentication. Entering a new decade, intelligent infrastructure (II) services seamlessly incorporated into the Internet of Things (IoT) stand at the forefront of technological progress. The authors [6] discussed the majority of existing public key cryptography (PKC) methods are theoretically susceptible to potential attacks carried out by quantum computers. Post-quantum cryptography (PQC) provides remedies to counter these threats. In the current age of the internet, sensitive information is often sent over insecure channels. Given the substantial advancements in quantum computing, there's a demand for absolute security in safeguarding confidential data. Quantum key distribution protocols are established as secure as long as all the devices involved are flawless [24]. Therefore, quantum key distribution protocols based on Heisenberg uncertainty principle such as BB84, BB92, SARG04, Six-State protocol and quantum entanglement-based such as E91 protocol, COW and DPS protocols are discussed by Kumar et al. [24].

Considering the necessity of quantum cryptography for future internet-based security, Chawla et al. [25] recently proposed a survey to establish a structured guide in the realm of quantum-secured IoT communication within the context of 5G. This guide will encompass the latest research on 5G-enabled IoT, including its pivotal enabling technologies, the potential risks faced by 5G-enabled IoT applications and the most advanced quantum-based solutions and initiatives available. In our work, an in-depth analysis of quantum cryptography for resource-constrained IoT considering the 5G-enabled IoT security concerns, an in-depth analysis of classical authentication-based schemes, a thorough investigation of quantum cryptography for IoT framework, an analysis of post-quantum cryptography and the comparison of quantum cryptography and post-quantum cryptography has been provided.

30.6.2 Quantum Authentication Protocols

We provide insight to the readers about how QC has proven to be a prominent technology for solving security concerns in IoT. Currently, IoT objects are extensively used and provide many benefits to the user. Kalaivani [26] discussed quantum-based authentication for securing healthcare data. Such data is generated from wireless body sensor networks (WBSN). The authors proposed the enhanced BB84 quantum cryptography protocol. The proposed protocol is based on the following few steps: (a) By converting the random numbers into the required set of qubits. (b) Sharing the qubits between sender and receiver by considering the random basis. (c) Comparison is performed to generate the final key. The proposed protocol is proven secure compared to classical authentication schemes such as RSA, ECC and AES. However, the authentication in healthcare scenario is not thoroughly investigated. To secure the banking transactions the authors proposed the idea of a quantum cheque [27]. The authors secure the cheque amount by securely generating and distributing the BB84 quantum key. The authors proposed the quantum locker for securing the messages. The proposed scheme is based on generating the quantum password. The scheme lacks in providing the authentication model for securing data generated from entities. Considering the drawbacks of authentication protocol Chawla et al. [20] protocol comprises five steps such as (a) ***Preparation phase***: During this stage, each participant within the communication network is assigned a distinct identifier. Furthermore, these entities must have the capability to generate both asymmetric and symmetric keys. (b) ***Gen Algorithm***: In this step, entities create maximally entangled states among themselves, including GWN, medical server, and professionals, as in Eq. (30.3):

$$\left|\psi^{\pm}_{0,0\oplus r,0\oplus u}\right\rangle_{ABC} = \frac{1}{\sqrt{2}}(|0, 0 \oplus r, 0 \oplus u\rangle + |1, 1 \oplus r, 1 \oplus u\rangle) \tag{30.3}$$

(c) ***Registration Algorithm***: In this step, the patient initiates registration with the medical server using their mobile device (GWN). Medical professionals also complete their registration with the medical server through the gateway. The GWN validates the registration information for authentication and subsequently shares this data with the medical server to ensure secure future logins.

1. **Patient Registration**: In this the patient's secret credentials are encoded as in Eq. (30.4):

$$|\phi\rangle_{\mathrm{GWSN}} = Q2 \otimes |\emptyset\rangle_{\mathrm{GHZ}} \tag{30.4}$$

   where $Q2 = h(\mathrm{ID}_p||\mathrm{ID}_{\mathrm{SN}_i}||\mathrm{ID}_{\mathrm{GW}}||\mathrm{ID}_{\mathrm{MP}})$. Such information is securely transferred using a quantum teleportation channel to authenticated entities.
2. **Medical Professional Registration**: Each medical professional must register onto the medical server by sharing the credentials such as $\mathrm{Q3} = h(\mathrm{ID}_p||\mathrm{PW}_p)$. The medical professional gets registered on the medical server by generating the Quantum Green Health Card (QGHC).

(d) ***Login Algorithm***: During this step, the authorized patient securely accesses the system to request medical advice from a certified physician.

(e) ***Authentication Algorithm***: In this stage, the patient, medical server (MS), and medical professional (MP) mutually authenticate themselves over a secure quantum channel. Each user must be securely login into the system by decrypting $|\phi\rangle_{\mathrm{GWSN}} = \mathrm{Q2} \otimes |\emptyset\rangle_{\mathrm{GHZ}}$. To secure the data from any unauthorized entities modified quantum key approach has been adopted as in Eq. (30.5)

$$\mathrm{QK} : k_{MP} \oplus k_{P} \oplus k_{CMS}. \tag{30.5}$$

The proposed protocol is for resource-constrained IoT devices. Quantum gates such as X and Z, Hadamard and CNOT are used for effective realization of the proposed scheme. The secure generation of GHZs states allows only authenticated

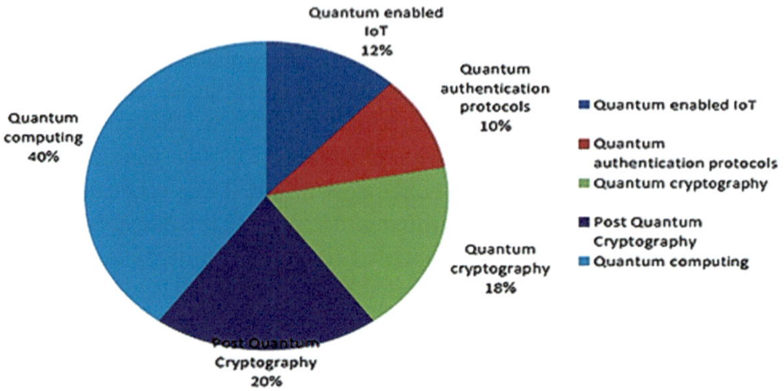

Fig. 30.2 Ratio of available quantum trends on databases (Web of Science, Science Direct, and IEEE Explore)

entities to retrieve the data. Figure 30.2 shows the ratio of recently available content of quantum cryptography and quantum-enabled IoT schemes.

30.7 Quantum Authentication Protocols: Performance Analysis

The performances of the quantum protocols are evaluated in terms of key generation time and success percentage of quantum schemes. The key generation time is evaluated by comparing the quantum schemes with classical cryptographic protocols as represented in Fig. 30.3. The successful quantum-based schemes can be evaluated in terms of total computation cost as in Fig. 30.4. The computation cost is computed by considering the total number of hash operations, messages exchanged, the number of entities involved and sacrificed qubits. As per the analysis of Chawla et al. [20], the number of entities is more, and; the total number of hash operations, messages exchanged, number of entities involved and sacrificed qubits are less compared to other schemes. Therefore, the proposed schemes by Chawla et al. [25] are considered successful as in this with more number of entities less computation is involved.

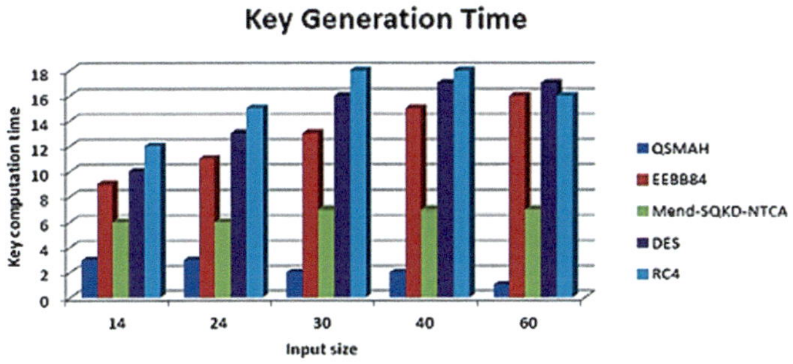

Fig. 30.3 Comparative analysis of key generation time

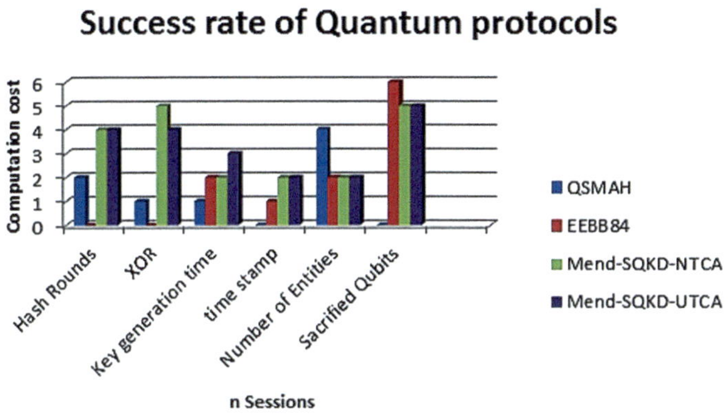

Fig. 30.4 Success percentage of quantum schemes

30.8 Technical Challenges

30.8.1 Cost Efficiency

The expense linked to quantum computers is significant because they cannot be maintained at room temperature. The sole method of accessing them is via a cloud-based platform. Nevertheless, if the reliance on cloud services grows following the application's requirements, it will also drive up costs due to a pay-as-you-use model.

30.8.2 Quantum Bit Error Rate

It represents the relationship between the error rate and the key rate, serving as an indicator of the presence of a potential eavesdropper. When there is an adversary in the quantum channel, extra errors are introduced into the raw key material, resulting in a higher rate of information leakage to the eavesdropper. Nevertheless, various methods have been suggested in existing literature, including the use of dual quantum channels and two-way classical communication.

30.8.3 Quantum Key Size

It represents the relationship between the error rate and the key rate, serving as an indicator of the presence of a potential eavesdropper. When there is an adversary in the quantum channel, extra errors are introduced into the raw key material, resulting in a higher rate of information leakage to the eavesdropper. Nevertheless, various methods have been suggested in existing literature, including the use of dual quantum channels and two-way classical communication.

30.8.4 Heterogeneous IoT Devices

Implementing a new level of security according to different end-user requirements is challenging, if the number of pre-shared EPR pairs is exponential in the size of the messages of the protocol.

30.9 Security Analysis

In this segment, we examine and deliberate on the security attributes achieved by contemporary classical authentication protocols and quantum authentication protocols. We scrutinize how the quantum-enabled protocol QSMAH fulfils nearly all the prerequisites for securing resource-constrained IoT communication, while others fall short in achieving some aspects, as detailed in Table 30.3.

30.9.1 Security Features

(a) **GWN Impersonation**: To thwart an attacker's attempt to intercept transmitted messages from a gateway node (GWN), it is imperative to employ an authenticated gateway node.
(b) **Denial of Service (DoS)**: A Denial of Service (DoS) attack can be illustrated by envisioning a situation in which an assailant inundates the target with excessive traffic, to seize and entirely exhaust its memory resources.

Table 30.3 Comparison of the security feature of quantum with other authentication protocols

Security consideration	Moulick et al. [28]	Mehra et al. [29]	Kumar Dwivedi et al. [30]	Khan et al. [31]	Mehra et al. [15]	Anusuya Devi and Kalaivani et al. [26]	QSMAH [20]
Replay attack	✓	✓	–	–	–	–	✓
MITM attack	–	✓	–	–	✓	✓	✓
DoS attack	–	✓	–	–	–		✓
Key exchange attack	–	–	–	✓	–	✓	✓
GWN impersonation	✓	–	–	–	–	–	✓
Traffic analysis	–	–	–	–	–	–	✓
Sensor impersonation	✓	–	–	–	–	–	✓
User impersonation	–	✓	–	✓	✓	–	✓
Mutual authentication	–	–	–	✓	–	–	✓

(c) **Traffic Analysis**: The assailant eavesdrops on network communications with the intention of discerning patterns in application behaviour, the configuration of routing, the positions of critical nodes, and the locations of base stations.

30.10 Conclusion and Future Work

IoT communication uses sensors that gather and share data. Classical cryptographic algorithms based on complex mathematical primitives such as integer factorization are used to secure IoT communication. Such cryptographic primitives, if broken by an attacker, will seriously impact user privacy, confidentiality and authentication of data shared in the IoT framework, as quantum computers can break classical cryptographic primitives in milliseconds by searching for weaknesses within the network. In our work, we identified IoT application-specific concerns and provided an in-detail analysis of quantum-based schemes. We reviewed the various quantum-enabling technologies. Our work considers performance analysis and security comparison of various quantum-enabled schemes with classical authentication protocols for a secure IoT communication framework. Quantum cryptography is at its peak. A lot of research has been done on simulations of quantum experiments. Quantum based error correction; quantum-portfolio risk optimization techniques presently have high expectations and will continuously evolve in the next ten years.

References

1. Sharma, J., Mehra, P.S.: Secure communication in IOT-based UAV networks: a systematic survey. Internet of Things **23**, 100883 (2023). https://doi.org/10.1016/j.iot.2023.100883
2. Cheng, C., Lu, R., Petzoldt, A., Takagi, T.: Securing the Internet of Things in a Quantum World. IEEE Commun. Mag. **55**, 116–120 (2017). https://doi.org/10.1109/MCOM.2017.1600522CM
3. El-Latif, A.A.A., Abd-El-Atty, B., Mazurczyk, W., Fung, C., Venegas-Andraca, S.E.: Secure data encryption based on quantum walks for 5G Internet of Things scenario. IEEE Trans. Netw. Serv. Manage. **17**, 118–131 (2020). https://doi.org/10.1109/TNSM.2020.2969863
4. Langenberg, B., Pham, H., Steinwandt, R.: Reducing the cost of implementing the advanced encryption standard as a quantum circuit. IEEE Trans. Quant. Eng. **1**, 1–12 (2021). https://doi.org/10.1109/tqe.2020.2965697
5. Ralegankar, V.K., Bagul, J., Thakkar, B., Gupta, R., Tanwar, S., Sharma, G., Davidson, I.E.: Quantum cryptography-as-a-service for secure UAV communication: applications, challenges, and case study. IEEE Access **10**, 1475–1492 (2021). https://doi.org/10.1109/access.2021.3138753
6. Malina, L., Dzurenda, P., Ricci, S., Hajny, J., Srivastava, G., Matulevicius, R., Affia, A.A.O., Laurent, M., Sultan, N.H., Tang, Q.: Post-quantum era privacy protection for intelligent infrastructures. IEEE Access **9**, 36038–36077 (2021). https://doi.org/10.1109/ACCESS.2021.3062201
7. Broadbent, A., Schaffner, C.: Quantum cryptography beyond quantum key distribution. Des. Codes Cryptogr. **78**(1), 351–382 (2016). https://doi.org/10.1007/s10623-015-0157-4
8. Gupta, S., Dutta, C.: Internet of Things security analysis of networks using quantum key distribution. Indian J. Sci. Technol. 9, 48 (2016). https://doi.org/10.17485/ijst/2016/v9i48/105551
9. Arul, R., Raja, G., Almagrabi, A.O., Alkatheiri, M.S., Chauhdary, S.H., Bashir, A.K.: A quantum-safe key hierarchy and dynamic security association for LTE/SAE in 5G scenario. IEEE Trans. Industr. Inf. **16**, 681–690 (2020). https://doi.org/10.1109/TII.2019.2949354
10. Mitchell, C.J.: The impact of quantum computing on real-world security: a 5G case study. Comput. Secur. **93**, 1–11 (2020). https://doi.org/10.1016/j.cose.2020.101825
11. Routray, S.K., Jha, M.K., Sharma, L., Nyamangoudar, R., Javali, A., Sarkar, S.: Quantum cryptography for IoT: a perspective. In: IEEE International Conference on IoT and its Applications, ICIOT 2017 (2017). https://doi.org/10.1109/ICIOTA.2017.8073638
12. Bhatt, A.P., Sharma, A.: Quantum cryptography for internet of things security. J. Electron. Sci. Technol. **17**, 213–220 (2019). https://doi.org/10.11989/JEST.1674-862X.90523016
13. EL-Latif, A.A.A., Abd-El-Atty, B., Venegas-Andraca, S.E., Mazurczyk, W.: Efficient quantum-based security protocols for information sharing and data protection in 5G networks. Fut. Generation Comput. Syst. 100, 893–906 (2019). https://doi.org/10.1016/j.future.2019.05.053
14. Gupta, P., Raj, P., Tiwari, S., Kumari, P., Mehra, P.S.: Energy efficient diagonal based clustering protocol in wireless sensor network. SSRN Electron. J. (2020). https://doi.org/10.2139/SSRN.3565781
15. Mehra, P.S., Doja, M.N., Alam, B.: Codeword authenticated key exchange (CAKE) light weight secure routing protocol for WSN. Int. J. Commun. Syst. **32**, 1–27 (2019). https://doi.org/10.1002/dac.3879
16. Nagdewani, H., Mehra, P.S.: A complete Internet of things based home security system. In: 2022 Second International Conference on Computer Science, Engineering and Applications (ICCSEA), pp. 1–5. IEEE (2022). https://doi.org/10.1109/ICCSEA54677.2022.9936168.
17. Chawla, D., Mehra, P.S.: A survey on quantum computing for Internet of Things security. Proc. Comput. Sci. **218**, 2191–2200 (2023). https://doi.org/10.1016/j.procs.2023.01.195
18. Dwivedi, A.K., Sharma, A.K., Mehra, P.S.: Energy efficient sensor node deployment scheme for two stage routing protocol of wireless sensor networks assisted IoT. ECTI Trans. Electr. Eng. Electron. Commun. 18, 158–169 (2020). https://doi.org/10.37936/ECTI-EEC.2020182.240541

19. Mehra, P.S., Mehra, Y.B., Dagur, A., Dwivedi, A.K., Doja, M.N., Jamshed, A.: COVID-19 suspected person detection and identification using thermal imaging-based closed circuit television camera and tracking using drone in Internet of Things. Int. J. Comput. Appl. Technol. **66**, 340–349 (2021)
20. Chawla, D., Mehra, P.S.: QSMAH: a novel quantum-based secure cryptosystem using mutual authentication for healthcare in the Internet of Things. Internet of Things 100949 (2023). https://doi.org/10.1016/j.iot.2023.100949
21. Mehra, P.S., Doja, M.N., Alam, B.: Energy efficient self organising load balanced clustering scheme for heterogeneous WSN. In: 2015 International Conference on Energy Economics and Environment (ICEEE), pp. 1–6 (2015). https://doi.org/10.1109/EnergyEconomics.2015.7235092
22. Mehra, P.S., Doja, M.N., Alam, B.: Low energy adaptive stable energy efficient (LEASE) protocol for wireless sensor network. In: 2015 International Conference on Futuristic Trends on Computational Analysis and Knowledge Management (ABLAZE), pp. 484–488 (2015). https://doi.org/10.1109/ABLAZE.2015.7155044
23. Musliyana, Z., Arif, T., Munadi, R.: Security enhancement of Advanced Encryption Standard (AES) using time-based dynamic key generation. ARPN J. Eng. Appl. Sci. **10**, 8347–8350 (2015)
24. Kumar, A., Garhwal, S.: State-of-the-art survey of quantum cryptography. Arch. Comput. Methods Eng. (2021). https://doi.org/10.1007/s11831-021-09561-2
25. Chawla, D., Mehra, P.S.: A roadmap from classical cryptography to post-quantum resistant cryptography for 5G-enabled IoT: challenges, opportunities and solutions. Internet of Things **24**, 100950 (2023). https://doi.org/10.1016/j.iot.2023.100950
26. Anusuya Devi, V., Kalaivani, V.: Enhanced BB84 quantum cryptography protocol for secure communication in wireless body sensor networks for medical applications. Pers. Ubiquit. Comput. **27**(3), 875–885 (2021). https://doi.org/10.1007/s00779-021-01546-z
27. Behera, B.K., Banerjee, A., Panigrahi, P.K.: Experimental realization of quantum cheque using a five-qubit quantum computer. Quantum Inf. Process. **16**, 1–12 (2017). https://doi.org/10.1007/s11128-017-1762-0
28. Moulick, S.R., Panigrahi, P.K.: Quantum cheques. Quantum Inf. Process. **15**, 2475–2486 (2016). https://doi.org/10.1007/s11128-016-1273-4
29. Mehra, P.S.: LBECR: load balanced, efficient clustering and routing protocol for sustainable internet of things in smart cities. J. Ambient Intell. Human Comput. **14**, 10493–10515 (2023). https://doi.org/10.1007/s12652-022-03706-2
30. KumarDwivedi, A., Sharma, A.K., Singh, S., Mehra, P.S.: SCSZB: Sensor congregate stable zonal-based routing protocol designed for optimal WSN. In: Agrawal, R., Kishore Singh, C., Goyal, A. (eds.), Advances in Smart Communication and Imaging Systems. pp. 139–149. Springer, Singapore (2021). https://doi.org/10.1007/978-981-15-9938-5_15
31. Khan, M.K., Kumari, S.: An improved user authentication protocol for healthcare services via wireless medical sensor networks. Int. J. Distrib. Sensor Netw. **10**, 347169 (2014). https://doi.org/10.1155/2014/347169

Chapter 31
E-Healthcare Record System Using Blockchain

Parthasarathi Pattnayak, Arpeeta Mohanty, Tulip Das, and Sanghamitra Patnaik

Abstract The term "Electronic Healthcare Record" (EHR) refers to an electronic recording of the medical record of a patient that is maintained by the system over time and can include all relevant, authoritative clinical information important to the user's care. All of the patient's medical data (also called PHI) about the diagnosis of disease along with diagnostic reports is saved and communicated across the system's users within a contemporary, stored in the cloud E-healthcare system. But a significant concern is the protection of these private documents from misuse and unauthorized access. In this situation, the blockchain-based record-sharing solution might be crucial for safe medical data exchange. In this paper we discussed the session data will be encrypted using intelligent deals with between planetary file solution (IPFS)-based secured healthcare picture security solution. For improved accountability, it also logs each health care interaction in the blockchain. For healthcare services, this cutting-edge paradigm offers significant adaptability and EHR accessibility.

31.1 Introduction

Healthcare records (EHRs) are a vital organization for collecting health records that support the electronic availability of paper-based individual healthcare records on the internet [1]. The COVID-19 pandemic is being addressed by the global healthcare management system, which is supported by the e-healthcare system. The improvement within social healthcare interactions, from small practices to massive events, is expected to be facilitated by health data creativity, particularly EHR. Based on these potential benefits, the Health Information and Technology in Clinical and Economic

P. Pattnayak · A. Mohanty
School of Computer Applications, KIIT Deemed to Be University, Bhubaneswar, Odisha, India

T. Das (✉)
Faculty of Emerging Technologies, SRI SRI University, Cuttack, Odisha, India
e-mail: tulipdas6387@gmail.com

S. Patnaik
School of Law, KIIT Deemed to Be University, Bhubaneswar, Odisha, India

P. Nanda et al. (eds.), *Intelligent System and Data Analysis*, Smart Innovation, Systems and Technologies 398, https://doi.org/10.1007/978-981-97-5200-3_31

Health (HITECH) Act supports the adoption of EHRs. Many emergency clinics are working to integrate these systems and show their usefulness in the recovery process. This framework is used to give patients the freedom to create, manage, and share their electronic health records (EHRs) with loved ones, friends, social insurance providers, and other authorized information buyers [2]. The advancement programme of the medical services arrangement must also be completed, given that the social service researcher as well as suppliers of such aid get to these EHRs via the on boarding. However, under the current situation, patients share their electronic health records (EHRs) across several areas during life events, causing the EHRs to transfer from a single specialist co-op database to the next. As a result, the patient may no longer have authority over the current medical management information, however the pro association largely maintains the necessary management. The e-healthcare record system built on blockchain will foster strong inter-organizational trust. EHRs are now managed by both hospitals along with their suppliers, with patients being denied the chance to unrestrictedly control their own EHRs. The development of the blockchain for electronic health records (EHRs) is made possible by the use of blockchain innovation [3]. The technology also keeps track of every exchange's assessment data in a permanent distributed record, ensuring honesty and transparency during the information-transfer process. So, the patient gives the clinic permission to monitor health care data and professional analytical information.

The manual method is the traditional methodology used in the majority of medical clinics to maintain data. Hospitals have access to each of information because it is kept within a database in an organized centralized way [4]. The individual receiving treatment must transport the documents and files that make up the medical records. The existing method has the drawback that patient access to medical information is intermittent and dispersed. Therefore, since the patient must constantly speak with hospital administration, acquiring past records or other significant details regarding the past could prove challenging to them. The other issue with the present system is that, despite the significance it is to maintain patient privacy and ensure the safety of their medical histories, clinicians along with doctors can still access information that has been saved as shown in Fig. 31.1 [5, 6].

The architecture of electronic healthcare records, which is totally controlled with practices compared to patients, is shown in the above image. This makes it difficult to seek physician guidance at various hospitals. While treatment or examining the individual, physicians or administrators of the hospital upload data/records on a server in this method. By the aforementioned method, individuals must request permission to their health history/record from the healthcare facility administration each time because only the medical administration gets access to it. Under this method, the patient may find it complicated to obtain their health record from the hospital. The main problem with this approach is losing data because of the centralized structure. The major objective aims to employ "Blockchain Technology" to categorize durable medical documents. One of the most crucial fundamentals of medicine is the ability to tolerate data uncertainty or classification. In addition to being an issue of ethical decency, protecting a patient's private information is essential to building confidence among the patient and the physician. The purpose of this study was to

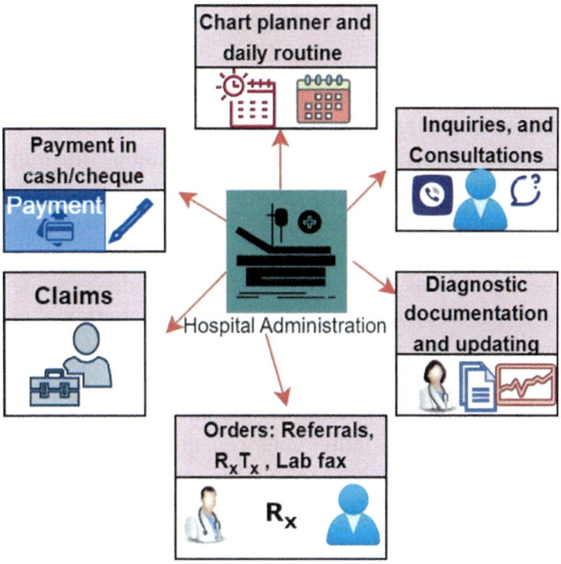

Fig. 31.1 Centralized EHR system

determine if emergency facilities with partially or no completed EHR structures could be suggested to quiet findings compared to medical clinics with fully achieved EHR setups after adjusting for other key patient and clinic characteristics. This research provides fresh information on the relationship among an EHR system's operation and, consequently, the type of social services provided in a hospitalized context.

31.2 Related Work

Patients must carry the information in a variety of files and papers since the historic technique, which is still used by the majority of clinics to manage EHR, is centralised [7]. Dealing with extensive data retrieval and storage represents a few of the system's main limitations [6]. However, in contemporary times, individuals now find that managing medical records is a simple process because of the decentralized method [2]. Protecting individuals' vital records is another big problem in the conventional system; nonetheless, people may hold doctors and hospitals liable if they handle patients' records carelessly, manipulate them, or compromise confidentiality [8]. By this time, ensuring privacy is a key component in the system, which allows individuals to control access to their records by physicians and hospitals [9]. By implementing a decentralised e-healthcare system, health records can be shared easily and swiftly across user with medical sectors [5].

31.3 Proposed System

The state-of-the-art healthcare framework uses computerized methods to maintain patient health records. Human services organizations generally employ an integrated method for preserving patient information, illustrative reports, and expert advice. Given that patients have no control over personal information and that exchanging the recordings provides a time-consuming and expensive process, it is a system with a high risk of data being leaked or exploited for a variety of causes. In order to address these problems, the system proposes the idea of switching from an isolated structure to a decentralized structure using blockchain technology. Individual medical data frequently remain through the blockchain network using their unique IDs and public keys. Associations for social security of a patient's unknowable information can be accessed through careful agreements using only the intriguing IDs' coordinates. While necessary, individuals can give social security insurance organizations the general open key, but without the individual key, the information would always be unrecognizable. As a result, clever contracts are essential to guaranteeing the clients' willingness that share/recover the information (Fig. 31.2).

A. *Overview on Blockchain*

Blockchain includes a network of blocks that are linked sequentially and constantly expanding as a result of storing transactions within rectangles. This level makes use of a decentralized methodology that enables data dissemination and shares ownership of every bit with appropriated data, or often known as information [10]. A distributed ledger offers advantages including protection, privacy, and data consistency without outside interference [8]. Due to the rise in creativity within the health care sector, security of individuals medical data has become an essential requirement [18], making it a cost-effective way to keep patient's medical records secure. A number of researchers have also noted how applying the technology of blockchain in the healthcare industry could represent a workable solution. A typical blockchain technology deal technique, through which the sender may produce an operation within a block, is shown in Fig. 31.3. Using hashing with cryptography, the aforementioned block can be shared and verified. Additionally, producers carry out, reward, and send the shared hashed activity to the person who will receive it.

B. *The Structure of the System Flow*

The overall system's flow is shown in Fig. 31.4, and this comprises three primary modules: administration, patient care, medical care, and hospitals. Besides these components, the suggested system's operational process operates as follows:

- The administrator must first log in and is then in charge of creating physician and hospital profiles using their accurate information.
- The patient must first log in and register with the system.
- The patient's first significant responsibility is to create a healthcare ID, which should be employed as an individual's identity.
- The patient is able to provide their medical and personal records.

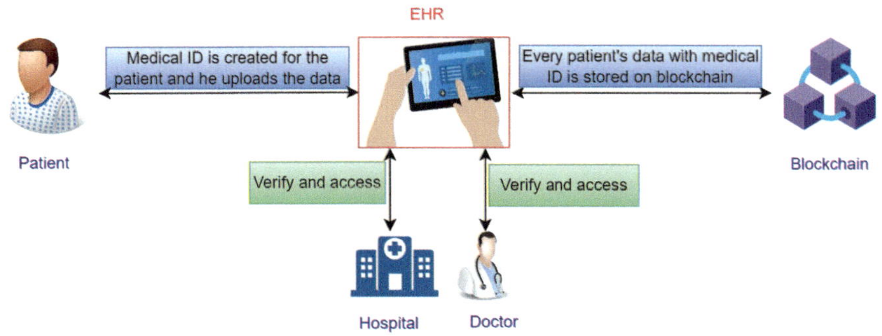

Fig. 31.2 Description of the EHR system based on blockchain

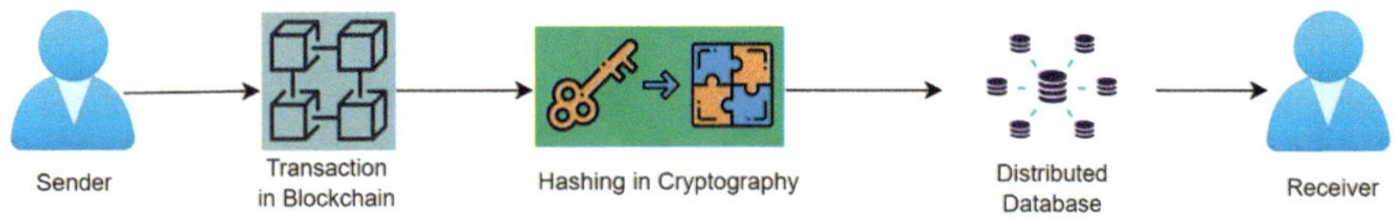

Fig. 31.3 Blockchain process

- The person receiving treatment has the power to give or revoke the right to access documents based on security considerations.
- Doctors and hospitals must log into the system to look for patients using unique medical identifiers, similar to administrators and patients.
- The patient must respond to a record accessibility request before the doctor or hospital can view their records.
- Both the doctor and hospital are able to read the records if a patient gives approval for access.

C. *System Architecture*

According to Fig. 31.5, the blockchain relies e-healthcare record is believed to exist in the cloud phase where medical data are gathered from a network of adjacent hospitals and stored on the open cloud for sharing with social security providers. Electronic health records may include patient-provided personal information and medical history. Individuals may generate their own personal healthcare ID, upload data to the blockchain relies cloud, and enable hospitals and physicians gain access to those documents. The powered by blockchain cloud is an essential element of this framework, which also contains the following critical elements.

1. *EHRs Manager:* Throughout the sharing of data framework, the EHRs manager has a crucial role to play. On the blockchain system, it is reliable to manage all client exchanges, including information access and information storage procedures [8]. Smart agreements and rigorous client tactics enable the administrative abilities of EHRs administrators.

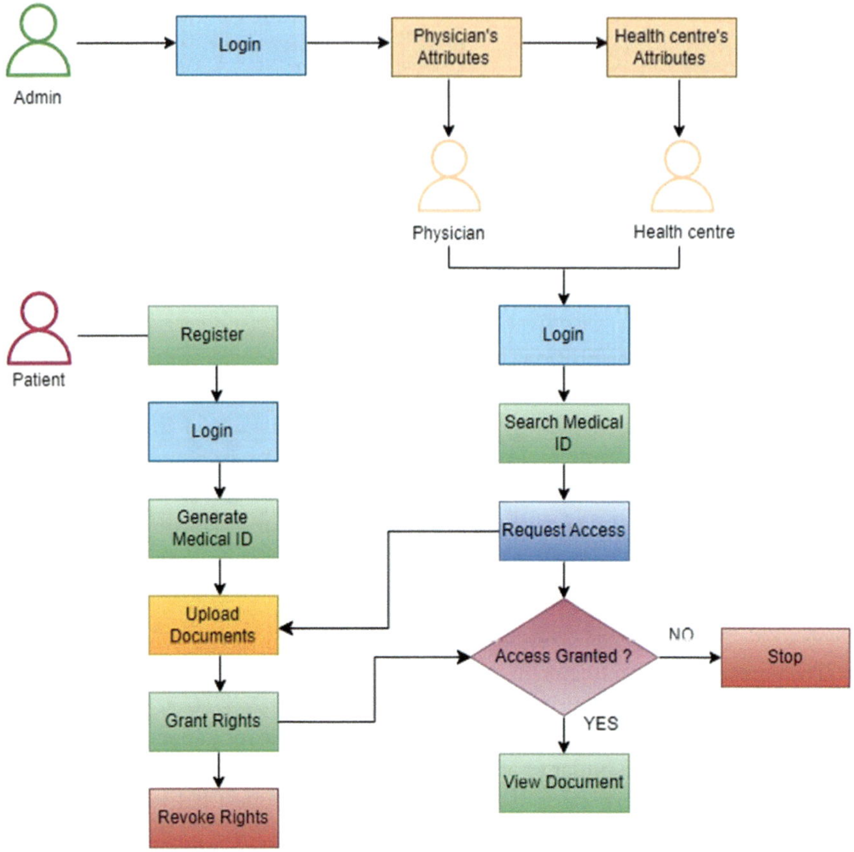

Fig. 31.4 System's flow

2. Admin: This is used to manage the cloud services and operations through techniques for adding, modifying, or rescinding permission acknowledges [8]. The individual in charge is aware of sending smart contracts and an essential component with the ability to update or amend terms in smart contracts.
3. *Smart Contracts:* These outline all actions allowed by the structure for admission control. By using the contract's address, clients can contact keen contracts [11]. By triggering transactions or communications, astute contracts may determine, evaluate advertisement, along with issue get-to authorizations for therapeutic clients. All blockchain components are welcome to participate in the clever agreement's actions. In the e-health system, it is regarded as centre programming.
4. *Decentralized Storage:* The blockchain's data capacity poses a fundamental problem, namely its storage capacity limit [11]. Everybody who has access to the blockchain may see the information, rendering it useless—clearly not the desired outcome for a decentralized system. The information stored on a blockchain

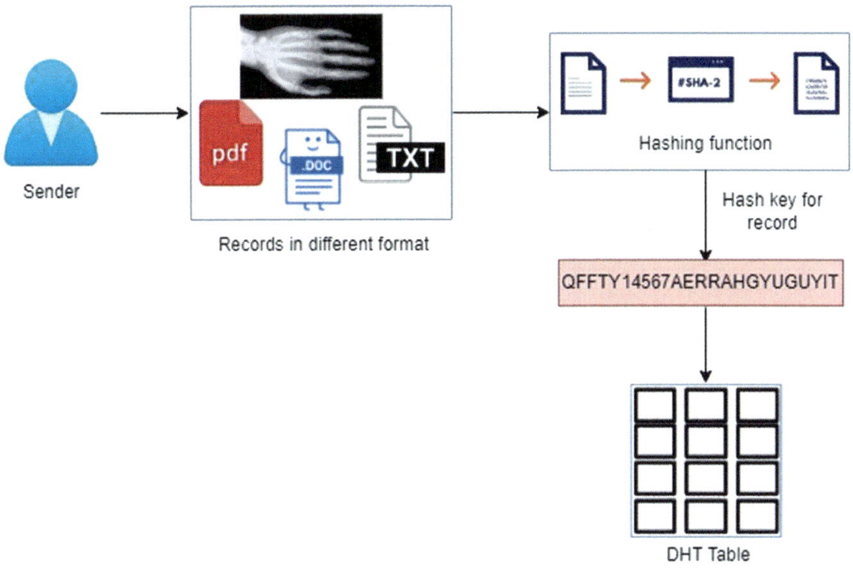

Fig. 31.5 Process of IPFS

would include extensive medical files, tests, X-beams reports, MRI results, and a variety of other reports. The blockchain's capacity limit would be significantly impacted by the storage of all of this extensive information. Off-chain stores will be employed to address this problem.

5. *Storage Off-Chain:* It refers to the practice of storing data or information outside of the main blockchain network. Storage off-chain for non-transactional data cannot be modified or removed in the blockchain because it is too massive to be efficiently sorted there. Off-chain data type samples are shown in Fig. 31.5. Utilizing distributed file systems like interplanetary file system (IPFS) along with the distributed hashing table (DHT) decentralized distributed file structure are used to store such a large amount of data.

31.4 Interplanetary File System (IPFS)

A distributed system is used by the IPFS conventions to store data. Since information stored on IPFS is shielded from modification, it allows for safe information storing. Because any attempt to change the data stored on IPFS must be completed through modifying the identifier, it uses a cryptographic identification to protect the data from modification [11]. Figure 31.6 explains the IPFS storing process. Every information record stored on IPFS contains a cryptographically generated hash value. It is unique and is used to distinguish the evidence of stored information records on the IPFS. The

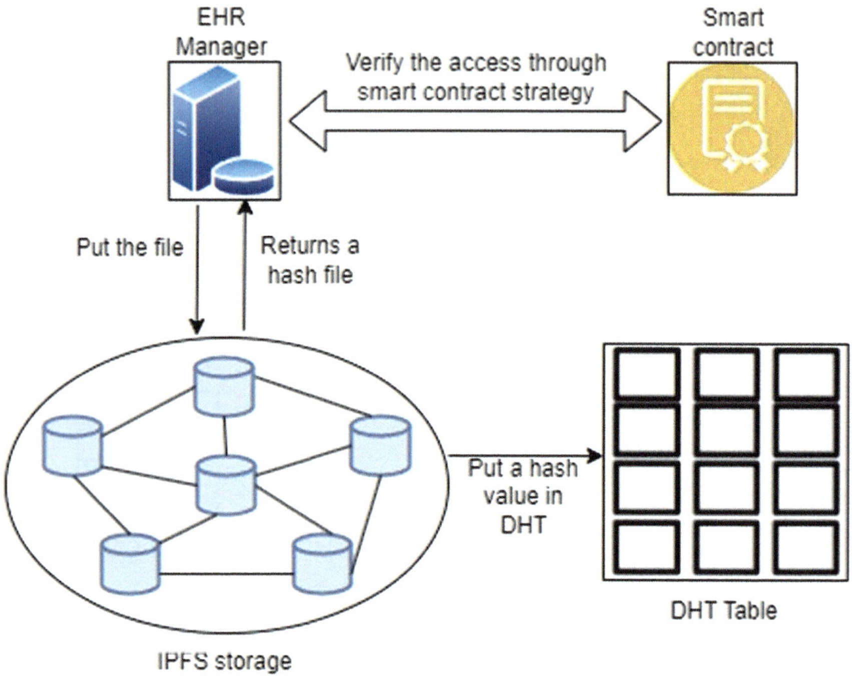

Fig. 31.6 IPFS storage for the system

IPFS convention's secure archiving architecture makes it a wise choice for preserving sensitive and fundamental information. The generated cryptographic hash could be stored on a decentralized application to cut down on intensive computing tasks on the blockchain. Contrary to the internet as a whole, data is spread across multiple nodes rather than being stored on a single server. IPFS uses a DHT mechanism that is similar to Bit Torrent to distribute this stuff. The function for the IPFS storage in the suggested system is depicted in Fig. 31.8. Every record in the peer-to-peer IPFS protocol maintains a collection of hashed files that have been checked by the EHR management utilizing smart contract law. A decentralized protocol that uses content-based addressing is IPFS. Once a record's hash value has been computed, the encrypted record and its unique key are added into a decentralized hash tables.

31.5 Distributed Hash Tables (DHTs)

Almost like a hash table, a hash table that is the peer-to-peer storage technology known as distributed (DHTs) offers search and storing techniques using content [8]. In a DHT, each node is in charge of both the key fields and the transposed values. Any node is capable of successfully recovering the value linked to a certain key.

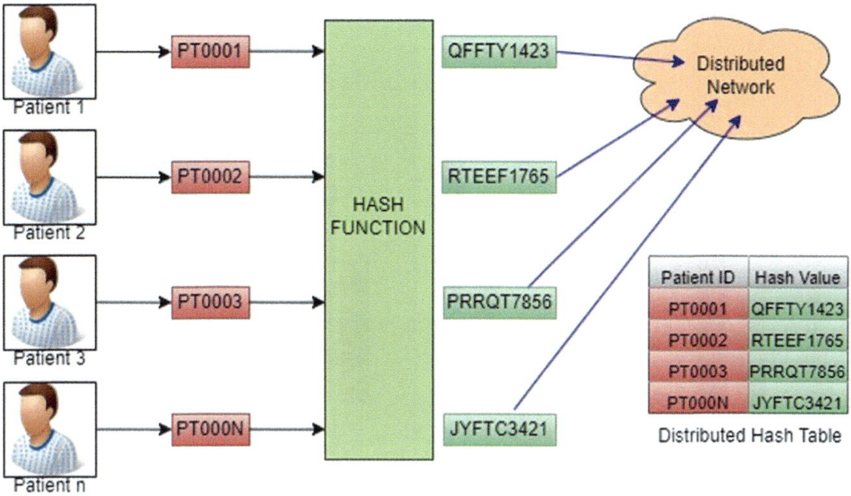

Fig. 31.7 Decentralized hash table (DHT) structure

DHTs, which are spread out, a portion of the DHT is responsible for each "peer" (or "node"). When a friend receives a request, they either respond to it right away or send it forward to another friend until a friend who can respond to it is identified. Depending on its execution, a request that is not answered by the main hub can be transmitted in one of the following ways:—from dispersed, having the final peer getting by mentioning peer—through exchanged, with the necessary response sent later in a similar fashion—responded with the contact information of a hub that offers better opportunity to reply. The IPFS employs this technique. A DHT structure example is shown in Fig. 31.7.

31.6 Analysis of Results

To verify the framework's practicality for actual usage scenarios, the suggested decentralized system is compared to the conventional structure, that adopts a centralized strategy, under a variety of efficiency criteria. The proposed system along with the centralized EHRs system's characteristics and levels are shown in Table 31.1. Assessment of the suggested system was carried out using the contrast shown in Fig. 31.8.

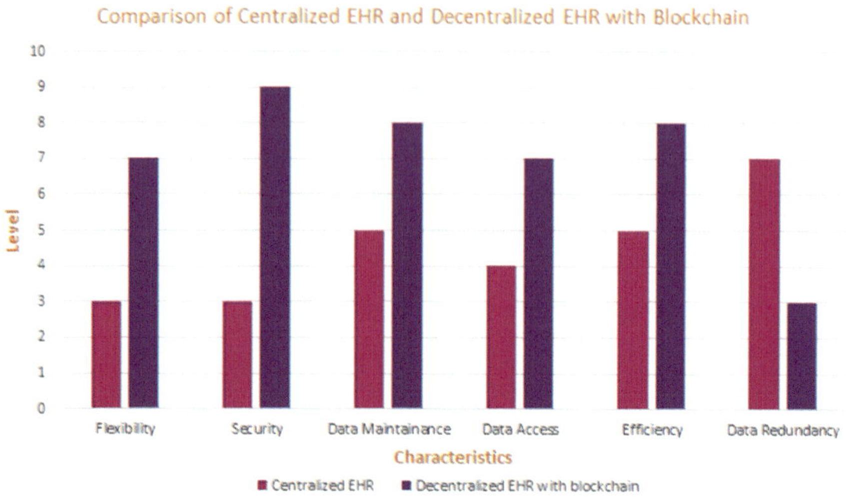

Fig. 31.8 Centralized EHRs and decentralized EHRs comparison

Table 31.1 Performance traits of centralized and decentralized electronic health records

Features	Blockchain used for decentralized EHR	Centralized EHR
Data access	7	3
Data redundancy	6	3
Efficiency	8	4
Data maintenance	8	5
Security	9	3
Portability	7	3

31.7 Conclusion

The entire world has moved approaching continuously driven interoperability, where people grant requests for access to their medical records. The individual receiving treatment operates as the sole owner of his medical records under the existing system, and he or she has the choice of what information should be released along with whom. Encryption is a key issue in this situation as well because medical information shouldn't be disclosed or should only be accessible with the patient's consent. All medical records and transactional details will be kept on the distributed ledger in ledgers, which means that Ethereum plays a significant role in the security of the data. Although the person in charge has authorization to view the information, the network will ensure that system data is not released. The patient will be required to provide their medical ID to any medical facility, such as a hospital, so that the facility can access all the information on their medical ID from any place. The suggested structure combines safe record storage with particular entry controls for those documents. It creates a structure that is easier for clients to use and understand. In addition, because it makes use of IPFS's off-chain hoarding feature, the architecture suggests ways to ensure that the framework addresses the problem of data storing. When both parties accept on the set of rules, the method also sends clever contracts, resulting in a code that runs automatically. Because one of this system's constraints, the proposed system does not contain checks and settlement of claims processes, although it does allow patients to submit insurance-related records alongside other records. The creation for a smartphone-based EHR programme is one of the workable options because the recommended system is an application that is web-based and delays in providing or cancelling authorization from the side of patients can be another issue associated to the system.

References

1. Peng, W., Yi, L., Fang, L., XinHua, D., Ping, C.: Secure and traceable copyright management system based on blockchain. In: IEEE 5th International Conference on Computer and Communications (ICCC) (2019)
2. Radhakrishnan, B.L., Joseph, A.S., Sudhakar, S.: Securing blockchain based electronic health record using multilevel authentication. In: 2019 5th International Conference on Advanced Computing and Communication Systems (ICACCS) IEEE, pp. 699–703 (2019)
3. Mikula, T., Jacobsen, R.H.: Identity and access management with blockchain in electronic healthcare records. In: 2018 21st Euromicro Conference on Digital System Design (DSD) IEEE, pp. 699–706 (2018)
4. Harshini, V.M., Danai, S., Usha, H.R., Kounte, M.R.: Health record management through blockchain technology, In: 2019 3rd International Conference on Trends in Electronics and Informatics (ICOEI) IEEE, pp. 1411–1415, (2019).
5. Novikov, S.P., Kazakov, O.D., Kulagina, N.A., Azarenko, N.Y.: Blockchain and smart contracts in a decentralized health infrastructure, In: 2018 IEEE International Conference Quality Management, Transport and Information Security, Information Technologies (IT&QM&IS). IEEE, pp. 697–703 (2019)
6. Azim, A., Islam, M.N., Spranger, P.E.: Blockchain and novel coronavirus: towards preventing COVID-19 and future pandemics. Iberoamerican J. Med. **2**(3), 215–218 (2020)
7. Kang, A.: Bitcoin's growing pains: intermediation and the need for an effective loss allocation mechanism. Mich. Bus. Entrepreneurial L. Rev. **6**, 263 (2016)
8. Nguyen, D.C., Pathirana, P.N., Ding, M., Seneviratne, A.: Blockchain for secure EHRs sharing of mobile cloud based e Health systems. IEEE Access **7**, 66792–66806 (2019)
9. Wang, Y., Zhang, A., Zhang, P., Wang, H.: Cloud-assisted EHR sharing with security and privacy preservation via consortium blockchain. IEEE Access **7**, 136704–136719 (2019)
10. Ma, Z., Huang, W., Gao, H.: Secure DRM scheme based on blockchain with high credibility. Chin. J. Electron. **27**(5), 1025–1036 (2018)
11. Novikov, S.P., Oleg, D.K., Kulagina, N.A., Azarenko, N.Y.: Blockchain and smart contracts in a decentralized health infrastructure. In: 2018 IEEE International Conference Quality Management, Transport and Information Security, Information Technologies (IT&QM&IS). IEEE, pp. 697–703 (2018)

Chapter 32
Navigating the Landscape: A Comprehensive Survey on Healthcare Recommendation Systems

Megha Gupta, Shilpa Singhal, and Dr.Abhinav Tomar

Abstract A recommender system (RS) is become an essential part of present day specially in healthcare systems (HS) as professionals performed maintenance of health through treatment of diseases, diagnosis, and other mental and physical destruction. To provide solutions to the health-related issues includes massive amount of data of patients so healthcare recommender system has become a growing platform. Doctors believe that these systems provide recommendations based on clinical experiences as patients past records are also being used. For RS in health sector includes many types of filtering like content-based, collaborating, knowledge-based, hybrid, etc. Machine learning (ML) algorithms and deep learning (DL) algorithms are used to do the same. The purpose of this paper is to review the usage of RS to solve many problems of patients and provide best solutions to them based on recommendations given by different professionals. The article discussed recent market trends, limitations and possibilities related to HRS, and new strategies.

32.1 Introduction

As we know that current health issues are frequently linked to adaption of our modern lifestyle. High blood pressure, elevated glucose levels, and physical inactivity are some of the factors that are connected to a modern lifestyle caused due to unhealthy food choices, persistent stress, and perhaps a heavy consumption of energy-dense foods such as recreational drugs [1]. Consequently, most people make unfavorable health-related decisions due to their hectic lifestyles, a variety of alternatives, and

M. Gupta (✉) · D. Tomar
Computer Science and Engineering, Netaji Subhas University of Technology, Dwarka, New Delhi 110078, Delhi, India
e-mail: meghagupta12502@gmail.com

D. Tomar
e-mail: abhinav.tomar@nsut.ac.in

S. Singhal
Computer Engineering, Marwadi University, Rajkot-Morbi Road, Rajkot 360003, Gujarat, India

P. Nanda et al. (eds.), *Intelligent System and Data Analysis*, Smart Innovation, Systems and Technologies 398, https://doi.org/10.1007/978-981-97-5200-3_32

a lack of knowledge that leads to the chances of health risks for people. So, health recommender system (HRS) helps them to find a validated health-related information that can lead to improved health status and minimize the health risks.

Health recommender systems (HRSs) have the potential to persuade and engage users in changing their behavior and also giving them better choices and useful information based on their observed behavior.

The major objective of the HRS is to give people the resources they need to keep track of and improve their health through technology-assisted, customized recommendations.

As we know that current health issues are frequently linked to adaption of our modern lifestyle. High blood pressure, elevated glucose levels, and physical inactivity are some of the factors that are connected to a modern lifestyle caused due to unhealthy food choices,persistent stress, and perhaps a heavy consumption of energy-dense foods such as recreational drugs [1]. Consequently, most people make unfavorable health-related decisions due to their hectic lifestyles, a variety of alternatives, and a lack of knowledge that leads to the chances of health risks for people. So, health recommender system (HRS) helps them to find a validated health-related information that can lead to improved health status and minimize the health risks.

The possible benefits of health recommender systems (HRSs) include persuading and engaging users in changing their behavior and giving those better choices and useful information based on their observed behavior.

The major objective of the HRS is to give people the resources they need to keep track of and improve their health through technology-assisted, customized recommendations.

Today, hospitals have collected a massive quantity of information based on a patient's prior medical history, present health condition, medicines, symptoms, previous treatments, and other health markers. As a result, an efficient technique is required for medical professionals to make effective use of this information. Moreover, many treatments recommendations are available today based on patient's symptoms to certain health problems, for instance evidence-based medicine and clinical pathways [2].

A proper nutritional intake is seen as a significant aspect in improving overall well being [3]. Even though most people understand how important it is to maintain good eating habits, they frequently fail to act in the right way due to their busy schedules or due to their unwillingness to put up the effort to plan meals. These challenges make it difficult for consumers to consume healthy meals. As a result, recommender systems are being investigated as a viable technique of supporting consumers in improving their eating habits and making healthier food choices.

The healthcare industry places a lot of importance on the use of recommender systems and big data analytics in terms of making health-related decisions for patients [4]. The application of big data analytics in a variety of fields is best exemplified in the field of healthcare.

As technology advances, new approaches to implement these designed treatments into practice are being suggested, and accessible tools are required for officials and experts to assess the quality and reliability of the proposed designs [5]. The use of recommender systems (RSs) is one of these cutting-edge strategies in computer-based health treatments.

We can also improve market sales productivity by using this recommender system technique. Since we know that through online e-commerce platforms, millions of patients purchase different products and post their thoughts and remarks regarding it on the appropriate web forum after purchasing it. Thus, every entrepreneur's primary objective is to generate revenue. While consumer choices in one area can be characterized as minimal-risk, other choices might have a more detrimental impact on their patient. Due to concerns on patient immunity and safety, decisions in the healthcare sector can be difficult. In the context of healthcare, the recommender system should interact in instantaneously via a streamline server to monitor patients, and dispense therapy as required in addition to supporting decision-making and preventing risks or failures [6].

Consider a real-world example "A drug recommender system" powered by AI assists healthcare providers in prescribing the best medications for their patients. The system considers a patient's medical history, allergies, and current circumstances to recommend the most effective and safe medications, minimizing the chance of bad reactions and drug interactions.

Accordingly, we define our research questions (RQs) for this study as follows:

RQ1: Which recommendation methods are applied across several HRSs?
RQ2: What are the major uses of the most recent HRS and what do they suggest?
RQ3: Do end users contribute to the evaluation of HRSs and how are they evaluated?
RQ4: How users are ensured that their data is secure?

The remaining part of the paper is structured as follows. Section 32.2 addresses the related work for the health recommender system. Section 32.3 compares several techniques and framework of HRS. Section 32.4 discusses the findings and upcoming directions. Section 32.5 discusses the conclusions and summary of this paper.

32.2 Literature Review

In [7], the author proposed a system in which recommendation will be given on diabetic datasets using deep machine learning technique. An automated systems learning approach is used to solve the hazardous diseases like hepatocellular carcinoma which is a type of cancer. Q Learning and Deep Q network-based techniques are also employed in this study to forecast several diseases so overall SHRS-M3DP model is used for diabetes recommender system.

In [8], in this paper a review is given on healthcare recommendation systems that medical experts can take patient's-oriented decisions. The author raised awareness of a variety of circumstances such as prescription recommendation, diet recommendation, health status prediction, and healthcare professional advice so that patients will have better solution for their problems and all this can be done by many machine learning algorithms like decision tree, classification, clustering, etc.

In [9], in this author thoroughly explains or analyzes the state-of-the-art literature review which is basically used to extract the open challenges or issues. In this recommendation system with blockchain technology is used so the security will be increased and user privacy factors will be increased which will give users transparent system.

In [10] the author includes predicting patients' health status by analyzing their diet, medical health information, and social activities. The suggested deep learning method (RBMCNN) produces fewer mistakes than existing approaches in terms of root square mean error (RSME) and mean absolute error (MAE) and restricted Boltzmann machine CNN deep learning method. High-level privacy is not well-defined in this research.

In [11] the author reviewed various studies and said that tele-machines and diagnosis are the main points which will help in recommendations. In order to get self-examinations and for health marketing electronic health records play a very important role. The main issue occurs is security as many cyber-attacks are done nowadays. Table 32.1 gives a summary of all their search findings related to health recommender system. It mentions the various technologies and dataset that are used along with the research gaps and future directions.

32.3 Fundamental Concepts of the Recommender System

32.3.1 Techniques of Recommender System

There are different types of recommender system approaches discussed in this section as shown in Fig. 32.1.

Collaborative Filtering (CF) is a technology that recommends items to a user based on the concept described below:"If users previously shared similar interests, their tastes would be consistent" [18]. This type of approach is based on information gathered and compiled from users. It consists of several model-based and memory-based techniques. In model-based collaborative filtering several types of machine learning techniques are used to find the user rating, whereas in memory-based approaches nonparametric ML approaches are used to find the user and item-based ratings. This strategy may be viewed as follows in the context of HRS: "Patients would receive similar treatments and healthcare services if they have similar disease profiles and/or health problems."

Content-Based Filtering (CB): This strategy searches for products that are comparable to previous user preferences and reflect the user profile [19]. This technique in HRS recommends quality healthcare that are appropriate for the patient's sickness or health state and are like those previously provided to him or her. According to

Table 32.1 Summary of literature review

Author	Year	Dataset	Technology	Research findings	Limitations
[7]	2023	Patients profile	HRS approaches well-explained	Novel methods for cold start problem explained well	There is no implementation
[8]	2023	Electronic health records, wearable devices, and social media	Filtering approaches	Concept of big data in HRS is well-explained	There is no implementation
[12]	2023	Dataset is created using platform and features as columns	Fabric and neo of blockchain are used	Proposed method is to develop smart contracts for detecting healthcare insurance fraud	Machine learning techniques integrated in blockchain toward privacy-preserving machine learning models is not done
[9]	2022	Health-related images, videos, etc.	Subject-oriented approaches	The health information being selected on the internet	Various challenges faced like cyber-attacks, data mining issues
[10]	2022	Patients profile, hospital locations, and doctors' profile	Methodological and collaborative architecture is used	Proposed algorithm gives better results in terms of medical facilities	The dependency on the Google directions API service to obtain the estimations and approximations for the arrival time to the medical facilities
[11]	2022	Patients profile	HRS many approaches and algorithms are used	Different methods and phases explained well	There is no applied tool presented
[13]	2021	Diabetes datasets for Pima Indians and the Hospital Frankfurt in Germany	SHRS-M3DP model	The proposed model accuracy is more when compared with many classifiers model used by all	SHRS-M3DP model only used for diabetic patients
[14]	2021	Consisting of name, age, etc.	Various filtering techniques like collaborative filtering	review of all the concerned parameters of HRS	There is no implementation of algorithms so no comparison will be taken
[15]	2021	Patients profile	Encryption and decryption	Patients' medical data privacy and protection is focused	Blockchain give better results
[16]	2019	Patients' details	modern approaches	Study about the various modern RS techniques and blockchain usage for storage	Study about the various modern RS techniques and blockchain usage for storage
[17]	2013	Patients profile-related data	RBM-CNN deep learning method is used	Based on the "4Vs," determine the significance of big data analytics in healthcare	Data security and privacy issue occur

this method, a suggestion is based on patient profiles that discuss the various product qualities as well as the patient's prior purchasing behavior. Patients express their preferences by giving evaluations that might be favorable, negative, or neutral. Positively rated products are suggested to the patient according to their choices [20].

Knowledge-Based Recommendation (KB): This method is used when the customer wishes to specifically state their needs for things (e.g., since I am allergic to dairy products, the meal should not contain cheese.) Or when there are very few item ratings accessible (as in the case of vehicles, residences, and financial sectors). With this method, suggestions are made based on knowledge of the items, explicit user preferences, and a set of constraints outlining the relationships between the preferences of the users and the qualities of the item. Knowledge-based recommender system mainly focuses on two main approaches: case-based and constraint-based recommendations. In case-based recommendations, the primary challenge is similarity evaluation and not the suggestions, whereas constraint-based recommendation takes several explicit constraints into consideration, such as filter constraints or mismatch constraints [21].

Hybrid Recommendation (HyR): The aim behind this strategy is to integrate the previous recommendation strategies in order to take use of one approach's advantages while addressing the limitations of another [22]. Any of the following methods can be used to implement the hybrid filtering technique: combining the two recommender systems into one of the mentioned methods; incorporating some collaborative filtering into a content-based technique, as well as using some content-based filtering in both the collaborative and content-based methods. According to its operations, this methodology uses a variety of hybrid strategies, including balanced hybrid, sequence hybrid and switching hybrid.

32.3.2 Framework of Health Recommender System

In collaboration with patients, physicians, surgeons, and other medical staff, we require a framework for creating an HRS. The framework's architecture is organized mainly into three sections: data gathering, data analysis, and data visualization as shown in Fig. 32.2. In first section data is gathered in structures and unstructured form from various data sources such as patients past records, laboratory results, data generated from devices such as sensors, hospital records, and clinical data. The second section consists of data analysis process in which several analytical methods are used to check the patient's condition and determine the sort of illness that they are experiencing such as Hadoop. The third section consists of data visualization process that determines how the items recommended can be presented in appropriate form to . the end patients. Several visualization and knowledge representation techniques are used for this process. The general health recommender system (HRS) framework is divided into the following phases.

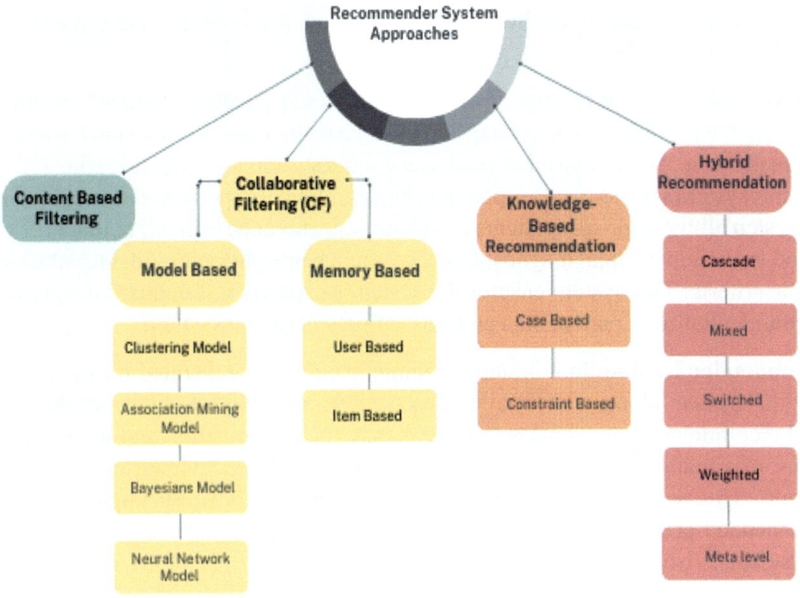

Fig. 32.1 Recommendation systems approaches

1. Data Collection
2. Data Processing
3. Patient Profile generation
4. Data Analysis
5. Recommender Engine
6. Visualization

Data Collection: This process consists of collecting the health-related data from various data sources. These data can be collected in the form of structured and unstructured data. The various data sources form which data can be collected includes patient's feedback, clinical reports, previous medicine descriptions, and other hospital record.

Data Processing: Once the data is collected next phase consists of data processing. In this phase data selection, data transformation, and outlier detection on data are done using various data mining concepts. In data selection process appropriate data type, source is selected among the all collected data. Then it is transformed into suitable form for further processing.

Patient Profile Generation: After the data is extracted in suitable form a patient profile is generated for each patient and comprises a variety of data such as patients' previous health record. If the new patient is registered, the entire procedure reverts to its initial state and begins with the data processing and development of a new patient medical file. In the case of an already existing person, the system changes the record according to the needs.

Data Analysis: This phase consists of data analysis steps once the patient's profile is generated. It uses various data mining and big data techniques such as Map Reduce to analyze the data. At this step all the end users of the system are identified who will be using this system such as doctors, patients, nurses, clinical experts, and other persons. At this step all the high-risk patients are identified who need critical treatment. This method speeds up medical diagnostics process and helps physicians identify the best suitable criteria. This system ensures the system's capacity to safeguard the privacy and confidentiality of patient data and trust of patient on the system.

Recommender Engine: In this step recommendations are generated for the patient using this system along with detailed descriptions and suggestions about the generated recommendation. These suggestions may include remedial and preventative actions, explanations of the disease's etiology, or a further course of therapy. Visualization: This section is composed of components that have an impact on how recommended products are presented. To communicate the derived content to the final patients, visualizations and knowledge representation approaches are deployed in this process. This process works on the representations of Health records, Health insurance plans, medical report and suitable medicine recommendations to the end patient.

Visualization: This section is composed of components that have an impact on how recommended products are presented. To communicate the derived content to the final patients, visualizations and knowledge representation approaches are deployed in this process. This process works on the representations of Health records, Health insurance plans, medical report and suitable medicine recommendations to the end patient.

32.3.3 Different Outlines of Recommender Systems

HRS give various types of suggestions that will give better result for any patient and for this there are the following outlines are discussed.

Recommendation of Food: Due to intake of inappropriate food items many people are suffering from health-related diseases so, by recommending appropriate food items HRS helps people to stay healthy and fit. HRS also suggests nutrients to users so that they can take proper nutrients by giving them comparisons of unhealthy things with healthy things. In this context, HRS motivates people to change their eating habits and stay fit.

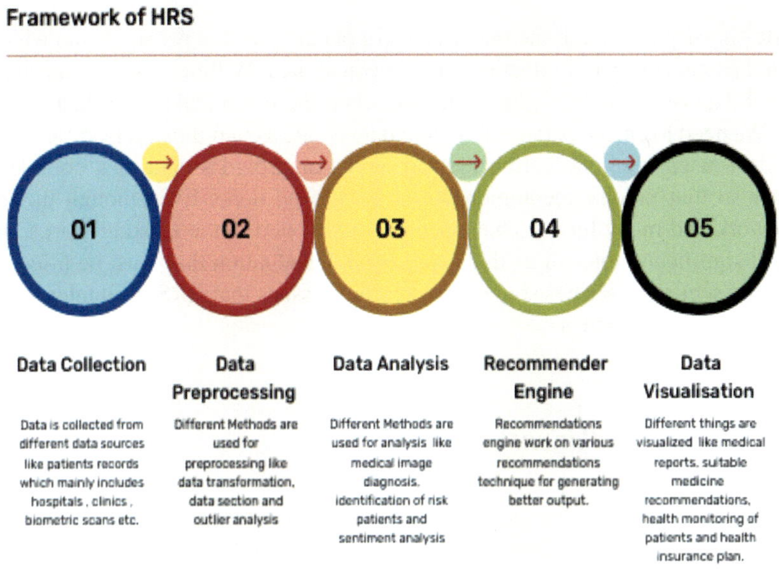

Fig. 32.2 Framework of HRS

Recommendation of Drugs: Medication plays an important role in the field of healthcare. HRS recommends user a proper medicine as medicinal error is very risky for patients' life. It also helps doctors who don't have much knowledge about medicines to suggest best treatment to patients by using records which are already stored in database.

Prediction of Health Status: Predicting risks about any disease is very important for the patients so HRS have been developed to find symptoms related to diseases based on stored information. For this purpose, the system mainly used collaborating filtering technique.

Recommendation by Professionals: Sometimes patients are confused that which doctor is best and how they are deciding that which professional will give the best treatment to users so, HRS recommends professionals related to their inputs given in the system so that patients will have proper treatment and this will create a good relationship between doctors and patients.

Physical Activity Recommendation: Physical activities are another focus of HRS as they help to decrease the possibility of becoming infirm of patients and prevent them from future problems. They also suggest patients to do daily physical activities to burn calories so that they become healthy and fit.

32.4 Open Issues And Future Scope

As HRS is very useful but the major concern occurs, i.e., the privacy issue which is referred as the ability to protect patients' medical data. Willingness of users sharing medical data over HRS may be decreased when there is a leakage of data occur in HRS. We need to ensure patients that their data is safe as in future a privacy technique should be used, i.e., encryption that preserves or provides data in a confidential manner so that precise recommendations should be done. Even though the HRS framework and many features have been demonstrated in the existing works, there are still significant holes in its development and evaluation that must be filled. The following section describes some of the research issues that HRS encounter, as well as answers to these issues.

32.4.1 Creating Patient's Profiles

Aside from the user choices that are commonly used in recommender systems additional user data such as statistical profile, existing health state, illnesses, previous treatment options, physical activities, caloric requirements, diet and lifestyle, moods, and past interactions should also be gathered to create relevant, varied, and accurate suggestions in HRS. Although there are numerous sources for gathering this information, documenting it might include several errors [23]. Therefore, in order to guarantee the accuracy of the information collected it is critical to impose a standard for data formats, the legitimacy of information sources, and regular update intervals. To provide optimal suggestions in terms of balancing user enjoyment and well-being, data in the user profile must be thoroughly and precisely analyzed. For example, a person with a strong risk of developing diabetes should not be advised to eat any product that contains Trans fats, even if it goes against his choice.

32.4.2 Early Detection of Disease

Many studies reveal that people with chronic diseases such as cancer are commonly misunderstood about their diagnosis or possible treatments until the disease reaches to the end states. Late diagnosis of such diseases reduces the likelihood of complete cure, and in some situations, this may endanger patients' lives [24]. So, it is necessary to provide proper knowledge regarding the identification of symptoms to patients so that diseases could be prevented at the early stages [25]. The author designed such a learning application that offers intuitive learning materials and makes individualized learning unit suggestions to the patients and provides them useful information about disease symptoms and effects. For critical and life-threatening disorders, due to early diagnosis of diseases several injurious consequences and difficulties in treatment process can be reduced.

32.4.3 Plausible Recommendations

To make good changes to users' schedules and practices the main aim of HRS is to track everyday activities. However, changing patterns that have grown so firmly ingrained over time appears to be a significant task [26]. As a result, researchers recently focused on developing persuading systems that employ various methods and convincing principles to encourage users to acquire and sustain positive habits and practices [27]. The author examined argument-based techniques in which encouraging arguments are developed in order to influence users' eating habits in a healthy manner. According to the findings of this study, it is critical to develop compelling arguments based on user attributes such as age, gender, or behavior. Although these studies suggest beneficial consequences on users' behavior altering, full approval of changes is not yet assured. Argumentation-based strategies have been shown to be adequately beneficial to people in the later phases of the condition, but they have less of an impact on those in the beginning phases. This leads to the issue of how to produce arguments that are persuasive, appropriate, and plausible enough to result in significant changes for people in the early stages of health hazards.

32.4.4 Legal Considerations

Aside from ethical considerations, there must be fundamental legal rules for the development and use of artificial intelligence in medicine and healthcare sector. On the other hand, medical AI has a massive effect on a patient's physical and mental health, and any breakdown could have serious repercussions. As a result, it is a very relevant area of AI in terms of constitutional rights. In addition to the training sample, the current performance of AI must be regularly checked for prejudice, to ensure that it is free from any bias. When healthcare AI is applied to subgroups of patients that were not sufficiently covered in the sample, the usefulness of the results must be evaluated. To prevent drawbacks, patient variety must constantly be considered while developing software or using applications [28].

32.4.5 Evaluation Metrics

The evaluation of healthcare recommender system should not only focus on the accuracy of the system but it must also include the quality of the system.

Trust: While evaluating the health recommender system trust is one of the most significant variables to be considered as it is important for convincing the patient to use HRS. To improve this functionality, the system can also include explanations that go into further detail about why a specific recommendation is made [29].

Robustness of a recommender system is associated to its credibility. In HRS, end users are sometimes unable to distinguish themselves from prospective attackers, resulting in a loss of faith in the system's integrity and trustworthiness. In order to provide customers with secure HRS, future research should simulate such threats and investigate how they might affect recommendation algorithms [30].

Privacy is described as the capability of health recommender system to secure patient's health records and clinical data. The disclosure of such information elevates patients' concerns, reducing their desire to provide important health information with HRS. Data encryption is the most used method to overcome the privacy issue that ensured at a security while allowing the users' data to be used to provide exact suggestions. Despite all the research to enhance the data encryption techniques, some of them continue to struggle with the system's low performance [31].

User Satisfaction: Depending on user variability, customer satisfaction with suggestions may vary. Significant research has been done to take a close look at modeling customer satisfaction with the goal of developing a satisfaction model [31]. Investigating the association between clinical suggestions and people' satisfaction from various user groups, such as patients, clinicians, pharmacists, and research scientists, might make sense in HRS. Uncertainty: There is an inability to identify a consistent structure like data that is insufficient, the user choices and preferences are not always adequately captured.

Uncertainty includes the inaccurate predictions made by user and related to potential dangers. These dangers can lower the patient's life expectancy. Due to this, when creating HRS, system developers should think about how to depict ambiguity in a list of suggestions so that users can fully assess the choice before making a choice [30].

32.5 Conclusion

Health recommender systems have become an apparent tool to help patients and doctors to make health-related decisions in a better way. In this paper, we've provided details on the possibilities for recommender system in healthcare sector. Various recommendation approaches such as collaborating filtering, content-based filtering, and hybrid filtering are discussed. We have given awareness into recommendation outlines given by these systems, such as physical activity recommendation, healthcare professional recommendation, food recommendation, drug recommendation, and health status prediction and framework for recommendation system. Although the proposed HRS already provides solutions for many health-related terms and issues, there are still many issues that need to be resolved for these technologies to evolve more effectively in the future.

References

1. Stevens, G.A., Mascarenhas, M.N., Mathers, C.D.: Global health risks: progress and challenges. Bull. World Health Org. **87**(9), 646 (2009)
2. A literature review on medicine recommender systems. Int. J. Adv. Comput. Sci. Appl. **10**(8) (2019). https://doi.org/10.14569/IJACSA.2019.0100802
3. Pinxteren, Y., Geleijnse, G., Kamsteeg, P.: Deriving a recipe similarity measure for recommending healthful meals, 105–114 (2011). https://doi.org/10.1145/1943403.1943422
4. Sahoo, A.K., Pradhan, C., Barik, R.K., Dubey, H.: Deepreco: deep learning based health recommender system using collaborative filtering. Computer **7**, 25 (2019)
5. Barros, M. , Couto, F.M. , Pato, M. , Ruas, P.: Creating recommender systems datasets in scientific fields, 4029–4030 (2021). https://doi.org/10.1145/3447548.3470805
6. Lafta, R., Zhang, J., Tao, X., Li, Y., Tseng, V. : An intelligent recommender system based on short-term risk prediction for heart disease patients, 102–105. In: 2015 IEEE/WIC/ACM International Joint Conference on Web Intelligence and Intelligent Agent Technology Workshops, WI-IAT Workshops 2015; Conference date: 06-12-2015 Through 09-12-2015 (2016). https://doi.org/10.1109/WI-IAT.2015.47
7. Etemadi, M., Bazzaz Abkenar, S., Ahmadzadeh, A., Haghi Kashani, M., Asghari, P., Akbari, M., Mahdipour, E.: A systematic review of healthcare recommender systems: open issues, challenges, and techniques. Expert Syst. Appl. **213**, 118823 (2023). https://doi.org/10.1016/j.eswa.2022.118823
8. Karthick, G.S., Sridhar, M.: Intelligent healthcare recommender systems for advanced healthcare informatics: a data fusion perspective, 24 (2023)
9. Sezgin, E. , Özkan, S. : A systematic literature review on health recommender systems, 1–4 (2013). https://doi.org/10.1109/EHB.2013.6707249
10. Torres-Ruiz, M.J., Quintero, R., Guzmán, G., Chui, K.T.: Healthcare recommender system based on medical specialties, patient profiles, and geospatial information. Sustainability (2022)
11. Etemadi, M., Abkenar, S.B., Ahmadzadeh, A., Kashani, M.H., Asghari, P., Akbari, M., Mahdipour, E.: A systematic review of healthcare recommender systems: open issues, challenges, and techniques. Expert Syst. Appl. **213**, 118823 (2022)
12. Ismail, L., Kaafarani, R., Zahwe, O.: Automatic recommender system for smart-contracts-based healthcare insurance fraud detection development platform: design, implementation, and performance evaluation (preprint) (2023). https://doi.org/10.2196/preprints.50730
13. Ihnaini, B. , Khan, M.A. , Khan, T.A., Abbas, S., Daoud, M.S., Ahmad, M.: A smart healthcare recommendation system for multidisciplinary diabetes patients with data fusion based on deep ensemble learning. Comput. Intell. Neurosci. **2021** (2021)
14. Tran, F.A.T.C.H.A., Trang, T.N.: Recommender systems in the healthcare domain: state-of-the-art and research issues. J. Intell. Inf. Syst. **57**, 171– 201 (2021)
15. Sharma, R., Rani, S.: A novel approach for smart-healthcare recommender system, 503–512 (2021)
16. Mekouar, L., Iraqi, Y., Damaj, I.W., Naous, T.: A survey on blockchain-based recommender systems: integration architecture and taxonomy. Comput. Commun. **187**, 1–19 (2022)
17. Sahoo, A.K., Pradhan, C., Barik, R.K., Dubey, H.: Deepreco: deep learning based health recommender system using collaborative filtering. Comput. **7**, 25 (2019)
18. Jannach, D., Zanker, M., Felfernig, A., Friedrich, G.: Recommender systems: an introduction (2010)
19. Sánchez-Bocanegra, C.L., Sanchez-Laguna, F., Sevillano, J.L.: Introduction on health recommender systems. Methods Mole. Biol. **1246**, 131–46 (2015).
20. Pinxteren, Y., Geleijnse, G., Kamsteeg, P.: Deriving a recipe similarity measure for recommending healthful meals (2011)
21. Felfernig, A., Burke, R.: Constraint-based recommender systems: technologies and research issues (2008)
22. Ricci, F., Rokach, L., Shapira, B.: Recommender systems handbook. In: Recommender Systems Handbook, 1– 35 (2010). https://doi.org/10.1007/978-0-387-85820-3_1

23. Mika, S., Erlangen-Nuremberg, F.-A.-U.: Challenges for nutrition recommender systems (2011)
24. Schäfer, H., Hors-Fraile, S., Karumur, R.P., Valdez, A.C., Said, A., Torkamaan, H., Ulmer, T., Trattner, C.: Towards health (aware) recommender systems. In: Proceedings of the 2017 International Conference on Digital Health (2017)
25. Stettinger, M., Tran, T.N.T., Pribik, I., Leitner, G., Felfernig, A., Samer, R., Atas, M., Wundara, M.: Knowledgecheckr: intelligent techniques for counteracting forgetting (2021). arXiv:abs/2102.07825
26. Tran, T.N.T., Atas, M., Felfernig, A., Stettinger, M.: An overview of recommender systems in the healthy food domain. J. Intell. Inf. Syst. **50**, 501–526 (2018)
27. Thomas, R., Masthoff, J., Oren, N.: Adapting healthy eating messages to personality. In: 12th International Conference on Persuasive Technology, PERSUASIVE 2017; Conference date: 04-04-2017 Through 06-04-2017, 119–132. https://doi.org/10.1007/978-3-319-55134-0_10
28. Stöger, K., Schneeberger, D., Holzinger, A.: Medical artificial intelligence: the European legal perspective. Commun. ACM **64**(11), 34–36 (2021). https://doi.org/10.1145/3458652
29. Valdez, A.C., Ziefle, M., Verbert, K., Felfernig, A., Holzinger, A.: Recommender systems for health informatics: state-of-the-art and future perspectives (2016)
30. Verhaert, D., Nateghizad, M., Erkin, Z.: An efficient privacy-preserving recommender system for e-healthcare systems (2018)
31. Ziarani, R.J., Ravanmehr, R.: Serendipity in recommender systems: a systematic literature review. J. Comput. Sci. Technol. **36**(2), 375–396 (2021). https://doi.org/10.1007/s11390-020-0135-9

Chapter 33
Multi-class Generative Adversarial Network for Hyperspectral Image Classification for Addressing Class Imbalance Problem

Deepak Kumar and Dharmender Kumar

Abstract Class-imbalanced datasets are commonly observed in hyperspectral image classification and deep learning models often tend to demonstrate a bias toward the dominant class, resulting in very low efficiency for minority classes. Data augmentation has been the most widely used strategy to mitigate class-imbalanced challenge. Therefore, in this paper, the authors have introduced a data augmentation approach utilizing a multi-class generative adversarial network (MCGAN) to deal with this issue. In the proposed approach a generative adversarial network is utilized to specifically generate fake minority class samples and integrating these samples with the existing class samples. Further, a convolutional neural network (CNN) has been employed for classification of hyperspectral image samples. A series of experiments, utilizing the proposed approach, has been conducted with two most popular and publicly available datasets. The experiment results obtained are then compared with four state-of-the-art data augmentation techniques. The comparison results prove the supremacy of the proposed approach in terms of the quality of the generated samples and the classification accuracy over the state-of-the-art techniques.

33.1 Introduction

The issue of class imbalance emerges when a dataset is characterized by a preponderance of samples from a single class, typically the negative class. This phenomenon is observed across diverse domains, encompassing fields like security, banking, and medicine. It manifests in both multi-class classification and binary tasks [1]. Deep learning (DL) models trained utilizing imbalanced datasets often exhibit a bias toward the dominant class. Mitigation strategies for this concern are typically implemented either at the algorithmic level [2] or the data level. At the data level,

D. Kumar (✉) · D. Kumar
Department of Computer Science and Engineering, Guru Jambheshwar University of Science and Technology, Hisar, Haryana, India
e-mail: drdeepakjanghu@gmail.com

P. Nanda et al. (eds.), *Intelligent System and Data Analysis*, Smart Innovation, Systems and Technologies 398, https://doi.org/10.1007/978-981-97-5200-3_33

techniques involving the resampling of data, such as under-sampling and oversampling, are employed to ensure a more equitable representation of instances among various classes. From an algorithmic perspective, solutions encompass modifying the learning objective to ensure equal involvement of all classes during the training phase. Another prevalent strategy is data augmentation, which generates supplementary training data by introducing artificial variations. This approach not only helps mitigate biases in data collection but also addresses issues related to class imbalance. This technique finds application across various domains, including scenarios where imbalanced classes are encountered. For instance, within the realm of image data, augmentation methodologies encompass a spectrum ranging from basic techniques like random cropping [3], image flips and noise injection [4], to more advanced approaches such as image pairing [5] and PCA-based color augmentation [3]. The utilization of data augmentation can yield a twofold benefit: it can act as a source for augmenting training data [6] and concurrently function as a regularizer [5], thus enhancing the model's generalization capabilities. These techniques have demonstrated their efficacy in addressing the challenges of learning from datasets characterized by class imbalance. However, under circumstances of extreme class imbalance, the application of augmentation to a limited subset of samples might fall short in generating the requisite variations necessary to rebalance the dataset effectively. Moreover, the intricacy deepens in multi-class scenarios, as attempts to improve the performance of one class could inadvertently impact the performance of others [7]. Furthermore, it should be noted that existing augmentation techniques may not seamlessly translate to the domain of deep learning [8].

More recently, there has been a growing utilization of generative adversarial networks (GANs) to create images characterized by a high level of visual authenticity [9]. The researchers have demonstrated, the generated images as supplementary training data, to enhance various procedures, including classification tasks [6, 10]. GAN models are adept at generating samples of commendable quality that possess the essential variations resembling those found in the original training dataset. Numerous GAN models have been introduced in prior research endeavors for the purpose of data augmentation [1, 6, 11–13]. Additionally, GANs have been employed to address the challenges posed by imbalanced data as an augmentation technique to enhance the accuracy of image recognition, as seen in Antoniou et al. [11]. While our approach shares certain parallels with these aforementioned investigations, it diverges in that it centers on a distinct GAN model situated within the domain of image classification. Moreover, our focus is directed toward multi-class classification tasks characterized by an imbalanced training dataset. In the presence of infrequent minority classes, generating images becomes a complex task, as the augmentation samples must be both plausible and diverse, while also aligning with the requisite minority class [11, 12].

To address the issue, the authors have proposed a novel multi-class generative adversarial network (MCGAN) data augmentation technique in this work. This approach strategically leverages a generative adversarial network to create synthetic samples specifically belonging to the underrepresented minority class. These newly generated samples are seamlessly incorporated alongside the existing class samples.

Additionally, a convolutional neural network (CNN) has been adopted for hyperspectral image classification (HSIC). The primary contributions of this study can be outlined as follows:

1. Introduction of multi-class generative adversarial network (MCGAN), as a novel data augmentation technique.
2. A methodology to address the challenges posed by class-imbalanced datasets by generating synthetic samples for minority classes.
3. Investigation of performance of the proposed MCGAN in multi-class imbalanced scenarios across two most popular HSI datasets.

The subsequent sections of this paper are structured as follows. In Sect. 33.2, the authors have discussed about relevant HSIC literature. The proposed MCGAN has been presented in Sect. 33.3. In Sect. 33.4, the results and discussion have been presented. Finally, the conclusions have been conferred in Sect. 33.5.

33.2 Related Works

The problem of class imbalance has emerged as a vibrant field of research in hyperspectral image classification, yielding a repertoire of well-established techniques. In contrast, the treatment of class imbalance within the context of multi-class classification has received relatively scant attention [2, 14]. The intricacies associated with imbalanced classes in a multi-class classification setting often necessitate novel sampling strategies and distinct data preprocessing approaches [2], setting them apart from those traditionally employed in binary classification scenarios. Prevailing strategies designed to tackle this issue encompass approaches such as the utilization of class rectification loss (CRL) [8], multi-class decomposition [7] and the employment of mean squared false error [15]. Furthermore, the resampling methodologies, including under-sampling and oversampling, hold sway in this domain. However, oversampling runs the risk of inducing overfitting, while under-sampling might result in the elimination of critical data points [16]. In [16], Buda et al. demonstrated the notable decline in CNN performance attributed to data imbalance. Conversely, Wang et al. [15] introduced modifications to the learning algorithm with the purpose of mitigating class imbalance. They achieved this by imposing penalties on misclassified instances from the minority class, effectively utilizing cost-sensitive techniques. However, the implementation of such methods necessitates meticulous calibration of the cost matrix settings, a challenge that can prove intricate within real-world problem contexts [2].

Traditional approaches like the Synthetic Minority Over-Sampling Technique (SMOTE) have proven to be ineffective when grappling with extreme class imbalance scenarios (as seen in datasets with substantial imbalances), consequently leading to a degradation in the algorithm's performance within such situations [2]. Furthermore, it's worth noting that SMOTE can induce an inclination toward over-generalization, accompanied by heightened variance [15]. In deep models, such as convolutional neural networks (CNNs), strategies like class rectification loss (CRL) [8] have been employed to address class imbalance. The CRL algorithm, within each batch for the minority class, employs a rigorous mining technique compelling the model to define boundaries for each minority class by imposing strict thresholds for both positive and negative instances. In a similar vein, authors in [17] utilize class clustering, large margin local embedding (LMLE), to maintain the structure of minority data. However, these techniques can incur substantial computational costs, particularly in domains characterized by extensive datasets [8]. Within the realm of deep models, data augmentation techniques have progressively integrated as pivotal components for classification tasks. For instance, Dosovitskiy et al.'s CNN [18] introduced exemplar to augment on the CIFAR-10 dataset, a method systematically augments data and exhibited improved performance. Ali et al. [4] utilized noise distortion and leveraged affine transformations to generate additional samples across classes effectively reducing the influence of class imbalance. Data augmentation wields significant influence in managing datasets characterized by class imbalance. Nevertheless, conventional augmentation might fall short when grappling with scenarios marked by extreme class imbalance or situations where ample data is lacking. Furthermore, domain-specific characteristics related to orientation can impose limitations on the applicability of straightforward augmentation approaches [12]. Consequently, the researchers have introduced mixup [19] and image pairing [5] as more sophisticated augmentation approaches. Furthermore, the authors in [20] introduced the BGWOHCNN framework HSIC, addressing challenges in CNN-based models due to limited training samples where binary gray wolf optimization (BGWO) has been employed for spectral band selection. Also, in [21] a SS-3D-ConvCapsule network model for HSIC was introduced addressing challenges posed by limited training samples and the need for efficient utilization of spectral-spatial features.

In recent times, the application of generative models for sample generation has garnered significant success. Particularly, generative adversarial networks (GANs) have demonstrated prowess in effectively producing and encapsulating data [9]. The GANs have been mainly utilized to mitigate imbalanced dataset issues and the emphasis rests on generating samples specific to individual classes. In this regard, the author in [22] introduced conditional GAN (C-GAN), a supervised GAN model as viable solutions. Nonetheless, AC-GAN [23] and vanilla GAN [24] have been proposed as GAN frameworks to conquer the issue of class-imbalanced datasets by falling short in generating the requisite minority samples [12, 25]. Notably, deep convolutional GAN (DCGAN) [26] was utilized on liver lesion images to fabricate synthetic images, in [6] exhibiting a noteworthy performance. This accomplishment was facilitated by oversampling the training set employing the combination of conventional augmentation methods. Correspondingly, Baur et al. [13] harnessed MelanoGAN (a DCGAN variant combined with Laplacian GAN [27]) to craft high-resolution skin lesion images from a limited 2 k sample dataset. The ResNet-50 [28] model, ameliorating the impact of class-imbalanced data, was subsequently leveraged to generate additional skin lesion samples, for classification. These instances underscore the efficacy of straightforward data augmentation techniques in addressing class imbalance-related challenges [6, 13]. Nevertheless, it is crucial to recognize that these instances were applied within binary datasets that did not possess orientation-dependent features or unclear class boundaries. Other strategies involve the fusion of GAN models with additional generative processes, like auto-encoder training. In this methodology, the features acquired from the auto-encoder are subsequently utilized to set the initial parameters of both the generator and discriminator in the GAN model. This might necessitate an additional training step [12] or simultaneous training [25] to execute conditional adversarial training. Various approaches such as fine-grained multi-attribute GAN (FM-GAN) [25], balancing GAN (BAGAN) [12] and data augmentation generative adversarial networks (DAGAN) [11] have employed analogous strategies to generate augmented samples. Another avenue is image refinement, a technique that upholds the image class while generating diverse synthetic samples. In a task centered on emotion recognition, Zhu et al. [10] utilized a reference sample to fabricated minority samples through image translation in the context of two closely related classes. Alternatively, certain methodologies reconfigure adversarial training through the introduction of supplementary losses or more stringent conditions during the generation process. The proposed framework drives the process of learning and generating minority samples, as exemplified in DeliGAN [29]. The Gaussian mixture model (GMM) has been utilized in DeliGAN to parameterize the latent space and the parameters for both, i.e., GAN and DeliGAN have been learnt in parallel. To encapsulate, resampling techniques exhibit suboptimal performance in scenarios characterized by significant class imbalance. Conventional data augmentation methods remain prevalent; however, they possess limitations, frequently failing to generate sufficient data variation, particularly in exceedingly imbalanced contexts. Notably, GAN-based approaches emerge as a more pragmatic solution for generating data samples and effectively addressing class imbalance concerns, encompassing both multi-modal techniques [11, 12] and image translation strategies [10].

33.3 Multi-class Generative Adversarial Network (MCGAN)

The hyperspectral images classification (HSIC) has been confronted by the challenge of imbalanced data distribution across different classes. Typically, a fraction of the entire dataset is allocated for the model's training, a practice susceptible to the adverse effects stemming from class imbalance. Indeed, the majority of existing HSI datasets exhibit a pronounced imbalance, with even randomly sampled training sets displaying a skewed distribution. Consequently, this situation imparts a bias in the training process of any DL model toward the prevalent classes. Put differently, the network inadequately learns the minority classes, a result of their feeble representation owing to the limited number of available samples. Therefore, the authors in this paper have introduced the MCGAN framework and demonstrate the ability of its generators to produce authentic images, even under conditions of class imbalance.

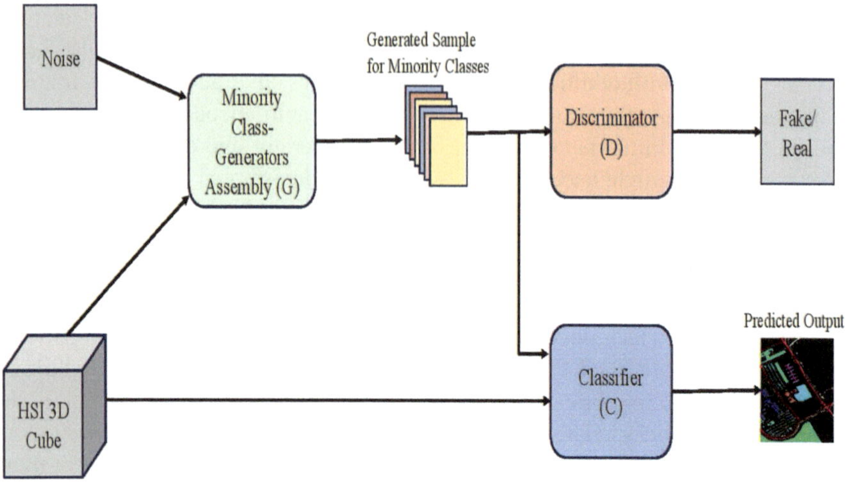

Fig. 33.1 Framework for proposed MCGAN

In instances where the real data distribution is imbalanced, generators equipped with conditional class distributions tend to struggle in generating minority classes due to the discriminator networks' inclination toward majority classes [1]. To surmount this challenge, the proposed MCGAN adopts an assembly of generators that operate within the confines of distinct real class distributions, effectively addressing the issue.

The MCGAN framework is composed of three neural networks: a classifier (C), a discriminator (D) and an assembly of minority class-generators (G). The generator configuration is presented as a summation encompassing each class generator along with its associated authentic class distribution as illustrated in Fig. 33.1.

The assembly of generator outputs results in generated samples operating within the realm of each authentic class data distribution. Given the domain limitations, MCGAN possesses the capability to produce samples from both minority and majority classes. The generator function G, represented by a differentiable CNN parameterized by, takes class-specific categorical information and a Gaussian normal distribution to engender a lifelike data distribution corresponding to the specified conditional class. The discriminator network (D), characterized by parameter, labels authentic and counterfeit instances to segregate real data from the generated realistic distribution data, respectively. The algorithm for MCGAN is outlined in Algorithm 1.

Algorithm 1 MCGAN Algorithm

Start

1. **For** each sample from random noise **do**
2. **For** each sample from HSI batch **do**
3. Generate samples using Gaussian normal distribution and Class-specific categorical information
4. Update discriminator based upon Eq. 33.1
5. Update classifier based upon Eq. 33.2
6. Update generator assembly based upon Equation
7. **End for**
8. **End for**

Stop

However, the selection of an appropriate generator holds heightened significance and hence, the authors have employed conditional categorical class information within the class, linked to the conditional authentic data to align the generated samples in a specific class within the authentic distribution domain.

Furthermore, the classifier network (C) has been utilized to classify class-conditioned real data alongside generated realistic data with fine-tuned parameter. From a classifier's standpoint, the G network has been employed as a data augmentation mechanism, producing credible data for both minority and majority classes in order to enhance classification accuracy further.

The objective function for the entropy loss in the proposed framework is formulated as follows:

$$L_D = P_i^{\text{Real}} E_{y \sim p^{\text{Real}}} [\log D(y)] + L_G \tag{33.1}$$

$$L_G = P_i^{\text{Gen}} E_{G\left(\frac{Z}{\text{Class}\, i}\right) \sim p^{\text{Gen}}} \times \left[\log\left(1 - D\left(G\left(\frac{Z}{\text{Class}\, i}\right)\right)\right)\right] \tag{33.2}$$

$$\begin{aligned} L_C = {} & P_i^{\text{Cfier}} E_{y \sim p^{\text{Real}}} \left[\log C(y)\right] \\ & + P_i^{\text{Cfier}} E_{G\left(\frac{Z}{\text{Class}\, i}\right) \sim p^{\text{Gen}}} \\ & \times \left[\log C\left(G\left(\frac{Z}{\text{Class}\, i}\right)\right)\right] \end{aligned} \tag{33.3}$$

Here L_D, L_G and L_C are the entropy-loss objective function for discriminator, generator and classifier, respectively. And P_i^{Real}, P_i^{Cfier} and P_i^{Gen} are conditional probabilities for real, classifier and generated fake data for ith class and Z denoted the Gaussian normal distribution.

33.4 Results and Discussion

The authors have examined the efficacy of the proposed MCGAN framework with four state-of-the-art HSIC models. This examination encompasses the utilization of two widely recognized imbalanced hyperspectral datasets, viz., Indian Pines (IP) and Pavia University (PU). The comprehensive performance evaluation of the proposed MCGAN in comparison with conventional methods and state-of-the-art techniques has been illustrated in Tables 33.1 and 33.2. All experiments were executed utilizing Python 3.8 on hardware comprising an NVIDIA Corporation TU104GL (Quadro TRX 5000) Graphics Processing Unit and an AMD 7662 central processing unit with 64 cores, 256 threads, and 512 GB of random-access memory. These computations were performed on the Ubuntu 20.04.2 LTS (64-bit) operating system. The effectiveness of the proposed framework has been evaluated on two real data distributions, i.e., 10% and 15% for both the hyperspectral datasets. In this evaluation, MCGAN is pitted against three neural network models (i.e., CNN-based), viz. 2D-CNN, 3D-CNN [30] and HSGAN [31] and a detailed comparative evaluation of the classification performance of MCGAN against state-of-the-art methodologies has been carried out. The evaluation has been based on three widely recognized performance metrics, i.e., average accuracy (AA), Kappa coefficient (K) and overall accuracy (OA) [31].

Further, the MCGAN model has superseded for PU dataset as well and rendered 88.46% kappa value, 87.22% average accuracy and an overall accuracy of 89.73% as delineated in Table 33.2. Further, HSGAN has achieved the second-highest position for both datasets, in terms of classification performance. SVM model has showcased the poor performance among all the HSIC models.

In addition, 15% of the total samples have been employed for training in the second set of experiments. Here again, proposed MCGAN model has manifested

Table 33.1 Performance comparison of HSIC models for IP dataset

Sample size	Perf. Ind.	SVM	2DCNN	3DCNN	HSGAN	MCGAN
10%	OA	67.32	74.66	75.1	81.4	88.45
	K	65.01	71.43	73.72	80.07	87.6
	AA	63.45	68.82	71.83	79.88	86.34
15%	OA	69.56	78.21	79.37	87.12	93.41
	K	67.11	75.02	77.92	85.67	92.61
	AA	65.39	72.34	76.07	84.42	91.44

Table 33.2 Performance comparison of HSIC models for PU dataset

Sample size	Perf. Ind.	SVM	2DCNN	3DCNN	HSGAN	MCGAN
10%	OA	68.74	75.81	76.73	82.47	89.73
	K	66.23	72.57	75.22	81.38	88.46
	AA	64.83	70.69	73.11	80.51	87.22
15%	OA	69.53	79.16	80.93	88.92	94.06
	K	67.18	76.68	79.11	87.27	93.38
	AA	65.62	73.84	78.25	86.63	92.49

improved performance metrics when compared with all other models. The MCGAN model has achieved substantial improvements of 6.94% in kappa, 7.02% in AA and 6.29% in OA and 6.11% in kappa, 5.86% in AA and 5.14% in OA in case of IP and PU datasets, respectively, compared to the second-best performer HSGAN model.

Furthermore, HSGAN along with other HSIC models have showcased improved performance when 15% of the total samples are utilized for training. Like, HSGAN model has demonstrated 88.92% of OA, 87.27% of kappa and 86.63% of AA for PU dataset. In case of IP dataset, the model has ill-treated 87.12% of OA, 85.67% of kappa and 84.42% of AA.

It is noteworthy that in the classification performances of all the compared HSIC models are better in case of PU dataset when compared with the IP HSI dataset. Also, the classification performances are improved when more samples are employed for training of the models. The HSIC models SVM, 2D-CNN and 3D-CNN have demonstrated 67.32%, 74.66%, 75.1% and 68.74%, 75.81%, 76.73% OA accuracies for IP and PU HSI datasets, respectively, when 10% of total samples were utilized during training.

Furthermore, in second sequence, (i.e., 15% of total samples were employed for training) for IP and PU datasets, these models have illustrated 69.56%, 78.21%, 79.37% and 69.53%, 79.16%, 80.93% OA accuracies, respectively.

The classification performances of all the HSIC models, employing performance indicators (i.e., AA, Kappa and OA), have been compared and depicted in Figs. 33.2a, b and 33.3a, b for both the HSI datasets. It can be inferred that the proposed MCGAN model outperforms all the other models for IP and PU datasets, especially when very limited samples are utilized for training. All the performance indicators reveal the supremacy of proposed framework over the other contemporary HSIC models during both the sequences of experiments. The MCGAN model has attained apex values 94.06%, 93.38% and 92.49% of OA, Kappa and AA for PU datasets in second sequence of experiments. Further, the MCGAN along with other HSIC models have achieved better performance with PU dataset when compared with IP dataset during both the sequence of experiments.

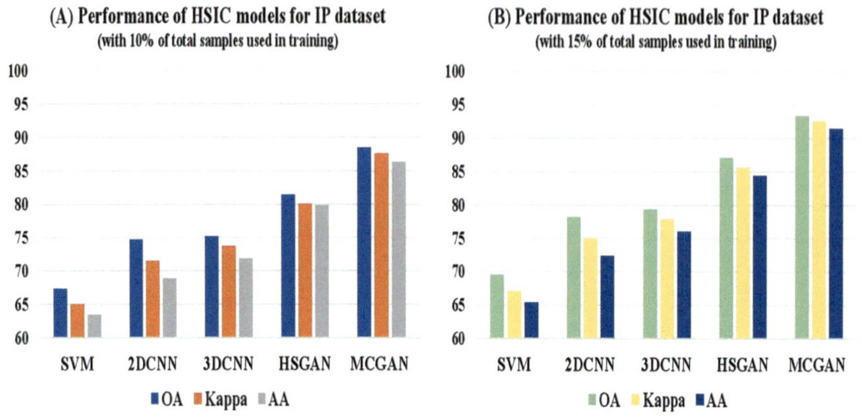

Fig. 33.2 Classification performance of HSIC models for IP dataset

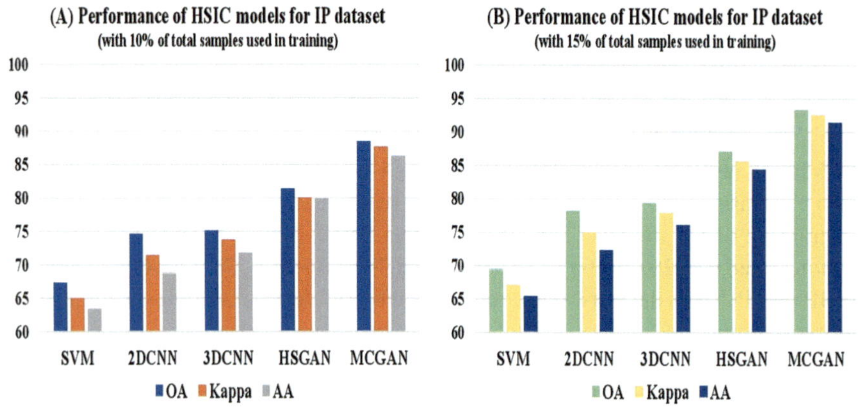

Fig. 33.3 Classification performance of HSIC models for PU dataset

33.5 Conclusion

In the current manuscript, the authors have proposed an innovative framework MCGAN to deal with the issue of class-imbalanced. The proposed model is designed to generate both minority and majority class samples, thereby enhancing classifier performance on datasets with imbalanced class distributions. To investigate the effectiveness of the proposed model, the authors have conducted two sequences of experiments employing only 10% and 15% of the total number of samples during the training, for the Indian Pines and University of Pavia datasets. The classification of the proposed MCGAN model has been compared with four state-of-the-art HSIC models, viz., 3D-CNN, 2D-CNN, SVM and HSGAN. The experiments result clearly demonstrates that MCGAN outperformed all the other models in terms of classification performance. The obtained results are very encouraging and will lead the future research endeavors to explore the integration of GAN-based techniques to enhance the classification performance in HSIC further.

References

1. Douzas, G., Bacao, F.: Effective data generation for imbalanced learning using conditional generative adversarial networks. Expert Syst. Appl. **91**, 464–471 (2018)
2. Krawczyk, B.: Learning from imbalanced data: open challenges and future directions. Prog. Artif. Intell. **5**(4), 221–232 (2016)
3. Krizhevsky, A., Sutskever, I., Hinton, G.E.: Imagenet classification with deep convolutional neural networks. In: Pereira, F., Burges, C.J.C., Bottou, L., Weinberger, K.Q. (eds.) Advances in neural information processing systems 25, pp. 1097–1105. Curran Associates Inc. (2012)
4. Ali-Gombe, A., Elyan, E., Jayne, C.: Fish classification in context of noisy images. In: International Conference on Engineering Applications of Neural Networks (2017)
5. Inoue, H.: Data augmentation by pairing samples for images classification. arXiv preprint arXiv:1801.02929
6. Frid-Adar, M., Klang, E., Amitai, M., Goldberger, J., Greenspan, H.: Synthetic data augmentation using GAN for improved liver lesion classification. arXiv preprint arXiv:1801.02385
7. Fernández, A., López, V., Galar, M., Del Jesus, M.J., Herrera, F.: Analysing the classification of imbalanced data-sets with multiple classes: binarization techniques and ad-hoc approaches. Knowl. Based Syst. **42**, 97–110 (2013)
8. Dong, Q., Gong, S., Zhu, X.: Class rectification hard mining for imbalanced deep learning
9. Karras, T., Aila, T., Laine, S., Lehtinen, J.: Progressive growing of GANs for improved quality, stability, and variation. arXiv preprint arXiv:1710.10196 ICLR2018
10. Zhu, X., Liu, Y., Qin, Z.: Data augmentation in classification using GAN. arXiv preprint arXiv:1711.00648
11. Antoniou, A., Storkey, A., Edwards, H.: Data augmentation generative adversarial networks. arXiv preprint arXiv:1711.04340
12. Mariani, G., Scheidegger, F., Istrate, R., Bekas, C., Malossi, C.: BAGAN: data augmentation with balancing GAN. arXiv preprint arXiv:1803.09655
13. Baur, C., Albarqouni, S., Navab, N.: MelanoGANs: high resolution skin lesion synthesis with GANs. arXiv preprint arXiv:1804.04338
14. Kumar, D., Kumar, D.: Hyperspectral image classification using deep learning models: a review. J. Phys. Conf. Ser. **1950**(1) (2021)
15. Wang, S., Liu, W., Wu, J., Cao, L., Meng, Q., Kennedy, P.J.: Training deep neural networks on imbalanced data sets. In: 2016 International Joint Conference on Neural Networks (IJCNN), pp. 4368–4374. IEEE (2016)
16. Buda, M., Maki, A., Mazurowski, M.A.: A systematic study of the class imbalance problem in convolutional neural networks. arXiv preprint arXiv:1710.05381
17. Huang, C., Li, Y., Change Loy, C., Tang, X.: Learning deep representation for imbalanced classification. In: Proceedings of the IEEE Conference on Computer Vision and Pattern Recognition, pp. 5375–5384 (2016)
18. Dosovitskiy, A., Springenberg, J.T., Riedmiller, M., Brox, T.: Discriminative unsupervised feature learning with convolutional neural networks. In: Advances in Neural Information Processing Systems, pp. 766–774 (2014)
19. Zhang, H., Cisse, M., Dauphin, Y.N., Lopez-Paz, D.: mixup: beyond empirical risk minimization. arXiv preprint arXiv:1710.09412

20. Kumar, D., Kumar, D.: A Binary Grey Wolf Optimization based Hybrid Convolutional Neural Network (BGWOHCNN) framework for hyperspectral image classification. Multimed. Tools Appl. **83**(4), 10091–10114 (2023)
21. Kumar, D., Kumar, D.: A spectral–spatial 3D-convolutional capsule network for hyperspectral image classification with limited training samples. Int. J. Inf. Technol. **15**(1), 379–391 (2023)
22. Mirza, M., Osindero, S.: Conditional generative adversarial nets. arXiv preprint arXiv:1411.1784
23. Odena, A., Olah, C., Shlens, J.: Conditional image synthesis with auxiliary classifier GANs. In: International Conference on Machine Learning, pp. 2642–2651 (2017)
24. Goodfellow, I., Pouget-Abadie, J., Mirza, M., Xu, B., Warde-Farley, D., Ozair, S., Courville, A., Bengio, Y.: Generative adversarial nets. In: Advances in Neural Information Processing Systems, pp. 2672–2680 (2014)
25. Wan, L., Wan, J., Jin, Y., Tan, Z., Li, S.Z. et al.: Fine-grained multi-attribute adversarial learning for face generation of age, gender and ethnicity (2018)
26. Radford, A., Metz, L., Chintala, S.: Unsupervised representation learning with deep convolutional generative adversarial networks. arXiv preprint arXiv:1511.06434
27. Denton, E.L., Chintala, S., Fergus, R. et al.: Deep generative image models using a Laplacian pyramid of adversarial networks. In: Advances in Neural Information Processing Systems, pp. 1486–1494 (2015)
28. He, K., Zhang, X., Ren, S., Sun, J.: Deep residual learning for image recognition. In: The IEEE Conference on Computer Vision and Pattern Recognition (CVPR) (2016)
29. Gurumurthy, S., Sarvadevabhatla, R.K., Radhakrishnan, V.B.: Deligan: generative adversarial networks for diverse and limited data. In: The IEEE Conference on Computer Vision and Pattern Recognition (CVPR), vol. 1 (2017)
30. Chen, Y., Jiang, H., Li, C., Jia, X., Ghamisi, P.: Deep feature extraction and classification of hyperspectral images based on convolutional neural networks. IEEE Trans. Geosci. Remote Sens. **54**(10), 6232–6251 (2016). https://doi.org/10.1109/TGRS.2016.2584107
31. Zhan, Y., Hu, D., Wang, Y., Yu, X.: Semisupervised hyperspectral image classification based on generative adversarial networks. IEEE Geosci. Remote Sens. Lett. **15**(2), 212–216 (2018)

Chapter 34
A Comparative Review of Resource Optimization and Provision Techniques in Cloud Computing

Sanjit Bhagat, Punit Gupta, and Dinesh Kumar Saini

Abstract Cloud computing is playing the major role to manage the user data virtually either an enterprise or an individual. Nowadays, almost everyone is using cloud directly or indirectly. Managing user data while satisfying the SLA with maintaining QoS is quite challenging. It is very complex to maintain the service with low cost and low power consumption. Primary challenge for every resource provider is to manage cost, power, and time while maintaining SLA. With all this, proper resource utilization is also needed to make it more efficient. In this paper, we have discussed various resource provisioning mechanisms with respect to the power, cost, and time. We present the current methods being used at various service levels such as SaaS, IaaS, and PaaS. Literature review shows some of the related works that have been done so far and what goals are achieved. Various static and dynamic approaches have been presented with their achievements and drawbacks. Cost is one of the prime factors for any organization, either cloud user or service provider. In recent years, migration of data and network has been significantly increased over cloud. Migrating and managing data over cloud network is very expensive. There are major factors affecting cost such as size of data, type of service, resources, duration, power consumption, maintenance, and many more. All these factors are affecting cost weather it is an individual user or an enterprise even service provider. In this article, we are going to discuss different approaches worked for the improvement of cost with better power efficiency and resource utilization. At last, we have concluded the overall challenges and future work that can be done.

S. Bhagat · D. K. Saini
Manipal University Jaipur, Jaipur, Rajasthan 303007, India

P. Gupta (✉)
University College Dublin, Dublin, Ireland
e-mail: punitgupta07@gmail.com

P. Nanda et al. (eds.), *Intelligent System and Data Analysis*, Smart Innovation, Systems and Technologies 398, https://doi.org/10.1007/978-981-97-5200-3_34

34.1 Introduction

Cloud computing has been playing major role in the organization in managing and storing high amount of data. Cloud can provide services from an individual to an enterprise based on their requirements. This on-demand capability of providing services makes it more in demand. Based on its services, it is categorized as SaaS, PaaS, and IaaS. The increasing amount of data on the cloud makes it more complex to handle. With the evolution of new technologies, cloud is capable to maintain its services (Fig. 34.1). Effective cloud computing can be obtained by proper resource management, power management, virtualization, load balancing, scaling, etc.

A resource can be anything used in cloud like servers, hardware, and operating systems. Proper management of resource makes cloud cost-efficient, reliable, and reusable. Development of new technologies making it possible, although a lot of research is still going on to make cloud more reliable, effective, power efficient, and scalable. Resource management is key factor for cloud utilization. Resource provisioning, resource scheduling, and monitoring are components of cloud resource management. A resource provider considers three basic services [1–3].

- Finding appropriate resource from all cloud resources available, which satisfies the consumer requirements.
- Proper monitoring of the resources for maximum and effective utilization of resources.
- Workload balancing to reduce the over-utilization or under-utilization of resources.

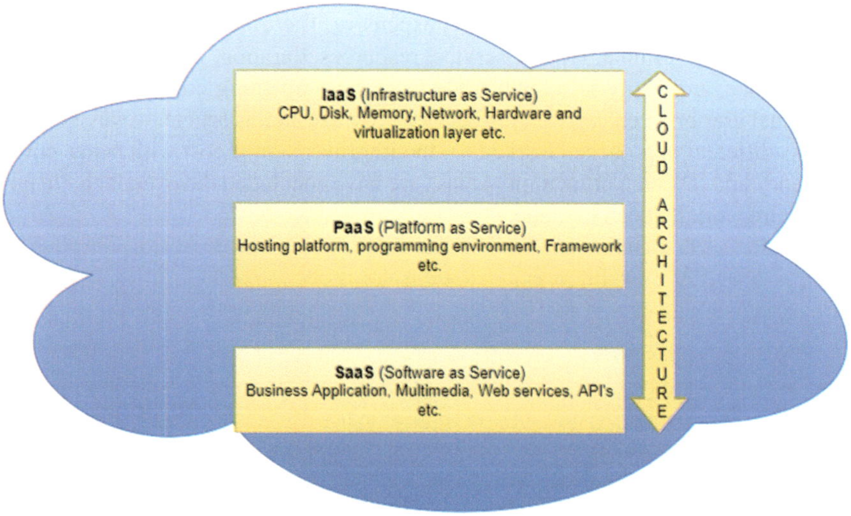

Fig. 34.1 Cloud architecture based on service [1]

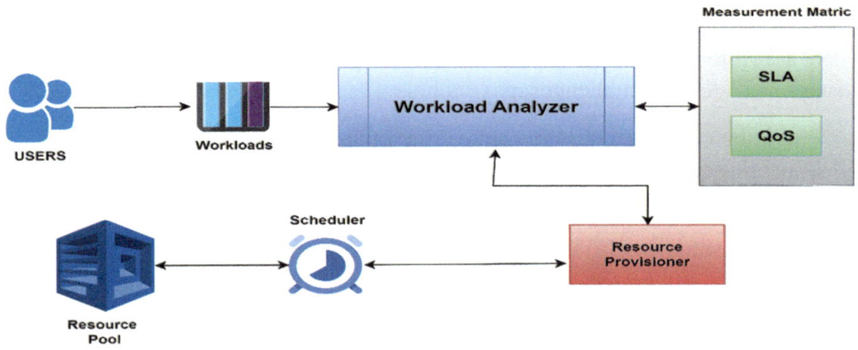

Fig. 34.2 Resource management process [4]

The major challenge for all the approaches is to provide high quality resource utilization at the service providers end and high quality and smooth services at the user end (Fig. 34.2). In this article, we have compared various resource provisioning techniques to achieve the better QoS while maintaining SLA and resource utilization. We have categorized the methods according to their achievements, i.e., cost, time, power, and other parameters.

34.2 Need of Resource Provisioning

Resource provisioning finds and provides proper resource for workloads in time. It is essential for cost and time management and also for proper resource utilization [5]. To maintain level of service, it is essential to provide proper amount of resource for workloads with least maintenance and maximum throughput. Proper resource detection and allocation are important for performance over heterogeneous or homogenous environments.

34.3 Cloud Resource Provisioning

A service provider provides resource among multiple consumers maintaining the maximum utilization of resources with minimum operating cost and minimum SLA violation. Resource provisioning is an important task to make full utilization of cloud, so it is important to analyze the task and provide resources efficiently and optimally. The process of deploying the application services on the resources provided by the cloud is called cloud resource provisioning. It involves two steps:

1. Physically hosting VM's over the cloud.
2. Applying consumer requests on the VM's using proper load balancing techniques.

For better utilization of resources, resource provisioning is a challenging task. There are various considerable parameters while providing resources, i.e., load, time, cost, power, etc. The challenges may include

1. Providing only the required resources and schedule the resources accordingly.
2. Managing resources when there is exponential increase in the requirements as it may consume more power, cost, and time if not managed properly.

Load balancing is also challenging while resource provisioning as the data is stored dynamically on the cloud server. Although so much work has been done so far, and it improved the resource provisioning and utilization in terms of cost, time, power, there is still much more to find and work onto. In the below section, we have discussed some related work in terms of resource provisioning.

34.4 Related Work

A lot of work has been done on resource optimization techniques in cloud computing, and some of the approaches are described below with different parameters and goals.

Gupta et al. [6] proposed bio-inspired GA-ANN technique that performed efficiently in a scalable environment when compared to other approaches, i.e., GA-EXE, BB-BC, etc. It can perform well in real, fabricated, and scalable environment. This approach schedules task in a virtual machine, and it can be extended for virtual machine scheduling.

Filali et al. [7] proposed particle swarm optimization (PSO) to achieve better resource scheduling and provisioning in cloud IaaS. Although it was suitable for smaller workflow, the cost increased with workflow size.

Pooyan Jamshidi et al. [8] provided integer programming mechanism for resource provisioning in grid. It was able to maximize revenue but needed to implement in different cloud environments.

Vijayakumar et al. [9] proposed match making technique for autonomic resource provisioning. It provided better QoS, but energy consumption was too high.

Rajiv et al. [10] proposed peer-to-peer technique for cloud resource provisioning and load balancing. This approach improved the performance and scalability.

Choi et al. [11] provided CPU allocation algorithm for dynamic resource provisioning which reduced the time and cost but still needed to implement of different cloud environments.

Wu et al. [12] proposed customer-driven heuristic algorithm to improve resource optimization at SaaS level, and this reduced the cost and SLA violations but was only applicable for SaaS.

Katyal et al. [13] provided min–min/max–min algorithm for effective resource provisioning which minimized the total execution time and improved the performance. Still dependencies on the tasks can be optimized.

Agarwal et al. [5] proposed efficient resource allocation (ERA) algorithm to improve effective resource provisioning. It reduced the response time, but it was only applicable on reserved instances.

Chase et al. [9] presented stochastic programming mechanism to optimally reserve the resources at different level with dynamic demands. The mechanism was able to retain the accuracy while reducing the cost. Network delay and VM migration factors should also be considered to improve the computational performance.

Varatharajan et al. [14] proposed modified cloud resource provisioning (MCRP) algorithm with cost, execution time, and memory as parameters. Based on the results proposed, it was able to reduce time and cost with low memory usage.

Li et al. [15] presented optimal cloud resource provisioning (OCRP) for SaaS level, and it was aimed to minimize the cost at user end and maximize the profit at SaaS provider by meeting the maximum QoS requirements. This algorithm was able to meet the QoS at SaaS level; further, it can be extended to PaaS or IaaS level of cloud.

Ghobaei-Arani et al. [16, 17] used biogeography and proposed a workload clustering-based resource provisioning mechanism. The proposed approach reduced the cost, and energy consumption also there was less delay and SLA violation. The reliability and failures need to be evaluated at different cloud scenarios.

Xiao et al. [12] used Dirichlet multinomial model (DMM) to determine the parameters for resource optimization. They proposed reputation-based resource provisioning mechanism that reduced the overall cost and meets the QoS requirements.

Bellavista et al. [18] presented a novel method for adaptive replication by reducing the fault tolerance when load balancing. This method reduced resource consumption with minimum information loss when load balancing.

Iqbal et al. [19] a prototype system for automatic identification mechanism for automatic identification and resolution of over resource provisioning in multi-tire cloud applications. It provided a better response time and reduced the cost by optimizing the resource allocation. Meeting to the SLA is still challenging over n-tier architecture.

Gupta et al. [20] presented a hybrid ANN and whale optimization model focused on increasing the fault tolerance and improving the load balancing and optimizing the resources to improve overall performance.

Gupta et al. [21] proposed hybrid harmony search and ANN (HS-ANN) approach to provide optimal solution and reduce task failure and improve reliability in a faulty environment. Execution time, task failure and power consumption (Kwh) were taken as parameters. This approach achieved a better solution in a feasible time.

Qian and Medhi [22] proposed an optimization model to optimize operational cost over dynamic time duration. They considered that the demand can be change during any time duration, thus turning off/on data center according to the demand. The proposed method was able to reduce operational cost in comparison of static capacity allocation, although this experiment was on a small dataset but can be applied on large scale datasets with heuristic algorithms.

Shahidinejad et al. [17] have proposed hybrid ICA-K-Means to cluster the workload according to the QoS. They have performed their work under NASA and

FIFA real workload traces. In their experiment, there was a significant amount of improvement in terms of cost, CPU utilization, response time, and elasticity.

Buyya et al. [23] proposed a resource management framework SCOOTER that has ability to manage resources automatically. This framework can self-heal, self-optimize and minimize execution cost and execution time. It performs better than other existing autonomic resource management techniques.

Patel et al. [24] worked on pCNN-LSTM, a hybrid prediction approach to predict CPU utilization at multiple consecutive time. It was able to learn different CPU loads at different scales. This model achieved up to 16% improvement in load prediction over other existing techniques.

34.5 Resource Optimization Techniques

Cloud optimization is referred as eliminating the waste of cloud resource by selecting best element from the set of available resources. Optimization techniques helps in the selection of resource provided by the cloud service provider. The optimization techniques can be classified into two broad categories: (I) static technique and (II) dynamic techniques (Fig. 34.3).

Resource optimization [25] followed while providing resource can be static or dynamic based on the requirements and availability of resources.

1. Static Approach: In this approach, the services are allocated in advance to the consumer. Consumer has to define the capacity of resource before its use. The static approach of optimization defines how efficiently static allocation can be done.
2. Dynamic Approach: Service provider provides the resources on requirement basis that can be scaled up or down in future according to the change in requirements. Genetic algorithms, stochastic algorithm, and other dynamic methods are used for optimized resource allocation.

    a. Genetic Algorithm: It is commonly used to generate solutions based on bio-inspired parameters such as mutation, crossover, and parent selection. GA is difficult to implement on complex and dynamic dataset.

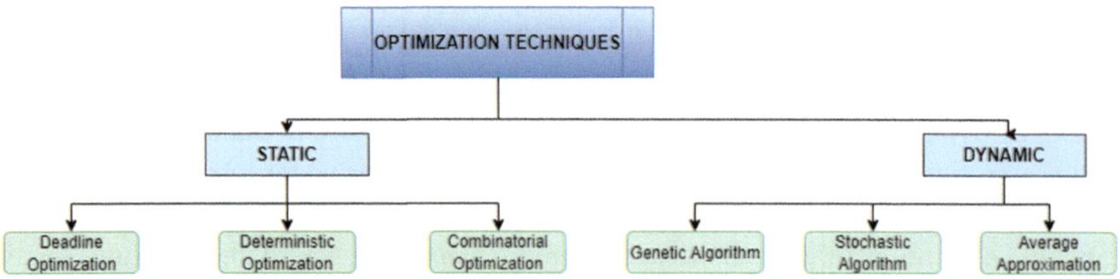

Fig. 34.3 Static and dynamic optimization techniques [2]

b. Stochastic Approach: At first, original formulation is derived, and then the core formulation is transformed into the deterministic equivalent formulation (DEF) that can be processed with any optimization tool.

In Table 34.1, we have compared the different approaches used to provide optimal solution for cloud resource provisioning. The comparison is done based on different parameters and mechanisms.

34.6 Resource Provisioning Techniques

The dynamic nature of cloud computing makes the optimization more challenging. Machine learning with nature-inspired algorithms is used to provide a better optimization with low cost and highly efficient mechanism. Table 34.1 provides various optimization techniques based on different parameters, i.e., time, cost, QoS, SLA, and heterogeneity.

Table 34.1 Resource optimization techniques comparative study

Problem	Proposed technique	Parameters considered	Merits	Challenges
Resource Scheduling on Cloud IaaS	Particle Swarm optimization approach	Cost and QoS	Suited for smaller workflow size	High cost for large workflow
Autonomic Cloud Resource Optimization Problem	Match-making technique		Provides QoS	Power efficiency is low
Resource Provisioning for Cloud Data Streaming Applications	CPU Allocation algorithm		Reduce processing time and cost	Need to work on different scenarios
Resource Provisioning for Software-as-a-Service Applications	Customer-driven heuristic algorithms	Heuristics and time	Less cost and reduce SLA violation	Only applicable at SaaS level
To manage dynamic resources in hybrid cloud	Deadline Priority Provisioning algorithm		Reduce total execution time	Only applicable for Hybrid Cloud
An effective algorithm for resource provisioning based on selection	Min–min or max–min algorithm		Minimize make span	Dependencies among tasks can be modeled
Cloud resource provisioning, application management, and load balancing	Hash Table—for routing and discovery Decentralized Peer-to-peer technique		Improvised the scalability and decreased the overall response time	Power efficiency is low
Efficient resource optimization in fog computing	Efficient resource allocation (ERA) Algorithm		Reduces the response time	Applicable for only reserved instances
Reduce task failure and improve overall performance	Hybrid Harmony Search and ANN (HS-ANN)		Increased overall performance and reduced failures in faulty environment	Can be improvised for more complex cloud structures
To reduce operational cost in limited time duration	Metaheuristic optimization model [6]		Overall cost reduction over dynamic time duration and demand	Experiment was performed on a small dataset, so it is questionable over large dataset
To maintain SLA, cost, CPU utilization, and elasticity	Hybrid ICA-K-means approach		Improvement in terms of cost, CPU utilization, response time, and elasticity	Energy efficiency, overall power consumption is still challengeable

(continued)

Table 34.1 (continued)

Problem	Proposed technique	Parameters considered	Merits	Challenges
Automatic Resource Management [26]	Resource management framework SCOOTER		This framework can selfheal, self-optimize, and minimize execution cost and execution time	Power efficiency is considerable
CPU utilization at multiple consecutive time	pCNN-LSTM, a hybrid prediction approach		It was able to learn different CPU loads at different scales	Cost and resource utilization can be improved

References

1. Patil, A., Rekha, P.: An analysis report on green cloud current trends and future research challenges (2020)
2. Sumalatha, K., Anbarasi, M.S.: A review on various optimization techniques of resource provisioning in cloud computing. Int. J. Electr. Comput. Eng. **9**(1), 2088–8708 (2019)
3. Sood, S.K., Singh, K.D.: SNA based resource optimization in optical network using fog and cloud computing. Opt. Switch. Netw. **33**, 114–121 (2019)
4. Tantar, A.-A., Nguyen, A.Q., Bouvry, P., Dorronsoro, B., Talbi, E.: Computational intelligence for cloud management current trends and opportunities. In: 2013 IEEE Congress on Evolutionary Computation, pp. 1286–1293. IEEE (2013)
5. Agarwal, S., Yadav, S., Yadav, A.K.: An efficient architecture and algorithm for resource provisioning in fog computing. Int. J. Inform. Eng. Electron. Bus. **8**(1), 48 (2016)
6. Rawat, P.S., Dimri, P., Gupta, P., Saroha, G.P.: Resource provisioning in scalable cloud using bio-inspired artificial neural network model. Appl. Soft Comput. **99**, 106876 (2021)
7. Filali, A., Hafid, A.S., Gendreau, M.: Adaptive resources provisioning for grid applications and services. In: 2008 IEEE International Conference on Communications, pp. 186–191. IEEE (2008)
8. Vijayakumar, S., Zhu, Q., Agrawal, G.: Dynamic resource provisioning for data streaming applications in a cloud environment. In: 2010 IEEE Second International Conference on Cloud Computing Technology and Science, pp. 441–448. IEEE (2010)
9. Chase, J., Niyato, D.: Joint optimization of resource provisioning in cloud computing. IEEE Trans. Serv. Comput. **10**(3), 396–409 (2015)
10. Jamshidi, P., Ahmad, A., Pahl, C.: Autonomic resource provisioning for cloud-based software. In: Proceedings of the 9th International Symposium on Software Engineering for Adaptive and Self-Managing Systems, pp. 95–104 (2014)
11. Choi, Y., Lim, Y.: Optimization approach for resource allocation on cloud computing for IoT. Int. J. Distrib. Sens. Netw. **12**(3), 3479247 (2016)
12. Xiao, Y., Lin, C., Jiang, Y., Chu, X., Shen, X.: Reputation-based QoS provisioning in cloud computing via Dirichlet multinomial model. In: 2010 IEEE International Conference on Communications, pp. 1–5. IEEE (2010)
13. Katyal, M., Mishra, A.: Application of selective algorithm for effective resource provisioning in cloud computing environment. arXiv preprint arXiv:1403.2914 (2014)
14. Suresh, A., Varatharajan, R.: Competent resource provisioning and distribution techniques for cloud computing environment. Clust. Comput. **22**(5), 11039–11046 (2019)

15. Li, C., Li, L.Y.: Optimal resource provisioning for cloud computing environment. J. Supercomput. **62**(2), 989–1022 (2012)
16. Ghobaei-Arani, M.: A workload clustering based resource provisioning mechanism using biogeography based optimization technique in the cloud based systems. Soft. Comput. **25**(5), 3813–3830 (2021)
17. Shahidinejad, A., Ghobaei-Arani, M., Masdari, M.: Resource provisioning using workload clustering in cloud computing environment: a hybrid approach. Clust. Comput. **24**(1), 319–342 (2021)
18. Bellavista, P., Corradi, A., Kotoulas, S., Reale, A.: Adaptive fault-tolerance for dynamic resource provisioning in distributed stream processing systems. In: *EDBT*, pp. 85–96 (2014)
19. Iqbal, W., Dailey, M.N., Carrera, D., Janecek, P.: Adaptive resource provisioning for read intensive multi-tier applications in the cloud. Futur. Gener. Comput. Syst. **27**(6), 871–879 (2011)
20. Gupta, P., Bhagat, S., Saini, D.K., Kumar, A., Alahmadi, M., Sharma, P.C.: Hybrid whale optimization algorithm for resource optimization in cloud e-healthcare applications. CMC-Comput. Mater. Cont. **71**(3), 5659–5676 (2022)
21. Gupta, P., Bhagat, S., Rawat, P.: Fault aware hybrid harmony search technique for optimal resource allocation in cloud. J. Intell. Fuzzy Syst. (Preprint), pp.1–13 (2022)
22. Qian, H., Medhi, D.: Server operational cost optimization for cloud computing service providers over a time horizon. In: Hot-ICE (2011)
23. Gill, S.S., Buyya, R.: Resource provisioning based scheduling framework for execution of heterogeneous and clustered workloads in clouds: from fundamental to autonomic offering. J. Grid Comput. **17**, 385–417 (2019)
24. Patel, E., Kushwaha, D.S.: A hybrid CNN-LSTM model for predicting server load in cloud computing. J. Supercomput. **78**(8), 1–30 (2022)
25. Verma, V.K., Gupta, R.K., Jain, T., Mundra, A., Sachdev, A.: Hybrid approach for optimized resource allocation load balancing in cloud infrastructure. Int. J. Adv. Sci. Technol. **28**(12), 224–235 (2019)
26. Kousiouris, G., Menychtas, A., Kyriazis, D., Gogouvitis, S., Varvarigou, T.: Dynamic, behavioral-based estimation of resource provisioning based on high-level application terms in cloud platforms. Futur. Gener. Comput. Syst. **32**, 27–40 (2014)
27. Zhan, Z.H., Liu, X.F., Gong, Y.J., Zhang, J., Chung, H.S.H., Li, Y.: Cloud computing resource scheduling and a survey of its evolutionary approaches. ACM Comput. Surv. (CSUR) **47**(4), 1–33 (2015)
28. Chase, J., Kaewpuang, R., Yonggang, W., Niyato, D.: Joint virtual machine and bandwidth allocation in software defined network (SDN) and cloud computing environments. In: 2014 IEEE International Conference on Communications (ICC), pp. 2969–2974. IEEE (2014)
29. Chakravarthi, K.K. and Vijayakumar, V., 2018. Workflow Scheduling Techniques and Algorithms in IaaS Cloud: A Survey. *International Journal of Electrical & Computer Engineering (2088–8708)*, *8*(2).
30. Tchernykh, A., Schwiegelsohn, U., Talbi, E.G., Babenko, M.: Towards understanding uncertainty in cloud computing with risks of confidentiality, integrity, and availability. J. Comput. Sci. **36**, 100581 (2019)
31. Kaur, A., Kaur, B., Singh, D.: Optimization techniques for resource provisioning and load balancing in cloud environment: a review. Int. J. Inform. Eng. Electron. Bus. **9**(1), 28 (2017)

Chapter 35
The Future of Artificial Intelligence in Fashion: Innovations, Challenges, and Implications

Deepak Sharma, J. P. Sampath Kumar, and K. R. Shylaja

Abstract The fashion business is striving to improve product quality and efficiency to stay competitive in the current global market, which is characterised by intense competition and unpredictable demand. Consequently, extensive efforts have been undertaken to examine and enhance the use of artificial intelligence (AI) methods in diverse applications within the fashion sectors. Apparel organisations are swiftly integrating AI-powered tools and solutions to improve all facets of the industry, including design, production, marketing, and consumer experiences. This article presents a comprehensive analysis of the current state of AI in fashion, examines emerging trends and technologies, and provides insights into the potential impact of AI on the industry's future. Through an examination of existing literature and case studies, this article aims to inform stakeholders about the transformative role of AI in fashion and its implications for businesses, consumers, and the society at large.

35.1 Introduction

The fashion industry is experiencing continuous transformation due to technological advancements and evolving consumer preferences. In recent years, there has been a growing interest in incorporating artificial intelligence (AI) into the fashion business as a viable approach for transformation. AI technologies have revolutionised the fashion industry by bringing significant advancements in production, design, marketing, and supply chain management. These technologies encompass several

D. Sharma (✉)
Fashion Design Department, Manipal University, Jaipur, India
e-mail: deepak.ds3210@gmail.com

J. P. Sampath Kumar
School of Planning and Design, Manipal University, Jaipur, India
e-mail: sampath.kumar@jaipur.manipal.edu

K. R. Shylaja
Department of CSE, Dr. Ambedkar Institute of Technology Bangalore, Bangalore, India
e-mail: Shylaja.kr.cs@drait.edu.in

P. Nanda et al. (eds.), *Intelligent System and Data Analysis*, Smart Innovation, Systems and Technologies 398, https://doi.org/10.1007/978-981-97-5200-3_35

ways that allow robots to replicate human intelligence. This article examines the past background and current condition of artificial intelligence in the fashion industry. It provides a clear objective for the study and outlines the methodology employed to determine the impact of AI on the sector.

Artificial intelligence possesses the capacity to revolutionise the fashion industry and cater to the evolving demands of modern customers. According to Giri et al. [1], AI has the capacity to enhance several areas of the fashion business, such as design, production, promotion, and supply chain management. Fashion businesses may use the capabilities of AI to enhance operations, save expenses, and deliver personalised experiences to consumers through the utilisation of data analysis, pattern recognition, and decision-making algorithms. The integration of AI into the fashion sector has the potential to disrupt traditional business models and foster innovation across the whole value chain.

The aim of this research article is to investigate the prospective role of artificial intelligence (AI) in the fashion business. This article seeks to offer a complete overview of how AI is altering the fashion industry by analysing the present state of AI applications, identifying new trends and developments, and evaluating the difficulties and ramifications. This article seeks to educate on the prospective influence of AI on the future of fashion by examining current research, case studies, and expert comments.

To achieve the research objective, a comprehensive evaluation of the relevant literature, encompassing academic articles, business reports, and case studies, being conducted. The initiative will analyse case studies and real-world instances of start-ups and fashion companies employing AI technology. The study findings will be synthesised to provide insights into the transformative potential of artificial intelligence in the domains of design, production, marketing, and supply chain management.

35.2 AI Applications in Fashion

35.2.1 AI in Design and Creative Processes

Integrating artificial intelligence (AI) technology in design and creative processes aims to enhance and simplify the design and creative aspects of the fashion industry. Artificial intelligence (AI) possesses the capability to generate novel and inventive ideas, assist in virtual prototyping and 3D visualisation, and evaluate vast amounts of data to identify emerging trends and client preferences. Designers may utilise AI-powered tools to generate ideas, develop patterns, and personalise items, hence enhancing the speed and efficiency of their creative processes.

The fashion industry has experienced a transformation due to the integration of AI-generated designs and trends, which merge computational capabilities with human creativity. Several researches, like those conducted by Obe et al. [2], demonstrate the ability of AI to generate novel and innovative fashion concepts. Deep neural networks in AI can assess vast amounts of fashion data and patterns, enabling them to generate unique designs that are influenced by prior styles and trends. AI-generated designs push the limits of traditional design methods by promoting creativity while also offering designers a wealth of inspiration and opportunities for experimentation.

The utilisation of artificial intelligence (AI) in conjunction with three-dimensional (3D) visualisation and virtual prototyping has fundamentally transformed the approach that fashion industry designers employ to actualise their creations. In order to enhance the design process, Lee et al. [3] propose an integrated framework that combines artificial intelligence with virtual reality. Designers may now utilise virtual environments to conceptualise their garments and create virtual prototypes, enabling rapid modifications, and improvements before physical production. This technology enables designers to effortlessly explore various materials, colours, and patterns, fostering creativity and enhancing efficiency.

35.2.2 AI in Supply Chain and Production

"AI in fashion supply chain and production" is the term used to describe the application of artificial intelligence (AI) technology to enhance and streamline different stages of apparel manufacture. AI algorithms play a crucial role in demand forecasting, optimising inventory management, and enhancing supply chain logistics. Artificial intelligence has the capability to predict consumer demand and facilitate data-driven decision-making by analysing historical data and market trends, hence reducing instances of excessive inventory and shortages.

AI plays a crucial role in enhancing supply chain management by accurately predicting demand and efficiently managing inventories. The utilisation of AI methodologies for enhancing the precision of fashion retail demand prediction is emphasised by Ren et al. [4] and Thomassey et al. [5]. In order to forecast future demand, machine learning algorithms and neural networks analyse historical sales data, market trends, and external factors. Consequently, fashion manufacturers have reduced stock outs, less waste, and enhanced consumer satisfaction, leading to cost reduction. AI enables accurate demand forecasting, ensuring timely availability of the necessary goods in the appropriate quantities.

The integration of AI in smart manufacturing and automation has brought about a significant revolution in production processes within the garment industry. Wong et al. [6] and Çelik et al. [7] go through how AI-driven automation and intelligent robotics improve productivity and quality control in the garment industry. Robotics and artificial intelligence (AI) technologies have the capability to efficiently and precisely perform repetitive tasks like as cutting, sewing, and quality inspection. Furthermore, this not only reduces the need for human labour, but also enhances productivity, reduces mistakes, and decreases time-to-market. The utilisation of AI-driven automation allows the fashion sector to fulfil customer demands for personalised products and quick fashion production, while maintaining exceptional quality standards.

35.2.3 AI in Marketing and Personalisation

Artificial intelligence (AI) in fashion marketing and personalisation involves utilising AI technology to enhance marketing strategies and provide tailored customer experiences. Fashion companies may enhance their understanding of customers' preferences, behaviour, and buying habits by utilising AI-powered data analytics and consumer insights. By utilising this data, recommendation systems driven by artificial intelligence may provide tailored product recommendations, marketing strategies, and promotions that are highly pertinent to the consumer. This ultimately results in increased customer satisfaction and involvement.

Fashion companies may enhance their understanding of clients' interests and behaviours by leveraging AI, which utilises big data analytics to provide valuable consumer insights. Madsen et al. [8] argue that data-driven analytics play a vital role in the fashion industry. In order to identify emerging trends, focus on specific consumer segments, and enhance marketing strategies, AI systems analyse vast amounts of data, including social media trends, purchasing behaviour, and customer feedback. Fashion firms may enhance their client experiences, implement targeted marketing strategies, and cultivate stronger customer connections by doing thorough research on consumer preferences.

AI-powered recommendation systems, which offer tailored product suggestions and enhance consumer satisfaction, have become an essential component of the fashion shopping experience. The efficacy of AI in providing fashion advice is stressed in research such as McAuley et al. [9] and Bhatnagar et al. [10]. AI algorithms may provide customised suggestions by analysing customer data and past purchases, taking into account unique likes, styles, and sizes. This improves the shopping experience while simultaneously increasing revenue and customer loyalty. AI-driven recommendation systems enable fashion enterprises to develop tailored and pertinent product offerings, hence enhancing client engagement and profitability.

35.3 Emerging Trends and Innovations

35.3.1 Computer Vision and Image Recognition

Computer vision and image recognition technologies have revolutionised the fashion industry by enabling tasks such as item identification, segmentation, and classification. Deep learning algorithms have exhibited remarkable proficiency in understanding fashion items and their aesthetic attributes. Ge et al. [11] produced DeepFashion2, which is a widely used dataset for analysing images of clothes. This dataset enables advancements in activities such as detection, posture estimation, classification, and re-identification. It serves as a valuable resource for training and evaluating computer vision algorithms.

Visual searching and similarity matching are revolutionising the way consumers discover and explore fashion goods. Veit et al. [12] proposed an approach for visual compatibility modelling that captures the visual relationships between fashion goods. This method enables accurate visual search and recommendation systems by analysing visual signals such as colour, pattern, and style. A convolutional neural network system (CNN) that uses style and attribute information to categorise fashion clothes. This approach utilises convolutional neural networks (CNNs) to differentiate between local and global styles, streamlining visual exploration and classification tasks.

The advent of augmented reality (AR) and virtual try-on technology has revolutionised the way shoppers see and interact with fashion. Virtual try-on allows consumers to visualise the appearance of items on themselves without physically putting them on, by simulating the fitting procedure. Mohammadi et al. [13] and Ghodhbani et al. [14] conducted a comprehensive investigation on virtual try-on techniques, which encompassed 3D body modelling, garment simulation, and picture-based virtual try-on. These advancements enhance the simplicity of online shopping by reducing returns and increasing consumer satisfaction. Kang et al. [15] investigated the potential applications of augmented reality (AR) in the fashion industry, highlighting its benefits in terms of consumer interaction, interactive purchasing, and virtual try-on.

35.3.2 Natural Language Processing and Chatbots

The advent of natural language processing (NLP) and chatbots is causing a shift in how consumers engage with fashion businesses. AI-driven virtual assistants provide fashion advice, address customer queries, and give customised recommendations. In their study, Husak et al. [16] developed a chatbot that utilises collaborative filtering and natural language processing (NLP) to offer tailored fashion suggestions. This technique enhances the conversational buying experience by tailoring recommendations based on client preferences. In a similar study, Landim et al. [17] proposed a technique to enhance chatbot suggestions for tailored shopping experiences.

AI-powered virtual assistants are crucial for crafting personalised shopping experiences. These assistants utilise machine learning and natural language processing algorithms to understand consumer preferences, provide fashion recommendations, and aid with the shopping experience. Kang et al. [18] examined the potential application of generative adversarial networks (GANs) in the field of customised garment creation. Their research demonstrated the potential of GANs to generate tailored apparel designs according to individual client preferences, hence promoting collaborative design and customisation.

Personalisation has become an essential element in the fashion industry, and advancements in technology enable customised shopping experiences. AI-driven systems may analyse customer data and preferences to produce customised suggestions, bespoke offers, and tailored content. In 2020, Li et al. developed a personalised recommendation architecture that integrates user attributes and fashion style preferences. Enhancing the accuracy of fashion suggestions with this method enhances customer engagement and pleasure.

35.3.3 Generative Adversarial Networks (GANs)

The fashion industry has seen the potential of generative adversarial networks (GANs) due to its ability to generate authentic and diverse fashion images. These models have the capability to generate unique patterns, textures, and designs, hence fostering innovation and creativity. Zerfu et al. [19] proposed FashionGAN, a model based on generative adversarial networks (GANs) that is capable of learning and generating new fashion designs. The research conducted by Singh et al. [20] demonstrated the potential of generative adversarial networks (GANs) in facilitating novel avenues for customisation and collaborative creativity within the realm of fashion design.

The advent of AI-generated fashion photography is causing a transformation in the creative process of the fashion business. By combining aesthetic transfer techniques and semantic data, AI models are capable of generating diverse and lifelike fashion photographs. Zhao et al. [21] suggested a method for generating fashion photos by employing semantic analysis and style adversarial networks. Their study showcased the capacity of AI to generate fashion images, hence facilitating the exploration of emerging trends and styles.

Customisation and co-creation have become essential elements in the fashion industry, allowing individuals to express their distinctiveness and actively participate in the creative process. Lee [22] employed generative adversarial networks (GANs) to provide a customised method for fashion design. Their research demonstrated the use of user preferences and criteria to create tailored fashion designs through the application of GANs, hence promoting collaborative design and individualisation.

The fashion industry is undergoing transformation as a result of emerging trends and advancements in computer vision, image recognition, augmented reality, chatbots, chatbot-driven virtual assistants, personalised purchasing experiences, generative adversarial networks, AI-generated fashion imagery, customisation, and co-creation. These advancements enable collaborative creation of possibilities, enhance visual search and similarity matching, and facilitate virtual try-ons. The fashion industry is currently experiencing a technological revolution due to the incorporation of various technologies. This revolution brings new prospects for innovation, customisation, and client engagement.

35.4 Challenges and Limitations

In recent years, the integration of artificial intelligence (AI) in the fashion industry has brought forth numerous advancements and opportunities. However, alongside its potential benefits, there are significant challenges and limitations that need to be addressed. This examines three key areas of concern: data privacy and ethical concerns, bias and diversity in AI systems, and technical limitations and implementation hurdles (Table 35.1).

Table 35.1 Challenges and limitations of AI in fashion industry

Challenges and limitations	Research directions	Key references
Data privacy and ethical concerns	Data breaches and unauthorised access	Dilmaghani et al. [23]
	Non-transparent data handling practices	Elliott et al. [24]
	Gathering and storage of personal data of wearable technology	Kapoor al. [25]
Data privacy and ethical concerns	Biases in facial recognition systems	Buolamwini and Gebru [26]
	Word embedding	Caliskan et al. [27]
Technical limitations and implementation hurdles	Creating AI systems for fashion industry with small dataset	Wong et al. [28]
	Clothing simulation and realistic textile rendering	Wang et al. [29]
	Integrating data and system compatibility as per fashion industry	Jin et al. [30]
	Model adaption to shifting fashion trends	Guo et al. [31]

35.4.1 Data Privacy and Ethical Concerns

Data privacy and ethical concerns provide substantial challenges for the integration of AI in the fashion sector. Dilmaghani et al. [23] emphasise the potential dangers associated with data breaches and illegal access. The preservation and safeguarding of this data are of utmost importance, given that fashion companies collect and analyse vast quantities of personal data. Elliott et al. [24] argue that it is crucial to have fair standards for data usage, obtain informed consent, and maintain transparent practices when processing data. The utilisation of wearable technology in the fashion business, as discussed by Kapoor et al. [25], also gives rise to concerns around the collection and retention of personal data. To mitigate these hazards, it is imperative to implement robust privacy rules, ensure secure data retention, and provide users control over their personal information.

35.4.2 Bias and Diversity in AI Systems

The fashion industry has significant challenges due to prejudice and an absence of inclusivity in AI systems. Buolamwini and Gebru [26] emphasise the presence of biases in facial recognition systems, contending that these biases might lead to unjust and inaccurate outcomes. Caliskan et al. [27] examine biases in word embeddings and their impact on natural language processing systems. These biases may perpetuate racial and gender disparities in the realm of fashion guidance and preferences. The requirement of building algorithms takes into account individual diversity and prevents from spreading stereotypes. It is vital to regularly monitor and review AI systems for biases in order to solve these concerns, as well as to have more varied and representative training datasets.

35.4.3 Technical Limitations and Implementation Hurdles

The apparel sector has challenges in effectively implementing AI due to technical limitations and operational complexities. Wong et al. [28] address the technological challenges associated with developing AI systems for the fashion sector. Challenges in this context encompass limited dataset sizes, absence of domain-specific training data, and the complexity in capturing diverse design elements. Wang et al. [29] enumerate the limitations of AI-driven virtual try-on systems, including the challenges of achieving precise garment simulation and lifelike textile representation. Technical challenges occur when faced with variations in body sizes and forms. Jin et al. [30] analyse the challenges associated with incorporating AI into the fashion retail business, specifically focusing on the integration of data and system compatibility. Successful implementation also necessitates continuous model adjustment to evolving fashion trends said by Guo et al. [31].

Cooperation among stakeholders is essential for surmounting these barriers and limitations. Dilmaghani et al. [23] emphasise the necessity of implementing laws and standards throughout the whole industry to ensure ethical data practices and protect consumer privacy. Kapoor et al. [25] propose the development of ethical frameworks to govern the process of collecting data, its utilisation, and the decision-making process using algorithms. Collaboration between fashion enterprises, AI experts, and ethicists can facilitate the creation of AI systems that are both responsible and equitable. In order to minimise biases in AI systems, it is crucial to encourage diversity and inclusion in the creation of datasets and the development of algorithms [27].

35.5 Implications of AI in Fashion

The integration of artificial intelligence (AI) in the fashion industry has significantly influenced several business domains. This analysis examines the impact of artificial intelligence on the fashion industry, specifically in terms of changes to business models, shifts in customer expectations and experiences, disruptions to the workforce and the need for skill development, as well as the implications for sustainability and the circular economy.

35.5.1 Transformation of Business Models

An important outcome of incorporating AI into the fashion industry is the alteration of business models. Casciani et al. [32] emphasise the fashion industry's adoption of cutting-edge business strategies, such as subscription-based services and personalised product recommendations. This modification allows firms to enhance client connection and provide personalised experiences, eventually leading to heightened consumer loyalty and revenue. In a similar vein, Mohiuddin Babu et al. [33] propose a data-driven business model for fashion shops that use artificial intelligence (AI) to enhance operational efficiency, optimise the supply chain, and enhance consumer satisfaction.

35.5.2 Consumer Expectations and Experiences

AI has had a significant influence on the fashion industry, particularly in relation to customer expectations and experiences. Casciani et al. [32] discuss the enhancement of the personalised shopping experience through the utilisation of virtual try-on technology and AI-driven tailored product suggestions. This level of customisation enhances customer satisfaction and fosters a stronger relationship between clients and organisations. In addition, Hsu et al. [34] examine the impact of AI-powered fashion chatbots on consumer satisfaction and loyalty. The study highlights the need of precise and efficient AI interactions to meet customer expectations and enhance their overall shopping experience.

35.5.3 Workforce Disruption and Skills Development

The integration of artificial intelligence (AI) in the fashion sector has caused significant changes in the labour force and has made it imperative to enhance skill sets. Gangoda et al. [35] examine the impact of AI on employment in the fashion sector and emphasise the importance of acquiring new skills and improving existing ones to adapt to changing job roles. As AI automates repetitive tasks in the fashion sector, workers may focus on creative and strategic duties, leading to increased innovation and production [36].

35.5.4 Sustainability and Circular Economy

Sustainability and circular economy are crucial elements in the fashion industry, and AI has significant ramifications in both domains. Oosthuizen et al. [37] and Wong et al. [28] discuss the potential for enhancing supply chain efficiency and reducing waste by employing predictive demand forecasting to advance sustainability objectives. Fashion firms may foster a more sustainable and ecologically conscious industry by reducing overproduction and enhancing resource utilisation. In this study, Akram et al. [38] examine the potential of AI-based solutions to facilitate the implementation of circular economy principles in the fashion industry, therefore enhancing resource efficiency and reducing environmental harm.

The fashion industry's embrace of AI leads to substantial consequences. Conventional corporate frameworks undergo a metamorphosis, enabling customised interactions and enhancing client satisfaction. To effectively respond to evolving employment requirements, workers must acquire new skills and enhance existing ones due to the implementation of artificial intelligence (AI). AI contributes to the circular economy and sustainability by minimising waste and optimising resource allocation. To ensure a long-lasting and customer-centric future in the fashion sector, it is imperative for stakeholders to effectively handle challenges and capitalise on possibilities presented by AI technology.

35.6 Case Studies and Success Stories

Artificial intelligence (AI) has emerged as a disruptive force in the fashion industry, profoundly transforming several aspects of the value chain. Candeloro [39] emphasises the significance of artificial intelligence (AI) in luxury fashion brands, fashion start-ups driven by AI, and AI's impact on e-commerce and online marketplaces. Presented below are a selection of case studies and success stories. AI is utilised to enhance consumer experiences, foster innovation, and optimise corporate operations through examination of these real-life instances.

35.6.1 The Role of AI in Luxury Fashion Brands

Luxury fashion producers have used artificial intelligence (AI) to provide their clients with distinctive and customised experiences. Burberry is a notable firm that integrates artificial intelligence into its shops. Burberry uses AI technology to offer personalised product suggestions to clients. The recommendations are tailored to each customer's individual preferences and fashion sensibilities [40]. This level of personalisation enhances the consumer experience by increasing engagement and fostering brand loyalty.

In a similar fashion, Gucci has implemented an AI-powered chatbot called "Guccio" that engages with consumers and provides guidance on style choices [41]. The chatbot employs machine learning and natural language processing to provide interactive and personalised shopping experiences. Gucci's AI-driven strategy enables it to effectively address the specific needs and preferences of individual consumers, resulting in heightened satisfaction and increased sales.

35.6.2 AI-Driven Fashion Start-ups

AI-driven fashion enterprises have transformed the industry by employing AI technology to deliver distinctive and personalised experiences. Stitch Fix, an online personal styling business, utilises AI algorithms to produce customised clothing recommendations for its subscribers [42]. Stitch Fix's AI engine enhances the dependability of its suggestions by continuously studying client profiles and feedback, resulting in increased client happiness and retention.

The Yes is a renowned fashion company that utilises artificial intelligence to offer personalised product suggestions through the application of machine learning and natural language processing [43]. The Yes enhances the shopping experience by offering customised recommendations according to user input and individual preferences, enabling shoppers to discover clothing items that perfectly align with their own fashion sense.

35.6.3 AI in E-commerce and Online Marketplaces

Online marketplaces and e-commerce platforms have utilised artificial intelligence (AI) to enhance the client buying experience. Amazon, the largest e-commerce company globally, extensively employs artificial intelligence across its operations. The company's recommendation engine utilises advanced artificial intelligence algorithms to analyse user browsing and purchase trends and provide relevant product recommendations [44]. This customised approach enhances consumer engagement and stimulates sales.

Alibaba, a leading global e-commerce company, developed a virtual assistant named "FashionAI" that utilises artificial intelligence (AI) technology [45]. FashionAI employs computer vision and deep learning algorithms to aid consumers in identifying suitable clothing products by analysing body types, preferences, and prevailing fashion trends. This AI-powered virtual assistant enhances users' online shopping experience by providing personalised style suggestions and allowing them to virtually try-on clothes.

The aforementioned case studies and success stories demonstrate the significant influence that AI has had on the fashion industry. High-end fashion firms, AI-driven fashion start-ups, and e-commerce giants are utilising AI technology to enhance consumer experiences, stimulate innovation, and optimise company operations. Artificial intelligence assists luxury fashion brands such as Burberry and Gucci in providing tailored experiences that cater to individual preferences through the provision of customised recommendations. AI-powered fashion companies such as Stitch Fix and The Yes enhance the shopping experience and boost consumer satisfaction and loyalty by providing tailored product suggestions. E-commerce platforms such as Amazon and Alibaba utilise AI algorithms to provide customised suggestions and virtual assistance, hence enhancing customer engagement and improving conversion rates.

35.7 Conclusion and Future Outlook

The fashion industry has substantially profited from the implementation of artificial intelligence (AI) in terms of advancement, customisation, and operational effectiveness. The subsequent discussion delves into the use of artificial intelligence (AI) in several domains including as design, supply chain management, marketing, and customer experiences. Additionally, it addresses the challenges and limitations associated with using AI in these areas. Based on these observations, we analyse the potential future paths and possibilities for artificial intelligence in the fashion industry and provide suggestions for those involved.

35.7.1 Summary of Findings

This article elucidated many ways in which AI is revolutionising the fashion industry. The utilisation of AI-generated designs and virtual prototyping is fundamentally transforming the creative process within the design sector, enabling designers to actively explore novel possibilities and iterate at an accelerated pace. Augmented reality and AI-powered virtual try-on now allow customers to preview and personalise items prior to making a purchase. Artificial intelligence (AI) is facilitating demand forecasting, inventory management, and intelligent manufacturing across the supply chain and production processes, resulting in cost reduction and enhanced operational efficiency. The use of AI-driven marketing approaches, such as data-driven insights from consumers and AI-driven recommendation systems, is enhancing sales.

It is essential to acknowledge the challenges and limitations that artificial intelligence in the fashion industry encounters. In order to facilitate the proper and transparent use of consumer data, it is imperative to address and surmount key challenges pertaining to data privacy and ethics. In order to mitigate the perpetuation of societal injustices, it is imperative to address the issues of prejudice and diversity in AI systems with utmost care and attention. The effective use of AI technologies in the fashion industry may face obstacles due to technological limitations and implementation difficulties, such as the need for skilled personnel and infrastructure.

35.7.2 Recommendations for Stakeholders

In order to fully use the capabilities of AI in the fashion industry while addressing its problems, stakeholders should take into account the following recommendations:

1. Stakeholders should give high importance to data protection and enforce strict ethical standards while collecting, keeping, and using customer data. Data practices should give priority to openness and consent in order to empower consumers with control over their personal data.
2. It is important to constantly do audits, vary the training data, and include different teams in the creation of AI in order to actively eradicate bias and ensure diversity in AI systems. Long-term biases can be identified and rectified by continuous monitoring and evaluation.
3. Efforts should be focused on improving infrastructure and implementing training programmes to provide the workforce with the necessary knowledge and skills to effectively deploy and manage AI technology. Academic cooperation and industry ties can potentially eliminate the skills gap.
4. Collaboration and Knowledge Exchanging: To facilitate the exchange of best practices, important insights, and innovative ideas, stakeholders should actively promote collaborative efforts and venues for exchanging knowledge. Industry-wide collaboration can expedite the adoption and advancement of artificial intelligence in the fashion sector.

35.7.3 Future Directions and Opportunities

The future of AI in fashion appears to be quite promising. There are other prospects available for more investigation and advancement:

35.7.3.1 Enhanced Customisation

AI may achieve a higher level of customisation by incorporating user choices, contextual information, and real-time feedback. In order to enhance the level of service provided to each consumer, it is possible to enhance bespoke suggestions, flexible size options, and personalised purchase experiences.

35.7.3.2 Improved Sustainability

Through the optimisation of supply chain logistics, reduction of waste, and advocacy for circular economy concepts, AI has the potential to bolster sustainability efforts. Fashion companies may mitigate their environmental impact by leveraging artificial intelligence to guide their decision-making.

35.7.3.3 Better Collaboration

AI can facilitate to collaborative design processes, enabling designers, consumers, and AI systems to collaborate in the creation of unique fashion goods. By including customers in the design process, brands can cultivate a feeling of ownership and create goods that specifically cater to their target market.

35.7.3.4 AI with AR VR

The integration of artificial intelligence (AI) with augmented reality (AR) and virtual reality (VR) may provide clients with immersive and captivating experiences. Customers may utilise virtual try-on and shopping settings to simulate real-world encounters, enabling them to explore and engage with items in innovative ways.

35.7.3.5 AI in Design Evaluation

It can assist designers and manufacturers in evaluating the acceptance or rejection rate of a design prior to its manufacturing. This can assist them in creating precise manufacturing quantities for any design and increasing their profitability.

To summarise, AI has the capacity to revolutionise the fashion industry in several ways. By adopting AI technology and addressing associated challenges, stakeholders may unlock fresh opportunities for innovation, customisation, and sustainability. In order to uphold transparency, fairness, and inclusivity, it is imperative to employ artificial intelligence in an ethical and responsible manner. Collaboration, information exchange, and ongoing research will be crucial in shaping the trajectory of AI in the fashion industry as it progresses.

References

1. Giri, C., Jain, S., Zeng, X., Bruniaux, P.: A detailed review of artificial intelligence applied in the fashion and apparel industry. IEEE Access **7**, 95376–95396 (2019). https://doi.org/10.1109/ACCESS.2019.2928979
2. Obe, O., Egwuche, O.S.: Genetic algorithm approach for fabric pattern generation in textile industries. In: International Conference in Mathematics (2020)
3. Lee, M.J., Sohn, H.S., Kim, J.J.: A study on representation of 3D virtual fabric simulation with drape image analysis II-focus on the comparison between real clothing and 3D virtual clothing. J. Fashion Bus. **15**(3), 97–111 (2011)
4. Ren, S., Patrick Hui, C.L., Jason Choi, T.M.: AI-based fashion sales forecasting methods in big data era. In: Artificial intelligence for fashion industry in the big data era, pp. 9–26 (2018)
5. Thomassey, S., Zeng, X.: Introduction: artificial intelligence for fashion industry in the big data era. Springer, Singapore, pp. 1–6 (2018)
6. Wong, W.K., Chan, C.K.: An artificial intelligence method for planning the clothing manufacturing process. J. Textile Inst. **92**(2), 168–178 (2001). https://doi.org/10.1080/00405000108659606
7. Çelik, H.I., Dülger, L.C., Topalbekiroglu, M.: Application of neural networks (NNs) for fabric defect classification. Artif. Neural Netw. Models Appl. (2016). https://doi.org/10.5772/63427
8. Madsen, D.Ø., Silva, E.S., Hassani, H.: The application of big data in fashion retailing: a narrative review. Int. J. Manag. Concepts Philos. **13**(4), 247–274 (2020)
9. McAuley, J., Targett, C., Shi, Q., Van Den Hengel, A. Image-based recommendations on styles and substitutes. In: Proceedings of the 38th International ACM SIGIR Conference on Research and Development in Information Retrieval, pp. 43–52 (2015)
10. Bhatnagar, S., Ghosal, D., Kolekar, M.H.: Classification of fashion article images using convolutional neural networks. In: 2017 4th International Conference on Image Information Processing, ICIIP 2017, vol. 2018(1), pp. 357–362 (2018). https://doi.org/10.1109/ICIIP.2017.8313740
11. Ge, Y., Zheng, H., Zhang, Y., Wang, X.: DeepFashion2: a versatile benchmark for detection, pose estimation, segmentation and re-identification of clothing images. In: Proceedings of the IEEE/CVF Conference on Computer Vision and Pattern Recognition, pp. 533–542 (2019)
12. Veit, A., Kovacs, B., Bell, S., McAuley, J., Bala, K.: Learning visual clothing style with heterogeneous dyadic co-occurrences. Proc. IEEE Int. Conf. Comput. Vis. **2015**, 4642–4650 (2015)
13. Mohammadi, S.O., Kalhor, A.: Smart fashion: a review of AI applications in virtual try-on and fashion synthesis. J. Artif. Intell. **3**(4), 284 (2021)
14. Ghodhbani, H., Neji, M., Alimi, A.M.: Image-based virtual try-on system: a survey of deep learning-based methods. University of Sfax, National Engineering School of Sfax (ENIS), Sfax (2021)
15. Kang, J.Y.M., Kim, J.E., Lee, J.Y., Lin, S.H.: How mobile augmented reality digitally transforms the retail sector: examining trust in augmented reality apps and online/offline store patronage intention. J. Fashion Market. Manage. Int. J. **27**(1), 161–181 (2023)

16. Husak, V., Lozynska, O., Karpov, I., Peleshchak, I., Chyrun, S., Vysotskyi, A.: Information system for recommendation list formation of clothes style image selection according to user's needs based on NLP and Chatbots. In: COLINS, pp. 788–818 (2020)
17. Landim, A.R.D.B., Pereira, A.M., Vieira, T., B. de Costa, E., Moura, J.A.B., Wanick, V., Bazaki, E.: Chatbot design approaches for fashion E-commerce: an interdisciplinary review. Int. J. Fashion Des. Technol. Educ. **15**(2), 200–210 (2022)
18. Kang, W.C., Fang, C., Wang, Z., McAuley, J.: Visually-aware fashion recommendation and design with generative image models. In: 2017 IEEE international conference on data mining (ICDM), pp. 207–216. IEEE (2017)
19. Zerfu, M., Tilahun, T.: Application of generative adversarial networks techniques in creative fashion design: the case of Ethiopian Dress Discussion (2016)
20. Singh, M., Bajpai, U.V.V., Prasath, S.: Generation of fashionable clothes using generative adversarial networks: a preliminary feasibility study. Int. J. Cloth. Sci. Technol. **32**(2), 177–187 (2020)
21. Zhao, Z., Ma, X.: A compensation method of two-stage image generation for human-AI collaborated in-situ fashion design in augmented reality environment. In: 2018 IEEE International Conference on Artificial Intelligence and Virtual Reality (AIVR), pp. 76–83. IEEE (2018)
22. Lee, Y.K.: How complex systems get engaged in fashion design creation: using artificial intelligence. Thinking Skills Creativ. **46**, 101137 (2022)
23. Dilmaghani, S., Brust, M.R., Danoy, G., Cassagnes, N., Pecero, J., Bouvry, P.: Privacy and security of big data in AI systems: a research and standards perspective. In: 2019 IEEE International Conference on Big Data (Big Data), pp. 5737–5743. IEEE (2019)
24. Elliott, D., Soifer, E.: AI technologies, privacy, and security. Front. Artif. Intell. **5**, 826737 (2022)
25. Kapoor, V., Singh, R., Reddy, R., Churi, P.: Privacy issues in wearable technology: an intrinsic review. In: Proceedings of the International Conference on Innovative Computing and Communications (ICICC) (2020)
26. Buolamwini, J., Gebru, T.: Gender shades: Intersectional accuracy disparities in commercial gender classification. In: Proceedings of the 1st Conference on Fairness, Accountability and Transparency, pp. 77–91 (2018)
27. Caliskan, A., Bryson, J.J., Narayanan, A.: Semantics derived automatically from language corpora contain human-like biases. Science **356**(6334), 183–186 (2017)
28. Wong, C., Guo, Z.X., Leung, S.Y.S.: Optimizing decision making in the apparel supply chain using artificial intelligence (AI): from production to retail. Elsevier, Amsterdam (2013)
29. Wang, Z., Wang, J., Zeng, X., Sharma, S., Xing, Y., Xu, S., Liu, L.: Prediction of garment fit level in 3D virtual environment based on artificial neural networks. Text. Res. J. **91**(15–16), 1713–1731 (2021)
30. Jin, B.E., Shin, D.C.: Changing the game to compete: innovations in the fashion retail industry from the disruptive business model. Bus. Horiz. **63**(3), 301–311 (2020)
31. Guo, Z.X., Wong, W.K., Leung, S.Y.S., Li, M.: Applications of artificial intelligence in the apparel industry: a review. Text. Res. J. **81**(18), 1871–1892 (2011)
32. Casciani, D., Chkanikova, O., Pal, R.: Exploring the nature of digital transformation in the fashion industry: opportunities for supply chains, business models, and sustainability-oriented innovations. Sustain. Sci. Pract. Policy **18**(1), 773–795 (2022)
33. Mohiuddin Babu, M., Akter, S., Rahman, M., Billah, M.M., Hack-Polay, D.: The role of artificial intelligence in shaping the future of Agile fashion industry. Prod. Plan. Control 1–15 (2022)
34. Hsu, C.L., Lin, J.C.C.: Understanding the user satisfaction and loyalty of customer service chatbots. J. Retail. Consum. Serv. **71**, 103211 (2023)
35. Gangoda, A., Krasley, S., Cobb, K.: AI digitalisation and automation of the apparel industry and human workforce skills. Int. J. Fashion Des. Technol. Educ. **16**(3), 319–329 (2023)
36. Pettinger, L.: Brand culture and branded workers: service work and aesthetic labour in fashion retail. Consum. Mark. Cult. **7**(2), 165–184 (2004)
37. Oosthuizen, K., Botha, E., Robertson, J., Montecchi, M.: Artificial intelligence in retail: the AI-enabled value chain. Australas. Mark. J. **29**(3), 264–273 (2021)

38. Akram, S.V., Malik, P.K., Singh, R., Gehlot, A., Juyal, A., Ghafoor, K.Z., Shrestha, S.: Implementation of digitalized technologies for fashion industry 4.0: Opportunities and challenges. Sci. Program. (2022)
39. Candeloro, D.: Towards sustainable fashion: the role of artificial intelligence--H&M, Stella McCartney, Farfetch, Moosejaw: a multiple case study. ZoneModa J. **10**(2), 91–105 (2020)
40. Burberry: Fashion meets technology. https://www.burberryplc.com/news/corporate/2020/burberry-uses_gaming-technology-to-streamlinefashiondesignproces. Accessed 02 Jul 2023
41. Gucci: Introducing Guccio, a virtual assistant. https://www.gucci.com/us/en/st/stories/article/virtual-assistant (2023/07/02).
42. Stitch Fix: How it works (n.d.). https://www.stitchfix.com/how-it-works/. Accessed 20 Mar 2023
43. The Yes: Meet The Yes (n.d.). https://www.theyes.com/about. Accessed 25 Jun 2023
44. Amazon: Personalize the customer experience. https://aws.amazon.com/personalize/ Accessed 20 Jun 2023 (n.d.)
45. Alibaba: Alibaba's "FashionAI" provides smarter fashion advice. https://www.alibabagroup.com/en/news/article?news=p180406. Accessed 20 Jun 2023

Chapter 36
A Comprehensive Review on Deep Learning-Based Pansharpening Approaches

Gurpreet Kaur, Manisha Malhotra, and Dilbag Singh

Abstract Pansharpening involves merging a high-resolution panchromatic image with a low-resolution multispectral image to create a fused image that is more informative for remote sensing applications. Various techniques have been developed to improve fusion results by leveraging the spatial and spectral characteristics of the input images. In recent years, deep learning has significantly advanced pansharpening, offering promising results through its strong feature extraction and reconstruction capabilities. However, there is a lack of comprehensive analysis of the latest deep learning techniques across different fusion applications. This paper provides a review of various Pansharpening methods employing deep learning, comparing them based on optimal parameters. It also discusses the limitations, challenges, issues, and future directions in Pansharpening approaches.

36.1 Introduction

Remote sensing has evolved rapidly in recent years, with the development of sophisticated satellite sensors and imaging systems that can capture data with higher resolution, accuracy, and speed. This has opened up new opportunities for researchers and professionals in various fields to use remote sensing to address complex problems and challenges [1–3]. Remote sensing images offer extensive information about Earth's observed scenes. Spatial and spectral resolutions are critical factors for assessing the quality of these images. Producing a high-resolution colored image with both high spatial and spectral details is impractical with a single sensor because of the physical and technical limitations inherent in different sensors. Consequently, a trade-off between spatial and spectral resolutions must be considered. High-resolution colored

G. Kaur (✉) · M. Malhotra
University Institute of Computing, Chandigarh University, Mohali, India
e-mail: gurpreetkdhiman@gmail.com

D. Singh
Center of Biomedical Imaging, Department of Radiology, New York University Grossman School of Medicine, New York, NY, USA

P. Nanda et al. (eds.), *Intelligent System and Data Analysis*, Smart Innovation, Systems and Technologies 398, https://doi.org/10.1007/978-981-97-5200-3_36

images need high cost and storage capacity to store, and it also slows down the data transmission process due to the large size of colored images [2].

Increasing bands and narrowing their width can improve spectral resolution. However, narrower bands will receive less energy if the incident light energy remains constant, leading to a reduction in the signal-to-noise ratio of the observed image. To improve signal-to-noise ratio, the size of the light-sensing components can be increased, but this solution will degrade spatial resolution as the spatial resolution can be improved by downsizing the light-sensing devices [1]. Table 36.1 illustrates that a panchromatic image possesses high spatial resolution but is limited to only one band. In contrast, multispectral images offer rich spectral resolution and consist of multiple bands. Images taken from different satellites in the table show that multispectral image possesses inferior spatial resolution when compared with the panchromatic image, but the former is a grayscale image. Color details are available in multispectral images.

Many applications like resource exploration [1], agriculture [4], medical diagnosis [5], and other visualizations [2] lack in interpreting the observed scene with less spatial or spectral details. Image fusion is introduced to integrate different images to interpret the observed scene in a better way. Image fusion helps in this situation since it adds to the additional information available from other sources. There have been numerous efforts to develop the best fusion techniques that incorporate cutting-edge modeling methodologies and image processing techniques that are suitable for remote sensing [5]. Image fusion, also known as Pansharpening, is the process of combining panchromatic and multispectral images to create a more accurate and higher-quality color image [1]. Figure 36.1 illustrates the merging process and presents the resulting Pansharpened image. Figure 36.2 comprises (a) a panchromatic image, (b) a multispectral image, and (c) the resulting Pansharpened image after applying a fusion technique. This Pansharpened image enhances both the spatial and spectral qualities of the observed area, providing more informative details.

Though Pansharpening has been an important area of research for a long time, it still requires further research due to advancements in remote sensing sensors. Also, various applications may require spatial and spectral details from other visual aspects using Pansharpening [2]. Over the past decades, multiple Pansharpening techniques are developed and these also provide promising results. Pansharpening techniques can be broadly classified into several categories, including component substitution methods, multi-resolution analysis methods, model-based methods, variational methods, hybrid methods, and deep learning-based methods [2]. The existing reviews like Meng et al. provided a large dataset that helped in creation of remote sensing images [6]. In addition, Vivone et al. give a Pansharpening review thorough comparison of various component substitution and multi resolution-based techniques on datasets from several satellites, which encouraged the standardization of many Pansharpening code implementations [7]. Kaur et al. presented a review of other mentioned categories of Pansharpening techniques [2].

It is very difficult now to visually distinguish the Pansharpened image produced by different Pansharpening methods. Numerous quality metrics are available to assess the quality of Pansharpened images [2]. These metrics fall into two main categories:

Table 36.1 Panchromatic, multispectral images and spatial, spectral resolutions of different satellites

Panchromatic (Grayscale image) Multispectral (colored image)	satellite	Spectral bands	spatial resolution
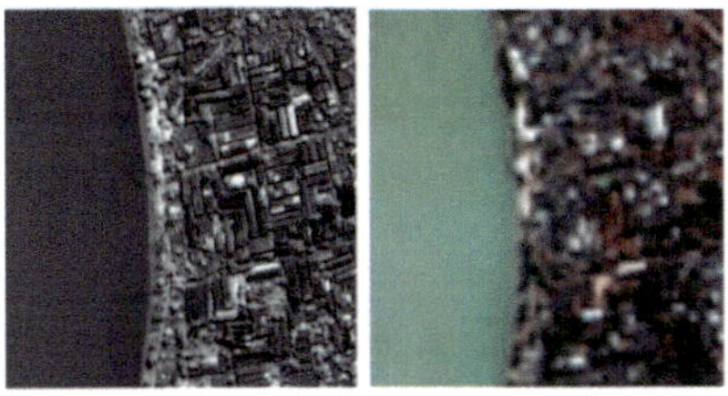	WorldView-2	Multispectral-8 Panchromatic-1	2 m 0.5 m
	WorldView-3	Multispectral-8 Panchromatic-1	1.24 m 0.31 m
	WorldView-4	Multispectral-4 Panchromatic-1	1.24 m 0.31 m
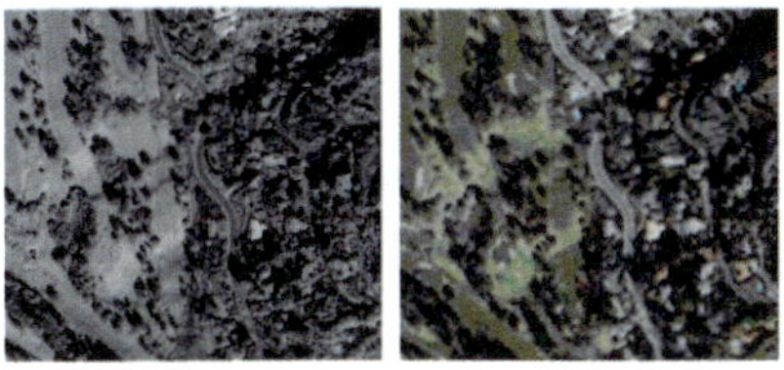	QuickBird	Multispecral-4 Panchromatic-1	2.44 m 0.61 m
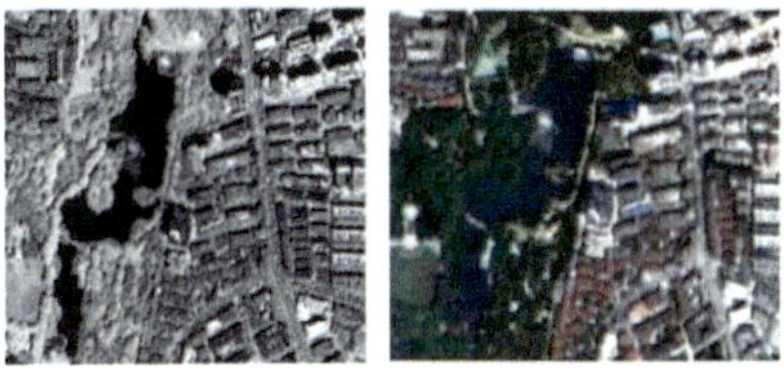	IKONO	Multispecral-4 Panchromatic-1	4 m 1 m
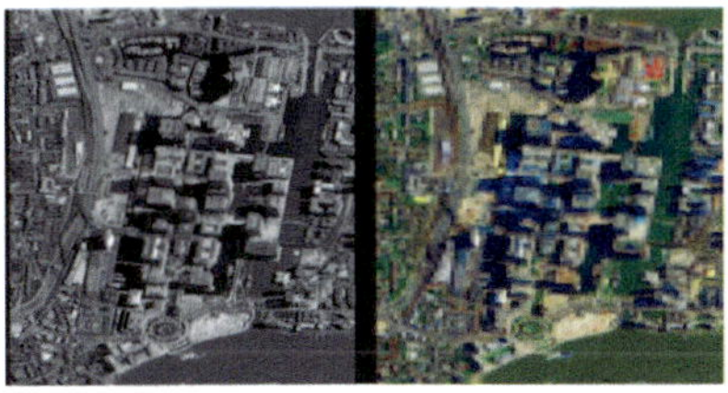	GeoEye-1	Multispecral-4 Panchromatic-1	2 m 0.5 m
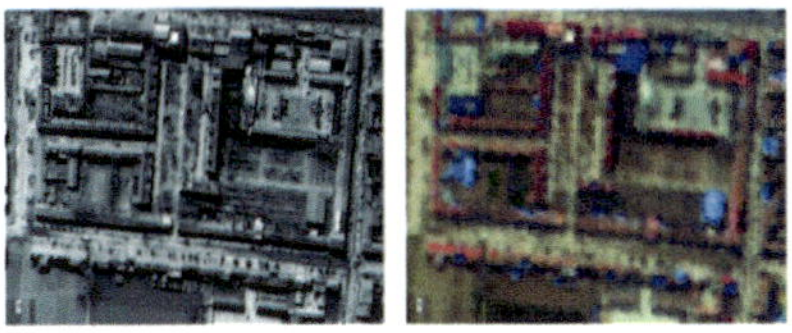	GF-1	Multispecral-4 Panchromatic-1	8 m 2 m
	GF-2	Multispecral-4 Panchromatic-1	4 m 1 m

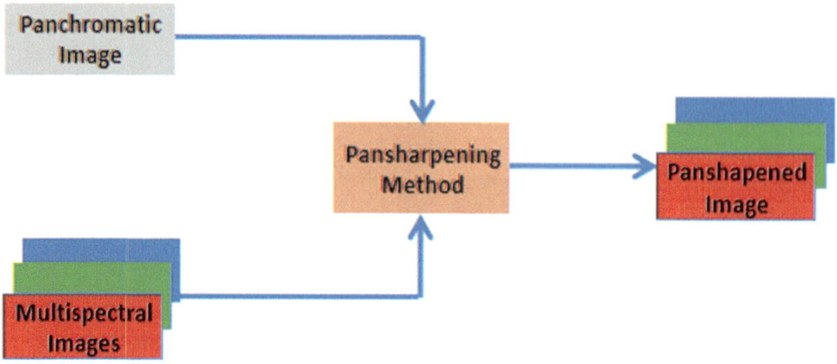

Fig. 36.1 Concept of Pansharpening

Fig. 36.2 **a** Panchromatic image, **b** multispectral image, **c** pansharpened image [2]

firstly, evaluating the fused image quality against a known reference image, and secondly, assessing the spectral and spatial details of the fused image in the absence of a reference image. Kaur et al. provide an overview of various quality metrics for evaluating Pansharpened image quality [2].

Table 36.2 gives an idea of different Pansharpening categories that show deep learning techniques provide us best spatial and spectral fidelity, but these lack in generalization to different satellites and have training and testing overheads.

36.1.1 Motivation

- Deep learning approaches may struggle to produce satisfactory results on full-resolution datasets because they are typically trained on artificially downscaled datasets. This downscaled training data can lead to complexities in mapping

Table 36.2 Features occupied by different Pansharpening categories

Feature	Component substitution methods	Multi-resolution methods	Variational methods	Deep learning methods
Spatial accuracy	✓✓	✓	✓✓	✓✓✓
Spectral accuracy	✓	✓✓	✓✓	✓✓✓
Generalization to different satellites	✓✓✓	✓✓✓	✓✓	✓
Running time	✓✓✓	✓✓✓	✓	✓✓
Interpretability	✓✓✓	✓✓✓	✓✓✓	✓

Performance—✓ [Poor], ✓✓ [Average], ✓✓✓ [Best]

between the source and fused images when applied to full-resolution data during generalization.

- Moreover, some deep learning methods are urged to utilize full-resolution data for training purposes due to challenges in obtaining and generating reduced-resolution training datasets [1].

36.1.2 Contribution

The following are the main contributions of this extensive study:

- This extensive study's primary goal is to classify recently proposed deep learning-based Pansharpening methods and an extensive review of these methods.
- The recent Pansharpening techniques based on the deep learning approach are compared through comparative reviews.
- Finally, opportunities and challenges for further deep learning-based Pansharpening studies are discussed.

The remainder of the paper is as follows: different categories of deep learning Pansharpening methods are presented in Sect. 36.2.

In Sect. 36.3, related research is reviewed and comparative assessment is discussed.

In Sect. 36.4, limitations, challenges, and issues in the field of Pansharpening are provided.

Section 36.5 covers future directions. In Sect. 36.6, a conclusion is given.

36.2 Categories of Deep Learning-Based Pansharpening Techniques

Pansharpening techniques play a crucial role in amalgamating valuable information from various source images to generate a single, information-rich image suitable for diverse applications. Traditional Pansharpening methods, such as component substitution class techniques outlined in Fig. 36.3, involve projecting the multispectral image into a different space where spatial and spectral details are separated. Subsequently, by replacing the spatial component with the panchromatic image, the multispectral image is sharpened. Reverting to the original space using inverse transformation completes the sharpening process. While these methods excel in enhancing spatial details, they often suffer from significant spectral distortion. However, they typically require less computational time for execution [1–3, 6–8].

The following formulation describes how the component substitution fusion process is simplified to achieve faster implementation under the assumptions of linear translation and the substitution of a unique component.

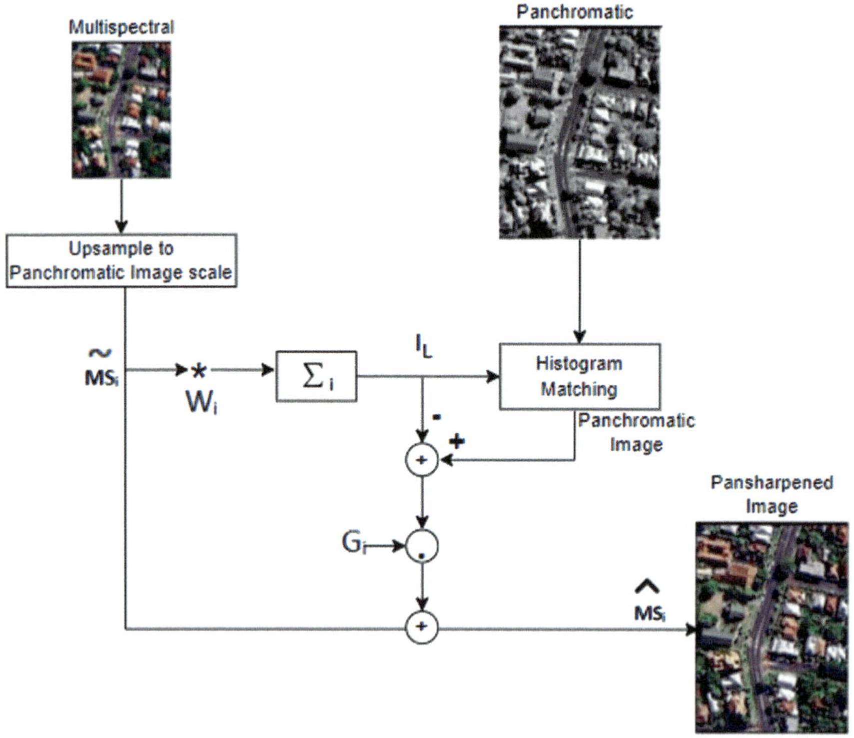

Fig. 36.3 Component substitution-based Pansharpening architecture [8]

$$\widehat{\mathrm{MSi}} = \widetilde{\mathrm{MSi}} + G_i(P - \mathrm{I_L}) \tag{36.1}$$

Here,

$\widetilde{\mathrm{MSi}}$—Multispectral image which is upsampled to panchromatic image scale.

G_i—Injection gain matrix.

P—Panchromatic image.

I_L—Intensity component created by averaging the weighted multispectral bands with W_i weights.

$\widehat{\mathrm{MSi}}$—Fusion of ith band.

Secondly, multi-resolution methods leverage the disparity between the panchromatic image (P) and its low-pass spatial variation, P_L, to extract spatial details. These methods employ the following formulation to obtain the fused image:

$$\widehat{\mathrm{MSi}} = \widetilde{\mathrm{MSi}} + G_i(P - P_\mathrm{L}) \tag{36.2}$$

Multi-resolution analysis techniques preserved good spectral resolution but suffer from high spatial distortion and stair-casing phenomena [1–3, 6–8]. The third category, model-based optimization delivered an impressive performance, but these are time-consuming and complex in implementation [2, 9]. Fourth are the variational techniques which preserved the spectral resolution of the Pansharpened image, but these produce spatial artifacts such as blurring and blocks in the final fused image and are time-consuming too [2, 6–8]. Hybrid methods are another type of Pansharpening techniques that have lots of blur issues because of the lack of spatial information added in the fused image. These also did not perform well for hyperspectral images [2]. Deep learning in Pansharpening has been motivated by the need to overcome the limitations of existing approaches [1–3, 5–8].

The development of deep learning methods has provided us the new possibilities for quality improvement of the fused image. Because of plenty of available datasets and robust learning capabilities, deep learning-based Pansharpening methods have proved successful [1–8]. However, the training datasets for the Pansharpening process must be prepared in accordance with the Wald protocol [10]. Deep learning-based approaches utilized distinct network branches for differential feature extraction, hence yielding more targeted features. Under the supervision of well-designed loss functions, deep learning-based approaches developed appropriate strategies for feature fusion [5]. Benefiting from this, deep learning encouraged enormous advancements in image fusion, achieving results better than all abovementioned conventional approaches.

Current deep learning-based image fusion algorithms primarily focus on "feature extraction, feature fusion, and image reconstruction" [5]. These techniques include autoencoder (AE)-based methods, convolutional neural network (CNN)-based methods, and generative adversarial network-based (GAN) methods [5, 8]. In most AE methods, an autoencoder is trained beforehand and then utilized to extract features and reconstruct the image. Intermediate feature fusion is typically achieved using standard fusion rules [5, 8], depicted in Fig. 36.4a. CNN methods employ two main approaches for integrating convolutional neural networks into image fusion. One approach involves loss functions and network structures for feature extraction and image fusion [5]. Alternatively, a trained CNN can be employed to establish fusion rules, as illustrated in Fig. 36.4b. GAN approaches depend on the adversarial game between the generator and the discriminator to assess the target's probability distribution for image fusion [5], as shown in Fig. 36.4c.

Deep residual learning, which uses the transformation $F(X)$, is an advanced approach to deal with Pansharpening. $F(X)–X$ is used in place of CNN(X). By enabling a skip connection between various layers in RES(X), it is possible to expand the network's layer structure and improve performance.

Wei et al. constructed a deep residual skip in the DRPNN with 10 layers and 7*7 kernel sizes on both sides of the convolutional filtering framework. Figure 36.5 explains that the multispectral bands are upsampled to a panchromatic scale and then fused with the panchromatic image. A restoration layer of convolution filters is used to get the Pansharpened image. Due to its usage of large size convolution kernels, specifically 7*7, which covers a larger region, the DRPNN can produce promising results.

However, the DRPNN has an excessive number of parameters because of these large kernels which increases the complexity of the method [8, 11].

36.3 Literature Review

Deep learning methods have produced several excellent outcomes for Pansharpening [9, 12–15]. In [16], Masi et al. used a CNN method named "PNN" with three layers to perform multispectral images of Pansharpening, achieving superior results compared to conventional methods.

In [17], Yang et al. introduced the "PanNet" method that uses a deep CNN approach and utilizes a high-pass filtering technique to merge panchromatic and low-resolution multispectral images, resulting in Pansharpening outcomes. Additionally, He et al. [18] developed the "DiCNN" method, a CNN approach for image fusion that employed a spatial detail injection architecture. Deng et al. proposed the "FusionNet" method in [19], which incorporates two network designs, CS-Net and MRA-Net, based on conventional CS and MRA fusion equations. To overcome these issues, they introduced FusionNet, an effective fusion network that gives better results. In recent years, GAN and unsupervised training have introduced novelty and effectiveness in the Pansharpening domain.

Ma et al. provided solutions for enhancing the efficiency of Pansharpening images by concentrating on two basic areas. The first is to create more generic datasets for training, and the second is to train networks using these datasets. The authors developed PanGan method that provides promising results [20]. Ozcelik et al. also used the same approach and developed TA-CNN method for Pansharpening [21]. Shen

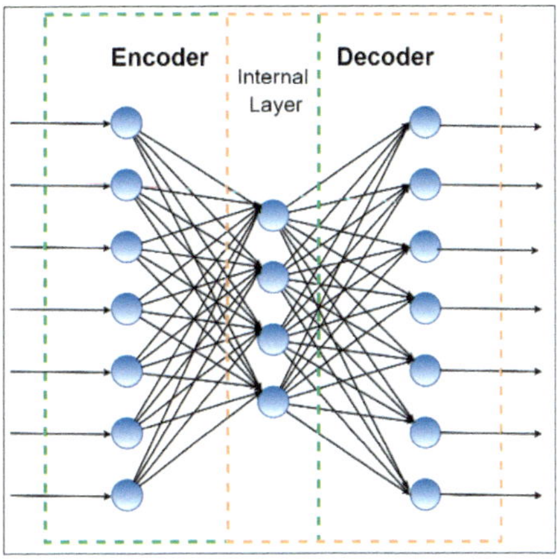

(a)

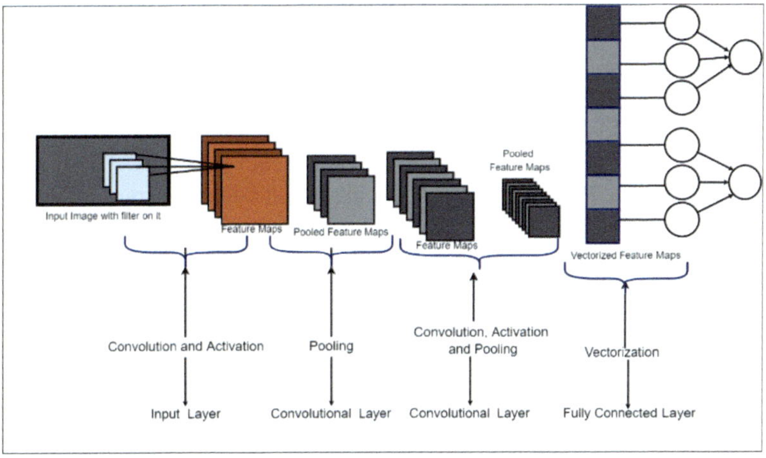

(b)

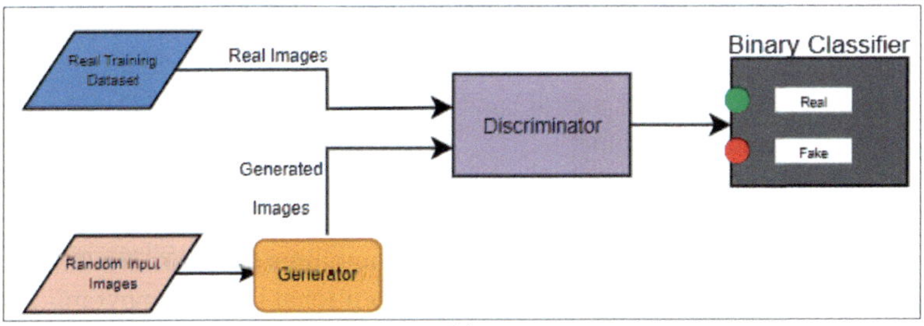

(c)

Fig. 36.4 **a** AE-based architecture [5], **b** CNN-based architecture [5], **c** GAN-based architecture [5]

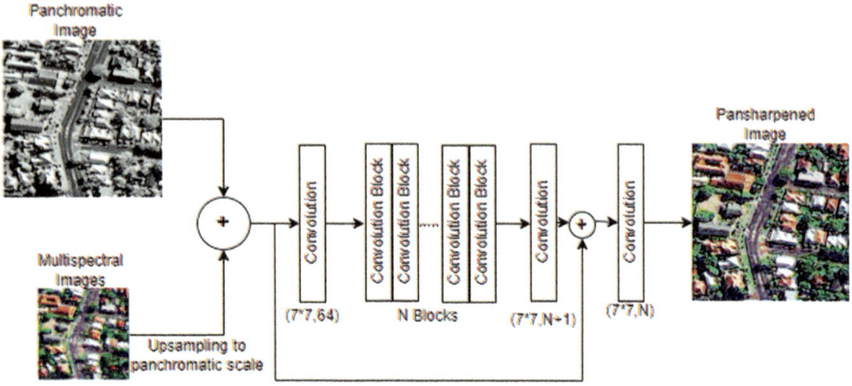

Fig. 36.5 Framework for DRPNN method [11]

et al. [22] focused on improving network design by integrating the aforementioned methodologies, using new cutting-edge modules, and incorporating mathematical knowledge into neural network training. There is still room for development in the following areas even though these approaches have achieved outstanding results. Firstly, multiscale feature is providing good results in enhancing image resolution [23]. So, multiscale structures for example "Laplacian pyramids"-based networks can be designed for Pansharpening. Secondly, the modulation transfer function is performing better in explaining the physical process of capturing the image. Hence, these can be used in simulating the datasets to enhance the fusion performance.

In [3], a Laplacian pyramid-based LPPN architecture was introduced as an improvement. This architecture considers both multiscale information and sensor modulation transfer functions (MTFs). It employs the Laplacian pyramid to decompose the original panchromatic image and the upsampled low-resolution multispectral image into multiple scales. Subsequently, it constructs relevant sub-networks for each image scale and combines them into a fusion convolutional neural network for multiscale information (FCNN). By incorporating multiscale sub-networks, the final loss function is computed. To reduce the number of network parameters, recursive blocks are integrated into the sub-networks. Unlike the Laplacian pyramid, which uses a single Gaussian kernel for all image channels, this solution utilizes separate Gaussian kernels for various spectral bands based on specific sensors' multi-transfer functions for each image channel.

This architecture demonstrated superior quantitative and qualitative performance compared to existing Pansharpening techniques. The LPPN architecture is the first attempt to handle the Pansharpening process by combining the recursive sub-network with the Laplacian pyramids. In contrast to the traditional Laplacian pyramid, which used a Gaussian kernel for all spectral channels, the Laplacian pyramid utilized in this study has taken advantage of the MTFs of the particular sensors for picture scale decomposition, which resulted in considerable improvements. Laplacian pyramids improved feature extraction spatial detail management in this study. Also, loss functions helped to recover image details by measuring information loss at each scale. The recursive block reduced network parameters in each sub-network. It also increased sub-network depth. The entire suggested architecture primarily consisted of several sub-networks that gradually merge and improve the spatial details. LPPN is a lightweight architecture for Pansharpening that has given significantly better results when compared with state-of-the-art methods [3].

Liu et al. proposed a SUFNet fusion network using a supervised network based on multiscale mechanisms, skip connections, dilated convolution, and an unsupervised network to improve spatial and spectral details in the fused image. The authors also advised using their proposed unsupervised network USSCNet as a universal network to enhance spatial and spectral resolution of different Pansharpening methods [24]. Xu et al. introduced UPanGAN method in which the method first generates initial difference image and then by using two optimized discriminators performed multi-level fusion process to generate full-resolution Pansharpened image. The authors proved their method as best by doing a comparison with other state-of-the-art methods [25].

Zhang et al. introduced the P2Sharpen Pansharpening method, which is based on deep learning and addresses two limitations of existing deep learning approaches. This method achieves improved results by balancing performance on both low and full resolutions. Additionally, a supervised network called STNet is employed to enhance the spatial texture of the fused image. The proposed method outperforms other compared methods in terms of results, as demonstrated in the paper [26]. Wei et al. proposed a framework that utilizes adversarial attacks to generate Pansharpened images. The authors incorporated label and shape loss for the attack task. To maintain the quality of the fused image, perceptual loss was employed to preserve spectral fidelity [27]. Beltran et al. devised a novel Pansharpening method based on deep learning named as W-NetPan that adopts W-shape architecture which integrated two U-Nets and those work sequentially. The first U segment helped in resolving the sensor deviation and the second helped in better data fusion. In this paper, the authors trained their method without any external data supervision using a joint loss formulation. The experimental results in the paper show the better results of the proposed method when compared with other state-of-the-art methods [28].

After identifying the drawbacks of existing deep learning techniques, Yang et al. introduced the DSCNN-RIE method, which utilizes two separate resolutions to extract spatial information through convolutional filters that transfer complementary information between these resolutions. Additionally, the method employs a residual information enhancement strategy to incorporate various residual levels into the outputs of cascaded convolutional blocks. By combining the retrieved features with the shallow feature data from the source images, the final Pansharpened image is generated [29]. Ciotola et al. developed a full-resolution training framework applicable to various deep learning-based Pansharpening techniques. This framework enables training in the full-resolution domain using only original data to prevent any loss of information. A two-component loss function is defined to ensure the preservation of spectral and spatial details. Target adaptive operating modality is used during testing to achieve good generalization with minimal computing complexity [30].

In order to promote a high-frequency detail prediction, Benzenati et al. presented DI-GAN, a detail injection GAN network for sharpening remote sensing images [31]. The scale-shift assumption in supervised learning-based Pansharpening algorithms is addressed by Shen et al. [32]. They presented the continuous learning guided training framework, which combines an unsupervised learning strategy to train the algorithm for full-resolution Pansharpening with a traditional supervised technique to train on reduced-resolution datasets. Hou et al. suggested multi-level feature fusion network (MLFNet) for hyperspectral Pansharpening that uses cross-layer guided attention mechanisms [33]. Using a mix of low-level details and deep level details, Wang et al. provided a technique to enhance the quality examination of full-resolution Pansharpened images [34].

Cheng et al. propose DMFANet, a multi-level dual branch aggregation network designed to efficiently extract spectral and spatial features. This network consists of two branches created through a residual framework. The goal of DMFANet is to thoroughly extract and fuse spectral and spatial features at multiple levels, enabling the utilization of complementary data for producing promising fusion results. This design facilitates the reconstruction of spatial details along with the spectral distribution to the reference image [35]. Wei et al. [11] developed the DRPNN method, which utilizes residual learning to create an extremely deep CNN and fully leverage the high nonlinearity of the deep learning models. Zhou et al. introduced an unsupervised GAN algorithm that learns from full-resolution images without relying on ground truth. This algorithm extracts image details from panchromatic and multispectral images, performs feature fusion, and incorporates a novel loss based on adversarial strategy and cycle consistency to enhance the quality of the fused image. The architecture has demonstrated reliability across various satellite sensors [36].

Table 36.3 gives a comparison of deep learning-based Pansharpening methods using optimal parameters. Here in the table, following abbreviations are used. L—Number of Layers, E—Epochs, Op—Optimizer, ILR—Initial Learning Rate, FS—Filter Size (for each layer), FN—Filter Number (for each layer), L—Loss Function used in the method, IBS-Images Batch Size, SGD—Stochastic Gradient Descent.

36.4 Limitations, Difficulties, and Challenges

We addressed the existing studies on deep learning-based Pansharpening. Despite significant research and good results for fused images, Pansharpening still faces multiple challenges. We discuss difficulties and present Pansharpening trends in this section.

- When we implement Pansharpening with deep learning, there is a requirement of plenty of datasets for the training of the deep neural network-based methods. But there is a lack of datasets available for comparing different Pansharpening techniques.

Table 36.3 Deep learning-based Pansharpening methods comparison using optimal parameters

Methods/Parameters	L	E	Op	ILR	FS	FN	L	IBS	Platform
DRPNN [36]	11	500	Adam	$2*10^{-4}$	7*7	32	L2	64	MatConvNet
UPanGan [25]	7	100	RMSProp	0.0002	3*3	32	Frobenius	32	Tensorflow
PNN [16]	3	12,000	SGD	0.0289*Bands	9*9,5*5	64,32	L2	64	Pytorch
PanNet [17]	10	450	Adam	0.0001	3*3	64	L2	32	Pytorch
SUFNet [24]	-	200	Adam	0.0001	–	–	L2	8	Pytorch
DiCNN [18]	3	1000	Adam	$2*10^{-4}$	3*3	32	Frobenius	64	Matlab
LPPN [3]	6	–	Adam	0.003	3*3	–	L2	32	Tensorflow
P2Sharpen [26]	4	600	Adam	$3*10^{-4}$	3*3	–	Frobenius	2	Matlab
Fusion Net [19]	10	400	Adam	0.0003	3*3	32	L2	32	Tensorflow
TSPSNet [27]	2	30	Adam	0.0001	–	–	L1	16	Pytorch

- Land cover types include farmlands, watery areas, grasslands, woodlands, rural areas, and urban areas. When a certain deep neural network is solely trained on some rural dataset, its fusion performance degrades when performed on a dataset of an urban area.
- The spectral response varies for different satellites due to their instrumental differences. So, for different satellite datasets, mapping relationships between panchromatic and multispectral images will be different. Therefore, if we use a deep neural network trained dataset of one satellite for example "QuickBird" to perform image fusion tasks on panchromatic and multispectral images of other satellites for example "GeoEye", satisfactory results may not be obtained.

36.5 Future Directions

- Due to misalignment of panchromatic and multispectral images, Pansharpened image suffers from spatial distortion. Geometric registration can be the solution, but this solution might not work for moving objects [2, 8].
- Most of the existing models have issues with hyper tuning parameters. Usually, the parameters used in Pansharpening models are only chosen on the basis of hits and trail method. So, it could limit the quality of the fused image [2].
- Deep neural network-based Pansharpening methods suffer from computational complexities and due to training of the neural networks, they require more time to get implemented. So, future Pansharpening methods can be proposed in this area to improve method implementation speed [1–3, 8].
- In future work, application-oriented Pansharpening techniques can also be developed [2].
- Deep learning methods encounter challenges in achieving optimal results on full-resolution datasets because they are typically trained on manually generated reduced-resolution datasets. This training approach complicates the mapping between the source image and the fused images when applied to full-resolution data during generalization. Thus, future research in deep neural network Pansharpening methods holds promise in training on full-resolution datasets to enhance fusion outcomes [1, 3].
- Unsupervised deep neural networks-based Pansharpening methods are not providing up to the mark fusion results when compared with supervised methods. So, spectral- and spatial model-driven deep neural networks may provide an excellent opportunity to enhance the performance of Pansharpening images [1, 25].
- In addition, there are lots of new deep neural networks popping up, like zero reference GANs, graph neural networks and transformers. These can make the fused image even better and provide additional solutions to the current challenges in this area [1].

36.6 Conclusion

While deep learning techniques have shown improved performance in Pansharpening, they exhibit reduced robustness across various satellite platforms, demand increased computational time for learning complex distributions, and are susceptible to image quality issues due to challenges in gathering comprehensive training data. Additionally, earlier DL Pansharpening methods, although capable of producing satisfactory fusion outcomes, often overlook the exploitation of complementary information between different resolutions and fail to fully leverage residual information at multiple levels. This oversight may result in detailed loss and spectral distortion.

References

1. Zhang, K., Zhang, F., Wan, W., Yu, H., Sun, J., Ser, J.D., Elyan, E., Hussain, A.: Panchromatic and multispectral image fusion for remote sensing and earth observation: concepts, taxonomy, literature review, evaluation methodologies and challenges ahead. Inf. Fusion (2023)
2. Kaur, G., Saini, K.S., Singh, D., Kaur, M.: A comprehensive study on computational Pansharpening techniques for remote sensing images. Arch. Comput. Methods Eng. **28**(7), 4961–4978 (2021)
3. Jin, C., Deng, L.-J., Huang, T.-Z., Vivone, G.: Laplacian pyramid networks: a new approach for multispectral Pansharpening. Inform. Fusion **78**, 158–170 (2022)
4. Jurado, J.M., López, A., Pádua, L., Sousa, J.J.: Remote sensing image fusion on 3D scenarios: a review of applications for agriculture and forestry. Int. J. Appl. Earth Obs. Geoinf. **112**, 102856 (2022)
5. Zhang, H., Xu, H., Tian, X., Jiang, J., Ma, J.: Image fusion meets deep learning: a survey and perspective. Inform. Fusion **76**, 323–336 (2021)
6. Meng, X., Xiong, Y., Shao, F., Shen, H., Sun, W., Yang, G., Yuan, Q., Fu, R., Zhang, H.: A large-scale benchmark data set for evaluating Pansharpening performance: overview and implementation. IEEE Geosci. Remote Sens. Mag. **9**(1), 18–52 (2020)
7. Vivone, G., Mura, M.D., Garzelli, A., Restaino, R., Scarpa, G., Ulfarsson, M.O., Alparone, L., Chanussot, J.: A new benchmark based on recent advances in multispectral Pansharpening: revisiting Pansharpening with classical and emerging Pansharpening methods. IEEE Geosci. Remote Sens. Mag. **9**(1), 53–81 (2020)
8. Deng, L.-J., Vivone, G., Paoletti, M.E., Scarpa, G., He, J., Zhang, Y., Chanussot, J., Plaza, A.: Machine learning in Pansharpening: a benchmark, from shallow to deep networks. IEEE Geosci. Remote Sens. Mag. **10**(3), 279–315 (2022)
9. Liu, C., Zhang, Y., Wang, S., Sun, M., Ou, Y., Yi, W., Liu, X.: Band-independent encoder–decoder network for Pansharpening of remote sensing images. IEEE Trans. Geosci. Remote Sens. **58**(7), 5208–5223 (2020)
10. Wald, L., Ranchin, T., Mangolini, M.: Fusion of satellite images of different spatial resolutions: assessing the quality of resulting images. Photogramm. Eng. Remote. Sens. **63**(6), 691–699 (1997)
11. Wei, Y., Yuan, Q., Shen, H., Zhang, L.: Boosting the accuracy of multispectral image Pansharpening by learning a deep residual network. IEEE Geosci. Remote Sens. Lett. **14**(10), 1795–1799 (2017)
12. Scarpa, G., Vitale, S., Cozzolino, D.: Target-adaptive CNN-based Pansharpening. IEEE Trans. Geosci. Remote Sens. **56**(9), 5443–5457 (2018)

13. Peng, J., Liu, L., Wang, J., Zhang, E., Zhu, X., Zhang, Y., Feng, J., Jiao, L.: PSMD-Net: a novel Pansharpening method based on a multiscale dense network. IEEE Trans. Geosci. Remote Sens. **59**(6), 4957–4971 (2020)
14. Xu, H., Ma, J., Shao, Z., Zhang, H., Jiang, J., Guo, X.: SDPNet: a deep network for Pansharpening with enhanced information representation. IEEE Trans. Geosci. Remote Sens. **59**(5), 4120–4134 (2020)
15. Zhang, H., Ma, J.: GTP-PNet: a residual learning network based on gradient transformation prior for Pansharpening. ISPRS J. Photogramm. Remote. Sens. **172**, 223–239 (2021)
16. Masi, G., Cozzolino, D., Verdoliva, L., Scarpa, G.: Pansharpening by convolutional neural networks. Remote Sensing **8**(7), 594 (2016)
17. Yang, J., Fu, X., Hu, Y., Huang, Y., Ding, X., Paisley, J.: PanNet: a deep network architecture for Pansharpening. In: Proceedings of the IEEE International Conference on Computer Vision, pp. 5449–5457 (2017)
18. He, L., Rao, Y., Li, J., Chanussot, J., Plaza, A., Zhu, J., Li, B.: Pansharpening via detail injection based convolutional neural networks. IEEE J. Select. Topics Appl. Earth Observ. Remote Sens. **12**(4), 1188–1204 (2019)
19. Deng, L.-J., Vivone, G., Jin, C., Chanussot, J.: Detail injection-based deep convolutional neural networks for Pansharpening. IEEE Trans. Geosci. Remote Sens. **59**(8), 6995–7010 (2020)
20. Ma, J., Yu, W., Chen, C., Liang, P., Guo, X., Jiang, J.: Pan-GAN: an unsupervised Pansharpening method for remote sensing image fusion. Inform. Fusion **62**, 110–120 (2020)
21. Ozcelik, F., Alganci, U., Sertel, E., Unal, G.: Rethinking CNN-based Pansharpening: guided colorization of panchromatic images via GANs. IEEE Trans. Geosci. Remote Sens. **59**(4), 3486–3501 (2020)
22. Shen, H., Jiang, M., Li, J., Yuan, Q., Wei, Y., Zhang, L.: Spatial–spectral fusion by combining deep learning and variational model. IEEE Trans. Geosci. Remote Sens. **57**(8), 6169–6181 (2019)
23. Lai, W.-S., Huang, J.-B., Ahuja, N., Yang, M.-H.: Fast and accurate image super-resolution with deep Laplacian pyramid networks. IEEE Trans. Pattern Anal. Mach. Intell. **41**(11), 2599–2613 (2018)
24. Liu, Q., Meng, X., Shao, F., Li, S.: Supervised-unsupervised combined deep convolutional neural networks for high-fidelity Pansharpening. Inform. Fusion **89**, 292–304 (2023)
25. Xu, Q., Li, Y., Nie, J., Liu, Q., Guo, M.: UPanGAN: unsupervised Pansharpening based on the spectral and spatial loss constrained generative adversarial network. Inform. Fusion **91**, 31–46 (2023)
26. Zhang, H., Wang, H., Tian, X., Ma, J.: P2Sharpen: a progressive Pansharpening network with deep spectral transformation. Inform. Fusion **91**, 103–122 (2023)
27. Wei, X., Yuan, M.: Adversarial Pansharpening attacks for object detection in remote sensing. Pattern Recogn. **139**, 109466 (2023)
28. Fernandez-Beltran, R., Fernandez, R., Kang, J., Pla, F.: W-NetPan: double-U network for inter-sensor self-supervised Pansharpening. Neurocomputing **530**, 125–138 (2023)
29. Yang, Y., Tu, W., Huang, S., Lu, H., Wan, W., Gan, L.: Dual-stream convolutional neural network with residual information enhancement for Pansharpening. IEEE Trans. Geosci. Remote Sens. **60**, 1–16 (2021)
30. Ciotola, M., Vitale, S., Mazza, A., Poggi, G., Scarpa, G.: Pansharpening by convolutional neural networks in the full resolution framework. IEEE Trans. Geosci. Remote Sens. **60**, 1–17 (2022)
31. Benzenati, T., Kessentini, Y., Kallel, A.: Pansharpening approach via two-stream detail injection based on relativistic generative adversarial networks. Expert Syst. Appl. **188**, 115996 (2022)
32. Shen, K., Yang, X., Lolli, S., Vivone, G.: A continual learning-guided training framework for Pansharpening. ISPRS J. Photogramm. Remote. Sens. **196**, 45–57 (2023)
33. Hou, S., Xiao, S., Dong, W., Qu, J.: Multi-level features fusion via cross-layer guided attention for hyperspectral Pansharpening. Neurocomputing **506**, 380–392 (2022)
34. Wang, Y., Liu, G., Wei, L., Yang, L., Xu, L.: A method to improve full-resolution remote sensing Pansharpening image quality assessment via feature combination. Signal Process. **208**, 108975 (2023)

35. Yadav, A., Saini, B., Verma, V.K., Pal, V.: A joint medical image compression and encryption using AMBTC and hybrid chaotic system. J. Discr. Math. Sci. Cryptogr. **24**(8), 2233–2244 (2021)
36. Zhou, H., Liu, Q., Weng, D., Wang, Y.: Unsupervised cycle-consistent generative adversarial networks for Pan sharpening. IEEE Trans. Geosci. Remote Sens. **60**, 1–14 (2022)

Chapter 37
Advancements and Applications of AI and ML Techniques in Breast Cancer Prognosis: A Comprehensive Survey

Anurag Jagetiya and Pankaj Dadheech

Abstract This paper provides an overview of the advancements and applications of machine learning (ML) techniques in breast cancer research. It discusses the use of ML algorithms for classifying cancers, predicting survival status, and forecasting recurrence. The survey reveals that machine learning approaches were employed for cancer classification, prognosis, and the identification of prognostic factors, while deep learning approaches were used for predicting patient survival subtypes, identifying concealed features, and predicting metastasis. Artificial neural network (ANN) approaches are effective in classifying cancer types, while logistic regression and K-nearest neighbours (KNN) perform poorly on certain datasets. SVM shows moderate performance. Feature ranking and filtering improve results and aid in prognosis. Random forest adapts well to multiple datasets. ML techniques are also applied to identify prognostic factors for breast cancer survival. It is found that genomic factors combined with clinicopathological data aid in early detection and recurrence prediction. ML models help identify genes responsible for recurrence. Various techniques, including survival analysis and mechanistic models, are employed. CNN shows promise in predicting breast cancer from microscopic images. LDA and random forest perform well in predicting metastasis. The survey emphasizes the need for validation on external datasets and the inclusion of larger datasets and varied data sources for improved accuracy.

37.1 Introduction

Human body comprises of billions of cells which follows different-different roles and responsibilities. This entire system is quite intricate and generally it works in a controlled manner, however, if anything goes wrong, some specific cells may behave unusually and grow abnormally. The spread of such cells in the body is termed as

A. Jagetiya (✉) · P. Dadheech
Department of Computer Science and Engineering, Swami Keshvanand Institute of Technology, Management & Gramothan, Jaipur, Rajasthan 302017, India
e-mail: anurag.mlvtec@gmail.com

P. Nanda et al. (eds.), *Intelligent System and Data Analysis*, Smart Innovation, Systems and Technologies 398, https://doi.org/10.1007/978-981-97-5200-3_37

cancer. Since, the cells are present in the entire human body; cancer can potentially affect any part of it. And, the bunch of such cells together is termed as a tumour. Cancer tumour, if not treated on time, may spread to other organs which is a very dangerous condition known as metastasis. Spread of the cancer cells to the adjoining cells is called regional metastasis and to other organs in the body is termed as distant metastasis. According to WHO, this cancerous metastasis is the root cause of deaths in human beings. In the year 2020, cancer is blamed to cause more than 10 million of deaths across the globe [1]. The lung cancer (1.8 million) caused highest number of deaths while highest number of breast cancer cases (2.26 million) was detected in the year 2020 [2]. The accurate disease prediction and its consequences is quite a challenging task in medical science. Machine learning and artificial intelligence methods are capable to discover and classify patterns and associations between intricate datasets and are able to effectively forecast results of any specific cancer. Machine learning prediction models excel in performance but face challenges in interpretation, generalization, and limited data variables, particularly for breast cancer recurrence due to class imbalances and data availability issues [3, 4].

In spite of advancements in medical science, the treatment of cancer is very costly and there are chances of relapse or recurrence after the surgery too. It is important to detect the cancer in early stages and perform careful prognosis to reduce the mortality rate. The emerging machine learning and deep learning-based approaches have found widespread applications in medical research in recent years and cancer prediction performance using AI has reached to a new height. Huang S. et al. discussed the prospects and barriers of AI in the clinical implementation of cancer prognosis [5].

Breasts are comprised of fatty tissues having tiny compartments termed lobules. The breast milk is generated in the lobules and reached at nipples through small tunnels termed ducts. Lymph nodes are also located in the nearby region of the breast and under the arm. Lymph nodes protect human body from dieses and remain connected with a network of lymph vessels. Similar to other cancers, the breast cancer occurs by the abnormal behaviour of breast cells. The cells grow out of order and form the tumours. Breast cancer that develops inside the ducts is known as ductal carcinoma (DCIS) in situ. But, if DCIS spreads into surrounding tissues, it is known as invasive ductal carcinoma (IDC). Irregular cells can also develop in the lobules of the breast; they are termed as lobular carcinoma in situ (LCIS). LCIS is not considered as cancer, but if the abnormal cells invade in the surrounding area, it is known as invasive lobular carcinoma. Generally, the cause of death due to cancer is the widespread metastasis. Lymph nodes located under the arm are the early indicators of the cancer spread aka metastasis [1].

Statistics indicates that the number of women diagnosed with breast cancer is around 2.3 million, while, nearly 685,000 died across the world due to breast cancer alone in the year 2020. Breast cancer witnessed highest number of cases in past five years among all other cancer types making it the most predominant type of cancer. Breast cancer affects every one in eight women of developed nations by the time she reached at her 85. The breast cancer can also be developed in males, but, such probability is very less. Every nation of this world is witnessing the severity of breast cancer. It can affect the women at any age but the risk increases with age. There are many existing techniques and ways like mammogram, breast ultrasound, biopsy, MRI, etc., to detect the breast cancer in medical science. During past 20 years, many developed counties implemented early cancer detection programmes along with suitable treatment to combat against aggressive nature of the cancer [1, 2]. Therefore, this survey work is targeting the machine learning or artificial intelligence-based research conducted to cancer prognosis, survival prediction, and recurrence.

37.2 Use of AI and ML in Breast Cancer Prognosis

Over the past two decades, statistical tools are being used in medical science too for the various analysis purposes, but, conventional statistic tools are not adaptive to identify new variable and generating an integrating visualization like ML algorithms [6].

37.2.1 Classification of Breast Cancer Using ML

Researchers who detected the severity of cancer tumour, generally, attempted to classify it as malignant or benign worked upon the Wisconsin Breast Cancer Dataset (WBCD) [7] and applied various ML algorithms with varied parameters and training strategies. Table 37.1 summarizes the performances of popular ML algorithms run on WBCD in the order of their accuracy level to detect a cancer as malignant or benign [8–10].

As shown in Table 37.1, multilayer perceptron (MLP) has shown highest accuracy provided hyper-parameters are manually allocated to produce the final outcome in all the classifiers. MLPs are neural network models used to approximate any continuous functions. However, the inclusion of cross validation techniques can improve the accuracy and help in obtaining optimal values for the hyper-parameter [8]. Gupta, P. et al. worked on six ML algorithms and increase the accuracy by incorporating a unique hyper parametric change. As given in Table 37.1, the highest accuracy is received by combining the benefits of adaptive gradient algorithm and root mean square propagation. The former suits to computer vision problems and the latter works well for nonstationary signals. In order to make the model faster and quick learner, the rectified linear unit (ReLU) function is applied. ReLU also helped to bypass a vanishing gradient problem. However, the best method suggested cannot be applied in classification of breast cancer using medical images [9]. Islam, Md. et al. concluded that ANN outperformed other algorithms while RD and LR show poorest performance. Confusion and scalar metrics along with ROC curves are used to measure the results and compare the performances. Additionally, Matthews correlation coefficient is also computed on the results of confusion metrics [10].

Table 37.1 Comparisons of accuracy of ML algorithms run on WBCD to detect cancer tumour as malignant or benign

Paper/year	Algorithm	Accuracy
7 [2018]	Multilayer perceptron (MLP)	99.04
	Softmax regression	97.65
	Support vector machine (SVM)	96.09
	Linear regression	96.09
	Nearest neighbour (NN) search	94.73
8 [2020]	Deep learning using Adam gradient descent learning	98.24
	Support vector machine with radial basis function-kernel	97.2
	Random forest	97.2
	Decision tree	95.8
	K-nearest neighbourhood	95.8
	Logistic regression	95.8
9 [2021]	Artificial neural networks	98.57
	Support vector machine	97.14
	K-nearest neighbours	97.14
	Random forests	95.71
	Logistic regression	95.71

Breast cancer can be diagnosed on the basis of either genomics data or histopathological data. Nassif, A. B. et al. applied deep learning algorithms upon the microscopic images of patient's histopathological data and found that convolutional neural network (CNN) performed better than the rest of the methods. It is observed that the attention mechanism along with CNN can be explored to refine the results and inclusion of feature selection can help focusing on relevant features only [11, 12]. During assessment of AI in mammography screening, Freeman, K. et al. found that image recognition methods of AI are not better than a radiologist analyzing the images and most of the research was conducted on small datasets, and cannot be generalized [13].

Osareh, A., and Shadgar, B. combined ML algorithms and feature selection methods to differentiate between the benign and malignant tumours and find out more relevant features for the clinical diagnosis of breast tumour. The highest accuracy level of 98.80% and 96.33% were achieved using fine-tuned SVM-RBF classifier upon two publically available benchmark breast cancer datasets containing specimens of fine needle aspiration of breast lesions/lump (FNAB) and gene microarrays, respectively. The inclusion of feature ranking and filtering helped measuring the prognostic risk of recrudescence and metastasis [14]. Li, Y., and Chen, Z. applied included Breast Cancer Coimbra Dataset (BCCD) [7] along with Wisconsin Breast Cancer Database (WBCD) and applied five diverse ML classification models to find out the nature of the breast cancer tumour, i.e. malignant or benign. The random forest found to be adaptive and performed better in terms of accuracy and other important metrics [15].

37.2.2 Survival Prediction Using ML

In disease management, prediction of survivability of patient is paramount. In order to find the factors influencing the survival status of the cancer patient in Asian setting, Ganggayah, M. D. et al. applied six ML algorithms on a Malaysia's big hospital-based breast cancer dataset having diagnosis information of past 23 years to find the factors influencing the survival status of the breast cancer patient as alive or dead. In the study, the accuracies obtained by various ML algorithms were quite closer to each other. However, the highest accuracy is obtained from random forest. Table 37.2 lists the six most significant variables identified in this study. However, a different selection of variable may lead to different outcomes based on the datasets of patients' lifestyle and geographical location [6].

Table 37.2 Comparison of cancer survival status prediction using ML methods

Paper/ year	ML method	Performance	Type of data	Influential variables
[6] 2019	Random forest	82.70	Patients' lifestyle and geographical location	Cancer stage classification, size of tumour, number of axillary lymph nodes removed, positive lymph nodes, primary treatment type, and method of diagnosis
	Neural network	82.00		
	SVM	81.80		
	Extreme boost	81.70		
	Logistic regression	81.70		
	Decision tree	79.80		
[16] 2019	DeepProg Framework	C-index 0.68–0.73	RNASeq, methylation, and miRNA data	Highest co-expressed genes related to cell proliferation, extracellular matrix (ECM) organization, and immunoregulation
[18] 2019	Decision support system based on ML	86	Demographic, clinical and biochemical data	Nottingham prognostic index, age, tumour stage and size, ER/PR/HER2, breast surgery status
		C-index 0.84		
[19] 2018	Random forest	72.9	Clinical and pathological features were considered	NPI score, age, tumour stage and size, ER/PR/HER2 status, and breast surgery status
	SVM	72.9		
	Gradient boosting	69.7		
	ANN	72.1		

The human biological data of genome, proteome, transcriptome, epigenome, metabolome, and microbiome are considered an essential mean to infer the survivability of patient. Since all of these parameters contained a postfix "ome", and biological analysis of these "omes" datasets in combination may lead to findings of new patterns, associations, and results. Therefore, such multiples "omes" data is termed as multi-omics data. Poirion, O. B. et al. suggested a computational framework called DeepProg that predicts patient survival subtypes on the basis of different-different types of omics data. Apart from measuring cancer survival status based on conventional parameters like cancer stage, tumour size, origin tissues, etc., DeepProg framework considered molecular signatures of the cancer patients and found that there is single type of omics data common in most of the survival types. DeepProg, therefore, developed a multi-omics-based integrated platform to gain more insightful information of coherent molecular signatures among various cancer types [16]. A similar framework called similarity network fusion (SNF) integrates the individual patient's data types acquired separately into a similarity network to perform the multi-level data analysis. The SNF approach generates accurate estimation of cancer survival due to integrative approach [17].

Ferroni, P. et al. suggested included demographic details along with clinical and biochemical data to measure progression-free survival of breast cancer patients. ML-based decision support system along with random optimization resulted in a concordance-index of 0.84 with an accuracy of 86%. However, validation in multicentre prospective studies is presently needed [18]. Due to intricate associations between numerous contributing factors, the accurate prediction of five-year survival of breast cancer patient is very difficult. To compare survival of predictions, Zhao, M. et al. considered genomic factors with clinicopathological data and the results favoured ensemble models. K-means clustering and KNN classification of gene expression profiles on training data points and validation data points resulted in a good way of dimensional reduction. The study can be used to find out high-risk patients and impact of CLCA2 gene in high-risk HER2-positive patients [19]. Tapak, L. et al. found that SVM outperformed in survival status prediction while linear discriminant analysis lead the results in finding metastasis. However, the study found to be prejudiced in measuring the parameters like sensitivity as it relies upon registry-based surveying work [20]. Few types of gene mutations enhance the risk of breast cancer to a great extent. And, early detection of such high penetration genes like BRCA1, BRCA2, and PALB-2 can significantly improve the effectivity of breast cancer treatment [21]. The similar approach of gene selections has been explored in lung cancers too. Raoof, S. S., et al. applied advanced ML techniques upon microarray gene expression data to find out genes having higher prognostic information to predict the early occurrence of lung cancer [22].

37.2.3 Cancer Recurrence Prediction Using ML

Cancer is a kind of disease that may recur after a few years of treatment. Therefore, it is equally essential to timely and accurately predict the recurrence of the cancer. The micro-level examination of numerous genes using ML algorithms can help to predict the cancer recurrence. Shoon Lei Win et al. applied ML methods over DNA gene level microarray data to figure out that any specific cancer will recur within a time frame of five years or not. An entropy-based gene selection and AODE with subsumption resolution approaches are used to select prominent genes responsible for recurrence and average accuracy of 98.9% over 3 datasets is achieved [23].

The death rate due to metastasis and recurrence of breast cancer is around 15–20% and the inclusion of clinicopathological features like sHER2 resulted in early prediction of recurrence by around 3 months with the help of random forest. However, the study is limited to the patient data of Chang Gung Memorial Hospital [24]. Ahmad, L. G. et al. studied the application of data mining techniques to develop predictive models for breast cancer recurrence in patients who were followed-up for two years. Analysis upon the Iranian Centre for Breast Cancer (ICBC) dataset showed that the SVM classification model leads in the prediction of breast cancer recurrence. However, the absence of some important data values like S-phase fraction and DNA index affected the end results [25]. Boeri, C. et al. applied six ML models to access the chances of recurrence (both loco-regional and systematic) or probability of death due to breast cancer and found that ANN and SVM performed more accurately. The study was conducted on the small dataset of an Italian hospital which can be explored on large and diverse datasets [26]. Macías-García, L. et al. used DNA methylation data which contains around thousands of features per sample, therefore, auto-encoding mechanism of machine learning is used to shortlist important features and remove noise. A specific set of genes responsible for recurrence is identified with the help of survival analysis and a gene ranking mechanism. However, the inclusion of large set of data including multi-omic data may improve the results [27].

Other than statistical or ML-based models, Nicolò, C. et al. suggest to mechanistic model time-event data that can be used to simulate the chances of metastasis after surgery in early stage breast cancers and advice the course of adjuvant therapy after. One of the important prognostic factors in metastasis is found as the size of the cancer tumour. However, inclusion of additional biomarkers and validation on external datasets may further improve the model [28].

37.3 Discussion

The paper discussed about the advancements and application of ML techniques and frameworks in classification of cancers as malignant or benign, prediction of survival status of cancer patient as alive or dead, and forecasting the recurrence of cancer in past years. Figure 37.1 shows the count of various ML and deep learning algorithms incorporated by various researchers cited in this work. Survey reveals that deep learning-based approaches accurately classified cancer as malignant or benign, whereas logistic regression and KNN performed relatively poor while working on WBC dataset. However, the performance of SVM was moderate during these studies [8–10]. Nevertheless, SVM outperformed KNN and PNN classifiers while working on two different datasets of FNAB and gene microarray. The inclusion of feature ranking along with filtering approaches not only improved the results but also paved the path for prognosis of cancer recurrence in the patients [14]. Random forest is found to be more adaptive while working on two different datasets simultaneously [15]. Other than cancer classification, the ML techniques are also applied to find out the important prognostic factors affecting the survival status of the breast cancer patients [6]. In order to find the common factors among multiple cancer types, DeepProg uses deep leaning and machine learning methods in a collaborative manner to predict the patient survival subtypes on the basis of types of omics data. The model was tested on 32 types of cancer and found that RNA-Seq has most of the concealed features while miRNA concealed features are common in all the cancer types [16].

Researchers combined various factors like genomic factors, demographic details, etc., with biochemical or clinical pathology data of patience to predict the risk. Combining genomic factors along with clinicopathological data resulted in the early detection of the presence of genes mutations responsible for enhancing the risk of breast cancer [21]. While, the inclusion of demographics details with biochemical data helped to measure the progression-free survival of patients. The use of ML

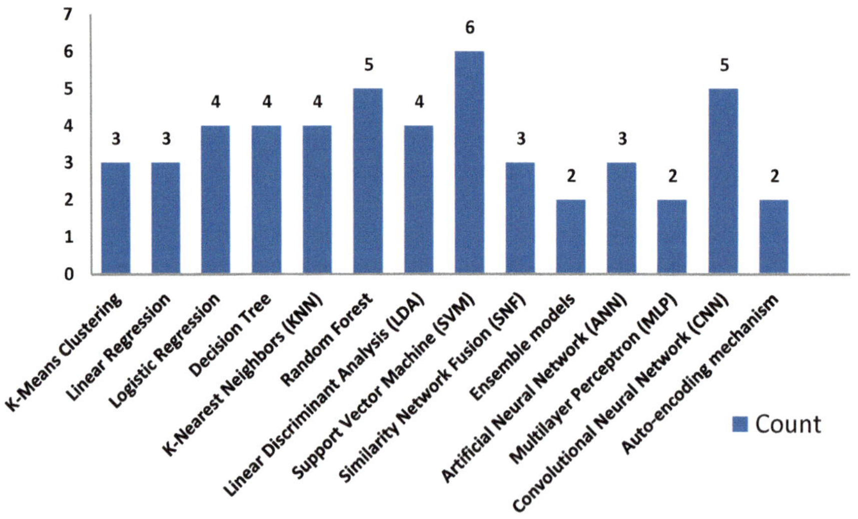

Fig. 37.1 Figure indicating the cumulative frequency of different machine learning and deep learning algorithms employed by the researchers whose work is referenced within this study. These algorithms are presented in a sequence, ranging from machine learning to deep learning

algorithms not only helped in the prediction of cancer recurrence but also in the identification of particular genes responsible for recurrence [18]. Cancer, generally, recurs after a certain period of time, therefore, it is essential to perform prognoses of the important factor responsible for the problem. One of the approaches used entropy-based selection of promising genes from DNA gene level microarray data to predict the recurrence within 5 years [23] and to develop predictive model for breast cancer recurrence, data mining techniques were applied and SVM outperformed DT and ANN [25]. Further, genes responsible for recurrence is identified with the help of survival analysis along with gene ranking mechanism [27]. To simulate chances of metastasis and course of adjuvant therapy after surgery, a mechanistic model is applied upon time-event data [28]. Deep learning mechanism like convolutional neural network (CNN) has shown good results in microscopic images to predict the breast cancer [11, 12]. While working upon different datasets, it is found that linear discriminant analysis [20] and random forest [24] performed better to predict metastasis.

The paper underscores the transformative potential of ML in cancer research and treatment, offering valuable insights into disease classification, prognosis, and the identification of crucial factors for patient outcomes. These findings contribute to the ongoing efforts to enhance cancer diagnosis and treatment, ultimately improving patient care and outcomes. The biotech company Aitia is applying Causal AI to advance drug discovery in oncology. They use multi-omic patient data, high-performance computing, and causal learning to uncover disease mechanisms. Aitia's approaches aid research in cancer, specifically in areas like multiple myeloma, prostate cancer, and pancreatic cancer [29].

37.4 Conclusion

Machine learning models heavily rely on the quality and representativeness of the data used for training. Biases or errors in the data can lead to biased or inaccurate predictions. In the case of breast cancer prognosis, if the training data is not diverse or representative enough, it may lead to biased predictions that disproportionately affect certain demographic groups or fail to capture important nuances of the disease. The choice of tuning parameters, optimization techniques, and lack of standard image acquisition methods affects the outcome of many studies. Many machine learning models, such as deep learning algorithms, are considered black boxes because they lack interpretability. The inability to understand the reasoning behind a model's predictions can hinder clinical adoption and trust. In the context of breast cancer prognosis, it becomes crucial to understand which features or factors contribute to a patient's predicted outcome and how they are weighted. Models trained on one dataset may not generalize well to different populations or healthcare settings. Variations in demographics, treatment protocols, and data collection practices can impact model performance.

The survey indicates that more accurate prediction of early stage cancers, cancer progression and recurrence and survivability using ML based models are still under explorations. The survey reveals that deep learning approaches have shown high accuracy in classifying cancer types, while traditional methods like logistic regression and K-nearest neighbours performed relatively poorly in specific datasets. Support vector machines (SVM) displayed moderate performance and even outperformed other classifiers in some scenarios. The combination of feature ranking and filtering methods has not only improved classification results but also paved the way for cancer recurrence prognosis. Random forest demonstrated adaptability when working with diverse datasets. ML techniques have also been instrumental in identifying important prognostic factors for patient survival, integrating genomic and clinical data for early detection of gene mutations and progression-free survival prediction. In conclusion, this discussion provides a comprehensive overview of the advancements in the application of machine learning (ML) techniques and frameworks for various aspects of cancer research and treatment. It has highlighted the versatility of ML in tasks such as classifying cancers as malignant or benign, predicting the survival status of cancer patients, and forecasting cancer recurrence.

References

1. WHO: Cancer. https://www.who.int/en/news-room/fact-sheets/detail/cancer. Accessed 1 Oct 2023 (2022)
2. Ferlay, J., et al.: Global Cancer Observatory: Cancer Today. International Agency for Research on Cancer, Lyon. https://gco.iarc.fr/today. Accessed Feb 2021 (2020)
3. El Haji, H., Souadka, A., Patel, B.N., Sbihi, N., Ramasamy, G., Patel, B.K., Banerjee, I.: Evolution of breast cancer recurrence risk prediction: a systematic review of statistical and machine learning-based models. JCO Clin. Cancer Informatics **7**, e2300049 (2023)
4. Howard, F.M., Dolezal, J., Kochanny, S., Khramtsova, G., Vickery, J., Srisuwananukorn, A., Pearson, A.T.: Integration of clinical features and deep learning on pathology for the prediction of breast cancer recurrence assays and risk of recurrence. NPJ Breast Cancer **9**(1), 25 (2023)
5. Huang, S., et al.: Artificial intelligence in cancer diagnosis and prognosis: opportunities and challenges. Cancer Lett. **471**, 61–71. https://doi.org/10.1016/j.canlet.2019.12.007 (2020)
6. Ganggayah, M.D., et al.: Predicting factors for survival of breast cancer patients using machine learning techniques. BMC Med. Inform. Decis. Mak. **19**(1), 1–17 (2019)
7. Dua, D., Graff, C.: UCI Machine Learning Repository. http://archive.ics.uci.edu/ml. University of California, School of Information and Computer Science, Irvine, CA (2019)
8. Agarap, F.M.: On breast cancer detection: an application of machine learning algorithms on the Wisconsin diagnostic dataset. In: 9th International Proceedings on Proceedings, pp. 1–2 (2018)
9. Gupta, P., Garg, S.: Breast cancer prediction using varying parameters of machine learning models. Procedia Comput. Sci. **171**, 593–601 (2020)
10. Islam, Md., et al.: Breast cancer prediction: a comparative study using machine learning techniques. SN Comput. Sci. **1**(5), 1–14 (2020)
11. Nassif, B., et al.: Breast cancer detection using artificial intelligence techniques: a systematic literature review. Artif. Intell. Med. 102276 (2022)
12. Saba, T.: Recent advancement in cancer detection using machine learning: systematic survey of decades, comparisons and challenges. J. Infect. Public Health **13**(9), 1274–1289 (2020). ISSN 1876-0341. https://doi.org/10.1016/j.jiph.2020.06.033

13. Freeman, K., et al.: Use of artificial intelligence for image analysis in breast cancer screening programs: systematic review of test accuracy. BMJ **374** (2021)
14. Osareh, A., Shadgar, B.: Machine learning techniques to diagnose breast cancer. In: 9th International Proceedings on Proceedings, pp. 1–2 (2010)
15. Li, Y., Chen, Z.: Performance evaluation of machine learning methods for breast cancer prediction. Appl. Comput. Math. **7**(4), 212–216 (2018)
16. Poirion, O.B., et al.: Multi-omics-based pan-cancer prognosis prediction using an ensemble of deep-learning and machine-learning models. medRxiv. 19010082 (2019)
17. Wang, B., et al.: Similarity network fusion for aggregating data types on a genomic scale. Nat. Methods **11**(3), 333–337 (2014)
18. Ferroni, P., et al.: Breast cancer prognosis using a machine learning approach. Cancers **11**(3), 328 (2019)
19. Zhao, M., et al.: Machine learning with k-means dimensional reduction for predicting survival outcomes in patients with breast cancer. Cancer Informatics **17**, 1176935118810215 (2018)
20. Tapak, L., et al.: Prediction of survival and metastasis in breast cancer patients using machine learning classifiers. Clin. Epidemiol. Global Health **7**(3), 293–299 (2019)
21. WHO: Breast Cancer. https://www.who.int/news-room/fact-sheets/detail/breast-cancer. Accessed 22 Feb 2022 (2021)
22. Raoof, S.S., Jabbar, M.A., Fathima, S.A.: Lung cancer prediction using machine learning: a comprehensive approach. In: 9th International Proceedings on Proceedings, pp. 1–2 (2020)
23. Win, S.L., et al.: Cancer recurrence prediction using machine learning. https://doi.org/10.5281/zenodo.4556482 (2021)
24. Tseng, Y.J., et al.: Predicting breast cancer metastasis by using serum biomarkers and clinicopathological data with machine learning technologies. Int. J. Med. Informatics **128**, 79–86 (2019)
25. Ahmad, L.G., et al.: Using three machine learning techniques for predicting breast cancer recurrence. J. Health Med. Informatics **4**(124), 3 (2013)
26. Boeri, C., et al.: Machine learning techniques in breast cancer prognosis prediction: a primary evaluation. Cancer Med. **9**(9), 3234–3243 (2020)
27. Macías-García, L., et al.: Autoencoded DNA methylation data to predict breast cancer recurrence: machine learning models and wet-lab validation. Int. J. Med. Informatics **151**, 104462 (2021)
28. Nicolò, C., et al.: Machine learning and mechanistic modeling for prediction of metastatic relapse in early-stage breast cancer. JCO Clin. Cancer Informatics **4**, 259–274 (2020)
29. Kolla, L., Gruber, F.K., Khalid, O., Hill, C., Parikh, R.B.: The case for AI-driven cancer clinical trials—the efficacy arm in silico. Biochim. Biophys. Acta Rev. Cancer **1876**(1), 188572 (2021). https://doi.org/10.1016/j.bbcan.2021.188572
30. Simpson, P.T., et al.: The diagnosis and management of pre-invasive breast disease: pathology of atypical lobular hyperplasia and lobular carcinoma in situ. Breast Cancer Res. **5**(5), 1–5 (2003)
31. Britt, K.L., Cuzick, J., Phillips, K.A.: Key steps for effective breast cancer prevention. Nat. Rev. Cancer **20**(8), 417–436 (2020)

Chapter 38
Perspective Analysis Toward Deep Learning Models for Wireless Capsule Endoscopy Images for Gastrointestinal Tract

Rakesh Sharma and C. S. Lamba

Abstract Analysis of medical image is new technique to solve the medical problems by examine the images that has been generated through some detailed clinical practice. For better clinical diagnosis, the goal is to extract information in a way that is both affective and efficient. Abdominal bleeding, ulcers, tumors, celiac disease, ulcerative colitis, and other gastrointestinal disorders are difficult to diagnose due to the uncertainty of accessing a volute condition in the human body. WCE is a non-invasive, painless, and patient-controlled gastrointestinal tract examination. Although using deep learning models to detect problems in WCE photos enhances detection accuracy, model training necessitates a large volume of labeled data. These deep models, on the other hand, are difficult to describe and do not incorporate expert input into the model decision-making process. Deep learning-based machine learning breakthroughs have improved the ability to categories, detect, and quantify patterns in medical images. Deep learning approaches are quickly being applied in medical imaging to increase performance in a variety of medical applications. Due to recent advancements in the field of biomedical engineering, medical image analysis has become one of the most important research and development areas. The use of machine learning algorithms for the processing of medical images is one of the reasons for this progress. Deep learning has been successfully applied as a machine learning technology, allowing a neural network to learn information automatically. This contrasts with methods that rely on traditional handcrafted elements. The selection and computation of these characteristics are a difficult task. Deep convolution networks, a type of deep learning approach, are widely employed in medical image processing. With these considerations in mind, the purpose of this research is to see if employing semi-supervised deep learning models over supervised deep learning methods for detecting and categorizing anomalies in WCE has any advantages. This paper provides a comprehensive overview of different deep learning algorithms for anomaly detection and localization methodologies utilized in WCE photographs in

R. Sharma (✉) · C. S. Lamba
Manipal University Jaipur, Jaipur, India
e-mail: srak.911@gmail.com

P. Nanda et al. (eds.), *Intelligent System and Data Analysis*, Smart Innovation, Systems and Technologies 398, https://doi.org/10.1007/978-981-97-5200-3_38

terms of performance, complexity, and dataset quality. The findings highlight existing research constraints in wireless capsule endoscopy image processing as well as future research prospects.

38.1 Introduction

ML algorithms have the possibility to be profoundly involved in all aspects of medicine, from drug development to clinical decision-making, dramatically transforming the practice of medicine. The recent success of machine learning algorithms in computer vision tasks comes at a good moment, as medical records become more digitalized. Office-based physicians in the USA increased their use of electronic health records (EHR) from 11.8 to 39.6% between 2007 and 2012 [1]. MRI, PET, CT, mammography, ultrasound, and X-ray have all been used to detect, diagnose, and treat illnesses throughout the last several decades [2]. Medical image analysis is a hotspot of machine learning research, thanks in large part to the well-structured and annotated data, and it's possible that this may be the first domain in which patients interact with genuine AI systems. For two reasons, this is significant. To begin, medical image analysis can be used as a barometer to see whether artificial intelligence systems can genuinely improve patient outcomes and survival in terms of actual clinical statistics. Second, it serves as a test bed for interpersonal interactions, with the goal of identifying how it should be done.

Patients will be receptive to nonhuman actors making or assisting them in making decisions that affect their health. Medical image processing has traditionally been done by human professionals such as physicians and radiologists in clinic science. Over the last few years, machine learning in medical imaging has gotten a lot of attention. Deep artificial neural networks began surpassing other known models on several important measures about 2009, fueling the boom. Medical images are an important aspect of a patient's electronic health record (EHR), and they are currently analyzed by human verification, whose speed, weariness, and experience limit them. Because it takes years and a lot of money to train a trained radiologist, many healthcare organizations use tele-radiology to outsource radiological reporting to moderate countries like India. The patient has received an inaccurate or delayed diagnosis. As a result, doing medical picture analysis with an accurate, reliable, and efficient machine learning system is a great idea.

WCE is a non-invasive way of detecting and assessing irregularities in the human gastrointestinal tract. The small intestine is the longest part of the digestive system and a vital organ for nutrition absorption. An illuminator, imaging sensor, optical dome, battery, and radio frequency transmitter are all contained in the Wireless Capsule, which has a length of 26 mm and a diameter of 11 mm (shown in Fig. 38.1). The patient consumes the full WCE, and the capsule gradually progresses it through the small intestine, photographing the entire GI tract in the process [3]. To address the existing approaches, Computer-Aided Detection (CAD) techniques were presented. Numerous studies have been conducted on detecting bleeding in WCE pictures [4].

Researchers, on the other hand, use a range of suggested methods to detect a particular condition such an ulcer, celiac disease, or polyp. To find anomalies in WCE pictures and videos, a variety of anomaly detection approaches have been explored. In the field of medical image processing and machine learning, CAD identification in wireless capsule endoscopy pictures has opened a new study path. Machine learning, image analysis, and computer vision are among the technologies used to discover and identify anomalies in various texture patterns and colors. ML-based models' overall performance is influenced by the features used to describe abnormalities and the techniques used to retrieve those characteristics from images.

In recent years, computer scientists have been particularly interested in medical imaging as they aim to develop automated diagnosis for diseases such as brain, skin, gastrointestinal, lungs, and many more. Health: 4.0, organizational issues, and software-defined radio are only a few of the related applications it includes to address medical sector challenges. Gastrointestinal issues impact most people worldwide, and a novel procedure called Wireless Capsule Endoscopy (WCE) was created to detect these issues. Every day, the global fatality rate from these gastrointestinal infections rises. Many reviews on the topics of DL design, performance, computational efficiency, and features used for anomaly representation in endoscopic image analysis have been published in recent years. In recent years, attention processing, expert knowledge integration in model decision-making, and generative deep model building have all advanced in computer vision using deep models. To uncover new research gaps in fields such as KNN, SVM, and random forest, this work investigates deep models in WCE images that do not use handcrafted features or classification methods.

38.2 Capsule Endoscopy and Standard Endoscopy

Both endoscopic procedures can quickly inspect the whole GI tract; however, they operate under different conditions. Manually identifying questionable spots is not time consuming because physicians or professionals may view the entire procedure in real time. The entire GI tract, as well as any conceivable pill-typed devices, flows through the human body like food or water in the case of CE [5]. The diagnostic procedure cannot be managed since the endoscopist is unable to move the camera manually. The video that was collected can be viewed at any time for inspection reasons. The process takes an average of 8 h to complete, making real-time viewing impractical. This technique generates 55,000 frames for each patient's evaluation and takes approximate 2 h for a physician for analyze the video. CE diagnostics require CAD tools to be swift, reliable, and accurate. If the computer system correctly identifies the sickness zone, the video inspection time can be greatly decreased. Based on the selected frames, the doctor will be able to assess the veracity of questionable parts afterward. Although the results aren't always precise or sufficient to detect all bleeding issues, the Suspected Blood Indicator is a tool for detecting red pixels in pictures [6].

38.3 Literature Review

In the last two decades, only a few assessments in the domain of WCE image processing have been conducted, with an emphasis on model designs and performance. In the fields of image processing and ML, medical image processing has advanced dramatically in recent years. The most well-known medical imaging specialties include lung nodule detection [5], brain cancer detection, skin cancer, and retinopathy. Artificial intelligence (AI) technology is rapidly evolving, with the potential to have a big impact on the medical industry. AI technology can improve the performance and consistency of patient treatment in gastroenterology and hematology since diagnostic imaging, endoscopy, and histopathology are so crucial. The author refines the progress of machine learning approaches to identification tasks, as well as the visual representation mechanism in deep learning, in this study. In gastrointestinal and hematological, several image analysis applications in endoscopy, radiography, and pathology are presented, suggesting the huge potential for AI in helping diagnosis [6, 7].

Reference [6] undertook a rigorous review of several computer-aided detection approaches in the field of WCE picture analysis. Color and texture feature extraction techniques in the spatial, geometric, and frequency domains, as well as other endoscopic treatments, were examined. The research also looks at and publishes datasets for various deep learning models for WCE image analysis that have been released. Furthermore, the survey reveals that no studies in the literature addressed dataset quality, dataset class imbalance, or model dependability.

The purpose of this work was to compare the accuracy and computational efficiency of two recent deep learning algorithms for autonomous cephalometric landmark detection. A total of 1028 cephalometric radiography photos were used as learning data for the YOLOv3 and SSD techniques. There were a total of 80 landmarks selected as targets. After the deep learning process, the algorithms were tested on a new test dataset of 283 photographs. To assess accuracy, spread graphs were utilized to illustrate the level error and success detection rate. The computation times for both strategies were also documented [4].

In medical imaging, semi-supervised, multiple instances, and transfer learning are covered, as well as their applications in diagnosis and segmentation. The author also talks about the connections between different learning contexts and future study options [36]. Deep learning algorithms, according to the authors, can analyze and identify lesions in collected photos in real time, allowing the capsule to take more images of a specific location, modify the focus level, or increase image quality. The authors also describe the challenges of building an automated capsule- endoscopy device that works.

The use of deep learning algorithms in the analysis of GI pictures is the subject of this review. Finally, the authors analyzed and contrasted the most recent published research on prevalent clinical digestive diseases, as well as the key uses of deep DL in GI image identification, classification, segmentation, recognition, and other tasks. Finally, the obstacles and future research targets for GI image analysis utilizing

DL will be discussed. This work created a computer-assisted automatic approach for detecting early gastrointestinal cancer lesion locations in gastroscopic pictures. To begin, the author devised a method for removing superfluous portions from a gastroscopic image in order to improve the accuracy of suspected lesion detection. Then, using the Canny-dilation technique, the image was segmented into smaller parts based on distinct image attributes. To describe each partitioned region, integrate different types of characteristics (such as color histogram, color moment, and rotation invariant local binary patterns). As a result, the ideal features can be chosen depending on the classification results of each feature between lesions and healthy regions. Finally, using clinical data from 407 gastroscopic pictures, they verified the proposed technique by detecting early gastrointestinal cancer lesions based on ideal features in a region. The authors present a historical overview of artificial intelligence research and analyze current studies that highlight major advancements in artificial intelligence for capsule endoscopy. Importantly, recent artificial intelligence advancements have yet to be implemented, making it premature to appraise the technology's potential based on current platforms. Artificial intelligence-based healthcare applications are projected to expand quickly as the remaining regulatory and standardization barriers are removed.

This study's authors provide a new feature descriptor for detecting frames containing polyps in WCE images automatically. The new approach is since discrete polyp pathological characteristics emerge when WCE images are divided into different resolution levels. They employed wavelet and emphasis feature extraction techniques as a result. The 2-D DWT, dual tree complex wavelet transform, GWT, and CT have all been examined to see which one is better at detecting polyps when paired with a probability distribution [41]. The whole set of receiver operating characteristic curves (ROCs) was plotted. There were 19 studies discovered out of a total of 45. All the research had already been completed. WCE was used to detect ulcers, polyps, celiac disease, hemorrhage, and hookworm, among other deep learning applications. The detection accuracy for most trials and illnesses was greater than 90%. This review paper reviews the most recent published material linked with widespread clinical GI disorders and discusses the crucial uses of DL in GI image detection, classification, segmentation, recognition, localization, and other tasks. Finally, based on DL [8], we highlight the challenges and potential research areas for GI image analysis.

This article outlines the different automated options for inspecting, identifying, and classifying various GI illnesses. This study also looks at content-based picture retrieval methods and automated endoscopic procedure summaries. In this study, the authors look at feature extraction approaches and deep learning procedures that are specifically designed for autonomous endoscopic video interpretation. We also separate features extraction methodologies into image processing domains, which are then divided into visual descriptions. The importance of WCE in medical imaging, the role of endoscopy for bleeding-related infections, and the scope of endoscopy are all discussed in this paper. The general steps have also been described, with the importance of each phase emphasized. There is a substantial discussion and future directions at the end. Deep model architectures and performance were emphasized over dataset quality and model dependability in the review literature on DL models for medical image processing that was assessed for evaluation. Without a benchmark dataset and consistent evaluation procedures for DL models, comparing the performance of different recommended models is pointless. This comprehensive evaluation considers all recent DL-based WCE image analysis research articles.

38.4 Research Method

The goal of this paper was to undertake a comprehensive evaluation of the literature on wireless capsule endoscopy image processing utilizing DL Models, which resulted in the creation of various research ideas. Several evaluations of WCE image analysis employing homemade feature engineering approaches have been undertaken during the last two decades. As a result, the scope of this review has been limited to deep learning-based studies in the disciplines of WCE, endoscopy, and colonoscopy image analysis published in the last five years (Table 38.1).

38.5 Findings and Scope

The quality of the labeled dataset, domain knowledge incorporation, and interpretability for model training, making decisions, and model understanding are all considerations in the examination of several deep models proposed for WCE anomaly detection and classification. In order to address the problems, the review was narrowed to focus on the potential of deep models for generating smart solutions that use unlabeled data, expert knowledge in deep model decision-making, and support systems for understanding model behavior. In terms of task-wise categorization, anomaly representation characteristics, the dataset used for assessment, and deep models utilized for model training.

38.6 Conclusion

Deep models for identifying anomalies in WCE pictures were investigated. The benefits and cons of using deep models for WCE image analysis are reviewed, as well as the dataset employed. Existing DL models are supervised learning models that are trained using a little amount of labeled data. The training data quality is found to be low due to the class imbalance. Domain information might be added to semi-supervised deep models in future. To improve model decision-making, and an attention mechanism could be employed to learn target-specific learning. Additionally, creating a decision tree support model aids in understanding model behavior and increases the model's interpretability.

Table 38.1 Existing WCE image analysis review articles in the areas of, deep model architecture, deep model performance, deep model trustworthiness, dataset quality, deep knowledge discovery model, YOLOv3, and single-shot detector

Article authors	Objective	DMA	DMP	DMT	DQ	DKDM	YOLOv3	SSD
Hao Chen (2020) [9]	AI's potential in gastroenterology and hepatology medical image analysis	✓	✓					
Ali, H. (2019) [10]	In WCE images, a survey of various CAD methods was conducted	✓	✓					
Ji-Hoon Park (2019) [11]	Comparisons of SSD and YOLOV3 model						✓	✓
Veronika Cheplyginab (2019) [12]	Study of different learning models					✓		
Jungmo Ahn (2018) [13]	Endoscopy image-based learning systems' challenges	✓	✓					
Du, W. (2019) [14]	Deep learning based on endoscopic image anomaly detection	✓	✓					
DY Liu (2015) [15]	Automatic diagnosis of early gastrointestinal cancer lesions from gastroscopic pictures using optimal feature extraction	✓	✓					
Roberto Trasolini (2020) [16]	DL for capsule endoscopy in the small intestine	✓	✓	✓				
Meryem Souaidi (2018) [17]	New features for polyp detection with WCE	✓	✓	✓				
Shelly Soffer (2020) [18]	Meta-analysis of DL for wireless capsule endoscopy	✓	✓			✓		

References

1. Hsiao, C.J., Hing, E., Ashman, J.: Trends in electronic health record system use among office-based physicians: United States, 2007–2012. Natl. Health Stat. Rep. **75**, 1–18 (2014)
2. Ciuti, G., Menciassi, A., Dario, P.: Capsule endoscopy: from current achievements to open challenges. IEEE Rev. Biomed. Eng. **4**, 59–72 (2011). https://doi.org/10.1109/RBME.2011.2171182
3. Schurischuster, S., Remeseiro, B., Radeva, P., Kampel, M.: A Deep Learning Approach for Red Endoscopies, vol. 1, pp. 553–561. Springer (2018). https://doi.org/10.1007/978-3-319-93000-8
4. Sainju, S., Bui, F.M., Wahid, K.A.: Automated bleeding detection in capsule endoscopy videos using statistical features and region growing. J. Med. Syst. **38**(4) (2014). https://doi.org/10.1007/s10916-014-0025-1
5. Xing, X., Yuan, Y., Jia, X., Meng, M.Q.: A saliency aware hybrid dense network for bleeding detection in wireless capsule endoscopy images. In: 2019 IEEE 16th International Symposium on Biomedical Imaging (ISBI 2019), pp. 104–107 (2019)
6. Fu, Y., Zhang, W., Mandal, M., Meng, M.Q.H.: Computer-aided bleeding detection in WCE video. IEEE J. Biomed. Health Inform. **18**(2), 636–642 (2014). https://doi.org/10.1109/JBHI.2013.2257819
7. He, J., Wu, X., Jiang, Y., Peng, Q.: Hookworm detection in wireless capsule endoscopy images with deep learning. IEEE Trans. Image Process. **27**(5), 2379–2392 (2018). https://doi.org/10.1109/TIP.2018.2801119
8. Khan, M.A., Sharif, M., Akram, T., Bukhari, S.A.C., Nayak, R.S.: Developed Newton-Raphson based deep features selection framework for skin lesion recognition. Pattern Recognit. Lett. **129**, 293–303 (2020)

Chapter 39
Exploring the Challenges and Software Risks of Agile Methodologies in Software Development

Neha Choudhary, Mahaveer Sain, and Sumitra Chopra

Abstract Software development model is the way that software development is organized and managed, which has a direct effect on the quality of software development. It is also one of the most significant topics of research in software engineering. It is necessary to take into account the software *development* process as well as its various phases in order to assure that the objectives are met effectively. Software engineering is concerned with the application of software systems to various system processes. Agile is a method of software development that emphasizes flexibility before its needed and pragmatic methods for delivering a final product. Agile software development requires a change in the way many organizations do things because it focuses on clean delivery of individual software parts instead of whole solution. Agile is an iterative approach to the management of projects as well as development of software. It makes it possible for teams to give value to customers more rapidly and with fewer complications than is possible with old methods. A "big bang" release is something an agile team strives to steer clear of in favor of smaller, more manageable components of work. First, deliver a quick introduction to the agile software development concepts. In this problem, lack of communication and collaboration among team members: Agile methodology relies heavily on communication and collaboration among team members. However, this can be challenging for teams that are geographically dispersed or have members working in different time zones. Afterward discussed different software development techniques including two techniques waterfall model and spiral model, agile software development, agile methodologies. After that, challenges in agile risk management and software risk have been discussed.

N. Choudhary (✉) · M. Sain · S. Chopra
Maharishi Arvind Institute of Science & Management, Jaipur, Rajasthan Technical University (RTU), Kota, India
e-mail: neha2861990@gmail.com

P. Nanda et al. (eds.), *Intelligent System and Data Analysis*, Smart Innovation, Systems and Technologies 398, https://doi.org/10.1007/978-981-97-5200-3_39

39.1 Introduction

Agile software development (ASD) is an iterative method for developing software that is both efficient and effective. The goal of an iterative process known as agile software development (ASD) is to produce high-quality software quickly. In order to improve one's own character and one's technical abilities, a strong motivator is crucial in the field of software development. Motivator is crucial to accomplishing project scope by elucidating corporate objectives. McHugh, Conboy, and Lang [1] analyzed the impact of demotivator and motivator on the three main agile principles. To achieve this objective, qualitative analysis was conducted. This systematic literature study will collect the current understanding of demotivators and motivators. In ASD, because of its iterative nature, proportion of failed projects is lower than in SDLC; yet, there is a need for demotivator and motivator variables affecting ASD in terms of individual personal and technical skills. These motivators and demotivators function as an umbrella activity over duration of project; therefore, it is necessary to manage demotivator elements in order to raise motivator factors afterward. The literature shows that good management is backbone of a successful project and can reduce failure rates and total costs by as much as a 70%. There are several different methods within ASD, all of which are committed to agile manifesto's ideal of continuous development across all stages [2]. Software development employs a wide variety of approaches. The following are some innovative software development methodologies:

- Component-based development.
- Agile software development.
- Reuse-based development.

All of these models are distinct and unable to serve the varied needs of software industry and its users. In order to achieve client pleasure, it is necessary to adopt a new method with extremely rapid creation and delivery. Thus, the ASD approach is utilized for this purpose. Moreover, ASD not only shortens the duration of software development, but also reduces the cost. ASD is capable of delivering products that meet market and customer requirements [3].

39.2 Problem Statement

The problem statement of software development in agile methodology can vary depending on the specific context, but some common challenges that organizations face when adopting agile methodologies include:

1. Lack of clear understanding and adherence to the principles and values of agile: Agile methodology requires a cultural shift toward collaboration, flexibility, and continuous improvement. However, some teams may struggle to fully embrace these principles, which can hinder the effectiveness of the methodology.

2. Inability to estimate project timelines and resource requirements accurately: Agile methodology often involves working in short sprints with flexible timelines. However, this can make it difficult to accurately estimate project timelines and resource requirements, which can impact the ability to deliver projects on time and within budget.
3. Difficulty in managing scope and priorities: Agile methodology places a strong emphasis on flexibility and adapting to changing requirements. However, this can make it challenging to manage project scope and priorities and can lead to scope creep and shifting priorities that can impact project timelines and deliverables.
4. Ineffective use of agile tools and techniques: Agile methodology relies on a range of tools and techniques, such as user stories, sprints, and retrospectives, to facilitate collaboration and continuous improvement. However, some teams may struggle to effectively use these tools and techniques, which can impact the effectiveness of the methodology.

39.3 Different Software Development Techniques

39.3.1 Waterfall

In 1970, Royce unnamed it when it was initially introduced. It sequentially summarizes essential software development activities (design, analysis, requirements, testing, programming, and operation).

Waterfall development was proposed to prevent dangers associated with code-and-fix techniques by putting requirements and analysis stages before coding stage. This ensures that the needs of end users are clearly expressed up front, cutting down on the need for reworked or entirely new pieces of code.

In traditional waterfall approach, errors propagate from one phase to the next until they are identified during testing. To eliminate this danger, it was advised that a review of the previous phase be conducted at the start of every phase to check that the previous stage was completed correctly. After that, he modified his original waterfall model by introducing localized iterations that feedback into earlier stages. Nonetheless, problems are still found during testing, even with these localized iterations, and these problems are usually result of design or need problems. Hence, to fix these mistakes, complex iterations were added to the requirements and design phases. These revisions need a lot of time, energy, and resources. In order to avoid the risks of operational constraints, Royce [4] suggested introducing a preliminary design stage between analysis and requirements stages to constrain analysts. This is accomplished by repeatedly cycling through processes of preliminary design analysis and design until an appropriate preliminary design is reached (Fig. 39.1).

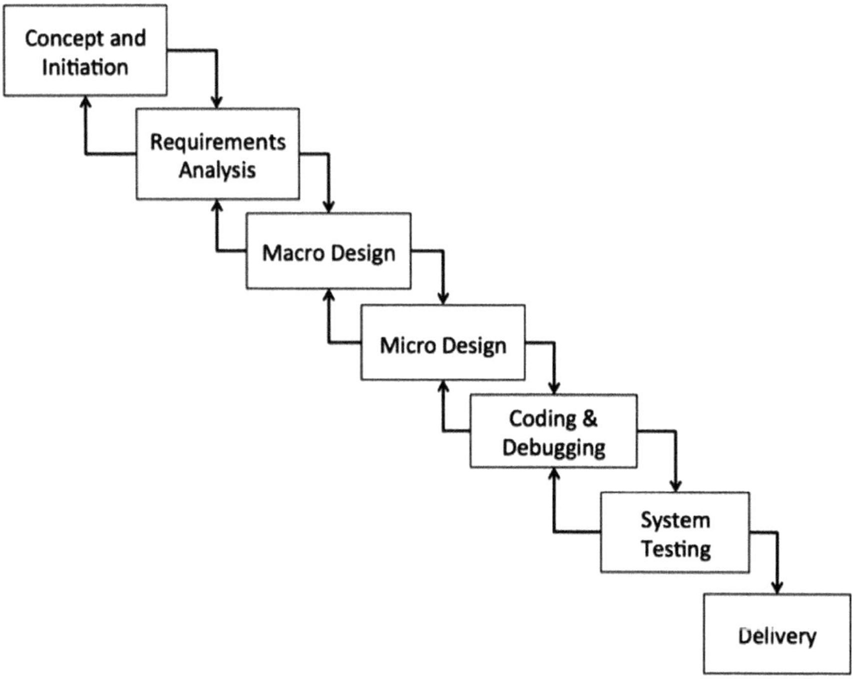

Fig. 39.1 Stages of waterfall model

39.3.2 Spiral Model

It manages risk during the stages of development. The amount of spiral loops varies according to the job. In first quadrant, customer requirements are gathered and potential solutions are identified. In second quadrant, hazards related to the solution have been identified and addressed. In third quadrant, solution is built and assessed. In the last section, the client evaluates completed version, and work on next iteration gets underway. Spiral model and its applications are utilized in a variety of contexts, including mobile application development, traffic management systems, and website development [5]. The purpose of this study is to compare spiral process with scrum methodology in terms of its strengths and weaknesses. Additionally, investigate the conditions for adopting either spiral process model or scrum technique when designing software [6]. The current investigation will make use of a number of different models for the development of mobile applications, most of which cannot be included in fundamental phases of software development life cycle (SDLC). The study will place a particular emphasis on cross-platform development, user interaction, and memory utilization. A model for mobile application development life cycle will also be suggested (Fig. 39.2).

Fig. 39.2 Spiral model

39.4 Agile Software Development

Agile implies "moving rapidly" or "being adaptable", and it also functions in this manner. In other words, agile development is a contemporary methodology that emphasizes inspection, monitoring, self-organization, rapid delivery of high-quality software, and alignment of customer needs with business objectives. All stages are reviewed throughout life cycle, hence incremental and iterative best characterize the agile approach to software development.

A. **Agile Manifesto**

The manifesto states, "we are discovering better methods for producing software by doing it ourselves and assisting others do it as well" [7]. The manifesto specifies four fundamental principles for excellent performance as follows [8]:

I. People and Interactions over Procedures and Instruments: There can't be any misunderstandings among members of the team if they want to perform above and beyond competition. Hence, inspect-and-adapt cycle aims to enhance collaboration and communication within agile technique.
II. Functional Software versus Extensive Documentation: The agile manifesto emphasizes providing the user with working software in small increments.
III. Partnership with Customers over Contract Negotiation: Customer participation in the software development process directly contributes to success. IV. Reacting to alterations rather than sticking to a plan: Agile methodology contains built-in criteria for modifying plans based on client feedback.

B. **Agile methods used for developing safety critical systems**

Extreme programming and SCRUM are two of many agile methodologies that have been used to create life-saving software.

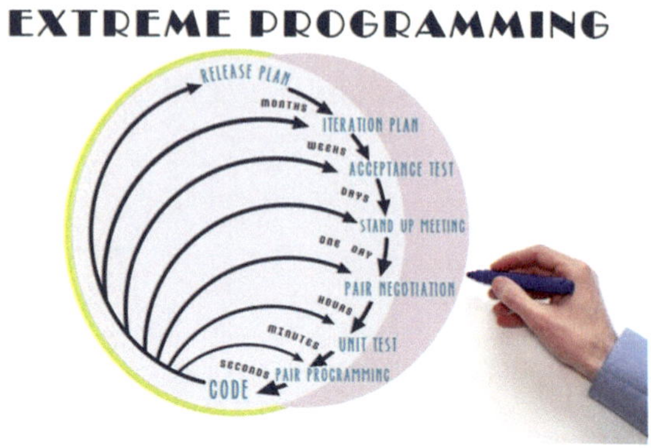

Fig. 39.3 Extreme programming

1. **Extreme Programming (XP)**

Extreme Programming was designed for use by small, colocated teams developing low-priority goods. A programmer can use it as a roadmap for the entire development cycle. The objective is to offer what the customer requires when it is required, with an emphasis on teamwork and a willingness to accept adjustments at any time. When requirements are unclear or changing and a project has a high level of risk, XP is right approach. Yet, there are issues when there is a sizable team, poor customer-developer communication, and a lack of projectwide testability. They used a subset of XP concepts for the creation of safety–critical systems and discovered an average quality improvement of 53% when contrasted to plan-driven software development initiatives [9]. It illustrates how XP methods are applied in a multinational corporation designing safety–critical systems. It proposes that some XP approaches, including simple designs linked with test-driven development and refactoring, work very well in safety–critical domain (Fig. 39.3).

Distinguishes features of XP are as follows:

Story Cards: Users provide needs as scenario-based stories that are subsequently shown as narrative cards. Developers then subdivide every story card into a series of smaller assignments. These smaller activities are then prioritized for implementation with the assistance of consumers [10].

- Simplicity: XP focuses on building the simplest product that suits users' fundamental needs. It adheres to the notion of developing only what is specified in given need. In response to consumer requests, product receives further capabilities.
- Feedback: At the conclusion of every release, appropriate consumer input is received, and subsequent level of iteration is depending on this feedback. In XP, for effective feedback, tiny design and implementation loops are created using pair programming and test-driven development methodology.

- Test-Driven Development: Extreme programming employs a test-oriented development strategy in which test cases are prepared in advance of proper code implementation. XP uses testing throughout entire process.
- Pair Programming: Pair programming is the defining characteristic of Extreme Programming, in which two programmers work collaboratively. This leads in significant time savings and decreased workload.

2. **Scrum**

This is one of the lightweight methods utilized for software development. Its guiding idea is which small cross-functional teams achieve positive outcomes. Scrum is more focused on revenue growth and software quality. Since it is lightweight, it can adapt to changing needs, and software is released in short cycles known as sprints. The responsibilities within scrum include scrum master, product owner, and team member. Scrum is an agile methodology for managing iterative and incremental software development. It challenges the "traditional, sequential approach" to product development and allows teams to self-organize through the promotion of physical co-location or close online collaboration of all team members and daily face-to-face communication between all team members and disciplines involved in the project, among other things [11] (Fig. 39.4)

3. **Adaptive Software Development (ASD)**

ASD, which was created by Jim Highsmith, is more accurately a culture than a software development process. ASD recognizes unpredictability and uncertainty in software development and offers a framework for delivering by releasing frequently and learning from errors. ASD is comprised of three nonlinear and overlapping phases [12] (Fig. 39.5).

4. **Feature-Driven Development (FDD)**

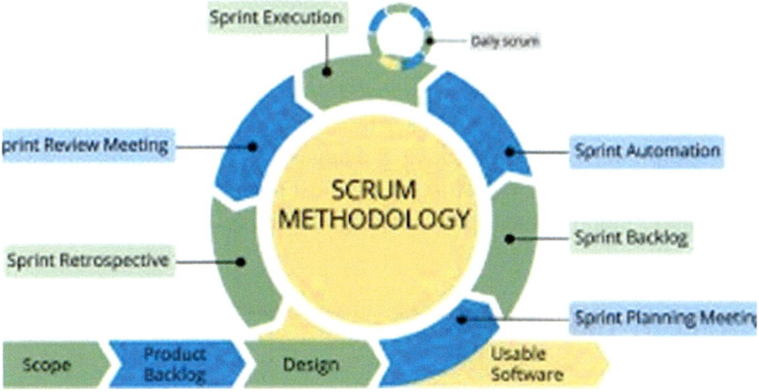

Fig. 39.4 Scrum process

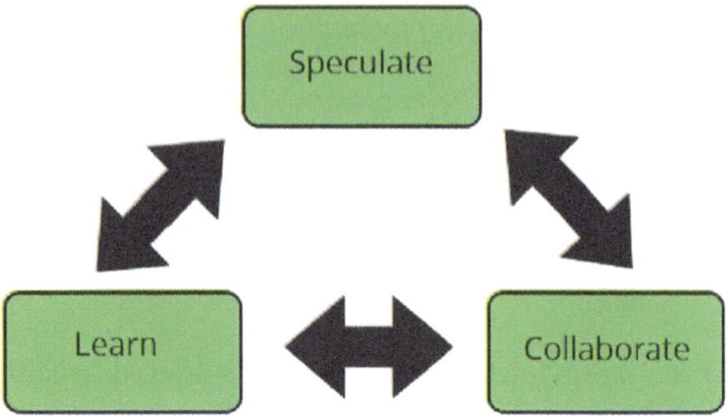

Fig. 39.5 Adaptive software development

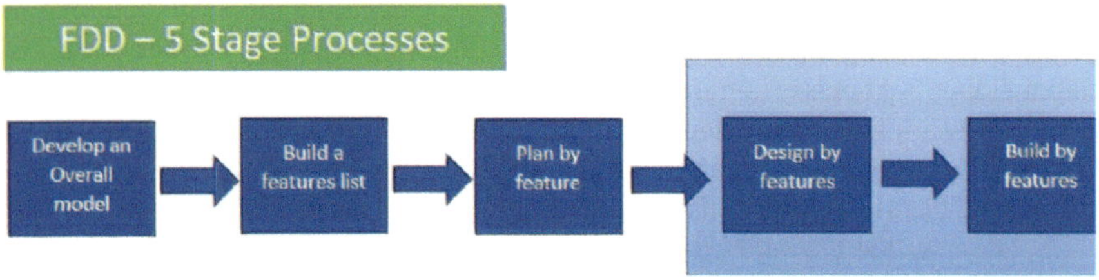

Fig. 39.6 Feature-driven

Feature-driven development (FDD) is an agile software development technique that prioritizes customer-requested features, groups them into areas, and iteratively designs and produces features. FDD contains project progress monitoring milestones and creates two-weekly builds to keep clients as informed as feasible.

Main features of FDD techniques (Fig. 39.6):

1. Developing by Feature: Implement only customer required features.
2. Feature Teams: Cooperating to discover a solution for a problem.
3. Individual Class (Code) Ownership: Every developer is tasked with creating and maintaining an allocated class.
4. Inspections: Covers unit testing as well as code and design reviews.
5. Domain Object Modeling: Development of a visual representation (class, UML) of the system. Every iteration updates the model with additional features.
6. Configuration Management: Tracking changes.
7. Regular Builds: Present to customers the most recent information possible.
8. The ability to monitor and view both progress and results.

39.5 Agile Methodologies

In early years of software development, the majority of customer needs were rather steady, and development closely followed the original intentions. Yet, as software development became more dependent on crucial and dynamic industrial initiatives, new challenges arose in tandem with the expansion of businesses. These difficulties include [13]:

- Evolving requirements: The expectations of customers are altering as a result of evolving company needs or legislative challenges. In the early phases, the majority of clients do not have a clear vision of their demands' specifications. Some consumers discover their true requirements only after using an application that does not adequately fulfill their expectations. During the course of one's development, one acquires experiences that serve as another source of change.
- Customer involvement: Lack of client interaction leads to a higher likelihood of project failure. Typically, firms devote minimal resources to client engagement.
- Deadlines and budgets: Customers frequently do not tolerate failure. On the other hand, firms typically give minimal budgets, short deadlines, and stringent requirements, all of which are the result of market competitiveness.
- Miscommunications: The miscommunication among developers and clients contributes to misinterpretation of requirements.

For instance, every side employs its own jargon, leading to a misunderstanding of customers' requirements.

Considering such issues, OO software development approaches cannot meet the aims of software development firms. In order to solve these issues, it is necessary to employ novel growth strategies.

A. **Limitations of Agile Methodology**

Nothing in the universe is without flaws or limitations. Here are a few of the limitations and drawbacks of agile approaches that we gleaned from research [14].

- Prolonged testing lead times and inadequate test coverage.
- A large number of teams demand a high level of coordination and communication from the project management.
- Development is emphasized more than design and the user. It focuses mostly on methods for obtaining needs and generating code while ignoring product design.
- Larger projects struggle to benefit from this method because of the many iterations required to achieve desired results. Too much time may be devoted to any single, small feature.
- As the application of an agile methodology relies so heavily on effective teamwork, the project manager needs to be actively involved in dynamics of team, which increases management overhead.
- Using agile methods on a massive undertaking may not be worth the potential loss of production when compared to other, more lucrative, or efficient endeavors.

39.6 Challenges in Agile Risk Management

A. **Software Development in Life Cycle**

This problem arises from differences in organizational frameworks of cybersecurity risk management, agile software development, and other related fields. Considering that cybersecurity risk management is frequently executed sequentially or "plan-driven" (controlling, analyzing, identifying, and monitoring risk) [15]. It is not necessarily compatible with lean, changeable, and adaptable characteristics of agile methodologies. Instead, cybersecurity risk management may overburden agile methodology due to the need to assess extra requirements introduced between sprints [16–18]. It has also been noted that time commitment involved in cybersecurity risk management operations may be stressful for practitioners who lack extensive background knowledge in the field [19]. This difficulty is referred to as C1 by us.

B. **Incremental Development**

This difficulty pertains to establishing a cybersecurity risk assessment that is consistent with incremental development of agile projects. Inconsistencies between development projects and management of cybersecurity risks can also be caused by constant refactoring and redesigning of code and other elements [20]. Code that has been evaluated for vulnerabilities and threats, for instance, may undergo significant modifications as project progresses, necessitating a new assessment [21]. This is Challenge Number Two or C2.

39.7 Software Risk

According to ISO Guide73:2002, risk is a "combination of likelihood and impact of an event" [Iso02]. It is built on three fundamental ideas: occurrence, frequency, and intensity. Yet, the primary emphasis is on unfavorable occurrences that represent a loss in a particular setting. Software risk is described as the probability of experiencing a loss, including customer unhappiness, poor quality, budget or schedule overruns, and passive customer involvement, as a result of an unfavorable occurrence and its repercussions throughout the project's life cycle.

- **Risk Event and Likelihood**

A risk event is the occurrence of potentially harmful circumstances or the discovery of evidence that suggests the existence of potentially harmful circumstances [22]. The event's certainty and/or uncertainty can be influenced by a single occurrence or a series of events. Probability refers to the likelihood that an event will occur. Hence, various mathematical concepts, like probability theory, model an occurrence based on likelihood of uncertainty. The expected utility theory (Hog87), Dempster–Shafer theory of evidence (Sha76), and the fuzzy set hypothesis (Zad99). These theories have been established for various objectives and represent distinct categories of uncertainty. One example might be the axioms of probability theory, which are [Fis67]:

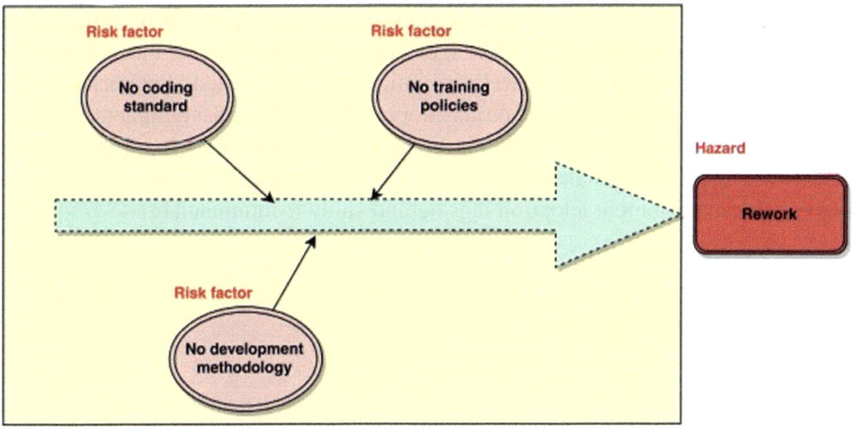

Fig. 39.7 Risks versus risk factors versus hazards

- Axiom I: For any random occurrence A, there is a specific integer P(A), known as probability of A, which meets inequality 0 P(A) 1.
- Axiom II: The probability of a particular occurrence is one, or $P(E) = 1$.
- Axiom III: If there are a finite or non-infinite number of mutually exclusive events, then the probability of all conceivable outcomes is equal to the sum of probabilities of those outcomes.
- **Risk Factor**

All hazards can be traced back to a set of underlying risk factors. Here we look at both internal and external elements that can have an impact on occurrence of a risk. If you take this view into account, risk is an aggregate of various potential threats. Thus, the risk statement consists of a condition and a consequence. Hence, risk is a complex and abstract notion, and one approach to make it more tangible is through risk factors, which serve to lessen its level of abstraction (Fig. 39.7).

A particular organization, for instance, produces software without following a development process, coding standard, or training regulations. These circumstances result in a hazard known as rework. In other words, the risk can be seen as a collection of factors that result in a certain danger. This condition is more effectively depicted in Fig. 39.7.

- **Risk management**

To accomplish the goals of a project, risk management is art and science of recognizing, evaluating, and dealing with risks at every stage of project's life cycle. The primary purpose of risk management is to recognize possible threats, analyze them to determine which ones are most likely to occur, classify which ones can have the greatest effect on project if they do occur, and then describe plans to mitigate or minimize the consequences or avoid risks, with ultimate goal of optimizing project objectives.

The majority of literature on software risk management that has been published focuses on detection and analysis of risk or risk evaluation. Despite the fact that risk management is a strategy that is well recognized but rarely used, there is sadly few research that specifically focuses on its adoption in software development projects. There is also a tendency to initiate its procedures, but despite being a vital aspect of software engineering management practices, it is disregarded over time. In other words, "risk management adoption lags behind study recommendations" [23].

39.8 Literature Review

The purpose of this work is to Beh et al. assess current performance assessment frameworks or models, especially agile ones. This study reviewed the literature on measurement metrics in agile software development. Agile-based project performance can be measured in a variety of ways, including organizational, process, team/project, and individual levels. Every dimension is interconnected in many ways rather than existing independently. More study is needed to acquire an understanding of metrics for every dimension, linkages among dimensions, and potential integration of many dimensions for organizations working on agile software development projects [24].

In this study, Kuhrmann et al. examine question: What makes an agile software development methodology?

Researchers present an empirical study based on a large-scale international survey with the objective of identifying software development methods that enhance or constrain agility. On the basis of 556 data points, they examine perceived level of agility in implementation of standard project disciplines and its relationship to the development methods and practices that are employed. Their research indicates that only a minority of project managers use exclusively conventional or agile methods (under 15%). In spite of this, the majority of project disciplines and practices exhibit a discernible upward trend in terms of agility. Compared to software development approaches, the selection of practices has a greater impact on degree of agility of a specific field. Ultimately, there are no specific procedures or practices which ensure or prohibit agility. They find that agility cannot be characterized only at the level of processes. When implementing or enhancing agility in a software organization, additional considerations must be taken into account. In conclusion, they explore the subject of software process-related research in light of their findings and suggest a research plan [25].

This study Ismail and Arviansyah examines agile issues in the external software development project as well as expert viewpoints for appropriate classification. This study intends to map obstacles associated with agile software development utilizing systematic literature review (SLR) and Delphi approach to fill gap. Utilizing SLR, previous research identified 19 issues and 5 categories. The conclusion of Delphi approach provides five types of issues and enhances the challenging concepts of prior research. In conclusion, this study might serve as a starting point for determining which set of difficulties significantly affects the information system sector [26].

The purpose of this study Lunesu et al. is to introduce and analyze a novel method for modeling major risk variables in agile development using software process simulation modeling (SPSM), which may supplement existing methods and whose application is particularly suited to agile development. Utilizing a simulator of agile development that they constructed for this purpose, they present a novel method for modeling many major risk indicators, including project duration, number of implemented problems, and key statistics of issue completion time. The technique comprises modeling the agile process, collecting data from tools used for project management, and doing Monte Carlo simulations of process in order to get insight into the projected time and effort required to finish project, as well as their distributions. Errors in effort assessment of the features to be developed, variances in developers' assignments to these features, and barriers linked to developers' availability and job completion are model factors that pose a risk. To test the simulator and show how the strategy may be implemented, they identified three open-source projects by collecting data from their JIRA repositories. They performed Monte Carlo simulations of these projects, demonstrating that the simulator is a good approximation of the actual project's development, before altering identified risk variables and statistically analyzing their influence on risk parameters. The suggested method is suitable for project managers because it can quantitatively evaluate risks, assuming process and project data are properly modeled and collected. As for the future, they are aiming to enhance their risk assessment approach by assessing it on other case studies and expanding the model from a single team to numerous teams working on one or more projects [27].

This work Afshari and Gandomani aims to establish a standard team structure and configuration for Agile teams in practice. Agile teams establish new ideas including on-site customer, leadership, self-organization, and cooperation due to differences between Agile and plan-driven software development methodologies. In addition, additional roles and duties, such as product owner, technical lead, architecture owner, and developer, have been specified for agile teams [28].

In this study, Tekbulut, Canbaz, and Kaya Participation in teams in retrospective meetings, which are an integral part of the agile product development process and focus on continuous improvement (Logo Agile Process Improvement System) LOGO's LAPIS, which was conceived in response to the lean manufacturing movement, features machine learning capability. The objective has been to increase the opportunity for teams to better their work and base it on data. Consequences of team-observed enhancements on the company, and their influence on software process measurement outcomes, have been investigated [29].

In this study, Sarikaya, Bagriyanik, and Gökalp, it has been determined if agile approaches that have gained popularity in software business over past 15 years and have same potential when employed as a teaching technique. Throughout the research, 49 engineering students from a Turkish university took part in the "Agile Software and Product Development" course. Learning experiences, curriculum design, student performance factors, and more outcomes have been presented.

In this course, three sprints are planned with eight agile project teams that are self-organized using scrum methodology. Constant team learning opportunities were reviewed, along with the obstacles experienced while applying the "agile" technique to students while they developed new ideas and impact of agile technique on learning performance. Considered to be beneficial for universities, agile coaches, faculty members, students, businesses, and professionals in the information and communication technology sector [30].

In this work, Paasivaara and Lassenius describe their findings from a case study of a big software development business that employed CoPs to assist their large-scale agile deployment and their transition from waterfall to lean and agile. In this instance, usage of CoPs was one of a number of crucial success factors. To promote both research and practice, they propose that future studies should focus more on CoPs in large-scale agile projects, such as scrum. By examining how CoPs should be cultivated within an organization, how they function in practice, what they are utilized for, as well as benefits and challenges of cultivating and implementing CoPs, they can better understand how they should be implemented [32].

This work by Doherty et al. this presentation presents an overview of a unique, large-scale, customized implementation of agile development known as the agile factory model (AFM), with the goal of emphasizing defects problems of particular concern to enterprises engaging in large-scale agile development. Using an extensive internal database of defects, this research examines cause, defect severity, and lifespan. Throughout development and testing, the bulk of AFM faults is of severity 3, whereas requirements' defects account for less than 0.7% of defects. Integration, acceptance, and performance (AIP) testing captures the greatest number of defects at 26.14%, while factory acceptance testing (FAT) captures an additional 17.28%; FAT and AIP are subsequent phases in AFM, respectively. These findings show that agile principles may successfully be applied in a remote development setting around the world [33].

39.9 Conclusion

During the 1970s, software development approaches changed. Agile approaches arose from the necessity for a lightweight approach to software development in order to adapt to an environment with fluctuating needs. Agile approaches include procedures that allow communication between developer and client, as well as develop-deliver-feedback cycles, in order to have a clearer understanding of requirements and be adaptable to any change. The primary objective of agile approaches is to provide what is required when it is required. Agile techniques consist of a collection of software development strategies. Despite their differences, they share the same fundamental ideas. The most prevalent agile approaches include SCRUM, XP, and agile modeling. XP is the testing and coding of customer-specified requirements. Agile modeling specifies a set of concepts, objectives, and practices that outline how to simplify modeling and documentation activities. In contrast, SCRUM promotes the management role in software development. Agile approaches are gaining favor in business sector despite the absence of risk management mechanisms. In past years, research on agile software development has indicated that agile approaches are efficient and applicable in a variety of circumstances and environments. Agile approaches are gaining acceptance in industry. In past years, research on agile software development has indicated that agile approaches are efficient and applicable in a variety of circumstances and environments. This paper examines recent research on agile software development. This paper examines some of the most recent research on agile software development. In this review paper, we discuss our study of agile methodology and its many methodologies or approaches. Several agile planning procedures with two objectives have been compared in the work. Reviewing planning procedures and requirements across different agile development methodologies is one goal, and deciding which agile development methodology(s) to use for planning procedures is the second. In the area of planning requirements and processes, we discovered that scrum is the top option among other agile methodologies.

References

1. McHugh, O., Conboy, K., Lang, M.: Using agile practices to influence motivation within IT project teams. Scand. J. Inf. Syst. (2011)
2. Ghayyur, S.A.K., Ahmed, S., Ali, M., Razzaq, A., Ahmed, N., Naseem, A.: A systematic literature review of success factors and barriers of Agile software development. Int. J. Adv. Comput. Sci. Appl. (2018). https://doi.org/10.14569/IJACSA.2018.090339
3. Singh, H., Singh, Y.: Review paper on software development. Int. J. Technol. Res. Eng. **3**(2) (2015)
4. Hijazi, H., Khdour, T., Alarabeyyat, A.: A review of risk management in different software development methodologies. Int. J. Comput. Appl. Technol. **45**(7), 8–12 (2012)
5. Doshi, D., Jain, L., Gala, K.: Review of the spiral model and its applications. Int. J. Eng. Appl. Sci. Technol. (2021). https://doi.org/10.33564/ijeast.2021.v05i12.053
6. Mathur, A., Acharya, A.: A comparative study on utilization of scrum and spiral software development methodologies: a review. Int. J. Eng. Res. (2015). https://doi.org/10.17577/ijertv4is110154
7. Ylimannela, V.: A model for risk management in agile software development. Commun. Cloud Softw. (2012)
8. Shah, R.: A literature review on agile model methodology in software development. Int. J. Adv. Res. Innov. Ideas Educ. **6**, 293–297 (2017)
9. Mwadulo, M.W.: Suitability of agile methods for safety-critical systems development: a survey of literature. Int. J. Comput. Appl. Technol. Res. (2016). https://doi.org/10.7753/ijcatr0507.1009
10. Nawaz, M., Nazir, T., Islam, S., Masood, M., Mehmood, A., Kanwal, S.: Agile software development techniques: a survey. Proc. Pak. Acad. Sci. Part A **58**(1), 17–33 (2021). https://doi.org/10.53560/PPASA(58-1)721
11. Khare, S.R., Shrivastava, R.: Literature survey on agile information systems development. Int. Res. J. Eng. Technol. (2015)
12. Balaban, S., Đurašković, J.: Agile project management as an answer to changing environment. Eur. Project Manag. J. **11**(1) (2021)
13. Hneif, M., Ow, S.H.: Review of agile methodologies in software development 1. Int. J. Res. Rev. Appl. Sci. **1**(1), 2076–2734 (2009)

14. Altameem, E.: Impact of agile methodology on software development. Comput. Inf. Sci. **8**(2) (2015). https://doi.org/10.5539/cis.v8n2p9
15. Lundgren, M., Bergström, E.: Dynamic interplay in the information security risk management process. Int. J. Risk Assess. Manag. (2019). https://doi.org/10.1504/IJRAM.2019.101287
16. Hauck, J.C.R., Vieira, M.: Towards a Guide for Risk Management Integration in Agile Software Projects (2021). https://doi.org/10.1007/978-3-030-85521-5_6
17. Tavares, B.G., Keil, M., Sanches da Silva, C.E., de Souza, A.D.: A risk management tool for agile software development. J. Comput. Inf. Syst. (2021). https://doi.org/10.1080/08874417.2020.1839813
18. Te Chiu, Y., Chen, H.G., Zhu, Y.Q.: Exploring IT/S Risk Management Agility (2018)
19. Wei, Y.C., Wu, W.C., Chu, Y.C.: Performance evaluation of the recommendation mechanism of information security risk identification. Neurocomputing (2018). https://doi.org/10.1016/j.neucom.2017.05.106
20. Tøndel, I.A., Jaatun, M.G., Cruzes, D.S., Moe, N.B.: Risk centric activities in secure software development in public organisations. Int. J. Secur. Softw. Eng. (2018). https://doi.org/10.4018/ijsse.2017100101
21. Ben Othmane, L., Angin, P., Weffers, H., Bhargava, B.: Extending the agile development process to develop acceptably secure software. IEEE Trans. Dependable Secur. Comput. (2014). https://doi.org/10.1109/TDSC.2014.2298011
22. Islam, S.: Software Development Risk Management Model—A Goal Driven Approach (2009). https://doi.org/10.1145/1595782.1595785
23. Bannerman, P.L.: A reassessment of risk management in software projects. In: Handbook on Project Management and Scheduling, vol. 2 (2015)
24. Beh, H.C., Jusoh, Y.Y., Abdullah, R., Hassan, S.: Dimensions in measuring performance of agile software development projects: a literature review. In: 2022 Applied Informatics International Conference (AiIC), pp. 83–87 (2022). https://doi.org/10.1109/AiIC54368.2022.9914025
25. Kuhrmann, M., et al.: What makes agile software development agile? IEEE Trans. Softw. Eng. **48**(9), 3523–3539 (2022). https://doi.org/10.1109/TSE.2021.3099532
26. Ismail, D., Arviansyah: A systematic literature review and Delphi study on agile software development challenges. In: 2021 6th International Conference on Management in Emerging Markets (ICMEM), pp. 1–6 (2021). https://doi.org/10.1109/ICMEM53145.2021.9869414
27. Lunesu, M.I., Tonelli, R., Marchesi, L., Marchesi, M.: Assessing the risk of software development in agile methodologies using simulation. IEEE Access **9**, 134240–134258 (2021). https://doi.org/10.1109/ACCESS.2021.3115941
28. Afshari, M., Gandomani, T.J.: A typical practical team structure and setup in agile software development. In: 2021 7th International Conference on Electrical, Electronics and Information Engineering (ICEEIE), pp. 483–487 (2021). https://doi.org/10.1109/ICEEIE52663.2021.9616743
29. Tekbulut, T., Canbaz, N., Kaya, T.Ö.: Machine learning application in LAPIS agile software development process. In: 2020 Turkish National Software Engineering Symposium (UYMS), pp. 1–6 (2020). https://doi.org/10.1109/UYMS50627.2020.9247069
30. Sarikaya, E., Bagriyanik, S., Gökalp, M.: Teaching agile software development using agile methods: a case study. In: 2020 Turkish National Software Engineering Symposium (UYMS), pp. 1–6 (2020). https://doi.org/10.1109/UYMS50627.2020.9247027
31. Braz, A., Rubira, C.M.F., Vieira, M.: Development of complex software with agile method. In: 2015 Agile Conference, pp. 97–101 (2015). https://doi.org/10.1109/Agile.2015.18
32. Paasivaara, M., Lassenius, C.: Deepening our understanding of communities of practice in largescale agile development. In: 2014 Agile Conference, pp. 37–40 (2014). https://doi.org/10.1109/AGILE.2014.18
33. Doherty, B., Jelfs, A., Dasgupta, A., Holden, P.: Defect analysis in large scale agile development: quality in the agile factory model. In: 2016 Joint Conference of the International Workshop on Software Measurement and the International Conference on Software Process and Product Measurement (IWSMMENSURA), p. 180 (2016). https://doi.org/10.1109/IWSMMensura.2016.034

Chapter 40
Guarding Databases: Unraveling SQL Injection Threats with Advanced Detection and Prevention

Harshit Soni, Ashish Tripathi, Rohit Kumar Gupta, and Pragati Gupta

Abstract Nowadays, web applications are generally utilized as a busybody between PC clients. Also, web applications are utilized by online business organizations, government agencies, research organizations, etc. that need to protect the database to keep data secure. SQL injection attack is among the most significant security vulnerability in web applications. It allows an attacker to intervene in the database through queries applicable to the application. This helps an attacker to view and access unauthorized data. This paper presents a common SQL injection attack, identification, and anticipation advancements.

40.1 Introduction

SQL injection is a significant issue, and various kinds of security breaches can be seen around us. Therefore, security personals are very much concerned about security threats. In this context, many websites' databases have been found vulnerable which causes a loss of data of individuals and industry [1]. The SQL injection attack is performed by attackers on entering the malicious query in the application which may result in resolved validation and access to the database. With access to the database, the attacker can pose a serious security threat to the owner of the data, which can even result in the theft of confidential data, its deletion, modification, and unauthorized access to systems, and accounts [2]. An individual injection attack comprised of small queries can compromise the entire network and data.

In an SQL injection attack, the web application is fooled into executing queries for which the application didn't expect to be as input [3]. For that data to be protected an

H. Soni
New York Institute of Technology, Vancouver, Canada

A. Tripathi · P. Gupta
SCSE, Galgotias University, Greater Noida, India

R. K. Gupta (✉)
Department of IT, Manipal University Jaipur, Jaipur, Rajasthan, India
e-mail: rohitkumar.gupta@jaipur.manipal.edu

P. Nanda et al. (eds.), *Intelligent System and Data Analysis*, Smart Innovation, Systems and Technologies 398, https://doi.org/10.1007/978-981-97-5200-3_40

inspection layer must be added to it, which includes running the submitted data information through mysql_real_escape_string (). This provides the protection ability that dangerous characters such as single quote ('), backslash (\), and other special characters are not passed to query in the data.

Validation is the most important aspect to be considered while having protection from SQL injections. While this validation occurs in two ways, the first is to blacklist dangerous characters and the second includes keeping the whitelisted characters only, these characters include only those characters which are allowed in the database for occurrence [4, 5]. The SQL injection attack [28] only occurs because of vulnerability in applications and databases. Those vulnerabilities are of four types:

- Error-based SQL injection: In this vulnerability, attackers can retrieve the data information which can include table names and content [6].
- Boolean-based SQL injection: When the submitted query fails, then at some moments the web page disappears and changes else the website fails to load, where such indication lets the attacker know where the input parameter is vulnerable and in knowing whether that allows data extraction.
- Time-based SQL injection: The attacker instructs the database to sleep for a defined time before responding and in the meanwhile, the page loads quickly then it is not vulnerable and if it takes a longer duration to load it informs the attacker that it is vulnerable [7].
- Out-of-band SQL injection vulnerability: If the injection does not occur directly then the attacker sends the data from the database server to a machine that controls the attacker, where the attacker accesses that database.

40.2 Background Details

40.2.1 MySQL

MySQL is the relational database management system software, which implements the client–server model. Whereas in communication between client and server, the RDBMS is used which only uses some specified query-based language called structured query language. These MySQL RDBMS using the structured query language are itself written in other programming languages where C and C++ are used in written MySQL [8].

For the working of MySQL [31], devices connect with the server through a network, where the client makes a request and, on that server, produces output, where in the process includes MySQL creating the database and defining the relationship of different tables. After that client makes a request by entering SQL queries in MySQL, from which the server application responds and the requested information is displayed to the client. The use of MySQL has been diverse as host-based verification and password encryption make it a highly secure feature.

As websites and web applications [27, 29, 30] become dependent on SQL-based databases, there is an increase in vulnerability to SQL injection because of flaws and improper implementation of websites and databases. These SQL vulnerabilities in websites are easy to detect using automated tools [24, 26] making it even more important to secure databases and websites as any website running MySQL can be vulnerable to SQL injection.

SQL injection attacks are too common mainly because of two reasons:

1. Significant prevalence for SQL injection vulnerabilities
2. The attacked SQL database contains every information and critical data of the website.

40.2.2 Cybersecurity

Cybersecurity is essential because it combines everything that pertains to protecting personal data, intellectual information, and governmental and sensitive industry data from unauthorized access, data theft, and damage that is attempted by cybercriminals.

As every organization and individual uses the internet, they are exposed to risks that can be used by any malicious actor on the internet. If hackers get personal information, they might sell that information on the Dark and Deep Web, which is a part of the internet that can only be accessed using some specially designed software; this internet part is not regulated by any government body which also makes it a digital black market where the people from all across the globe can share and also sell the private information data which is not legal.

40.2.3 SQL Injection

SQL injection (SQLi) is one of those attacks that allow malicious SQL statements to be executed. These commands control the databases that have private data of the web application. The attacker can exploit database vulnerabilities to bypass the security systems of web applications. They can bypass the authorization and authentication of the website and obtain sensitive information from the complete database. Also, they can use SQL injection to delete, manipulate, and add data to that database.

Any web application that uses an SQL database such as SQL Server, Oracle, or MySQL with SQL injection vulnerability [35] can be used to have unauthorized access to any sensitive information including personal data, customer data, intellectual property information data, and trade secrets. SQL injection attack [32] is among the most dangerous and oldest web application-based vulnerability. The Open Web Application Security Project (OWASP) has listed SQL injection as the first among all the data threats in web application security.

Any successful SQL injection attack [33] may lead to the access of unauthorized information including personal details, details of credit cards, passwords, and many sensitive information data. There have been many recent data breaches in recent years which have resulted from SQL injection only, resulting in the reputational damage of many companies. An attacker even gains the permanent backdoor of the organization's database, which leads to the compromise in the long term and goes unnoticed by the company.

Initially in an SQL injection attack [34], the task is to study the working of the targeted database, which is achieved by sending different random values to a query to see the way in which the server responds.

The attacker then uses this observation to create a query that the server interprets and executes as an SQL statement. For example, the database stores some information about their students with some student identification codes. Instead of searching for a specific Student ID, the attacker can put "StudentID = 10,000 or 1 = 1" in the input field. Because the 1 = 1 statement is always true, SQL will give all the available student IDs and all matching information, which allows an attacker to bypass the authentication and to have the administration access.

40.2.4 Need of Security

Security is a significant aspect because this includes everything that pertains in protecting intellectual property, governmental information, and organization data from damage, which can be used by attackers.

As every organization and individual uses the internet, they are exposed to risks that can be used by any malicious actor on the internet. If hackers get personal information, they may sell it on the Dark and Deep Web which is a worldwide network that can be accessed by using special software and is not in control of any organization or government acting as the digital open world where the people share any illegal information data and also can sell it.

40.3 Related Work

See Table 40.1.

40.4 Proposed Work

SQL injection attack permits the attackers to type SQL queries on a website and get a reaction from the targeted database, which allow the attacker to manipulate the current data present in the database and to spoof identities. In this work approach, Wireshark is being used, as a tool to analyze the network packets and traffic on the

Table 40.1 Related work of SQL injection

S. No.	Author	Method of implementation
1	Taran et al. [9]	The designing of a system that detects the login attempts on MySQL servers using the Honeypot technique
2	Shahriar et al. [10]	Identifying the potential security vulnerabilities in both NoSQL and SQL databases being used in the MOOC applications
3	Medeiros et al. [11]	Detecting SQL injection attacks and the related vulnerabilities present in the database
4	Mukherjee et al. [12]	The implementation of attacks by comparative analysis among different SQL injection techniques of prevention with effectivity
5	Qian et al. [13]	SQL injection attacks are being prevented by type-safe SQL parameters methods and input validation
6	Liban et al. [14]	The enhancement of MySQL vulnerability checker tool with the use of interface binary search algorithm for blind timing-based attacks
7	Wegerer et al. [15]	Using Honeypot technology in the detection of vulnerabilities in the database and defeating the adversary using deception
8	Hlaing et al. [16]	SQLIA techniques are being used for SQL injection attacks, detection, and prevention
9	Sharma et al. [17]	The anatomy of order-wise injection and blind injection attacks is done to objectify and to prevent SQL injection
10	AL-Malikia et al. [18]	Detection of SQLi to improve the security of web applications from client-side attacks
11	GUO Chun et al. [19]	Truncated key payload method to prevent the SQLi attack
12	Nasereddin et al. [20]	A systematic review of different detection and prevention methods in SQL injection
13	Falor et al. [21]	The use of deep learning in SQL injection detection by convolutional neural networks

network, on running Wireshark and opening some captured networks to view them for SQL injection attack against the SQL.

40.4.1 Comparative Analysis

In the detection of SQL injection earlier techniques used a proposed tool that identified the detection of SQL injection attack and vulnerability [4, 22]. Those techniques were not able to identify the much of different SQL attacks as the proposed tool had an inability to identify the attack on the network with SQL injection. It identified the SQL injection vulnerability that a website designer and developer can use to identify the vulnerability of SQL injection in almost every aspect of their website and web application for search and analysis where the network can be analyzed for miscellaneous activities, to manipulate with the data present in database [13]. The detection is based on analysis work rather than using the proposed tool to identify the vulnerability of the SQL injection.

40.4.2 Detection Technique

Following are the steps of SQL injection technique and the flowchart for the same is shown in Fig. 40.1.

1. Choosing the GET HTTP request and following the HTTP stream as the application layer leads to the testing of queries for SQL injection. On entering 1 = 1 union select database (), the page that needed to respond with login failure responded with:

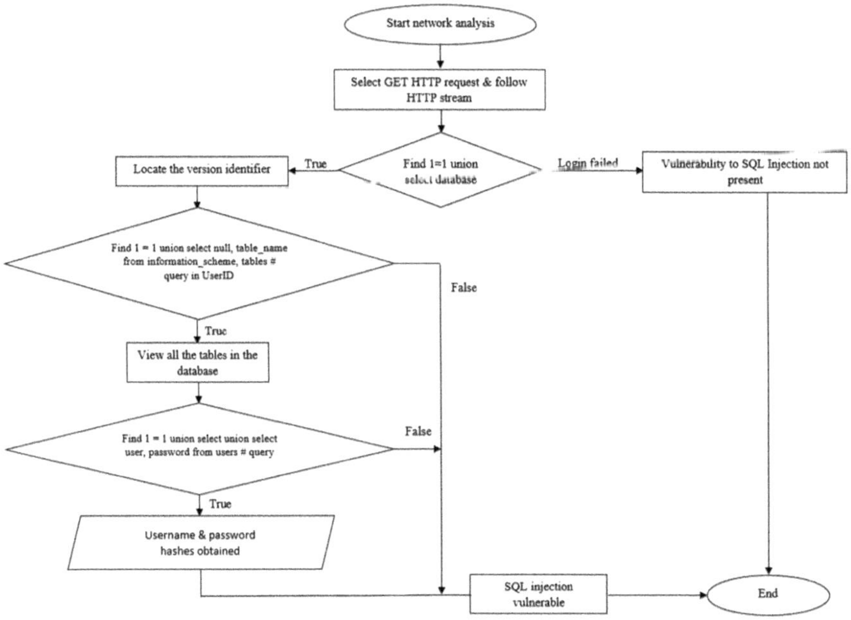

Fig. 40.1 Flowchart of SQL injection detection technique

```
.<div class="vulnerable_code_area">
..<form action="#" method="GET">
...<p>
....User ID:
....<input type="text" size="15" name="id">
....<input type="submit" name="Submit" value="Submit">
...</p>

..</form>
..<pre>ID: 1' or 1=1 union select database(), user()#<br />First name: admin<br />
>Surname: admin</pre><pre>ID: 1' or 1=1 union select database(), user()#<br />First
name: Gordon<br />Surname: Brown</pre><pre>ID: 1' or 1=1 union select database(), user
()#<br />First name: Hack<br />Surname: Me</pre><pre>ID: 1' or 1=1 union select database
(), user()#<br />First name: Pablo<br />Surname: Picasso</pre><pre>ID: 1' or 1=1 union
select database(), user()#<br />First name: Bob<br />Surname: Smith</pre><pre>ID: 1' or
1=1 union select database(), user()#<br />First name: dvwa<br />Surname:
root@localhost</pre>
.</div>

.<h2>More Information</h2>
.<ul>
..<li><a href="http://www.securiteam.com/securityreviews/5DP0N1P76E.html"
target="_blank">http://www.securiteam.com/securityreviews/5DP0N1P76E.html</a></li>
..<li><a href="https://en.wikipedia.org/wiki/SQL_injection" target="_blank">https://
en.wikipedia.org/wiki/SQL_injection</a></li>
Entire conversation (6532 bytes)
```

2. Attacker used "1' or 1 = 1 union select" query in an ID search box to locate version identifier. All these version identifiers are in the end of the output obtained before </pre>.</div> tag.

```
....
<div class="body_padded">
.<h1>Vulnerability: SQL Injection</h1>

.

.<div class="vulnerable_code_area">
..<form action="#" method="GET">
...<p>
....User ID:
....<input type="text" size="15" name="id">
....<input type="submit" name="Submit" value="Submit">
...</p>

..</form>
..<pre>ID: 1' or 1=1 union select null, version ()#<br />First name: admin<br />
>Surname: admin</pre><pre>ID: 1' or 1=1 union select null, version ()#<br />First name:
Gordon<br />Surname: Brown</pre><pre>ID: 1' or 1=1 union select null, version ()#<br />
>First name: Hack<br />Surname: Me</pre><pre>ID: 1' or 1=1 union select null, version
()#<br />First name: Pablo<br />Surname: Picasso</pre><pre>ID: 1' or 1=1 union select
null, version ()#<br />First name: Bob<br />Surname: Smith</pre><pre>ID: 1' or 1=1
union select null, version ()#<br />First name: <br />Surname: 5.7.12-0ubuntu1.1</pre>
.</div>

.<h2>More Information</h2>
.<ul>
Entire conversation (6548 bytes)
```

3. On using Ctrl+F and entering the users, search for entry. When the text is found, close the Find textbox.

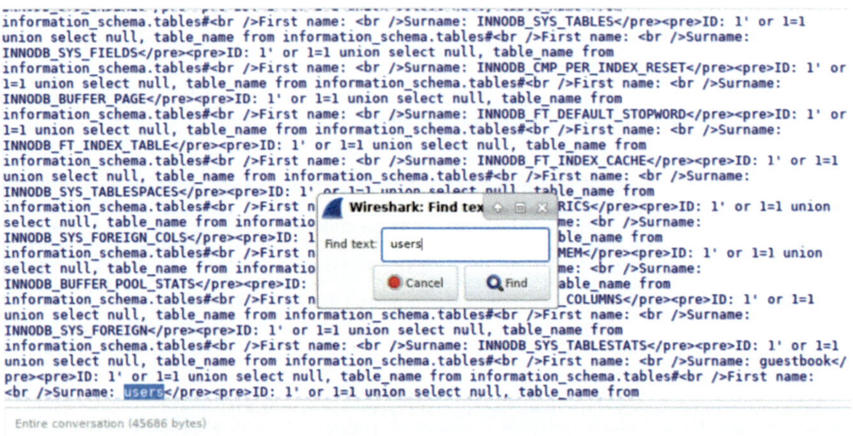

4. Attacker used “1’or 1 = 1 union select null, table_name from information_ schema. tables#” query in the ID search box to get tables present in database. This will provide the output of many tables, as attacker has specified null without the specification.

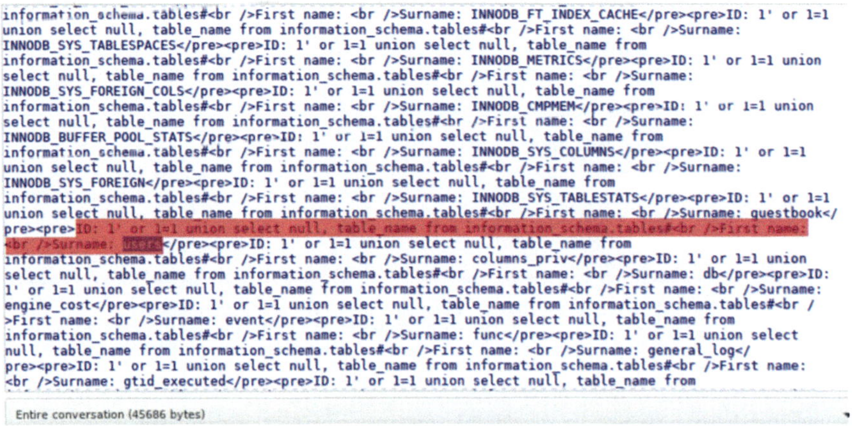

5. In Find box type 1 = 1 and search for the entry.

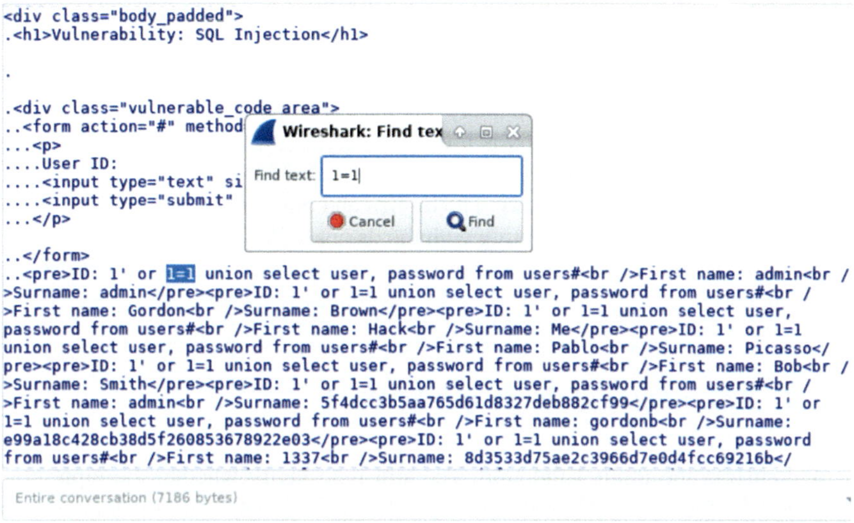

6. Attacker used "1'or 1 = 1 union select user, password from users#" query to the ID search box to obtain the user names with their password hashes.

40.4.3 Prevention Against the SQL Injection Attacks

The organizations can follow the following methods to protect the database against SQL injection attacks.

- ***Use prepared statements***: Always use prepared queries by generating SQL statements initially and then taking all the submitted data as the parameters [8].
- ***User input must not be trusted***: User input should always be verified before being used in the dynamic structured query language statement.
- ***Use regular expressions***: This is used in detecting a potentially mischief code and to remove it before the SQL query execution [12, 23].
- ***Database rights for user access***: The limited and only important rights to access must be given to users connecting with the database [9, 22].
- ***Stored procedures***: This will encapsulate SQL queries and take inputs as the parameter [25].

40.4.4 Advantages of the Proposed Work

- Whitelisting the Input Validation: The aim is to verify whether the input type entered by the user is allowed or not. Validating the input ensures that input is of acceptable length, type, and format. The values that are valid only gets processed.
- Use of Prepared and Parameterized Queries: They are the means to precompile a structured query to provide parameters to execute the statement. This method allows the database to identify and differentiate from the provided input data.
- Escaping All User Supplied Data: The system will not consider the external users provided by the attacker to get into database.
- Enforcing the Least Privilege of Users: The privilege to provide access and to make changes in database is restricted to some particular user only.
- Using Stored Procedures: Inside of using the query whenever the data is to be loaded, pre-generated stored procedure should be used.
- Deploy Web Application Firewall: To prevent the activity of outsider, a firewall must be deployed to prevent unauthorized access to database.
- Disable Unnecessary Database Features and Capabilities: Only the required features should be activated so the scope of making changes can get limited.
- Monitor SQL Statements: A proper system should be placed to monitor the activities going on, and if any suspicious activity is found, it can be stopped immediately.

40.5 Conclusion

SQL mixture ambush is among the highest certifiable security shortcomings in the web applications structure, an enormous part of these shortcomings are realized by the nonattendance of data endorsement and SQL limits use. These days, web applications are commonly used as a rubberneck between PC customers. Web applications are moreover used for the most part by online business associations, and these sorts of usage need a secured database in order to keep sensitive and grouped information.

Regular SQL mixture attacks and distinguishing proof and expectation progressions are introduced in the paper.

References

1. Ŝtamper, M.: Inferential SQL injection attacks. Int. J. Netw. Secur. **18**(2), 316–325
2. Sonakshi, Kumar, R., Gopal, G.: Case study of SQL injection attacks. Int. J. Eng. Sci. Res. Technol. 176–189
3. Yigit, G., Amavutoglu, M.: SQL injection attacks detection & prevention techniques. Int. J. Comput. Theory Eng. **9**(5), 351–356
4. Kilaru, D.: Improving Techniques for SQL Injection Defenses. University of Colorado Colorado Springs
5. Kumar, B.J.S., Anaswara, P.P.: Vulnerability detection and prevention of SQL injection. Int. J. Eng. Technol. 16–18
6. MySQL 5.5 Reference Manual, Loading Data into a Table. Available at: https://dev.mysql.com/doc/refman/5.5/en/loading-tables.html
7. Kumar, M., Indu, L.: Detection and prevention of SQL Injection attack. Int. J. Comput. Sci. Inf. Technol. **5**(1), 374–377 (2014)
8. Kumar, M., Indu, L.: Detection and prevention of SQL Injection attack. Int. J. Comput. Sci. Inf. Technol. **5**(1), 374–377
9. Taran, A., Silnov, D.S.: Research of Attacks on MySQL Servers Using Honeypot Technology
10. Shahriar, H., Haddad, H.M.: Security vulnerabilities of NoSQL and SQL databases for Massive Open Online Courses (MOOCs) applications. Int. J. Digit. Soc. (IJDS) **8**(1)
11. Medeiros, I., Beatriz, M., Neves, N., Correia, M.: Detecting injection attacks and vulnerabilities inside the DBMS
12. Mukherjee, S., Sen, P., Bora, S., Pradhan, C.: SQL Injection Sample Review
13. Qian, L., Zhu, Z., Hu, L., Liu, S.: Research of SQL injection attack and prevention technology. In: International Conference on Estimation, Detection and Information Fusion (ICEDIF 2015)
14. Ali, S., Rauf, A., Javed, H.: SQLIPA: an authentication mechanism against SQL injection. Eur. J. Sci. Res. **38**(4), 604–611. ISSN 1450-216X
15. Janot, E., Zavarsky, P.: Preventing SQL injections in online applications: study, recommendations and Java solution prototype based on the SQL DOM. In: OWASP Application Security Conference
16. Hlaing, Z.C.S.S., Khaing, M.: A detection and prevention technique on SQL injection attacks
17. https://owasp.org/www-project-top-ten/
18. AL-Malikia, M.H.A., Jasima, M.N.: Review of SQL injection attacks: detection, to enhance the security of the website from client-side attacks (2021)
19. Chun, G.U.O., Wenyan, C.A.I., Guowei, S.H.E.N., Xuemei, Z.H.O.U.: Research on SQL injection attacks detection method based on the truncated key payload. Netinfo Secur. **21**(7), 43–53 (2021)
20. Nasereddin, M., ALKhamaiseh, A., Qasaimehb, M.: A Systematic Review of Detection and Prevention Techniques of SQL Injection Attacks (2021)
21. Falor, A., Hirani, M., Vedant, H., Mehta, P., Krishnan, D.: A Deep Learning Approach for Detection of SQL Injection Attacks Using Convolutional Neural Networks (2021)
22. Kumar, P., Pateriya, R.K.: A survey on SQL injection attacks, detection and prevention techniques. In: IEEE International Conference on Computing Communications & Networking Technologies
23. Tang, J.: The security analysis of MySQL database system. Comput. Knowl. Technol. **6**(23), 6418–6420
24. Nunes, P., Medeiros, I., Fonseca, J., Neves, N., Correia, M., Vieira, M.: Benchmarking static analysis tools for web security. IEEE Trans. Rel. **67**(3), 1159–1175

25. Wei, K., Muthuprasanna, M., Kothari, S.: Preventing SQL injection attacks in stored procedures. In: Proceedings of the Australian Software Engineering Conference (ASWEC'06 IEEE)
26. Kie'zun, A., Guo, P.T., Jayaraman, K., Ernst, M.D.: Automatic creation of SQL injection and cross-site scripting attacks. In: ICSE, Vancouver, Canada. 978-1-4244-3452-7/09/$25.00 ©2009 IEEE
27. Halder, R., Cortesi, A.: Obfuscation-based analysis of SQL injection attacks. 978-1-4244-7755-5/10/$26.00 IEEE
28. Sadeghian, A., Zamani, M., Abdullah, S.M.: A taxonomy of SQL injection attacks. In: IEEE International Conference on Informatics and Creative Multimedia, pp. 269–273
29. Fonseca, J., Vieira, M., Madeira, H.: Vulnerability & attack injection for web applications. In: IEEE/IFIP International Conference on Dependable Systems & Networks, pp. 93–102
30. Wegerer, M., Tjoa, S.: Defeating the database adversary using deception
31. Zoratti, I.: mySQL Security Best Practices. The Institution of Engineering and Technology Conference on Crime and Security, IET
32. Bandhakavi, S., Bisht, P., Madhusudan, P., Venkatakrishnan, V.N.: CANDID: preventing SQL injection attacks using dynamic candidate evaluations. In: Proceedings of the 14th ACM Conference on Computer and Communications Security
33. Kumar, P., Pateriya, R.K.: A survey on SQL injection attacks, and detection
34. Boyd, S.W., Keromytis, A.D.: SQLrand: preventing SQL injection attacks. In: Proceedings of the 2nd Applied Cryptography and Network Security Conference, pp. 292–302
35. Sharma, C., Jain, S.C.: Analysis and Classification of SQL Injection Vulnerabilities and Attacks on Web Applications

Author Index

A
Akshaya, 75
Amit P. Singh, 309
Ananthi Sheshasaayee, 57
Anitha Mary, X., 343
Anshul Bamb, 407
Anuradha Chug, 309
Anurag Jagetiya, 529
Arjun Choudhary, 163
Arpeeta Mohanty, 445
Aryan Dokania, 407
Aryan Karande, 211
Asha Rani, 269
Ashim Saha, 107
Ashish Sharma, 31, 417
Ashish Tripathi, 163, 565
Ashok Kumar Saini, 325

B
Bhupathiraju S. S. M. Karthikeya Varma, 343
Biswaroop Nath, 253
Buen Bajrami, 149

C
Chetanya Kunndra, 163

D
Deepak Kumar, 471
Deepak Sharma, 493
Deepak Sinwar, 373
Deepika Agrawal, 359
Dharmender Kumar, 471
Dheeba, J., 407
Diksha Chawla, 427
Dilbag Singh, 511
Dinesh Kumar Saini, 483
Divya Gangwar, 325
Dr.Abhinav Tomar, 457

G
Gaurav Patil, 1
Gurpreet Kaur, 511
Gurram Aakash, 343

H
Harmunish Taneja, 269
Harshita Bhagat, 211
Harshit Soni, 565

I
Iram Fatima, 325

J
Jatinderkumar R. Saini, 391
Jeen Marseline, K. S., 195
Jyoti Parashar, 325

K
Kamal Upreti, 325
Kavita Taneja, 269
Kiruthika, R., 181
Kocherlakota Snigdha, 343
Kostandina Veljanovska, 149

P. Nanda et al. (eds.), *Intelligent System and Data Analysis*, Smart Innovation, Systems and Technologies 398, https://doi.org/10.1007/978-981-97-5200-3

Kotagiri Chaitanya, 283

L
Lamba, C. S., 541
Likhitha Rayana, 237

M
Madhu Babu Chunduri, 283
Mahaveer Sain, 549
Mahendra Parihar, 75
Makhan Singh, 137
Manisha, 31
Manisha Malhotra, 511
Manish Singhal, 417
Medhunhashini, D. R., 195
Meenakshi Punia, 163
Megha Gupta, 457
Mohamed Sirajudeen Yoosuf, 237

N
Neha Choudhary, 549
Nikhil Debbarma, 125
Nisha, N., 373
Nitai Shah, 297

P
Padmanabha Reddy, Y. C. A., 283
Pallav Jain, 15
Pankaj Dadheech, 529
Parthasarathi Pattnayak, 445
Parth More, 211
Pawan Singh Mehra, 427
Perala Venkata Akanksha, 283
Poonam Devi, 39
Poonam Jyoti, 137
Pragati Gupta, 565
Pramod Bide, 1
Prashant Shukla, 359
Prashant Vats, 325
Punit Gupta, 483

R
Radha, N., 181
Rajdeep Chatterjee, 253
Rakesh Sharma, 541
Ramanpreet Singh, 297
Ranjeetsingh Suryawanshi, 211
Rekha Jain, 31
Ritik Singh, 1
Riya Rawal, 309
Rohit Kumar Gupta, 565

S
Saikat Roy, 391
Sai Sathwik Kosuru, 283
Sai Suchith Kurra, 237
Sajimon Abraham, 223
Sampath Kumar, J. P., 493
Sanghamitra Patnaik, 445
Sanjay Patidar, 15
Sanjit Bhagat, 483
Sharmila Ghosh, 107
Shilpa Singhal, 457
Shivani Gupta, 359
Shrddha Sagar, 39
Shubham Shetty, 1
Shylaja, K. R., 493
Snehaja Parsi, 283
Sourabarna Roy, 91
Sreelakshmi, K. R., 223
Sudhakar Pandey, 359
Suganya, A., 57
Sumitra Chopra, 549
Swapan Debbarma, 91

T
Tannistha Pal, 91, 107, 125
Tina Debbarma, 125
Tulip Das, 445

V
Vaishali More, 211
Vijaypal Singh Dhaka, 373
Vijila Gnanaraj, 343
Vikas Kumar Kharbas, 163
Vimal Kumar Pathak, 297
Vinita Tiwari, 417

GPSR Compliance

The European Union's (EU) General Product Safety Regulation (GPSR) is a set of rules that requires consumer products to be safe and our obligations to ensure this.

If you have any concerns about our products, you can contact us on ProductSafety@springernature.com

In case Publisher is established outside the EU, the EU authorized representative is:

Springer Nature Customer Service Center GmbH
Europaplatz 3
69115 Heidelberg, Germany

Batch number: 10704305

Printed by Printforce, the Netherlands